# Composite Materials Handbook

# Composite Materials Handbook

**M.M. Schwartz**

*Chief of Metals and Metals Processing*
*Sikorsky Aircraft*
*Division of United Technologies*

**McGraw-Hill Book Company**

New York   St. Louis   San Francisco   Auckland   Bogotá
Hamburg   Johannesburg   London   Madrid   Mexico
Montreal   New Delhi   Panama   Paris   São Paulo
Singapore   Sydney   Tokyo   Toronto

*To Carolyn, my best friend,*
*30 years of love, appreciation,*
*encouragement, and emotional support.*

**Library of Congress Cataloging in Publication Data**
Schwartz, Mel M.
    Composite materials handbook.

    Includes bibliographies and index.
    1.Composite materials—Handbooks, manuals, etc.
I.Title.
TA418.9.C6S38    1983      670      82-21719
ISBN 0-07-055743-8

1234567890    HAL/HAL      89876543

ISBN 0-07-055743-8

The editors for this book were Harold B. Crawford and Elizabeth Richardson,
the designer was Mark E. Safran, and the production supervisor was Thomas
G. Kowalczyk. It was set in Garamond by University Graphics, Inc.

Printed and bound by Halliday Lithograph.

# Contents

# THREE

## Composite Designs and Joint Criteria, Properties, and Quality Assurance 3.1

# FOUR

## Processing 4.1

# FIVE

## Automated Fabrication Methods 5.1

## SIX

### Cutting, Machining, and Joining Composites                    6.1

## SEVEN

### Applications Development                    7.1

## EIGHT

### Future Potential of Composite Materials                    8.1

# Preface

Composite materials have been used for centuries—bricks reinforced with straw, laminated iron-steel swords and gun barrels, linoleum, plasterboard, and concrete, to name but a few. Today industrial innovation, improved energy planning, uncertain availability, and sky-rocketing costs have created a greater interest in composites. Now that increasingly severe performance requirements are taxing many conventional monolithic materials to the limit, the engineer's traditional approach of fitting the design to the properties is changing into one of finding materials with the right properties to meet the demands of design, service, and economics.

This book answers many practical questions that arise as engineers and technicians find more and more uses for composites, which, unlike metals, have developed so fast that reliable information is often hard to find. By collecting data from manufacturers' literature, reports of technical meetings, magazine articles, and government reports this book helps answer questions that usually take hours to research. Since composites encompass hundreds and thousands of materials, the book is devoted primarily to fiber- and metal-reinforced matrix composites.

It includes useful and practical information on the fibers and metals, their fabrication into a matrix, resin systems, composite design criteria, new fabrications methods and equipment to produce the end product, quality assurance, and details of machining, trimming, joining, and finishing. It also discusses applications in a multitude of industries and the future outlook for composites. A glossary defines common terms used in the industry.

I wish to acknowledge the able support of Alice Hardy, who assumed responsibility for typing the manuscript, and Nancy McGaan, who ably assisted in creating the illustrations and tables.

M. M. SCHWARTZ

Laminate, one type of composite.   (*Sikorsky Aircraft, Division of United Technologies.*)

# Introduction to Composites

What do we mean by a composite? The idea of combining chemical or structural elements can be productive on many different levels of matter, but for engineering purposes we must limit the concept so that we can apply it to today's problems. How can we classify the hundreds and thousands of composites? After outlining the general structural characteristics of composites we show how this outline can be the basis of a fairly simple scheme for descriptive classification and prediction of behavior.

The extra degrees of freedom composites give us (the mixtures of materials, forms, proportions, distributions, and orientations) cannot be exploited until one learns how to deal quantitatively with the same number of variables. Because the technology of producing and combining many of the constituents is new and imperfect, we must be more than usually concerned with processing and reliabilty.

That two or more materials judiciously combined will perform differently and often more efficiently than the materials by themselves is obvious and well known. But this simple concept offers a useful and even revolutionary way of thinking about the development and application of materials. Only with the emergence of a unified and interdisciplinary approach to materials have we begun to realize the full significance and huge potential of composite materials.

## DEFINITIONS

There is no universally accepted definition of composite materials. Definitions in the literature differ widely. The problem is the level of definition. In the dictionary and in everyday usage the term *composite* refers to something made up of various parts or elements. When we start to devise a definition for composite materials in accordance with this idea, we quickly discover that several definitions are possible. Since we can state a valid definition in terms of the constituents making up engineering materials at each of the several structural levels of matter, which materials are to be regarded as composites and which as monolithics depends upon the level chosen as the basis for definition.

*Elemental or basic level:* At this level, that of single molecules and crystal cells, all materials composed of two or more different atoms would be regarded as composites. They would include compounds, alloys, polymers, and ceramics. Only the pure elements would be excluded.

*Microstructural level:* At this level of crystals, phases, and compounds a composite would be defined as a material composed of two or more different crystals, molecular structures, or phases. By this definition many materials traditionally considered to be monolithic or homogeneous would be classified as composites. Of all the metallic materials only single-phase alloys, such as some brasses and bronzes, would be monolithic by this definition. Steel, a multiphase alloy of carbon and iron, would be a composite.

*Macrostructural level:* At this level, with which we shall be principally concerned, we deal with gross structural forms or constituents, e.g., matrixes, particles, and fibers, and think of a composite as a *materials system* composed of different macroconstituents.

The definition at the macrosectional level encompasses many but not all of the materials now commonly considered composites. To be more inclusive we must go beyond the forms of the constituents and include two other characteristics: (1) the individual constituents making up a composite are almost always different chemically, and (2) they are essentially insoluble in each other. A working definition of composite materials which takes into account both the structural form and composition of the material constituents follows:

> *A composite material is a materials system composed of a mixture or combination of two or more macroconstituents differing in form and/or material composition and that are essentially insoluble in each other.*

Even this definition needs clarification. To some engineers it will seem too broad because it includes many engineering materials that are not usually thought of as composites, e.g., precoated materials, filled plastics, concrete, and impregnated materials. Nevertheless, all such materials fall within the concept of composites and regardless of the common idea of them should be treated as composites. On the other hand, a strict application of this definition might exclude some materials now widely regarded as composites. For example, in many of the particulate-type composites, such as dispersion-hardened alloys and cermets, the composite structure is microscopic rather than macroscopic. Finally this definition does not draw the line between *composite materials* and *composite structures*. For example, should honeycomb be considered a structure or a material? Should one define as a composite structure or composite material a rocket nozzle constructed of five or six integrated layers or an automobile tire built up of several layers and a fabric-reinforced matrix? There are arguments on both sides. Some contend that the rocket nozzle and the tire are not composite materials because they are a finished structure, component, or product, but a finished structure or product can be regarded as a composite material even though it is also an integrated materials system.

Instead of trying to establish a distinction between composite materials and composite structures it is more useful to make a distinction between mill composites and specialty composites. Typical *mill composites* are nonmetallic laminates, clad metals, and honeycomb, produced in more or less standard lines and suitable for many different applications. Typical *specialty composites* are tires, rocket nose cones, and glass-reinforced plastic boats, which are designed and produced specifically for a given application and often are the product itself.

## CONSTITUENTS OF COMPOSITES

In principle, composites can be constructed of any combination of two or more materials, whether metallic, organic, or inorganic. Although the possible material combinations in composites are virtually unlimited, the constituent forms are more restricted. Major constituent forms used in composite materials are fibers, particles, laminae or layers, flakes, fillers, and matrixes (Fig. 1.1). The matrix is the *body constituent,* serving to enclose the composite and give it its bulk form. The fibers, particles, laminae, flakes, and fillers are

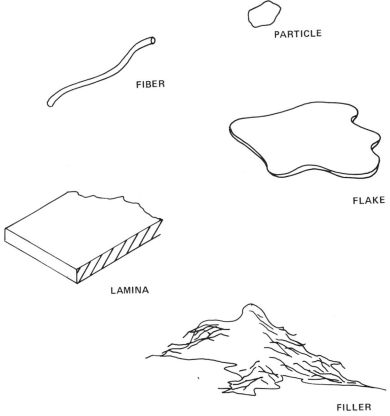

PARTICLE

FIBER

FLAKE

LAMINA

FILLER

**FIG. 1.1** The different constituent forms in composites.[16a]

the *structural constituents;* they determine the internal structure of the composite. Generally, but not always, they are the additive phase.

Perhaps the most typical composite is one composed of a structural constituent embedded in a matrix, but many composites have no matrix and are composed of one (or more) constituent form(s) consisting of two or more different materials. Sandwiches and laminates, for example, are composed entirely of layers, which, taken together, give the composite its form. Many felts and fabrics have no body matrix but consist entirely of fibers of several compositions, with or without a bonding phase.

## Interfaces and Interphases

Because the different constituents are intermixed or combined, there is always a contiguous region. It may simply be an interface, i.e., the surface forming the common boundary of the constituents. An interface is in some ways analogous to the grain boundaries in monolithic materials. In some cases, however, the contiguous region is a distinct added phase, called an *interphase.* Examples are the coating on the glass fibers in reinforced plastics and the adhesive that bonds the layers of a laminate together. When such an interphase is present, there are two interfaces, one between each surface on the interphase and its adjoining constituent (Fig. 1.2). In still other composites the surfaces of the dissimilar constituents interact to produce an interphase. For example, in some cermets bonding between the particles and matrix results from the small solubility at the surface of one or both these constituents.

## Distribution of Constituents

Constituents making up a composite can be distributed in two general ways. Perhaps the most common way is for the constituents to be present in a regular and repetitive pattern, with a relatively uniform cross section both in material and structure and uniform density. Matrix-particle and some matrix-fiber composites, in which the structural constituent is evenly distributed throughout the matrix, are of this homogeneous type. The second pos-

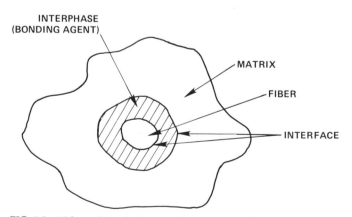

FIG. 1.2 Makeup of interface between fiber and matrix.[16a]

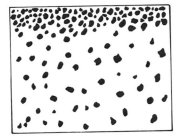

MATERIALS

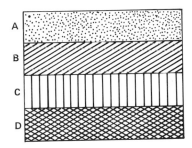

GRADIENT COMPOSITES

FIG. 1.3    Gradient composites.[16a]

sibility is a variable pattern of constituents that is nonrepetitive in either internal form or material. Materials of this type are termed *graded* or *gradient composites*. Laminated materials, which are composed of several different layers, belong in this category. Filament-wound composites can also be designed with variable fiber distribution (Fig. 1.3).

In both homogeneous and gradient composites the structural constituents (fibers or flakes) can be arranged in either an oriented or random fashion (Fig. 1.4).

## CLASSIFICATION

With this explanation of the nature and structure of composites, a working classification of composites can be set up. Several classification systems have been used, including classification by basic material combinations, e.g., metal-organic or metal-inorganic, (2) by bulk-form characteristics, e.g., matrix systems or laminates, (3) by distribution of the constituents, e.g., continuous or discontinuous; and (4) by function, e.g., electrical or structural. The classification system used in this chapter is based on the form of the structural constituents. This gives five general classes of composites (Fig. 1.5):

1. Fiber composites, composed of fibers with or without a matrix
2. Flake composites, composed of flat flakes with or without a matrix
3. Particulate composites, composed of particles with or without a matrix

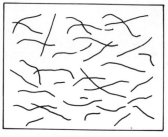

RANDOM

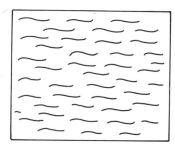

ORIENTED

**FIG.** 1.4   Orientation of fibers.[16a]

PARTICULATE COMPOSITE

FIBER COMPOSITE

LAMINAR COMPOSITE

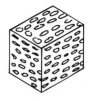

FLAKE COMPOSITE

FILLED COMPOSITE

**FIG.** 1.5   Classes of composites.[16b]

4. Filled (or skeletal) composites, composed of a continuous skeletal matrix filled by a second material
5. Laminar composites, composed of layer or laminar constituents

## PERFORMANCE

From the previous discussion of the nature and morphology of composites it is evident that the behavior and properties of composites are determined by:

The materials of which the constituents are composed

The form and structural arrangement of the constituents

The interaction between the constituents

Taking the most important of these factors first, it is obvious that the intrinsic properties of the materials of which the constituents are composed are of critical importance. They largely determine the general order or range of properties which the composite will display. Although, as we shall see, the interaction of constituents results in a new set of properties, these combined properties in the composite derive from those of the individual materials.

Structural and geometrical characteristics of the constituents also make important contributions to the composite's properties. The shape and size of the individual constituents, their structural arrangement and distribution, and the relative amount of each are important factors contributing to the overall performance of the composite. These many variables are what give composite materials much of their versatility.

Of far-reaching importance in the performance and use of composites are the effects produced by the combination and/or interaction of the constituents. Since composites are mixtures or combinations of constituents that differ either in material or form, the properties of their combination must always be different. The basic principle underlying the design, development, and use of composites is that we use different constituents to obtain combinations of properties and/or property values different from those of the individual constituents.

### Combination Effects

There are three general ways in which the new sets of properties offered by composites can differ from those of the constituents.

**Summation:** The most obvious way follows the simple mixture rule and results in a summation of the individual properties. This happens when the contribution of each constituent is independent. For example, the density of a composite is usually the sum of the volume fractions of each of the constituents times their individual densities. In laminar composites this simple summation rule generally also applies to electrical conductivity and heat-transfer properties.

**Complementation:** The second way is that each constituent will complement the other by contributing separate and distinct properties. Many clad materials and laminates, for example, are composed of a layer that provides corrosion resistance, attractive appearance, or other surface properties plus a layer or layers of a stronger material to provide the necessary structural strength. Metal-filled plastic, such as lead dispersed in vinyl sheet, provides sound-absorbing qualities.

**Interaction:**   The third way occurs when a given property or action of one constituent is not independent of the property or action of the other. One constituent supplements the other. The resulting composite properties are usually intermediate between those of the constituents or higher than those of both.

The most desirable supplementary effect is a significant property improvement; e.g., the addition of wood flour to a plastic resin provides a mixture which is stronger than either of the materials alone. A classic example is glass-fiber-reinforced plastics, in which the composite strength is considerably higher than the strength of either the resin or the glass fibers alone. In the wood flour-resin composite the flour inhibits plastic deformation of the plastic resin while the resin bonds the wood-flour particles together. In fiber-reinforced plastics increased strength results because the deformations of the two constituents are not independent of each other. The low-modulus, low-strength plastic deforms first and distributes the stress to the higher-strength glass fibers. In addition, when individual glass fibers fail, the stress is redistributed through the resin to the other fibers.

The nature and success of the interaction between the composite constituents depend to a significant degree on the interfaces. The interface is particularly critical in structural composites. What takes place at the interface depends on reactions between the surfaces of the constituents or between these surfaces and those of a bonding phase. These reactions involve such phenomena as chemical compatibility, adsorption characteristics, wettability, and stresses resulting from differences in expansion. Little is known about these phenomena, empirically or theoretically. Of course, when a bonding phase is added, the adhesion of the bonding agent and its ability to transfer stresses is of paramount importance.

## Quantitative Analysis

Our discussion of composites and the factors influencing their behavior has been largely qualitative, but, as with all engineering materials, making the optimum use of composites means being able to characterize their properties and behavior quantitatively. Thus, in order to design or develop composites and in order to design with them engineers need models and mathematical expressions that predict composite behavior under various service conditions. There are two general approaches to the quantitative analysis of composites. The simplest method, an empirical or "black box" approach, measures only the behavior (or output) of the composite as a whole when it is subjected to a given set of conditions, individual contributions of the constituents and the effects of their interaction not being considered. The analytical approach, on the other hand, gets inside the black box and attempts to predict composite performance on the basis of the behavior and interaction of the constituents. This approach, which is more complex but yields more useful information, is discussed in Chap. 3.

## PROCESSING VARIABLES

The many complex factors controlling the engineering properties of composites have already been touched upon. Perhaps as many variables, equally complex, control the quality and reliability or uniformity of composites. They relate both to the quality of the individual constituent elements and, often more important, to that involved when the constituents are combined into the finished composite. For example, to obtain the promised strength properties in glass-flake composites, flatness and freedom from surface and edge cracks in the flakes are essential, but a high degree of parallelism in the flake distribution is more important than uniformity of the particles themselves. In many fiber composites, uniformity in orientation of the fibers, control of matrix content, quality of the interfacial bonds, and many other processing variables influence the composite's performance directly.

In the past 20 years only a limited number of standards and test procedures for evaluating the quality and uniformity of composites have been established. The problem is complex. The field of composites is still quite new, and only limited effort has been devoted to measuring and controlling quality. In view of the history of monolithic materials and conventional forms, e.g., castings, moldings, and mill products, the problem of measuring and controlling quality and uniformity is not peculiar to composites. Over the years such troublesome quality variables as agglomeration, shrinkage, segregation, porosity, and chemical composition have had to be dealt with in the production and processing of established materials and forms. Thus the difficulty and vastness of the task of evaluating quality and obtaining uniform properties in composites are not as great as they seem. As the composite field advances, sufficiently accurate methods will be developed to measure changes in composite composition and structure which are responsible for significant changes in properties. These methods will probably be nondestructive and capable of determining the degree of uniformity of entire composites on a production basis.

## FIBER-MATRIX COMPOSITES

Of all composite materials, the fiber type (specifically the inclusion of fibers in a matrix) has evoked the most interest among engineers concerned with structural applications. Initially most work was done with strong, stiff fibers of solid, circular cross section in a much weaker, more flexible matrix, i.e., glass fibers in synthetic resins. Then development work disclosed the special advantages offered by metal and ceramic fibers, hollow fibers, fibers of noncircular cross section, and stronger, stiffer, and more heat-resistant matrixes.

Some textile composites consist solely of fibers, with no matrix, and bonded fiber structures to which a matrix is added. Since the fibrous form of most materials is many times stronger than its bulk form, engineers have long sought ways of making practical use of fibers as engineering materials. The most efficient method yet found is to combine a fibrous material of high tensile strength and high modulus of elasticity with a lightweight bulk material of lower strength and lower modulus of elasticity. In nature the most common example of this reinforcing principle is probably the bamboo pole. Among synthetic materials practically every type (plastic, rubber, ceramic, and metal) is now being reinforced with fibers; uses of these reinforced materials range from the low-performance cafeteria tray to the high-performance rocket-motor case.

Although this chapter is primarily concerned with fiber-matrix composites, fiber-fiber composites should not be overlooked. The polyester-and-wool textile used for men's suits is a familiar example, but there are many industrial textiles as well. Usually one fiber is chosen for its mechanical properties and the other for a secondary purpose, e.g., heat resistance. Conveyor belts of asbestos-metal-fiber composite are used to carry heavy hot materials. Many felts are also fiber-fiber composites in which the fibers are randomly oriented.

In fiber-fiber composites there is no matrix and little or no dependency on a bonding phase. Fiber-matrix composites have not only these two constituents but usually a bonding phase as well. We look first at the roles played by these three constituents and then at how they are combined in various classes of fiber-matrix composites.

## Fiber Factors

What factors contribute to the engineering performance of a fiber-matrix composite? Among the most important are the orientation, length, shape, and composition of the fibers; the mechanical properties of the matrix; and the integrity of the bond between fibers and matrix. Of these, orientation of the fibers is perhaps most important.

## *Orientation*

Fiber orientation (how the individual strands are positioned) determines the mechanical strength of the composite and the direction in which that strength will be the greatest. There are three types of fiber orientation: one-dimensional reinforcement, planar (two-dimensional) reinforcement, and three-dimensional reinforcement. The one-dimensional type has maximum composite strength and modulus in the direction of the fiber axis. The planar type exhibits different strengths in each direction of fiber orientation. The three-dimensional type is isotropic but has greatly decreased reinforcing values (all three dimensions are reinforced but only to about one-third of the one-dimensional reinforced value). The mechanical properties in any one direction are proportional to the amount of fiber by volume oriented in that direction. As fiber orientation becomes more random, the mechanical properties in any one direction become lower (Fig. 1.6).

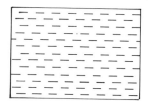

**ONE–DIMENSIONAL REINFORCEMENT**

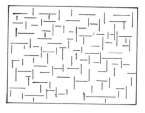

**PLANAR REINFORCEMENT**

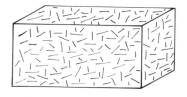

**THREE–DIMENSIONAL REINFORCEMENT**

**FIG. 1.6** Three types of orientation reinforcement.

## *Length*

The orientation of fibers in the matrix can be accomplished with either continuous or short fibers. Although continuous fibers are more efficiently oriented than short fibers, they are not necessarily better. Theoretically, continuous fibers can transmit an applied load or stress from the point of application to the reaction by a continuous load path. In practice this is not possible for two reasons: (1) manufacturing variables make it impossible to obtain continuous fibers with optimum tensile strength throughout their length, and (2) in the ideal

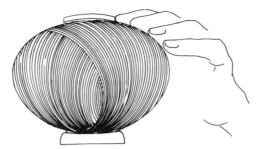

**FIG. 1.7**   Filament-wound bottle. *(United Technologies.)*

continuous-fiber structure the fibers will be stress-free or equally stressed, but this situation never obtains in reality. Instead, during fabrication some fibers will be highly stressed while others will be unstressed. As the composite is loaded to its ultimate strength, the prestressed fibers must support more of the load than the fibers with lower built-in stresses.

Composites made from shorter fibers, if they could be properly oriented, could have substantially greater strengths than those made from continuous fibers. This is particularly true of whiskers, which have uniform tensile strengths as high as 1500 kips/in² (10.3 GPa). Because the shorter fibers can be produced with few surface flaws, they come extremely close to achieving their theoretical strength.

Besides mechanical properties, fiber length has a bearing on the processability of the fiber composite. In general, continuous fibers are easier to handle but more limited in design possibilities than short fibers. Continuous filaments are normally incorporated by the filament-winding process, which wraps a continuous fiber (impregnated with a matrix material) around a mandrel the shape of the part, ensuring good distribution and favorable orientation of the fiber in the finished article (see Fig. 1.7). Since filament winding is limited chiefly to the fabrication of bodies of revolution, short-fiber composite materials are used to make many flat and irregular shapes. The numerous open- and closed-mold processes for these materials are less efficient in their use of fiber but generally have higher output and lower cost than filament winding.

### Shape

Practically all fibers presently being used have a circular cross section whether they are continuous or short; however, hexagonal, rectangular, polygonal, annular (hollow circle), and irregular cross sections appear to promise improved mechanical properties.

Solid circular fibers are easy to produce and handle. Glass, plastic, and metal fibers have been drawn and extruded in great quantity and in various sizes. Steel reinforcing rods for concrete are examples of thick fibers; some glass fibers have diameters as small as 0.0004 in (0.01 mm). In general, the smaller the diameter of a fiber the greater its strength, probably thanks to the elimination of surface flaws on the fiber.

The potential of high-strength cylindrical structures made of fibers of rectangular cross section is due to the development of thin, flatsided filaments. Rectangular filaments make it possible to obtain almost perfect packing; however, at a packing of about 85 vol % glass the composite's strength depends more on the bond of the glass to the matrix than on the glass itself.

Single crystal whiskers with hexagonal fibers make the strongest fibers, but handling them and incorporating them into a fiber composite is difficult.

Hollow glass fibers demonstrate improved structural efficiency for applications where stiffness and compressive strength are the governing criteria. The transverse compressive strength of a hollow-fiber composite is lower than the strength of a solid-fiber composite when the hollow part is more than half the total fiber diameter. Hollow fibers are also quite difficult to handle and to incorporate into a composite.

Metal fibers produced as shavings have no definite shape and are very inexpensive to produce since often they are a by-product of machining. Their irregularity prevents them from being used efficiently in designs based on tensile strength.

## Composition

Both organic and inorganic fibers are available for fiber composite materials. The organics, such as cellulose, polypropylene, and graphite† fibers, can be characterized in general as lightweight, flexible, elastic, and heat-sensitive. Inorganic fibers, such as glass, tungsten, and ceramic, can be generally described as very high in strength, heat-resistant, rigid, and low in energy absorption and fatigue resistance. While many organic fibers satisfy both the strength and elasticity requirements for structural composites, graphite in recent years has become the most popular. The inorganics, notably glass, dominate the field. Most of the other inorganic fibers, e.g., metallic whiskers and ceramic fibers, have not received the financial backing required for development into an accepted material.[1-3]‡

## The Matrix

The other major constituent in fiber composites, the matrix, serves two very important functions: (1) it holds the fibrous phase in place, and (2) under an applied force it deforms and distributes the stress to the high-modulus fibrous constituent. The choice of a matrix for a structural fiber composite is limited by the requirement that it have a greater elongation at break than the fiber. Thus, if the elongation at break in a fiber is 4%, the matrix must have an elongation of at least 5%, if not more. Also, the matrix must transmit the forces to the fibers and change shape as required to accomplish this, placing only tensile loading on the fibers. Furthermore, during processing the matrix must encapsulate the fibrous phase without excessive shrinkage, which can place internal strain on the fibers. In addition chemical, thermal, and electrical properties are often affected by the type of matrix used. Further discussion of properties appears in Chaps. 2 and 3.

## The Bonding Phase

Fiber composites are able to withstand higher stresses than either of their individual constituents because the fibers and matrix interact and redistribute the stresses. The ability of these two constituents to exchange stresses depends critically on the effectiveness of the coupling or bonding between them. Such bonding can sometimes be attained by direct contact of the two phases, but usually a specially treated fiber must be used to ensure a

---

†Graphite fibers are classified as organic because they are manufactured from organic precursors.
‡Numbered references appear at the end of each chapter.

receptive adherent surface. Much research on factors influencing these reactions has led to the development of types of fiber finishes, called coupling agents.

There are several theories on the effect of bonding materials, variously suggesting chemical or mechanical bonding as the primary function of coupling agents. The latest work suggests that both chemical and mechanical interactions occur.

Voids (air pockets) are harmful because portions of a fiber passing through the void are not supported by surrounding resin. Under load, the fiber may buckle and transfer the stresses to the resin, which readily cracks. Weak or incomplete bonding between the fiber and the matrix is another cause of early failure. The fiber-matrix bond is often in a state of shear when the material is under load. When this bond is broken, the fiber separates from the matrix and leaves discontinuities that may cause failure. Coupling agents can be used to strengthen these bonds against shear forces.

## TYPES OF FIBER MATRIX

For purposes of discussion fibrous composites can be grouped into four classes: organic fiber organic matrix, inorganic fiber organic matrix, inorganic fiber inorganic matrix, and organic fiber inorganic matrix.

### Organic Fibers in Organic Matrix

Since most organic materials, such as plastics and rubbers, have a low specific gravity, they are often used in weight-critical applications. Though strengthening lightweight materials is important, often flexibility must be retained. The automobile tire is a good example of how an all-organic composite functions. The tire resists stresses because of the nylon and rayon fiber cords bonded into the tire rubber. The rubber, though thoroughly reinforced with the fibers, remains flexible. There are many other composites of this type, e.g., plastic- and rubber-coated fabrics used as tarpaulins.

Wood (cellulose) fibers in matrixes of rigid thermosetting plastics are exceptions to the generally flexible organic composites. These two-phase materials offer the attractive combination of low cost, good strength-to-weight ratio, and easy processing. The finer size of wood-fiber-reinforced plastics which are similar to particle-board compositions, ensure better distribution and wetting of the reinforcement and consequently better strength. Cellulose fibers have been evaluated for reinforcing thermoplastics, e.g., polyethylene. These two-phase flexible materials combine the chemical resistance of the plastic with the paper-like qualities of the cellulose. Examples are a battery separator and a chemical filter cartridge.

An exception to the generally flexible nature of organic fiber composites is found in plastics containing graphite fibers to improve their heat resistance. Since the graphite fibers are good heat conductors, they dissipate the heat applied to the matrix.

### Inorganic Fibers in Organic Matrix

Probably the greatest potential for lightweight high-strength composites exists in the inorganic-fiber–organic-matrix composites. Although the choice of materials for the two constituents making up such fiber composites is wide, no combination of inorganic fiber and organic matrix has yet proved as successful as glass-fiber-reinforced plastic composites. Taken as a whole, glass-fiber–plastic composites have the following advantages:

1. Good physical properties, including strength, elasticity, impact resistance, and dimensional stability
2. High strength-to-weight ratio
3. Good electrical properties
4. Resistance to chemical attack and outdoor weathering
5. Resistance to moderately high temperatures [about 500°F (260°C)]
6. Suitability for fabrication by a variety of production methods
7. Adaptability to a wide range of sizes

Glass-fiber-reinforced plastics have been successfully used in the aerospace, transportation, recreational, appliance, electrical equipment, and tank and piping industries.

## Glass Fibers

One of the most popular of the commercially available glass fibers, E glass, has an ultimate tensile strength of about 500 kips/in$^2$ (3448 MPa) for single diameters of 0.00037 in (0.01 mm). It has a Young's modulus of 10,500 kips/in$^2$ (72.4 GPa) and a density of 0.092 lb/in$^3$ (2547 kg/m$^3$). Other glass compositions have different mechanical properties. For example, window or sheet glass, which is a high-alkali glass and has a tensile strength which is 50% lower than that of E glass, is used as a cheaper fiber for many low-performance applications. Research continues in attempts to increase the modulus or tensile strength of glass; some glass fibers have tensile strengths of 700 kips/in$^2$ (4827 MPa).

Glass fibers are available as continuous filaments, cut or chopped strands, rovings, and yarns, which can be used individually or in the form of cloths, mats, or tapes. This variety of forms allows glass fibers to be used in many different fabrication processes, such as filament winding, matched-die molding, lay-up, etc.

## Resin Matrixes

Although many types of plastic resins, both thermosetting and thermoplastic, are being reinforced with glass fibers, polyester resins are the most widely used, especially for low-performance applications. In the cured state polyesters are hard, light-colored, transparent materials, which may be rigid or flexible. They are resistant to water, weather, aging, and a variety of chemicals. They can be used at temperatures up to about 175°F (79°C) or higher, depending upon the particular resin or service requirements. Other principal advantages of polyesters are that they combine easily with glass-fiber reinforcements and can be used with all types of reinforced-plastic fabricating equipment.

Polyesters shrink 4 to 8 vol % on curing. This property is undesirable in composites because it often exposes the glass-fiber reinforcement, necessitating additional surface finishing. Shrinkage can also weaken the glass-resin bond and place stresses other than tension on the glass fibers.

Epoxy resins are more expensive than polyesters, but their high adhesion makes them useful in many high-performance applications. Epoxy resins also have excellent water resistance and low shrinkage during cure (about 3%), properties conducive to good fiber-matrix adhesion.

Continuous development to improve materials and utilize epoxy resin systems with graphite has examined the most feasible method of combining epoxy-impregnated graphite fibers[4] and/or polyimide-impregnated graphite fibers to form very-high-strength honeycomb. The end result is expected to be a graphite-fiber honeycomb material with a higher strength-to-weight ratio than conventional honeycomb having comparable high-temperature properties.

Several other thermosetting resins have also been used for fiber-reinforcement applications requiring high-temperature stability, notably silicone and phenolic resins.

When reinforced with glass fibers, thermoplastic resins, such as polystyrene, polycarbonate, and polyvinyl chloride, show improved mechanical properties, and these normally unstable plastics have excellent dimensional stability and improved heat resistance.

## Coupling Agents

Although many plastics can be reinforced with glass fibers, the permanence of the mechanical properties of the composites, especially their resistance to degradation upon environmental exposure, depends largely on the behavior of the bond between fibers and matrix. Because glass is a polar material, it has a strong attraction to water. Unprotected glass fibers will be coated with a number of molecular layers of water, which will have an adverse effect on the bond between the glass fiber and the resin matrix. The bond is improved by coating the fibers with a coupling agent to promote greater adhesion between the two materials. Coupling agents improve bond strength between the fiber and resin. The ideal coupling agent should provide a low-modulus flexible layer at the interface that will improve adhesive strength of the fiber-resin bond and reduce the number of voids in the material.

Two general types of chemical coupling agents are chrome complexes and silane compounds. The complex molecules of these agents can orient themselves in such a way that one group of ions can react with the silicon ions of the glass surface and effect a bond while the unsaturated organic group takes part in the reaction of the resin during its curing (see Fig. 1.8). While coupling agents normally increase the original bond strength, their usefulness becomes much more pronounced if the composite is subject to moisture. All plastics absorb water to a greater or lesser extent, but the strong chemical bonds between the glass and coupling agent prevent the water from having a catastrophic effect on the interface adhesion.

## Metal Fibers

Glass fibers are not the only fibers being used with plastics. Metal-fiber-reinforced plastics are not only mechanically strong and tough composites but can be designed to be good conductors of heat and electricity. The incorporation of short, random metal fibers in epoxy resins has led to systems with impact strengths, heat-distortion points, and thermal conductivities greatly exceeding those of glass-fiber-reinforced composites. Much of the work done with these materials has centered around their use in sheet-metal forming dies intended as prototypes or for short runs. In addition to their desirable mechanical properties, metal-fiber composites can be easily machined, thus reducing tooling costs. Another key property of metal-fiber–plastic composites is their wear resistance. The automotive industry has examined their use for clutch facings.

Continuous metal fibers have also been investigated as a reinforcement for plastics. Steel wire [0.004-in diameter (0.10-mm)] of high tensile strength [on the order of 600 kips/

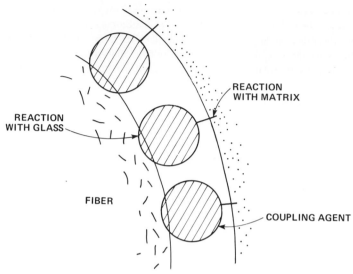

**FIG. 1.8** Mechanism of coupling agent.[16c]

in$^2$ (4137 MPa)] has been used to reinforce epoxy and polyethylene plastics. The advantages of this two-phase material are high strength, light weight, chemical resistance, and good fatigue resistance. Continuous metal fibers are also easier to handle than glass fibers. For some high-performance missile applications fiber-reinforced composites that can resist cycle loading without internal cracking have been developed. In addition to better flexural properties, the wire composites offer better modulus-to-weight ratios than glass-fiber composites. Metal-fiber-reinforced plastics have some limitations, especially their poor high-temperature properties. Furthermore, the difference in coefficient of thermal expansion between the metal fibers and the plastic matrixes can cause problems.

Other efforts to develop high-temperature fiber composites have centered around the use of ceramic and quartz fibers to reinforce thermosetting plastics. Refractory oxide fibers, e.g., zirconia, have been used as reinforcement for plastic ablatives. Aluminum phosphate and fused-silica fibers have been used for high-temperature [1300°F (704°C)] radome materials. Refractory-fiber-reinforced plastics also can be used to fabricate some parts which are impossible to produce out of pure metals or ceramics.[5]

## Inorganic Fibers in Inorganic Matrix

One of the major reasons for combining inorganic fibers and inorganic matrixes is to achieve high-temperature performance not possible with organic materials (Fig. 1.9). One of the most promising composites in this group is metal reinforced with alumina whiskers. Whiskers have outstanding tensile strengths: sapphire whisker has a tensile strength of 17,500 kips/in$^2$ (12.1 GPa). The tensile strength of a silver matrix with alumina whiskers at 1400°F (760°C) is nearly 45 kips/in$^2$ (310 MPa), which is far greater than that of pure silver and more than twice that of dispersion-hardened silver at 1000°F (538°C).

Ceramic whiskers have been of greatest interest since they do not deform plastically and have high elastic moduli. Alumina and silicon whiskers have been offered in commercial quantities. Silicon nitride whiskers have been grown in quantity and studied thoroughly in composites, but phase changes at relatively low temperature have limited interest in this system. Alumina whiskers have been grown by a method in which water vapor is passed over molten aluminum, forming a suboxide. Process improvements and elegant techniques have resulted in longer alumina whiskers.

One continuous method of whisker manufacturing relies on the controlled hydrolysis of aluminum chloride on an aluminum or alumina nucleus in a fluidized bed. Aluminum chloride, prepared by passing chlorine over molten aluminum, is passed into a fluidized-bed reaction chamber along with $CO$, $H_2$, and $CO_2$. This results in the formation of a variety of whisker types containing considerable debris, especially when grown on alumina nuclei.

Silicon carbide whiskers generally are formed by the pyrolysis of chlorosilanes in hydrogen at temperatures above 2552°F (1400°C). Selection of proper silanes, optimum reaction conditions, and suitable reaction chambers have led to most of the improvements in the production of silicon carbide whiskers (Fig. 1.10). When silicon carbide whiskers are used in nickel, the whiskers and metal react, so that it is necessary to coat the whiskers with a metal. A low-temperature vapor-deposition process such as radio-frequency sputtering or decomposition of metal carbonyls on the fiber is used.

Whiskers can be incorporated into composites by several techniques. Slip-casting and powder-metallurgy techniques have been used successfully to prepare metal-whiskers systems.[6]

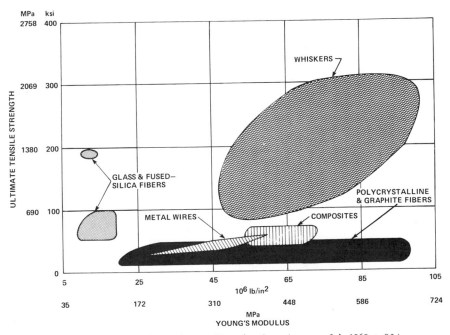

**FIG. 1.9** Strengths of various fibers and wires. (*Data from Space Aeronaut., July 1969, p. 93.*)

Strengthening metals at elevated temperatures with whiskers has been demonstrated in laboratories and limited prototype operations. Continued work in this area will open a vast potential for materials and applications.[7] If all or part of the strength of whiskers could be retained when they are incorporated into an engineering material, a superior product would result, but their very fine size (diameters on the order of 2 μm) makes whiskers extremely difficult to handle and fabricate into fiber composites. Before a usable product can be made, a great deal remains to be learned about handling, mixing, and sintering very fine whiskers with powders or liquid metals.

**FIG. 1.10**  SiC whiskers.[19]

Continuous metal fibers also have been used to reinforce metals. Tungsten-wire–copper-matrix composites are considerably stronger than either material in bulk form, and molybdenum-fiber–titanium-matrix composites have shown substantially better high-temperature strength than titanium alloys. A filamentary system of titanium alloy matrix reinforced with beryllium filaments holds promise for jet-engine compressor blades. Specific strength and specific modulus are about 2 to 2.8 times better than for solid-metal structures.[8]

Sintered-metal-fiber bodies have also been used to reinforce metal matrixes. These two-phase materials look and feel like metal but have less than 10% the density. They can also be designed to have both high density and high permeability so that gases or fluids pass through them readily. These two-phase materials are produced by *fiber metallurgy*, which is similar to powder metallurgy. Metal-fiber mats are produced by compacting the fibers into the desired form and then sintering. This produces a random interlocked mass of fibers bonded at each point of fiber contact. The pores of this metal felt are then filled with a second material, such as a low-melting metal, to form the fiber-matrix composite.[9-11] Some of the metal fibers which have been treated are lead, copper, iron, nickel, titanium, and molybdenum. Apparently all metals and alloys can be processed by fiber metallurgy, but those forming stable surface oxide films, e.g., aluminum, are difficult to sinter.

This type of fiber composite has been found to be useful in many ways. Composites made by infiltrating a sintered mat of steel fibers with a lower-melting alloy have useful high-temperature properties since remelting the lower-melting alloy does not necessarily cause the liquid to run out. Composites of this type have shown potential use as unlubricated bearings. Similarly, stainless-steel–copper systems have proved useful as bulk brazing materials.

In addition to reinforcing metals, metal fibers have been used with ceramic materials. The incorporation of a fine metal skeleton in a refractory ceramic can almost certainly improve the ceramic's strength, shock resistance, and thermal properties. Possibly metal-ceramic composites analogous to prestressed reinforced concrete can be developed for even higher strength.

## Organic Fibers in Inorganic Matrix

In addition to metal fibers, lightweight graphite fibers are also used to improve the thermal-shock resistance of ceramics. Since graphite is one of the best high-temperature materials known, it makes an ideal reinforcement for refractory ceramic materials. Although graphite fibers normally oxidize in air at temperatures above 800°F (427°C), encasing them in the ceramic matrix protects them.[12] The high-temperature resistance of graphite fibers is an exception to the rule for organic fibers. Some new plastic fibers can withstand temperatures up to about 700°F (371°C), but since most inorganic matrix materials require temperatures far above this in order to be worked, development of bulk fiber composites is limited.

## Fiber-Fiber Composites

Fiber-matrix composites doubtless receive most of the attention paid to fiber composites today, but a good deal of textile engineering is being done with fiber-fiber composites.

### *Organic-Organic*

Many synthetic fibers, with their good chemical resistance, mechanical properties, and electrical properties, are combined with natural organic fibers to form nonwoven, woven, and knit fabrics that are low in cost and very lightweight. Although in most of these textiles the fibers are not physically joined, some bonded fabrics have been investigated. Bonded either under pressure with an adhesive or (for thermoplastic fibers) under temperatures that cause the fibers to soften and interlock, these fiber composites have greater tensile strength than the unbonded materials (Fig. 1.11). The major limitation of the totally organic fiber composites is their low strength compared with that of systems using inorganic fibers. In addition, very few materials can be used at temperatures above 300°F (149°C). Organic systems are also affected by aging and weathering.

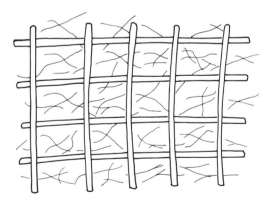

FIBER—FIBER COMPOSITE

**FIG. 1.11** Orientation of fiber-fiber composite.[16d]

### Organic-Inorganic

A number of textile and felt fiber composites use organic and inorganic fibers. Most of these materials depend upon strengthening and toughening of metal fiber and the resiliency, chemical resistance, and colorability of the organic fibers. Typical applications include conveyor belts, filters, gaskets, oil pads, tapes, and seals.

## FLAKE COMPOSITES

Although the strength and one-dimensional nature of fibers make them highly useful in composite structures, particularly where anisotropy is desirable, there are many applications where two-dimensional elements, or flakes, are preferable. These flat forms can be packed together more closely than other shapes. Embedded in a matrix and made parallel in a plane, flakes provide equal properties in all directions in the plane. While they offer little structural strength perpendicular to the plane, their overlapping in that direction is an effective barrier to fluid penetration of the matrix. It is not easy, however, to control the orientation of flakes precisely in a composite or to make suitable flakes; a limited number of materials are available, and development work on flakes continues.

Flake composites are perhaps the least known of the composites. They have been examined for structural applications. The ability of flakes to strengthen resinous materials was demonstrated in the sixties. A flake composite consists of flakes held together by an interface binder or incorporated into a matrix. Depending on the material's end use, the flakes can be present in a small amount or constitute almost the entire composite.

The special properties obtainable with flakes are due in large part to their shape. Being flat, they can be tightly packed to provide a high percentage of reinforcing material for a given cross-sectional area. The considerable amount of overlap between flakes in a composite means that they naturally form a series of barriers to the passage of liquid and vapors; they also reduce the danger of mechanical damage by penetration. If the overlap is such that the flakes actually touch, metal flakes can be used to provide electrical (and sometimes thermal) conductivity through the composite. With nonconductive flakes such as glass or mica it is possible to obtain good dielectric properties and resistance to heat.

Controlling the shape and orientation of the flakes can lead to special decorative effects. Flake aluminum, for example, is used in automobile paints and molded plastics to provide decorative color effects and various degrees of transparency.

### Flakes vs. Fibers

In structural applications flakes appear to offer several advantages over fibers, both real and speculative. For example, as long as the flakes are parallel, flake composites can provide uniform mechanical properties in the plane of the flakes. Although properties approaching isotropic can be obtained in continuous-fiber composites, special processing techniques such as filament winding are required (Fig. 1.12). Flake composites also have a higher theoretical modulus than fiber composites and can be packed closer and with fewer voids. Compared with fibers, flakes are relatively inexpensive to produce and can be handled in batch quantities. Although they are relatively easy to incorporate into composites, obtaining parallel orientation is not easy.

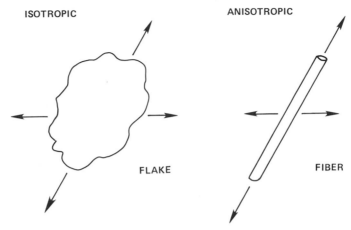

**FIG. 1.12** Difference between fiber and flake.[16c]

## Flake Materials

Flakes are easy to make, but not always easy to make in the desired shape and size. This does not arise with metal flakes, but mica and glass flakes are more critical with respect to shape. From the initial mining stage mica flakes and powders require special sorting and handling procedures. Even though glass flakes are synthetically produced, special care must be taken to avoid cracks or notches at the edges. Some manufacturing processes produce flakes with an inherent microscopic curvature, which are susceptible to cracking under transverse loads during fabrication into flat or molded shapes. However, many methods have been developed for producing glass flakes, and the shape problem is not insurmountable.

The number of materials used in flake composites is limited. Most metal flakes are aluminum, silver being used to a minor extent. The other important flake materials are mica and glass. Almost all flakes can be used with a wide variety of organic or inorganic binders or matrixes as long as the material has chemical, mechanical, and processing compatibility with the flakes.

### Glass-Flake Composites

Glass-flake-reinforced plastics and laminates are said to have originated during World War II. From the first investigations it appeared that small, thin films or flakes of glass offered important theoretical advantages over conventional glass fibers:

1. High flexural modulus since flakes are free to bend in only one plane
2. Uniform mechanical properties in the plane of the flakes (in conventional fiber-reinforced composites mechanical properties drop off at angles away from the fiber axes)
3. Higher strength because flakes can be packed to higher density
4. Lower moisture, liquid, and vapor transmission thanks to the labyrinthine structure
5. Higher dielectric strength and resistance to heat
6. Lower cost because flakes are less expensive to make than glass fiber or cloth

These advantages suggest several important areas of use. Such complex shapes as missile fins, ablative nose cones, windshields, exhaust nozzles, electronic gates, and electronic-circuit frames have been manufactured from 50 and 70% glass-epoxy and 50% glass-polyester resin premixes. The excellent combination of strength and electrical properties has resulted in the production of laminates for printed-circuit applications as well as molded insulators, polarized lighting panels, phosphorescent panes, and electrical potting mixes.

## Mica Flake Composites

Mica flakes are the most familiar and widely used flake materials. Although less versatile than glass flakes for structural and moisture-barrier applications, they are much more useful where a combination of dielectric strength and heat resistance is needed. Natural or synthetic mica flakes are available in an extremely large range of shapes and sizes. Some flakes are so small that they are essentially powders; some are so large that they are essentially thin films and sheets. The size and shape of the flake depend on the type of application, which also determines the amount of binder used. The binder may serve as a matrix making up the bulk of the composite or be used in an amount just sufficient to bind the flakes together. Because they have been available longer, natural mica flakes enjoy a considerably wider range of uses. In heater plates for domestic applicances, for example, mica splittings remain in their original position after the binder has volatilized and insulate perfectly as long as the unit is undisturbed and compression maintained.

## Metal-Flake Composites

The idea of using metals in flake form is attractive because of the special properties metals can contribute to the composite. For example, metal flakes can be used in a matrix to provide impermeable barriers and corrosion resistance, to impart thermal and electrical conductivity to the composite, or to provide a special visual or decorative effect.

Metal-flake production initially was confined principally to aluminum and silver. Probably the greatest deterrent to development of a wider variety of metal flakes has been the relative difficulty of producing them. Finding good flakes for reinforcing composites has been a problem. The composite properties of silicon carbide (SiC) are disappointing because of the low aspect ratio of SiC flakes and the outgrowths at the edge of the crystals. Flake-reinforced metals have been produced by unidirectional solidification of binary eutectic alloys such as $CuAl_2$ in Al, but the density of the flake material is high and control of flake concentration is poor.

A new flake material, aluminum diboride ($AlB_2$), has been successfully produced. $AlB_2$ flakes grown from aluminum melt are thin, flexible hexagons with a density [181 lb/ft$^3$ (2.9 g/cm$^3$)] less than one-third that of steel but a stiffness greater than that of steel (Fig. 1.13), good oxidation resistance to 900°F (482°C), and good thermal resistance to 1900°F (1038°C). Mechanical properties compare favorably with those of other reinforcement materials. $AlB_2$ flakes have been incorporated in epoxy and phenolic matrixes with good results (Fig. 1.14) and in aluminum matrixes as well. Table 1.1 compares the planar mechanical properties of $AlB_2$-flake-reinforced composites with those of unidirectional and pseudoisotropic boron and E-glass composites. This ability to reinforce in the planar direction, a major advantage of single-crystal flakes, is not possible with fiber- or whisker-reinforced composites. Flakes up to 0.39 in (10 mm) diameter with a width-to-thickness ratio of 500:1 have been produced. Binary composites listed in the table consist of 64 vol %

FIG. 1.13 Matrix of AlB$_2$ flake.[13]

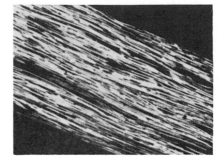

FIG. 1.14 Cross section of AlB$_2$ flake.[13]

AlB$_2$ flakes in an epoxy matrix. The ternary composites consist of alternate layers of flake-reinforced and glass-fiber-reinforced epoxy. As the table indicates, the flake composites have superior multidirection tensile and flexural modulus, the boron composites have the best compressive strength, and the glass-fiber-reinforced composites exhibit the highest tensile and flexural strength.

A new grade of beryllium made from high-purity electrolytic flake has better elevated-temperature creep strength than previous grades of beryllium. Temperature capability has been extended by 250 to 400°F (121 to 204°C) above previous upper limits; and the new grade resists recrystallization and grain growth up to 2300°F (1260°C).

### Graphite-Flake Composites

A cast iron with an intermediate strength of 5.1 kips/in$^2$ (35 MPa), good machinability, thermal properties and castability is called *compact-flake-graphite* (CFG) *iron.* Although the graphite in CFG resembles that of gray iron, the flakes are shorter, thicker, and more rounded at the edge. At a tensile strength of 5.1 kips/in$^2$ (35 MPa), CFG has nearly the same machinability and castability as conventional gray iron. With some refinements in process control left to be achieved, CFG iron appears certain to become a standard casting material with a different range of properties from either gray or nodular iron.

## PARTICULATE COMPOSITES

Particulate composites have an additive constituent which is essentially one- or two-dimensional and macroscopic. In some composites, however, the additive constituent is macroscopically nondimensional, i.e., conceptually a point, as opposed to a line or an area. Only on the microscopic scale does it becomes dimensional, i.e., a particle, and thus the concept

**Table 1.1** Mechanical Properties of Flake- and Filament-Reinforced Composites[13]

| Type[a] | Tensile strength ksi | Tensile strength MPa | Tensile modulus $10^6$ psi | Tensile modulus GPa | Compressive strength ksi | Compressive strength MPa | Compressive modulus $10^6$ psi | Compressive modulus GPa | Flexural strength ksi | Flexural strength MPa | Flexural modulus $10^6$ psi | Flexural modulus GPa | Interlaminar shear strength[b] ksi | Interlaminar shear strength[b] MPa |
|---|---|---|---|---|---|---|---|---|---|---|---|---|---|---|
| **Unidirectional** | | | | | | | | | | | | | | |
| Boron, 0° | 110 | 758 | 43 | 296 | 164 | 1131 | 42 | 290 | 220 | 1517 | 24 | 165 | 12 | 83 |
| 45° | ... | ... | ... | ... | 16 | 110 | 2.8 | 19 | 16 | 110 | 4 | 28 | | |
| 90° | 3 | 21 | 5 | 35 | 17 | 117 | 3 | 21 | 8 | 55 | 4 | 28 | | |
| E glass, 0° | 160 | 1103 | 8 | 55 | 113 | 779 | 5 | 35 | 137 | 945 | 5 | 35 | 12 | 83 |
| 45° | 25 | 172 | ... | ... | 19 | 131 | ... | ... | 11 | 76 | 1.3 | 9 | | |
| 90° | 4 | 28 | 1.4 | 10 | 17 | 117 | 1.2 | 8 | 6 | 41 | 1 | 7 | | |
| **Pseudo-isotropic[c]** | | | | | | | | | | | | | | |
| Boron, 0° | 23 | 159 | 10 | 69 | 89 | 614 | 18 | 124 | 52 | 357 | 12 | 83 | 3 | 21 |
| 45° | 25 | 172 | 10 | 69 | 86 | 593 | 14 | 97 | 42 | 290 | 9 | 62 | | |
| 90° | 26 | 179 | 8 | 55 | 84 | 579 | 15 | 103 | 47 | 324 | 10 | 69 | | |
| E glass, 0° | 60 | 414 | 2 | 14 | 54 | 372 | 3 | 21 | 69 | 469 | 3 | 21 | 5 | 35 |
| 45° | 46 | 317 | 2 | 14 | 50 | 345 | 3 | 21 | 63 | 434 | 2 | 14 | | |
| 90° | 50 | 345 | 2 | 14 | 51 | 350 | 3 | 21 | 62 | 427 | 2.5 | 17 | | |
| **AlB$_2$ flake composites** | | | | | | | | | | | | | | |
| Binary[d], 0° | 35 | 241 | 33 | 228 | 23 | 159 | 19 | 131 | 56 | 386 | 20 | 138 | 5 | 35 |
| 45° | 35 | 241 | 33 | 228 | ... | ... | 19 | 131 | 56 | 386 | 20 | 138 | | |
| 90° | 35 | 241 | 33 | 228 | ... | ... | 19 | 131 | 56 | 386 | 20 | 138 | | |
| Ternary[e], 0° | 70 | 483 | 22 | 152 | ... | ... | 17 | 117 | 80 | 552 | 16 | 110 | 4.5 | 31 |
| 45° | ... | ... | ... | ... | ... | ... | ... | ... | ... | ... | ... | ... | | |
| 90° | 24 | 165 | 17 | 117 | ... | ... | ... | ... | 33 | 228 | 11 | 76 | | |

[a] 0° = 0 rad, 45° = 0.79 rad, and 90° = 1.6 rad.  [b] Value for all orientations.  [c] Fiber axis aligned at 0, 45, and 90°.
[d] 64 vol % AlB$_2$ flakes in epoxy matrix.  [e] Alternate layers of flake-reinforced epoxy and glass-filament-reinforced epoxy.

of composites must come down to the microscopic level if it is to encompass all the composites of interest to engineers. *Particulate composites* differ from the fiber and flake types in that distribution of the additive constituent is usually random rather than controlled. Particulate composites are therefore usually isotropic. This family of composites includes dispersion-hardened alloys and cermets.

## Types and Properties

The size, nature, and function of the particles vary widely—from control of mechanical behavior at the atomic level to less sophisticated averaging of antithetical properties. Thus, the particulates pose some of the trickiest obstacles to developing a unified theory of composites, but it is this very breadth of application that makes particulate composites an exciting field for speculation and experiment. In terms of applications and industrial applicability to current problems, particulate composites range from some of the most widely used materials to a lengthy list of laboratory materials.

Strength of a dispersion-hardened composite is directly proportional to the hardness of the dispersed particle (particularly at elevated temperatures) since the particle must resist the stresses caused by dislocation pileups against it (Fig. 1.15). Coherency strains between particle phase and matrix also affect strength. Low energy at the particle-matrix interface implies good coupling, which is necessary if the particles are to act as barriers to dislocation flow. High interfacial energy, on the other hand, is equivalent to a hole surrounding a particle. This would not only be a poor barrier to dislocation motion (with a resulting lowering of strength) but would also act as a microcrack in the structure.

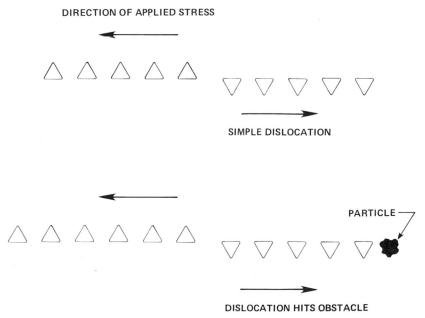

FIG. 1.15   Mechanism of dispersion-hardening composites.[17]

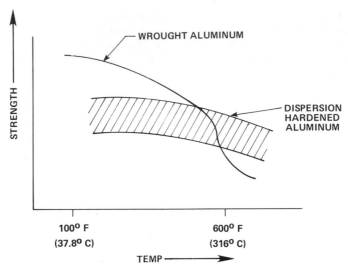

**FIG.** 1.16  Strength of dispersion-hardened aluminum compared with that of aluminum.[16f]

Because of the wide range of particle variables, particles combined in a matrix can produce composites with many unusual properties. Dispersions in metals can greatly expand the temperature range of use, as well as strengthen the matrix; e.g., the sintered-aluminum-powder alloys have usable strength ranging up to 1000°F (538°C), which is 400°F (204°C) higher than heat-treated aluminum alloys (Fig. 1.16).

The cermets are mixtures of ceramics and metals which offer the engineer an opportunity to combine the properties and take advantage of the inherent characteristics of both. For example, ceramics do not degrade at high temperature but lack ductility. Combining ceramic particles in a metal matrix results in a composite that has better ductility than the ceramics alone along with good high-temperature resistance.

Metal-plastic composites are among the most widely used groups of the particulate composites and have many interesting characteristics. Transparent plastics heavily filled with metal powder, for example, take on the color of the powder. Plastic-metal combinations can also reduce electrical resistivity of the plastic. Resistors and capacitors can be tailored to yield an infinite range of resistivity through combinations of metal powders and conductive carbon blocks.

### Cermets

Under the microscope a cermet has a structure composed of ceramic grains held in a metal matrix (or binder). The matrix usually accounts for up to 30% of the total volume. In the more successful cermets, bonding between constituents results from a small amount of mutual or partial solubility. On the other hand, some cermet systems, such as the metal oxides, generally exhibit poor bonding between constituents and require additions that are partially soluble in both the metal and the oxide. For example, the addition of titanium nitride to a nickel–magnesium oxide cermet provides a metal–metallic oxide interface type of bond. Cermet composites are formed by powder-metallurgy techniques and can achieve a wide range of properties, depending on the composition and relative volumes of the metal and ceramic constituents.

**Oxide-Base Cermets:**  The outstanding characteristic of oxide-base cermets is that the metal or ceramic can be either the particle or the matrix constituent. Thus, a wide range of property values is available and the composition can be tailored to service requirements. Properties depend strongly on the volume of the binder material. For example, while thermal shock resistance of a 28% $Al_2O_3$–72% Cr cermet is good, it can be greatly improved by reversing the proportions of the constituents. The same reversal reduces modulus of elasticity at 75°F (24°C) from 52.3 to 4.7 kips/in$^2$ (361 to 32.4 GPa). Numerous oxide-base cermets have been developed. $Al_2O_3$–Cr has already been mentioned, and $Al_2O_3$ has also been combined with stainless steel and 80Cr–20Mo to achieve good thermal conductivity and closely matched thermal-expansion characteristics.

The 6MgO–94Cr cermets reverse the roles of the oxide and chromium; i.e., the MgO is added to improve the fabrication and performance of the chromium. Chromium is not ductile at room temperature. Adding MgO not only permits press forging at room temperature but also increases oxidation resistance to almost 5 times that of pure chromium.

The oxide-base cermets have been extensively applied as a tool material for high-speed cutting of materials difficult to machine. Other applications include flow control pins (because of chromium-alumina's resistance to wetting and erosion by many molten metals and to thermal shock), thermocouple protection tubes (high thermal conductivity), mechanical seals (low coefficient of friction plus abrasion resistance), and gas-turbine flameholders (resistance to flame erosion).

Of the cermets, the oxide-base alloys are probably the simplest to fabricate. Normal powder metallurgy or ceramic techniques can be used to form shapes, but these materials can also be machined or forged.

**Carbide-Base Cermets:**  There are three major families of carbide-base cermets: tungsten carbide, chromium carbide, and titanium carbide, and the number of compositions available to the user in each of these families is large.

Tungsten carbide, widely used as a cutting-tool material, is high in rigidity, compressive strength, hardness, and abrasion resistance. For bonding, cobalt is added up to 35 vol %, and properties vary accordingly. As with the oxide-base cermets, increasing binder volume produces such property improvements as improved ductility and toughness. Compressive strength, hardness, and modulus of elasticity are reduced as binder content increases, but transverse rupture strength and density are increased. Structural uses of tungsten carbide–cobalt cermets include wire-drawing dies, precision rolls, gages, and valve parts. Higher-impact grades can be applied where die steels were formerly needed to withstand impact loading. Combined with superior abrasion resistance, the higher impact strength results in die-life improvements as high, in some cases, as 5000 to 7000%.

Chromium carbide offers phenomenal resistance to oxidation, excellent corrosion resistance, relatively high thermal expansion, relatively low density, and the lowest melting point of the stable carbides. Gages of chromium carbide make use of its rigidity, abrasion resistance, corrosion resistance, and ability to take a high polish. Also important is the similarity of its coefficient of expansion to that of steel. Valve parts make use of the material's compressive strength and erosion resistance. Valve liners, oil-well check valves, spray nozzles, bearing seal rings, and pump rotors are other uses. Bearings use the high compressive strength and rigidity to provide extremely high load-carrying ability. Harder grades can be run against softer ones and operate without lubricants up to 1800°F (982°C). Nonmagnetic compositions can be formulated for use where magnetic materials cannot be tolerated.

Titanium carbide, used principally for high-temperature applications, has good oxidation and thermal-shock resistance, good retention of strength at elevated temperatures, and a high modulus of elasticity. Because of the relatively poor oxidation resistance of cobalt at elevated temperatures, nickel is more commonly used as the binder. Titanium carbide cermets have been used in such applications as gas-turbine nozzle vanes and buckets, integral turbine wheels, torch tips, hot-upsetting anvils, hot-spinning tools, hot-mill roll guides, valve and valve seats, and thermocouple protection tubes.

**Other Cermets:**   Refractory cermets, such as tungsten-thoria and barium carbonate–nickel, are widely used in higher-power pulse magnetrons. The cermets resist poisoning agents, recover from sporadic arcing, and prevent excessive frequency drift by keeping deposits to a minimum. Cermets also serve as friction materials, where they combine the thermal conductivity and toughness of metals with the hardness and refractory qualities of ceramics. Roughness, the principal disadvantage of cermets in this application, which can lead to grabbing, can be overcome by careful formulation. In the nuclear field, cermets are used for fuel elements and control rods. Structural components such as valve seats, bearings, etc., have also been fabricated. As reactor operating temperatures increase, we can expect to see more cermet applications in this industry. Uranium oxide, the reactant in particle form in the fuel elements, is combined with several types of matrix material, e.g., stainless steel–uranium oxide. Cermets also play an important role in sandwich-plate fuel elements. For example, a relatively porous graphite–uranium oxide fuel plate is prepared and then siliconized to form beta silicon carbide. The finished element is a siliconized silicon carbide with a core containing uranium oxide. Other cermets used include chromium-alumina, nickel-magnesia, and iron–zirconium carbide. Control rods have been fabricated from boron carbide–stainless steel and rare-earth oxides–stainless steel.

## Dispersion-Hardened Alloys

Like cermet composites, dispersion-hardened alloys generally consist of a hard particle constituent in a softer metal matrix. They differ from cermets in that their dispersed-particle constituent is only a small proportion of the total, seldom exceeding 3 vol %, and the particles themselves are considerably smaller, often ranging below micrometer sizes (Fig. 1.17). Dispersion-hardened alloys differ from precipitation-hardened alloys in that the particle is added to the matrix, usually by nonchemical means. Precipitation-hardened alloys derive their properties from compounds that are precipitated from the matrix, not separately added.

**Particle Control of Properties:**   As with cermets, the characteristics of the particles largely control the properties of dispersion-hardened composites; this is particularly true of strength. The finer the interparticle spacing for a given alloy series the better the ductility and the higher the thermal and electrical conductivities. Spacing of 0.2 to 0.3 $\mu$m provides the best of these properties. Particle size is also important. The same interparticle spacing can be developed with a wide range of particle diameters. As particle size increases, less material is needed to achieve the desired interparticle spacing. The most desirable dispersion is the least soluble and most refractory. Refractory oxides offer the best potential,

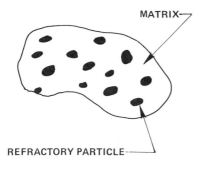

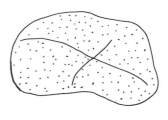

CERMET                    DISPERSION HARDENED MATRIX

FIG. 1.17   Difference between cermet and diffusion-hardened alloy.[16/]

although intermetallic compounds such as AlFe$_3$ have been used. Cold working is required to achieve high strength levels. Cold working the matrix combined with the dispersion of fine particles impedes dislocation movement and strengthens the dispersion-hardened metal.

**Production Methods:**   The method of producing dispersion-hardened alloys is usually governed by the materials involved; in turn, the method has considerable effect on the final characteristics of the particle constituent. Particles can be introduced into the matrix in several ways.

*Surface Oxidation of Ultrafine Powders:*   This method, used for sintered-aluminum-powder alloys (SAP), consists of controlled oxidation of aluminum powder, which causes each particle to be surrounded by a thin layer of oxide, and simultaneous grinding to weld the oxidized particles together and trap fine oxides in their interiors. Compacts are then sintered and cold-worked to form a strong dispersion-strengthened structure. The unique characteristic is the temperature dependence of strength (Fig. 1.18). Instead of softening at a temperature about half the melting temperature, recrystallization and grain growth is avoided, and an appreciable fraction of the material's lower-temperature strength persists up to 80% of its melting temperature.

*Internal Oxidation of Dilute Solid-Solution Alloys:*   These systems consist of a matrix containing a relatively large amount of a noble metal and a readily oxidizable solute element.

*Mechanical Mixing of Fine Metal Powders:*   Metal powders (less than 10 $\mu$m in diameter and for optimum properties $\mu$m and finer) are blended with much finer metal oxide powders. The normal procedure is to add 0.15 to 0.5 by vol %, which produces a series of alloys with increasing strength and decreasing ductility. This procedure is less efficient than either surface or internal oxidation.

*Direct Production from a Liquid Metal:*   Liquid metal is quenched and atomized to produce the powder. This method requires an alloying element which is soluble in the liquid and relatively insoluble in the solid. As solubility decreases, the structure becomes more stable at elevated temperatures. One advantage of this technique is that it produces coarse powders which are easier to handle than the fine powders normally produced by the first three methods.

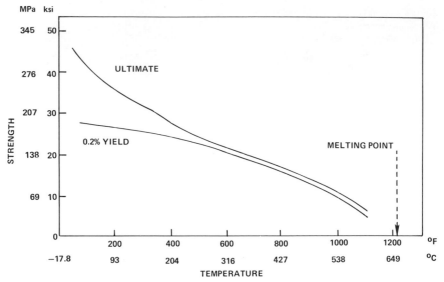

**FIG. 1.18**   Temperature dependence of the strength of SAP alloys.

*Colloidal Technique:*   Developed in 1962, this technique involves a patented codeposition technique which inherently controls the size and distribution of a dispersed thoria (ThO$_2$) phase. The codeposition process involves the encapsulation of submicrometer thoria particles with the matrix material by means of colloidal techniques. The process offers a distinct advantage over conventional powder metallurgical mixing techniques in that particle segregation is avoided during processing. Mixtures of nickel oxide, thoria, and chromium oxide (TDNiCr) as well as nickel oxide and thoria (TDNi) have been produced. The process essentially follows powder-metallurgy techniques. TD nickel has modest strength at room temperature but retains it nearly to the melting point (Fig. 1.19). Above

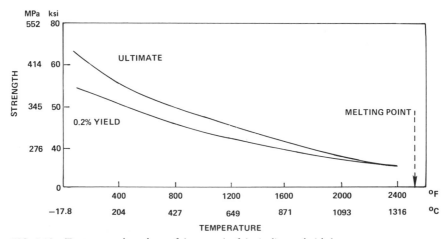

**FIG. 1.19**   Temperature dependence of the strength of thoria-dispersed nickel.

2000°F (1093°C) its strength is superior to many of the iron, cobalt, and nickel superalloys. The fine dispersion inhibits the grain growth that is a source of weakening at these temperatures in pure nickel and its alloys.

**Dispersion-Hardened Metal Systems:** Of the many dispersion-hardened alloy systems only a few have reached commercial significance, namely, aluminum, nickel, and tungsten; two others, copper and titanium, have proved successful in laboratory and developmental stages. The aluminum alloys represent the principal commercial use of dispersion hardening. There are two groups of commercial dispersion-hardened aluminum alloys: the original AIAG alloys, now called SAP alloys, and the APM alloys. SAP alloys are normally prepared by surface oxidation; APM alloys are produced by atomizing liquid aluminum and forming insoluble intermetallic compounds. Both SAP and APM alloys maintain usable strength up to 1000°F (538°C), compared with 600°F (316°C) for 2219-T6, a high-strength wrought-aluminum alloy. However, the ductility of the dispersion-strengthened alloys above 600°F (316°C) is lower than that of the wrought alloy. SAP alloys have been used for pistons and impellers in the aircraft industry.

Joining dispersion-hardened metals is one of the principal problems slowing their universal acceptance. Fusion methods cannot be used because the effect of straining is lost and the particles agglomerate. Successful techniques include cold and hot pressure welding, spot welding, flash butt welding, and ultrasonic welding.

TD nickel is 3 to 4 times stronger than pure nickel in the 1600 to 2400°F (871 to 1316°C) range, and oxidation resistance of the alloy is better than that of nickel at 2000°F (1093°C).

An alloy which falls within the realm of composite materials and has been in service for over 30 years as a lamp filament material is tungsten-thoria, still the most successful and widely used tungsten alloy. It is superior in strength to tungsten, and no commercially available tungsten alloys exceed it in strength at 3500°F (1927°C).

**Metal-in-Metal and Metal-in-Plastic Materials:** Metal-in-metal composites, composed of metal particles in a metal matrix, occupy an important place as industrial materials. They also constitute a means of producing ductile materials by combining an essentially brittle metal with a more ductile one. Steel- and copper-base alloys containing lead particles are characteristic of this group.

Metal-in-plastic materials include a number of useful particulate composites, which consist of metal particles in a plastic matrix. Such filled plastics may contain up to 90 vol % of metal particles. The range of metal-filled plastics is very broad. After looking at some of the filler materials used, we examine applications in various industries.

As a filler aluminum has applications ranging from a decorative finish to the improvement of thermal conductivity. Improved thermal or electrical conductivity is developed in castable thermosetting plastics. At 30 vol % aluminum improves thermal conductivity 300% (compared with an unfilled epoxy resin) and lowers electrical resistivity somewhat. One use of the improvement in thermal conductivity has been in plastic tooling. In molding polyester or phenolic resins and glass fibers high thermal conductivity is desirable. Though unsuitable for long production runs, the aluminum-powder–epoxy-resin composite is easily formed in less time and at lower cost than metal molds.

Unlike aluminum particles, when added to liquid resin polymers iron and steel tend to settle because of their higher density relative to the resin. Nevertheless, steel-filled plastics have been used successfully for many years on small-lot production tooling; parts range in weight from a few ounces to several hundred pounds. Other important uses for steel-particle-filled plastics include automotive body solders and hull-smoothing cements for shipbuilding.

Adding copper particles to liquid plastics poses the problem of rapid settling because of the large density differences between the two materials, but the problem becomes less pronounced as the volume percent of the copper filler is increased. A significant use for copper and its alloys is as a coloring material. Other uses include bearing and friction materials, magnetic materials, electrical encapsulants, and nuclear applications.

**Metal in Ceramic and Nonmetallic in Nonmetallic:**   Composites consisting of metal particles in a ceramic matrix have been used in industry for many years. Copper-graphite composites, used as motor brushes and commutator segments, are one example. Nonmetallic particles in a nonmetallic matrix include oxidation-resistant graphites containing zirconium diboride, boron, and silicon. The thermal-shock-resistant refractories combine graphite and zirconium carbide, silicon carbide and graphite, zirconium diboride and boron nitride, or aluminum oxide and zirconium oxide.


# FILLED COMPOSITES

In its simplest form a filled composite consists of a continuous three-dimensional structural matrix infiltrated or impregnated with a second-phase filler material. The filler also has a three-dimensional shape, determined by the voids in the matrix. The matrix itself may be an ordered honeycomb, a group of cells, or a random spongelike network of open pores. In most of the familiar filled composites the cellular structure of the matrix is rather fine, and the function of the filler is to seal it or to provide a more desirable and self-renewing surface. Thus metal-powder parts and castings that are impermeable or alloys that are self-lubricating fall into this category. Some of the newer filled composites, however, arrive at a solid structure consisting of two intertwining skeletons of different properties. In others, those having large cells, a major purpose of the matrix may be to confine the filler in discrete and limited volumes to control accumulation of strain.


## Constituents

In effect, both the matrix and the filler exist as two separate constituents that do not alloy and (except for a bonding action) do not combine chemically to any significant extent. The matrix is always continuous, but the filler may be either continuous (as in an impregnated casting) or discontinuous (as in a filled honeycomb). In most filled composites the matrix provides the framework and the filler provides the desired engineering or functional properties. Although the matrix usually makes up the bulk of the composite, the filler material is often used to such a large extent that it becomes the dominant material and makes a significant contribution to the overall strength and structure of the composite. In order to obtain the optimum properties in filled composites the two materials must be compatible and not react in a way that would degrade or destroy their inherent properties. Thus, it is important for the matrix and filler materials to exist as two separate constituents.

## Types of Skeletons

The skeletal structures most receptive to impregnation are an open honeycomb or group of cells and a spongelike network of open pores. In general, the open-pore structure presents more processing problems than the honeycomb because its random orientation usually prevents introducing filler materials in their solid state. The bond between the structure and filler is usually not as strong in honeycomb or cellular structures as in spongelike structures. The random orientation of the spongelike structure helps keep the filler material in place even if there is little or no adhesive action between it and the matrix. Because honeycomb cells have flat, unbroken areas, the bond strength between structure and filler is not likely to be high. However, the bond-strength factor is seldom of great importance in filled-honeycomb composites because they are generally not used in structural applications.

### Filled Honeycomb and Cells

There are many applications, especially at high temperatures, where a material will perform better if it is used in small sections rather than in bulk. For example, a thin plastic ablator of large area has little mechanical strength at high temperatures. Similarly, a large mass of ceramic material tends to have poor resistance to the stresses caused by large temperature differences between its surfaces. Such problems can be avoided by using the materials in smaller sections, i.e., incorporating them as fillers in a honeycomb or cellular structure.

**Metal-Ceramic:** A composite consisting of a metal honeycomb structure filled with ceramic can be useful where very high heat fluxes occur for short periods. Even though the heat may be high enough to vaporize or melt the surface of the ceramic, the segments can be prevented from rupturing if their cross-sectional area is kept small enough to avoid critical gradients.

**Nonmetallic-Ceramic:** Filled composites for high-temperature service can also be constructed with nonmetallic matrixes such as paper or wax. The matrix is subsequently burned out before or during service, leaving the ceramic "pencils" separated by a narrow gap.

**Metal-Nonmetallic:** The principal advantage of a nonmetallic-filled metal-honeycomb composite is that the metal matrix strengthens or reinforces the otherwise weak nonmetallic material. For example, in reentry applications the shear load experienced by a nonmetallic heat shield is high enough to erode mechanically a charring ablative material. Damage can be prevented by incorporating the material into a strengthened honeycomb. The composite can be processed more reliably than a large monolithic structure and presents fewer difficulties in fastening to the base surface. A metal honeycomb can also enable a nonmetallic material to withstand greater vibration and flexing. A metal-silica honeycomb, for example, can be flexed without having the silica fall out even though it has been cracked. Materials such as fiber-reinforced epoxy and silicone rubber can also be incorporated in a metal-honeycomb structure. Considerable improvements in the mechanical properties of silicone rubber have been obtained by confining the material to cells, where it can maintain its structural integrity.

## Filled Sponge and Pores

Most filled composites in use today consist of a matrix formed from a random network of open passages or pores. Unlike the matrix in a filled honeycomb, which is specifically designed to a given shape, the open matrix in a sponge or pore composite is formed naturally during processing. Typical of materials that have this kind of structure and lend themselves to filling are metal castings, powder-metal parts, ceramics, carbides, graphite, and foams. The open network of a spongelike structure can be filled with a wide range of materials, including metals, plastics, and lubricants, depending on the end properties being sought. Metal impregnants, for example, can be used to improve the strength of a matrix or to provide better bearing properties. Plastics can be impregnated into metals to make them pressuretight or to act as lubricants to provide special bearing properties. They can also be incorporated into porous ceramics and graphite as a structural binder.

**Metal-Metal:** Although metal-filled metal matrixes have been used to overcome the inherent disadvantages of a naturally porous structure, as in metal-powder parts, functional advantages can be gained by deliberately designing a porous metal structure that can be filled with another metal to provide a special characteristic such as transpiration cooling. The principal advantages of infiltrating metal-powder parts are to provide higher strength and hardness and uniform density in parts that are difficult to press uniformly.

Metal-filled metal composites also provide an ingenious approach to the problems of using metals at high temperature. If a high-temperature metal can be made porous, it should be possible to impregnate it with a lower-melting metal to provide a new set of characteristics not obtainable with either metal alone. For example, tungsten bodies with 80% initial density that are impregnated with copper or silver have good strength at room temperature, are relatively easy to machine, and provide transpiration cooling as well as good resistance to thermal shock. The cooling, which is useful in applications such as rocket nozzles, occurs because the copper or silver impregnant absorbs heat and is vaporized before the tungsten reaches its melting point. Further cooling is provided by the vapor layer at the surface. In addition to high thermal conductivity, the copper or silver has inherent ductility that helps stop crack propagation due to thermal shock. Silver-impregnated tungsten composites are feasible for other applications, such as bearings, where it is desirable to have a two-phase metal system.

Although the idea of a metal-filled-metal composite appears feasible for many applications, the two metals must be compatible. For example, silver and copper work well with tungsten because they are insoluble in it. On the other hand, nickel is soluble in tungsten and would probably destroy a tungsten matrix.

Of course, metals other than tungsten can be used for bearing applications. Automotive bearings can be prepared from a spongelike copper-base alloy that is brazed to a steel backing and then impregnated with babbitt metal. The babbitt provides good bearing properties and resistance to corrosion, and the matrix serves as a cushion between the bearing surface and the steel backing.

Impregnating one metal with another has also proved valuable for electric contacts. By impregnating molybdenum or tungsten with copper or silver it is possible to obtain an excellent combination of mechanical and electrical properties. For example, a filled composite of molybdenum and 20% or more silver is useful at high currents: the hardness and wear resistance of molybdenum combine with the conductivity of silver to produce a contact whose properties remain substantially unaltered by the heat generated from the passage of current. Similar properties can be obtained with tungsten-silver composites.

Metal fillers can also be used to make a pressuretight composite. Although the common practice is to seal castings with plastic or inorganic materials, engineers have found that some metal fillers are capable of running into the narrowest pores of ferrous and nonferrous castings by capillary action. Thus, molten combinations of zinc, copper, and aluminum powders can be used to provide good sealing action in metal castings in the event of surface wear.

**Metal-Nonmetallic:** Special engineering characteristics can often be imparted to a porous metal structure by impregnating it with a nonmetallic fluid lubricant or with a fluid resin that subsequently hardens into a solid mass. One of the most successful composites consists of a steel backing on which is sintered a lining of spherical bronze (89Cu–11Sn) impregnated with a mixture of tetrafluoroethylene and lead. Finally one of the most obvious advantages of filling a porous metal with plastic or other material is that it can make the metal pressuretight and improve its corrosion resistance.

**Nonmetallic-Nonmetallic:** The idea of filling one nonmetallic material with another is very attractive, an example being graphite pipe and structural shapes. Another interesting application is heat-shield composites for the thermal protection of reentry vehicles. These materials are resin-impregnated ceramic foams. Zirconia foam has exhibited better properties than alumina, and foams have been produced with about 90% porosity. The pore size and density can be varied within reasonable limits to obtain the desired balance between porosity, weight, and strength.

**Nonmetallic-Metal:** By infiltrating a porous cermet skeleton with molten metal several composites have been produced. These composites consist of a steel matrix containing titanium carbide, which can be machined and heat-treated (because of the steel) and have good wear resistance (because of the carbide). They thus combine the machining advantages of the tool steels with the long life of the carbides. Several different metals can be incorporated into the composite, depending on the type of service. A heat-treatable alloy-steel matrix, for example, is popular and can be used for punches, dies, and gages. High-speed steels and austenitic-steel matrixes can also be used to contain the titanium carbide.

Another family of composites, consisting of graphite impregnated with metallics such as babbitt, copper, bronze, or silver, is particularly useful in bearing applications, where the metal improves overall strength and heat dissipation. Metal-impregnated graphite composites also combine good electrical conductivity with high wear resistance. Because they can carry high currents without welding, they are particularly useful as contact materials for relays, interrupters, and circuit breakers, and electrical brushes in high-rotating-speed equipment.

## Microspheres

Microspheres are particulate fillers used in plastic and other materials. They can cut costs, add strength, lower density, or solve problems of warpage and shrinkage. They are excellent fillers for plastics, work well in both thermosets and thermoplastics, and can be molded in a variety of processes. By displacing significant amounts of resin microspheres can double resin volume and parts produced. The combination of good economics and improvement in properties is making microspheres a competitor with glass fibers in some areas. New developments find them being used in spray-up for parts for furniture and in sheet- and bulk-molding compounds.

Plastic microspheres are best known, and their widest use is as a component of syntactic foam. Microspheres made from glass are receiving the most attention as resin extenders. Solid spheres, which are 10 times heavier than hollow spheres, provide reinforcement and are tough enough to be extruded or injection-molded. Hollow spheres, on the other hand, are used for weight reduction because of their low density. They affect most mechanical properties adversely and cannot be used in high-pressure molding processes. The solid spheres are cheaper by the pound than hollow ones, but the difference in density makes them more expensive on a volumetric basis.

## Solid Spheres

Solid spheres are made from soda-lime glass. They range in size from 0.00015 to 0.2 in (0.004 to 5.0 mm) in diameter, although 0.00015 to 0.0015 in (0.004 to 0.038 mm) is said to be the optimum size. With a specific gravity of about 2.5, solid spheres increase the weight of a composite rather than reduce it. Their use results in marked improvements in mechanical properties.

For epoxy (Table 1.2) the addition of solid spheres (80 phr) almost doubles flexural modulus. This increase is 16% more than is gained by the use of 80 phr of calcium carbonate filler and 14% more than with 80 phr of clay filler. The spheres improve tensile, flexural, and compressive strength and lower elongation and water absorption more than the other fillers. The spheres also have an advantage in viscosity. They can be mixed in the resin easily at 120 phr. Calcium carbonate and clay are difficult to mix at 80 phr and cannot be used above that concentration.

Added to type 6/6 nylon, solid spheres increase compressive strength from 4.2 to 36.5 kips/in$^2$ (29 to 252 MPa) and reduce ultimate elongation from 60 to 2.5%. A compound of 60% 6/6 resin, 25% spheres, and 15% glass fiber provides a greater increase in tensile strength than compounds with 20 to 30% glass fibers. Among the benefits that spheres bring to other properties are lower deformation under load, increased abrasion resistance, higher melt index, and smoother surface. Because the glass is inert, it improves the corrosion resistance of a composite and decreases the flammability of resins.

Coating the spheres with a coupling agent is recommended (except with silicones and fluorocarbons) to obtain a good bond between spheres and resin. The type of coupling agent depends upon the type of resin. Coating with a coupling agent adds about 3 cents per pound (7 cents per kilogram) to the cost of the spheres.

A glass sphere is isotropic (no orientation) and provides uniform cure shrinkage without causing internal stresses. As a result, glass spheres can be used to control warpage and reduce shrinkage. For example, the addition of spheres to an automotive glove-box door made of reinforced styrene-acrylonitrile polymer eliminated a shrinkage problem caused by the anisotropic nature of the fibers. Glass spheres disperse evenly throughout a part, assuring that all the resin will be reinforced whatever the shape of the mold. Using spheres in a resin results in strong edges and corners. In terms of processability, a producer of solid spheres has shown that resins reinforced with spheres require less torque to process than

**Table 1.2** Effect of Fillers on Physical Properties of Epoxy Resins†

| Filler, phr‡ | Tensile strength | | Elonga-tion, % | Flexural strength | | Flexural modulus | | Compressive strength | | Water absorp-tion, % | Hardness, Rock-well M | sp gr |
|---|---|---|---|---|---|---|---|---|---|---|---|---|
| | ksi | MPa | | ksi | MPa | ksi | GPa | ksi | MPa | | | |
| None | 2.36 | 16.3 | 2.55 | 2.82 | 19.4 | 370 | 2.55 | 10.46 | 72.1 | 0.183 | 74 | 1.14 |
| Solid sphere, 40 | 2.37 | 16.3 | 1.98 | 6.61 | 45.6 | 525 | 3.62 | 10.18 | 70.2 | 0.058 | 73 | 1.29 |
| 80 | 3.38 | 23.3 | 2.08 | 7.7 | 53.1 | 733 | 5.05 | 10.37 | 71.5 | 0.061 | 76 | 1.39 |
| 120 | 3.49 | 24.1 | 1.88 | 6.8 | 46.9 | 810 | 5.59 | 10.51 | 72.5 | 0.034 | 87 | 1.48 |
| $CaCO_3$, 40 | 2.47 | 17.0 | 1.58 | 4.63 | 31.9 | 514 | 3.54 | 10.09 | 69.6 | 0.093 | 77 | 1.22 |
| 80 | 2.88 | 19.9 | 2.38 | 3.5 | 24.1 | 617 | 4.25 | 9.58 | 66.7 | 0.091 | 51 | 1.48 |
| Clay, 40 | 2.76 | 19.0 | 2.63 | 4.96 | 34.2 | 471 | 3.25 | 9.42 | 65 | 0.074 | 56 | 1.32 |
| 80 | 2.5 | 17.2 | 2.13 | 4.67 | 32.2 | 620 | 4.34 | 9.5 | 65.5 | 0.098 | 65 | 1.48 |

†Based on *Mater. Des. Eng.*, April 1975, p. 58.
‡Parts of additive by weight per 100 parts by weight of resin.

other fillers tested. In processing nylon resins, nylon filled with 25% spheres required about one-fifth the torque the unfilled nylon required. Ease in processing results in shorter cycle times. In addition to extrusion and injection molding, resin with solid spheres can be processed by blow molding, compression molding, and transfer molding.

### Hollow Spheres

Hollow glass microspheres are light enough to float on water; densities range from 5 to 50 lb/ft³ (0.08 to 0.8 g/cm³). They range in size from 0.00075 to 0.0075 in (0.02 to 0.2 mm) in diameter. Hollow spheres, as a component of syntactic foam, have found use in aerospace and deep-submergence applications. As an example of their capacity to reduce weight, the addition of 8% spheres to a 20% glass-reinforced polyester laminate reduces density from 82 to 64 lb/ft³ (1.31 to 1.03 g/cm³). The disadvantage of adding hollow spheres to the composite is that they cut flexural strength and modulus in half.

In a phenolic molding compound, the addition of 25% hollow spheres reduces mechanical properties but improves electrical and thermal properties and cuts density in half. The spheres change the dielectric constant of the compound from 4.56 to 1.76 and the dielectric strength from 325 to 435 kV/mil (12.8 to 17.1 MV/mm). Used in the core of syntactic foam composites, microspheres not only serve to make the sandwich structure lightweight but also increase stiffness and improve crack resistance. Because they give a polyester composite the workability of wood (they permit nailing and sawing), spheres are growing in use in the furniture industry. The spheres can be used in spray-up for polyester laminates and for making thermoformed parts rigid.

**Processing Hollow Spheres:** The strength of the standard hollow sphere, in terms of hydrostatic pressure required to reduce the volume of the spheres by 10%, is about 220 lb/in² (1.5 MPa). A much stronger hollow sphere, or bubble, developed recently will withstand a hydrostatic pressure of 2.2 kips/in² (15.2 MPa) while showing only a 10% loss of volume in water. This bubble is strong enough to take the pressures of compression molding and can be used in both sheet-molding compounds (SMC) and bulk-molding compounds (BMC). One molder using the new bubbles reports that they reduce the weight of reinforced parts by 25 to 35% while retaining almost equivalent physical properties and improving the flow characteristics of the SMC and BMC, making them easier to mold.

### Other Types of Microspheres

Microspheres can also be made out of plastic and other materials. Plastic microspheres are used mostly in the production of syntactic foams, although they also are used as bulk fillers. They have a lower service temperature than glass spheres. Although plastic microspheres made from polyvinylidene chloride are expensive [$5 per pound ($11 per kilogram) (1979)], they are excellent resin extenders, saving as much as 40% resin in spray-up laminate applications. They reduce density in polyester castings more than most glass spheres and can be used in combination with other fillers. Applications for polyvinylidene chloride spheres include polyester furniture and sandwich construction of boat hulls. Epoxy spheres can be used as low-density bulk filler for plastics and ceramics. Developed for use in submerged deepwater floats, they can withstand hydrostatic pressures of 10 kips/in² (69 MPa). Microspheres made of phenolic and filled with nitrogen are used for the production of polyester foam compositions and syntactic epoxy foams. Polystyrene microspheres have also been used in producing syntactic foams.

## LAMINAR COMPOSITES

Of all the composites consciously devised the laminar type is by far the oldest. It differs from the other types in that its distinguishing characteristic, the presence of layers, usually becomes evident at a grosser level of perception. Since it is dimensionally of a higher order of magnitude, it is not surprising that its constituents may themselves be composites. Laminar composites have a special appeal because, being made up of films or sheets as they so often are, they are easier to design, produce, standardize, and control than other types of composites. Many standard types have been developed to provide desirable combinations of bulk and surface properties. Perhaps the single most successful application of the laminar principle has been the development of sandwich materials. The most ingenious applications are undoubtedly the custom components that are carefully built up of varied layers to meet specific requirements.

### Composition and Properties

Laminar, or layered, composites consist of two or more different layers bonded together. The layers constituting a composite can differ in material, form, and/or orientation. For example, clad metals are made up of two different materials. In sandwich materials, such as honeycomb, the core layer differs in form from the facings, while the layer materials may or may not be different. Similarly, in plywoods, though the layers are often of the same type of wood, the orientations differ.

The great variety of possible combinations makes it difficult to generalize about laminar composites, but for the most part properties of laminar composites tend to be anisotropic, i.e., vary from one side of the composite to the other, and each layer may perform a separate and distinct function. Like the other types, laminar composites combine the properties of their component materials to provide properties not available in either material alone, or they make it easier or less costly to obtain certain properties. However, laminar composites are unique in their ability to incorporate the advantages of other composites.

Thus, the improvement in properties obtained from combining fibers in matrix, particles in matrix, or flakes in matrix can be used in combination with other materials in layered construction. For example, a reinforced-plastic sheet (fibers in matrix) may be clad with copper to provide a printed circuit (laminate) with a combination of strength, electrical conductivity, and electrical insulation. This laminate may in turn be bonded to another composite, say a sandwich, to provide additional strength, electrical conductivity, and electrical insulation. This laminate may in turn be bonded to another composite, say a sandwich, to provide additional strength, vibration damping, or thermal insulation for use in spacecraft. The sandwich itself may be a combination of composites. For example, the outer faces may consist of a dispersion-strengthened metal (particles in matrix), and the core may be a foam-filled honeycomb. As a result, the final composite may consist of several distinct layers, each performing a specific function, in which some or all of the layers are themselves composites.

## Laminates and Sandwiches

Laminar composites can be divided into laminates and sandwiches. Laminates are defined as composite materials consisting of two or more superimposed layers bonded together. Sandwiches, a special case of laminates, consist of a thick low-density core (such as honeycomb or foamed material) between thin faces of comparatively higher density. Although this distinction between laminates and sandwiches is more expedient than scientific, some general observations can be made. In sandwich composites, for example, a primary objective is improved structural performance, or more specifically, high strength-to-weight ratios. To this end, the core serves to separate and stabilize the faces against buckling under edgewise compression, torsion, or bending and provides a rigid and highly efficient structure. Other considerations, such as thermal insulation, heat resistance, and vibration damping, dictate the particular choice of materials used.

Laminates, on the other hand, are most often designed to provide characteristics other than superior strength. There are, of course, exceptions, the most notable being plastic laminates consisting of layers of resin-impregnated fabrics, paper, glass cloth, etc., with high strength-to-weight ratios. In general, laminates are designed to protect against corrosion and high-temperature oxidation; to provide impermeability; to facilitate fabrication; to cut costs; to improve appearance; to reduce thickness; to modify electrical and other properties; or to overcome size limitations. In other words, while the choice of materials is certainly an important consideration in sandwich composites, it is the nature of the structure that controls the essential properties. In laminates, however, properties depend much more upon the combination of materials used.

### Laminates

Theoretically there are as many different types of laminates as there are possible combinations of two or more materials. If we divide all materials into metals and nonmetals, and if we divide nonmetals into organic and inorganic, there are six possible combinations in which laminates can be produced: metal-metal, metal-organic, metal-inorganic, organic-organic, organic-inorganic, and inorganic-inorganic. In laminates containing more than two layers there are obviously considerably more possiblilities, and one or more of the layers may be a composite, making the combinations even more variable and complex.

**Metal-Metal:** There are three basic functional categories of metal-metal laminates: (1) those whose face is primarily decorative, (2) those whose face provides one or more important surface properties (other than appearance) and whose base makes the laminate cheaper and/or stronger than the equivalent face material alone, and (3) those which provide special bulk properties or properties resulting from a reaction between face and base. Within this framework the two major classes, based on methods of production, are precoated and clad metals.

*Precoated Metals:* In precoated metals the face is formed by building up the second constituent on a substrate to form a thin, essentially continuous film. This is usually done by electroplating or hot dipping, although chemical plating is also used. The best-known examples of preplated metals are electrolytic tinplate, used mostly for tin cans, and electrogalvanized steel, widely used for roofing. Another important surface property provided by precoated materials is corrosion resistance. In general, however, the preplated and precoated metals are suitable only for atmospheric corrosion. Severe chemical environments

**Table 1.3** Bend Properties of Outer Fibers of
Be–Al Laminated Composites†

| | |
|---|---|
| Yield strength (0.2% offset) | 55 kips/in$^2$ (379 MPa) |
| Fracture stress | 78 kips/in$^2$ (538 MPa) |
| Plastic strain | 0.9% |

†Based on *Mater. Des. Eng.*, January 1968, p. 56.

require the thicker clad metals. The usefulness of many precoated metals depends at least partly upon a chemical or electrochemical relationship between the two materials. Best-known examples are galvanized steel and the Alclads; in each case the face metal is anodic to the base and protects it by corroding preferentially.

*Clad Metals:* In clad metals the face is applied as a solid, wrought material. Clads are more suitable than precoated metals when the environment is more severe, calling for a thicker face. Cladding is accomplished in a variety of ways, including rolling, hot pressing, casting, extruding, brazing, and welding.

The face of a clad metal is usually considerably thicker than that of a precoated metal. A range of 5 to 20% of the thickness of the combination is most common, although the proportion may be as high as 90% in special cases. The great number of combinations of metals that have been clad include aluminum-clad uranium, copper-clad tungsten, and molybdenum clad with nickel, copper, aluminum, gold, silver, tin, or lead. Other combinations are titanium and steel and beryllium and aluminum. Laminated composites consisting of aluminum foil and beryllium sheet have better bend behavior and resistance to impact than solid beryllium. Delamination of the composite arrests crack propagation and channels it along longer fracture paths, increasing fracture energy. The bend properties of the laminate are shown in Table 1.3. Failure occurred layer by layer, and each layer failed at the failure values of solid beryllium. The extra work required to fracture the specimen because of its lamellar construction is shown in Fig. 1.20 (crosshatched area). A solid beryllium specimen would have fractured at 6.5° (0.11 rad) bend angle, but fracture was not complete in the lamellar specimen when the test was stopped at a bend angle of 24° (0.42 rad). A new gas-turbine material with promise uses laminar composites consisting of alternate layers of metal sheet or foil diffusion bonded together which have potential high

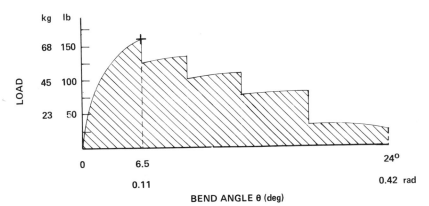

**FIG. 1.20** Load-deflection diagram of Be–Al composite in three-point bending test. Discontinuities represent independent failures of each beryllium layer.[18]

strength and are capable of withstanding high temperatures. Strengths are comparable to those of fiber-reinforced composites. Scientists have laminated Nichrome V (Ni–20Cr) alternately with tungsten, the reinforcing phase. Fabricated composites contain 50 vol % of each material as foil or sheets 0.001, 0.005, or 0.020 in (0.03, 0.13, or 5 mm) thick. Another laminar composite consisting of 77 vol % W–Re–Hf–C alloy sheet has been alternated with 25% Inconel 600. Tensile strength at 2000°F (1093°C) is 99 kips/in$^2$ (682 MPa), and projected 100-h stress-rupture strength[14,15] is 50 kips/in$^2$ (345 MPa).

Some clad materials have a thin layer of a third metal interposed between the two principal layers to prevent formation of an interfacial alloy layer that is brittle or may become so with heat treatment. A good example is ferritic stainless-clad steel, in which a layer of electrolytic nickel serves to keep the carbon in the steel from diffusing into the stainless and combining with the chromium.

Clad metals are particularly valuable for electrical applications requiring conductivity. Silver, platinum, and platinum-group metals clad to copper, brass, bronze, and nickel have found wide acceptance as contact materials. The superior electrical conductivity of silver can be used in high-frequency circuits by means of silver-clad copper wire or strip and silver-clad brass or aluminum waveguides.

Other surface properties obtainable are heat reflectivity (steel or copper clad with aluminum, nickel, or silver), radiation shielding (lead-clad steel); and ease of fabrication; both clad and precoated metals are often used where the principal function of the face metal is to facilitate joining, for example. Other interesting applications are those in which special combinations of properties are exploited, such as heat transfer and corrosion resistance in stainless-clad copper or aluminum for pots and pans.

**Metal-Organic:** The several types of metal-organic laminates can be conveniently divided into those with a thin face built up on a base and those with a considerably thicker face, usually formed by bonding on a solid layer. The best-known metal-organic laminate is probably prefinished or prepainted metal, whose primary advantage is the elimination of final finishing by the user. The most popular base metals are cold-rolled steel, tinplate, tin-mill black plate, hot-dipped and electrogalvanized steel, and standard aluminum alloys.

The second type are the plastic-metal laminates. Although there are many possible combinations of solid organic films and metals, plastic-metal combinations are probably most familiar. Vinyl-metal laminates probably account for 90 to 95% of all plastic-metal laminates now used. These laminates are commonly made by adhesive-bonding preprocessed vinyl sheet to metal.

Other applications include tetrafluoroethylene plastic bonded to aluminum or stainless-steel foil for use in flexible bladders for missile fuel-storage control systems where impermeability is the main objective. Other metal-organic laminates include such things as jacketed wire, metal-faced plywood, and a great variety of rubber-metal combinations such as rubber-covered steel rolls, rubber-lined tanks, etc.

**Metal-Inorganic:** The best known metal-inorganic laminates are probably porcelain-enameled steel or copper and other ceramic-coated metals. Others include metal-reinforced ceramic coatings for supersonic aircraft combustion chambers and glass-lined steel used in corrosive environments involving acids and alkalis at temperatures up to 450°F (232°C). Ceramic-lined steel has also been used for parts and provides excellent resistance to impact, abrasion, and thermal shock up to 1400°F (760°C).

**Organic-Organic:** The most obvious organic-organic laminate is glued-laminated wood (plywood is not included here because the differences in face and core densities and thicknesses place it more accurately in the category of sandwich materials). Other examples of organic-organic laminates include plastic-faced wood, laminated paper, rubber-fabric industrial belting, laminated layers of transparent plastic (such as acrylic) for glazing and light-transmitting applications, synthetic felts bonded into layers for a variety of uses, and self-adhesive materials consisting of a layer of adhesive and a carrier (usually plastic or paper).

Another important class of organic-organic laminates is the high-pressure thermosetting plastic laminates consisting of resin-impregnated paper, cotton fabric or mat, and nylon fabric. These laminates are used for a variety of mechanical and electrical applications. The higher-performance laminates generally use an inorganic constituent with a glass-fiber or cloth reinforcement.

**Organic-Inorganic:** High-pressure thermosetting plastic laminates (with glass or asbestos reinforcement) and glass-plastic laminates are the two most common forms of organic-inorganic laminates. High-pressure laminates are produced by impregnating a reinforcing material with a thermosetting resin, laminating the material into multiple layers, and curing with heat and high pressure to form a dense, hard solid with good mechanical strength. Outstanding characteristics of glass- and asbestos-reinforced plastic laminates are their electrical insulation, strength-to-weight ratios, and corrosion resistance.

Glass-plastic laminates usually consist of two or more layers of glass sheet and one or more layers of plastic, although other materials such as wood and fabric are sometimes used. The most common laminated glass is safety glass. Applications include auto and aircraft, which uses vinyl plastic 4 to 10 times the thickness of that used in automotive glass. Bulletproof glass is a composite structure consisting of multiple layers of plate glass bonded together with alternate layers of vinyl to provide even greater thickness and energy absorption. Another example of glass-plastic laminates is a combination of colored glasses with interposed transparent plastic layers to provide a variety of light filters.

**Inorganic-Inorganic:** There are not many all-inorganic laminates. Although inorganic materials provide important properties, they seem to be exploited best in monolithic materials or in laminates containing layers of other materials where the special properties of inorganics are an adjunct. The most obvious inorganic-inorganic laminates involve glass. Structural glass-glass laminates, for example, find wide application for partitions and for construction of tabletops, counter tops, and signs. Glass-glass laminates are also used for special lenses.

## COMBINATION LAMINATES

Not only monolithic materials but almost any composite material can be used as a layer to form a composite laminate. For example, a relatively simple combination laminate is a printed-circuit board consisting of a layer of silicone rubber bonded between two layers of glass-reinforced epoxy laminates.

The use of plastic laminates bonded to other materials is a solution to many materials problems. Laminates have thus far been successfully bonded to steel, aluminum, copper, rubber, cork, plastic film, vulcanized fiber, asbestos, gold, silver, and several other materials, including other plastic laminates. Specific design advantages include better strength-to-weight ratios, increased resistance to corrosion and chemicals, greater range of electrical characteristics, dimensional stability over a wide temperature range, increased rigidity and strength for soft sealing materials, improved bearing surfaces, greater range of frictional characteristics, improved fabrication characteristics, and reduced costs.

An example of a multifunctional laminate is an insulating system designed to reduce temperature from 5400°F (2982°C) at the hot face to ambient temperature on the cold side. The system consists of three composite layers: the first (on the hot side) is graphite fiber and resin binder cured, pyrolized, and graphitized at 5400°F (2982°C); the second is carbon fiber and resin binder cured and pyrolized at 1470°F (799°C); and the third is glass fiber and resin binder cured at 300°F (149°C).

Ablation systems and materials are another area where multifunctional laminates provide the best solution to high-temperature problems. One of the most successful, used in rocket nozzles, is a laminate which has withstood 6800°F (3760°C). One exotic combination laminate is used for the walls of large missile nozzles, which may comprise eight separate layers (not arranged directly above one another). From the inside out, the layers include graphite, asbestos-reinforced phenolic laminate, graphite-reinforced phenolic laminate, silica-reinforced phenolic laminate, silica fabric–phenolic tape, two layers of glass filament-wound epoxy, and a thick steel insert (Fig. 1.21).

The ski industry has exploited the advantages gained from multilayer laminates. Figure 1.22 illustrates a ski design which uses seven distinct layers in an interesting multilayer laminate. Two strips of high-carbon steel give the ski its basic camber and flexure. A layer of high-density wood particle board is added between the steel strips to dampen the spring action. Since steel adheres to aluminum better than to wood, aluminum strips are bonded to the top and bottom of the wood layer and then bonded to the steel strips. Layers of cotton fabric, applied to the aluminum, increase its bond strength to both the wood and steel. The bottom running surface is phenolic plastic, and a layer of phenolic plastic also forms the top sheet and sides of the ski.

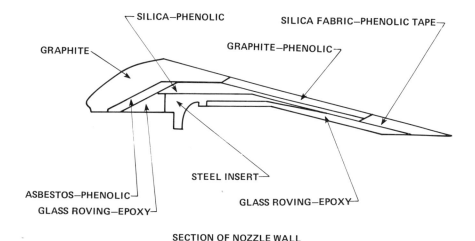

**SECTION OF NOZZLE WALL**

**FIG. 1.21** Breakdown of layers in a nozzle wall of a combination laminate.[16g]

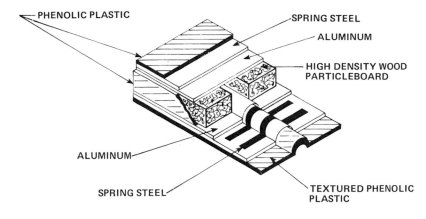

PHENOLIC PLASTIC

SPRING STEEL

ALUMINUM

HIGH DENSITY WOOD PARTICLEBOARD

ALUMINUM

SPRING STEEL

TEXTURED PHENOLIC PLASTIC

SKI COMPOSITE

**FIG. 1.22**   Multilayer laminate design for a ski.[16b]

# REFERENCES

1. Dannöhl, P. W.: MgO Fibre-Skeleton Composites, *Met. Mater.*, **6**(1):19 (1972).

2. Hurley, G. F., H. E. LaBelle, Jr., and E. G. Roberge: Sapphire Multiple Filament and Large Plate Growth Processes, *Tyco Laboratories, Inc., Final Tech. Rep. April 1970–March 1972*, AFML TR-72-190, Contr. F33615-70-C-1471, Air Force Materials Laboratory, Wright-Patterson AFB, Ohio, October 1972.

3. Evans, C. C.: "Whiskers," *M&B Monog.* ME/8, Mills and Boon, London, 1972.

4. Smith, M. F., "Carbon and Graphite," pt. 1, "Carbon and Graphite Fibers and Fiber Composites," Vol. 2, cited in E. Fitzer and M. Heyn, Chemical Fibers: The Outlook, *Chem. Ind. (Lond.)*, 16:663–676 (1976).

5. Economy, J., W. D. Smith, and R. Y. Lin: Preparation of Refractory Fibers by Chemical Conversion Method, *Proc. ACS Conf. High Temp. Flame-Resist. Fibers, New York, Aug. 28–29, 1972*.

6. Takahashi, S: Preparation of SiC Whisker–Reinforced Silver Composite Material on Skylab, *AIAA J.*, **16**(5):452–457 (May 1978).

7. Ordway, F., P. J. Lare, and R. A. Hermann: Silicon Carbide Whisker–Metal Matrix Composites, *Artech Corp. Final Tech. Rep., December 1969–September 1971*, TR-71-252, Contr. F33615-69-C-1187, Air Force Materials Laboratory, Wright-Patterson AFB, Ohio, March 1972, AD-752 589.

8. Goodwin, V. L., J. G. Theodore, and R. A. Foos: Characteristics of Beryllium-Reinforced Titanium Composites, *19th Refract. Compos. Work. Group Meet., Houston, Feb. 1–2 1972*.

9. Rudy, J. F.: Fiber Metal for Joining, *Weld. J.*, **42**(12):529s–534s (December 1963).

10. Fisher, J. I.: Fiber Metal Provides Controllable Properties, *Mater. Des. Eng.* October 1964, pp. 96–99.

11. West, P.: Metal Fabrics for High Temperatures, *Mater. Des. Eng.* January 1965, pp. 102–104.

12. Sambell, R. A. J., D. H. Bowen, D. C. Phillips, and A. Briggs: Carbon Fibre Composites with Ceramic and Glass Matrices, *J. Mater. Sci.*, **7**(6):663–681 (June 1972).

13. Economy, J., L. C. Wohrer, and A. A. Woskait: High-Strength, High-Modulus AlB$_2$ Flake Reinforced Composites, *Carborundum Co. Tech. Rep. March 1972–February 1973*, AFML-TR-72-74-Part 2, Contr. F33615-72-C-1496, Air Force Materials Laboratory, Wright-Patterson AFB, Ohio, November 1973, AD-917 950L.

14. Hoffman, C. A., and J. W. Weeton: Metal-Metal Laminar Composites for High-Temperature Applications, *NASA Lewis Res. Cent., Tech. Mem.* X-68056, 1972, N72-32541.

15. Hordon, M. J.: Development of Laminate Metal Matrix Composites, *National Research Corp. Final Rep,* Contr. N00019-71-C-0183, Naval Air Systems Command, May 1972.

16. *Mater. Des. Eng.,* September 1963: *a,* p. 83; *b,* p. 84; *c,* p. 90; *d,* p. 92; *e,* p. 93; *f,* p. 97; *g,* p. 117; *h,* p. 118.

17. *Mater. Des. Eng.,* August 1968, p. 50.

18. *Mater. Des. Eng.,* January 1968, p. 56.

19. *Ind. Res.,* February 1967, p. 88.

# BIBLIOGRAPHY

Baker, A. A.: Carbon Fibre Reinforced Metals: A Review of the Current Technology, *Mater. Sci. Eng.,* 17(2): 177–208 (February 1975).

Barnet, F. R., and M. K. Norr: Carbon Fiber Microstructure, *Nav. Ord. Lab. Tech. Rep.* NOLTR 72-32, March 1972, AD-740 315.

Benzel, J. F., J. K. Cochran, R. K. Feeney, J. W. Hooper, and J. D. Norgard: Melt-Grown Oxide-Metal Composites, *Ga. Inst. Technol. Sch. Ceram. Eng. Annu. Tech. Rep.,* July 1972, Contr. DAAH01-71-C-1046, AD-749 506.

Berghezan, A.: Precise Alignment and Uniform Distribution of Fibers in a Metal Matrix, *Composites,* 3(5): 200–210 (September 1972).

Bittence, J. C.: Pumping More Strength into Reinforced Plastics, *Mater. Eng.,* May 1978, pp. 27–30.

Brennan, J. J.: Development of Fiber Reinforced Ceramic Matrix Composites, *United Aircraft Res. Lab. 2d Q. Prog. Rep.,* Contr. N62269-73-C-0268, NADC, October 1973.

———— and M. A. DeCrescente: Fiber-Reinforced Ceramic Matrix Composites, *United Aircraft Res. Lab. 3d Q. Rep.,* Contr. N00019-72-C-0377, NASC, November 1972.

Burykina, A. L., Y. V. Dzyadykevich, and V. V. Gorskii: Compatibility of Boron Fibers with a Tungsten Substrate and a Titanium Matrix, *Sov. Powder Metall. Met. Ceram.,* 11:900–903 (November 1972).

Butler, I. G., W. Kurry, J. Gillot, and B. Lux: The Production of Metal Fibres and Wires Directly from the Melt, *Fibre Sci. Technol.,* 5(4):243 (October 1972).

Cooper, G. A., D. G. Gladman, J. M. Sillwood, and G. D. Sims: The Development of Composite Carbon Fibres of Large Diameter Jumbo Fibres, *Natl. Phys. Lab. (UK), Rep.* NPL-IMS-18, November 1972, N73-18580.

DeBolt, H., and V. Krukonis: Improvement of Manufacturing Methods for the Production of Low-Cost Silicon Carbide Filament, *AVCO Corp. System Div. Q. Rep.* IR-360-2(11), Contr. F33615-72-C-1177, September 1972.

————, ———— and F. E. Wawner, Jr: High-Strength, High-Modulus Silicon Carbide Filament via Chemical Vapor Deposition, *"Silicon Carbide 1973," Proc. 3d Int. Conf. Silicon Carbide, Miami Beach, Fla., Sept. 17–20, 1973,* University of South Carolina Press, Columbia, 1974, pp. 168–175.

DeLamotte, E., K. Phillips, A. J. Perry, and H. R. Killias: Continuously Cast Aluminum–Carbon Fiber Composites and Their Tensile Properties, *J. Mater. Sci.,* 7(3):346–349 (1972).

Diefendorf, R. J.: Fiber and Matrix Materials for Advanced Composites, *AGARD Lect. Ser. Composite Mater., Oslo, Copenhagen, Lisbon, June 1972.*

Directionally Solidified Eutectics: Promising Turbine Materials, *Mater. Eng.,* December 1972, p. 30.

Dittmer, W. D., and P. R. Hoffman: Boron Composites: Status in the USA, *Interavia,* 28:654 (June 1973).

Donald, I. W., and P. W. McMillan: Review: Ceramic-Matrix Composites, *J. Mater. Sci.* 11(5):949–972 (May 1976).

Economy, J.: Present Status and Future for High-Strength Fibers, *SAMPE J.,* 12(6):5–9 (November/December 1976).

Ellis, R.: Laminated Metallic Structure: An Approach to Fracture Control, pp. 329–335 in "Materials and Processes for the 70's—Cost Effectiveness and Reliability," *SAMPE 5th Natl. Meet., Kiamesha Lake, N.Y., Oct. 9–11, 1973.*

Fleck, J. N., and V. D. Linse: Advances in Boron/Aluminum Processing, *Proc. 6th St. Louis Symp. Composite Mater. Eng. Des., May 11–12, 1972,* pp. 251–256 (1973).

Giamei, A. F., E. H. Draft, and F. D. Lemkey: The Art and Science of Unidirectional Solidification, *New Trends Mater. Proc. Sem. Am. Soc. Met., Oct. 19–20, 1974,* Metals Park, Ohio, pp. 48–97 (1976).

Goddard, D. M., and E. G. Kendall: Fibreglass-Reinforced Lead Composites, *Composites,* 8(2):103–109 (April 1977).

Hart, P. E.: New Class of Ceramic Composites, *Ceram. Age,* 88(3):29–30 (1972).

Hashin, Z.: Theory of Fibre-Reinforced Materials, NASA CR-1974, March 1972.

Heldenfels, R. R.: Recent NASA Progress in Composites, *NASA, Langley Res. Cent. Tech. Mem.,* X-72713, August 1975, N75-29188.

Hilado, C. J.: Carbon Composite and Metal Composite Systems, *Mater. Techno. Ser.* 7, Technomic, Westport, Conn., 1975.

Hunter, R. L.: The Outlook for Advanced Filaments, Fibers, and Whiskers, *Proc. 1975 Int. Conf. Compos. Mater.,* vol. 1, pp. 16–21, The Metallurgical Society of AIME, New York, 1976.

LaBelle, H.E., Jr., G. F. Hurley, and A. Morrison: Spinel Ribbon, *Tyco Lab. Saphikon Div. 6th Q. Prog. Rep.,* Contr. DAAB05-72-C-5841, October 1973.

Lilholt, H., and H. Carlsen: A Study of Nickel Reinforced with Tungsten Wires, a Potential High-Temperature Material, *Proc. 1975 Int. Conf. Composite Mater.,* vol. 2, pp. 1321–1333, The Metallurgical Society of AIME, New York, 1976.

MacKenzie, J. D.: Preparation and Properties of Beta-Alumina/Glass Composites, *Univ. Calif., Los Angeles, Mater. Dept., Final Rep.,* June 1976, PB-255549.

McCreight, L. R., and H. W. Rauch, Sr.: Filaments for Advanced Composites: An Overview, pp. 43–50 in Advanced Materials: Composites and Carbon, *Am. Ceram. Soc. Symp., Philadelphia, Apr. 26–28, 1971,* American Ceramic Society, Columbus, Ohio, 1972.

Mehrabian, R., R. G. Rick, and M. C. Flemings: Preparation and Casting of Metal-Particulate Non-Metal Composites, *Metall. Trans.,* 5(6):1899–1905 (June 1974).

Metcalfe, A. G., and M. J. Klein: Compatible Alloys for Titanium Matrix Composites, *Titanium Sci. Technol.,* 4:2285–2297 (1973).

Morin, D.: A New Fiber for Making High-Modulus Composite Materials: Boron Carbide–Coated Boron Filament, *New Horiz. Mater. Process.,* 1973:599–618.

Moss, M., W. L. Cyrus, and C. B. Haizlip: Fabrication of Filament Reinforced Metals by Hot Roll Bonding, *Sandia Lab. Rep.* SC-DR-720117 (1972).

National Aeronautics and Space Administration: Composite Materials: A Compilation, *NASA Summ. Rep.* SP-5974, January 1976, N76-21294.

National Materials Advisory Board, National Academy of Sciences: Directionally Solidified Composites: Known Also as in Situ Composites, or Directional Solidified Eutectics, Final Rep. NMAB-301, Contr. DA-49-083-SA-3131, U.S. Army, April 1973.

National Materials Advisory Board: Metal-Matrix Composites: Status and Prospects, *Rep. ad hoc Comm. Met. Matrix Composites* NMAB-313, December 1974, AD-A005 774.

Signorelli, R. A.: Review of Status and Potential of Tungsten-Wire-Superalloy Composites for Advanced Gas Turbine Engine Blades, *NASA Lewis Res. Cent. Tech. Mem.* X-2599, September 1972, N72-30471.

Sippel, G. R., and M. Herman: Fibre Reinforced Titanium Alloy, *Proc. 1971 Fall Meet. Metall. Soc. AIME, Detroit,* pp. 211–251, The Metallurgical Society of AIME, New York, 1974.

Toth, I. J., W. D. Brentnall, and G. D. Menke: Fabricating Aluminum Matrix Composites, 1: A Survey of Aluminum Matrix Composites, *J. Met.,* 24(9):19–25 (September 1972).

Weeton, J. W.: Fiber-Reinforced Superalloys, Ceramics, and Refractory Metals and Directionally Solidified Eutectics (Heat-Resistant Composites), pp. 255–301 in *Proc. 1971 Fall Meet. Metall. Soc. AIME, Detroit,* The Metallurgical Society of AIME, New York, 1974.

Wickens, A. J.: Fibre-Reinforced Metals: The Chemistry of the Interface, *Chem. Ind.,* February 1977, pp. 147–151.

Wu, E. M., and T. T. Chiao: Fiber Composites: A Review and Assessment, *Energy Res. Dev. Admin. Summ. Rep.* UCRL-78146, Contr. W-7405-eng-48, April 1976.

Graphite fibers.   *(Celanese Corp.)*

# Fabrication
# of Composite
# Materials

The composite materials discussed in this chapter consist of various combinations of high-performance fibers or reinforcements with known properties in a suitable matrix to obtain properties superior in some cases to those of metals and commercial-grade fiber-glass–epoxy materials. The fibers, coated and uncoated, typically control the strength and stiffness characteristics, formability, and machining characteristics of the laminate. The matrix, either organic (resin) or metallic, determines the transverse mechanical properties, interlaminar shear characteristics, and service operating temperatures of the laminate; it also influences the selection of laminating processes and conditions and tool design.

## FIBERS

Fibers are one of the oldest engineering materials in use. Jute, flax, and hemp have been used for such products as rope, cordage, nets, water hose, and containers since antiquity. Other plant and animal fibers are still used for felts, paper, brush, or heavy structural cloth. The fiber industry is clearly divided between natural fibers (from plant, animal, or mineral sources) and synthetic fibers. Many synthetic fibers have been developed specifically to replace natural fibers, because synthetics often behave more predictably and are usually more uniform in size. Synthetic fibers are frequently less costly than their natural counterparts. In the garment industry, for example, acrylic and rayon fibers were developed to replace the more costly silk and wool. For engineering purposes, glass, metallic, and organically derived synthetic fibers are most significant. Nylon, for example, is used for belting, nets, hose, rope, parachutes, webbing, ballistic cloths, and reinforcement in tires.

Attempts to define inorganic artificial fibers run into difficulty because no common denominator runs through this large family of materials. Certain fibers, such as metallic yarns and those based on tungsten substrates, are clearly inorganic and artificial. Carbon and graphite fibers, based on rayon, polyacrylonitrile, or pitch, are also artificial, but the resulting carbon is neither organic nor inorganic. It is probably simplest to classify carbon and graphite fibers as organic because they are manufactured from organic precursors.

Metal fibers and single-crystal metal fibers, called *whiskers,* are used in high-strength high-temperature lightweight composite materials for aerospace applications. Fiber composites improve the strength-to-weight ratio of base materials such as titanium and aluminum. Anisotropic (directional) properties can be designed into a part made from a fiber composite by selectively aligning the fiber-base lay-up. Metal-fiber composites are used in turbine compressor blades, heavy-duty bearings, pressure vessels, and spacecraft reentry shields. Boron, carbon, graphite, and refractory oxide fibers and whiskers of alumina, silica, and SiC are common materials used in high-strength fiber composites. Among the strongest materials are metal fibers formed by controlled solidification and cold drawing. Some nonmetallic fibers such as $Al_2O_3$ and SiC are nearly as strong as metal fibers but have a higher modulus of elasticity. Fibers in a metal matrix combine the strength of the fiber with ductility or other characteristics of the matrix, making many combinations of properties possible.

## Selection of Specific Fibers

The structural potential of fiber-reinforced composites is being realized. Filament composites of boron are saving 350 lb (159 kg) in the tail sections of a fighter aircraft, and the same material has cut helicopter rotor blade weight by 25% and increased stiffness by 36%. What factors determined the selection of this specific fiber for these applications? Cracks and flaws in the surface of a material drastically reduce overall strength. Concentration of stress at these imperfections propagate their strength-reducing effect into the bulk of the material. Materials with high theoretical strengths and stiffness suffer most from this phenomenon, but decreasing the size and surface area reduces the probability of flaws and limits their effect. Thus, strengths approaching those theoretically predicted are possible with thin fibers. Table 2.1 illustrates the potential strength properties of fibers in various matrixes. High strength and stiffness are only a start in defining the ideal fiber for composite reinforcement.

### Density

Since the cost of putting hardware into orbit is about $1000 per pound ($2200 per kilogram) and the range of an aircraft depends largely on its weight, materials selection for aerospace structures is based on high strength-to-density and modulus-to-density ratio. Consequently, reinforcement fibers should be lightweight as well as strong and stiff.

## Producibility

Unfortunately, many high-modulus materials are not readily made into fiber, and when the necessary techniques are developed, they are expensive. Hence, some promising materials may cost $5000 to $10,000 per pound ($11,000 to $22,000 per kilogram). Despite recent advances in fabricating high-performance fiber reinforcement, it is not likely that they will be priced competitively for some time.

## Heat Resistance

High-performance fibers are generally considered for elevated-temperature applications. Thus, strength retention often is more significant than room-temperature mechanical properties. Figure 2.1 (page 2.17) shows temperature dependence of tensile strength for some typical fiber reinforcement materials. Besides high service temperatures, fiber reinforcement must withstand even higher temperatures during composite fabrication, particularly for metal-matrix composites.

## Chemical Compatibility

The chemical interaction between a reinforcement fiber and its matrix is critical. Many fiber-matrix combinations will not work because of chemical incompatibility. In some cases, reaction between the two degrades their properties; in other cases, there is so little reaction that fiber-matrix bonding is inadequate. The use of protective coatings, diffusion barriers, and adherence-promoting coatings helps in some cases.

## Others

In addition to the factors listed above, fibers in special applications may be required to meet specific tolerances in thermal expansion, radiation resistance, and optical behavior.

## Forms

The raw material, or starting stock, for various fabrication methods has been produced in a variety of forms for many fiber reinforcements and resin systems (Table 2.2). One of the most common forms, the prepreg, combines partially cured (B-stage) thermosetting resins such as epoxy or polyimide with high-strength reinforcing fibers. The resin, also called the *binder* or *matrix,* is applied as a syrup to the fibers, which are usually in the form of strands. By using a prepreg, the process of impregnating the fiber is separated from that of laying up the laminate or component. This avoids the problems inherent in laying up dry, uncooperative fibers in the presence of sticky liquid resins.

Combining the fiber with the resin as a separate process makes laying up very much simpler and quicker and results in laminates of much better quality. In particular, the proportion of resin to fiber is automatically kept constant within very close limits; this helps ensure optimum strength in the cured laminate and makes it possible to maintain strict control of weight distribution on small or large areas. Fiber orientation is also easily controlled, to give maximum strength and stiffness where they are needed and to minimize

**Table 2.1** Properties of Synthetic Inorganic Fibers†

| Fiber | Diameter | Density | Max service temp‡ | Tensile strength | Modulus of elasticity | Specific modulus of elasticity |
|---|---|---|---|---|---|---|
| | | | U.S. customary units | | | |
| | mils | lb/in$^3$ | °F | kips/in$^2$ | $10^6$ lb/in$^2$ | $10^6$ in |
| *Available in continuous or chopped form* | | | | | | |
| Carbon, 91% | 0.33 | 0.061 | ..... | 120 | 6 | 98 |
| 95% | 0.034 | 0.065 | | | | |
| From pitch | 0.035 | 0.072 | | 350–360 | 32–24 | 444–472 |
| Glass, D | ..... | 0.078 | 1420§ | 350 | 7.5 | 96 |
| E | 0.015–0.5 | 0.0917 | 1555§ | 500 | 10.5 | 114 |
| S | 0.38 | 0.089 | 1778§ | 665 | 12.4 | 139 |
| Graphite, intermediate | 0.31–0.32 | 0.065 | ¶ | 410 | 30–34 | 461–523 |
| High-strength | 0.29–0.31 | 0.0655 | ¶ | 400 | 34–37 | 519–564 |
| High-modulus | 0.28–0.29 | 0.0677 | ¶ | 340 | 50–55 | 738–812 |
| Silicon oxide | 1.4 | 0.079 | 3020 | 850 | 10.5 | 13.3 |
| *Available in chopped form* | | | | | | |
| Carbon steel | 4–17 | 0.28 | 800 | 135–400 | 29–30 | 103–107 |
| *Available in continuous form* | | | | | | |
| Aluminum oxide (polycrystalline) | ..... | 0.114 | 3000 | 300 | 25 | 219 |
| Beryllium | 4.9 | 0.66 | 1300 | 185 | 35 | 53 |
| Boron, on carbon | 4 | 0.081 | 600 | 475 | 53 | 654 |
| | 5.6 | 0.082 | 600 | 475 | 55 | 670 |
| On silicon oxide | 4 | 0.085 | 600 | 330 | 53 | 625 |
| On tungsten | 2 | 0.122 | ..... | 400 | 58–60 | 604–625 |
| | 4 | 0.093 | 600 | 500 | 58 | 623 |
| | 5.6 | 0.090 | 600 | 510 | 58 | 644 |
| | 8 | 0.089 | 600 | 530 | 58 | 651 |
| SiC-Coated | 4.2 | 0.096 | ..... | 350 | 58–60 | 604–625 |
| | 5.7 | 0.093 | ..... | 350 | 58–60 | 604–625 |
| Boron carbide–boron on tungsten | 4 | 0.085 | 600 | 390 | 62 | 730 |

| Material | | | | | | |
|---|---|---|---|---|---|---|
| Boron nitride | 0.27 | 0.069 | 2000 | 200 | 13 | 188 |
| Molybdenum | 0.98 | 0.369 | 3000 | 320 | 52 | 141 |
| René 41 | 0.98 | 0.298 | 1800 | 290 | 24 | 81 |
| Silicon carbide, on carbon | 0.35–0.43 | ...... | ...... | 600–700 | | 539 |
| On tungsten | 4 | 0.115 | 1110 | 450 | 62 | 548 |
| | 5.6 | 0.113 | 1110 | 450 | 62 | 93 |
| Stainless steel, type 304 | 0.47–098 | 0.29 | 1600 | 300 | 27 | 85 |
| Titanium boride | ...... | ...... | 4000 | 15 | 7 | |
| Tungsten | 0.5 | 0.697 | 5700 | 580 | 59 | |
| Zirconium oxide | ...... | 0.175 | 3500 | 300 | 50 | 286 |
| Available in whisker form | | | | | | |
| Aluminum oxide | 0.11–0.39 | 0.143 | 3000 | 3000 | 62 | 434 |
| Beryllium oxide | 0.39–1.17 | 0.103 | 3500 | 1900 | 50 | 485 |
| Boron carbide | ...... | 0.091 | 2000 | 2000 | 70 | 769 |
| Chromium | ...... | 0.260 | 1000 | 1290 | 35 | 134 |
| Copper | ...... | 0.322 | 500 | 475 | 18 | 56 |
| Graphite | ...... | 0.060 | ¶ | 2845 | 102 | 1700 |
| Iron | ...... | 0.283 | 1000 | 1900 | 29 | 102 |
| Nickel | ...... | 0.324 | 1000 | 560 | 31 | 96 |
| Silicon carbide | 0.03–0.11 | 0.116 | 1110 | 3000 | 70 | 608 |
| Silicon nitride | ...... | 0.115 | 1110 | 2000 | 55 | 478 |

†Based on *Mater. Eng.*, August 1975, p. 22. §Softening point.

‡If used in a resin matrix, the maximum service temperature depends on the matrix.

¶600°F (316°C) in air; 5000°F (2760°C) in inert atmosphere.

# Table 2.1 Properties of Synthetic Inorganic Fibers† (continued)

| Fiber | Diameter μm | Density kg/m³ | Max service temp‡ °C | Tensile strength MPa | Modulus of elasticity GPa | Specific modulus of elasticity Mm |
|---|---|---|---|---|---|---|
| | | | SI units | | | |
| *Available in continuous or chopped form* | | | | | | |
| Carbon, 91% | 8.4 | 1,700 | ..... | 827 | 41 | 2.5 |
| 95% | 8.6 | 1,800 | | | | |
| From pitch | 9 | 2,000 | | 2,413–2,482 | 221–234 | 11.3–11.9 |
| Glass, D | ...... | 2,160 | 771§ | 2,413 | 51.7 | 2.44 |
| E | 3.8–13 | 2,538 | 846§ | 3,447 | 72.4 | 2.9 |
| S | 9.6 | 2,464 | 970§ | 4,585 | 85.5 | 3.53 |
| Graphite, intermediate | 7.8–8.1 | 1,799 | ¶ | 2,827 | 207–234 | 11.7–13.3 |
| High-strength | 7.4–7.8 | 1,813 | ¶ | 2,758 | 234–255 | 13.2–14.3 |
| High-modulus | 7.2–7.4 | 1,874 | ¶ | 2,344 | 345–379 | 18.7–20.6 |
| Silicon oxide | 35 | 2,187 | 1660 | 5,861 | 72.4 | 3.38 |
| *Available in chopped form* | | | | | | |
| Carbon steel | 102–432 | 7,750 | 427 | 931–2,758 | 200–207 | 2.6–2.7 |
| *Available in continuous form* | | | | | | |
| Aluminum oxide (polycrystalline) | ........ | 3,156 | 1649 | 2,069 | 172 | 5.6 |
| Beryllium | 127 | 1,827 | 704 | 1,276 | 241 | 1.34 |
| Boron, on carbon | 102 | 2,242 | 316 | 3,275 | 365 | 16.6 |
| | 142 | 2,269 | 316 | 3,275 | 380 | 17.1 |
| On silicon oxide | 102 | 2,353 | 316 | 2,275 | 365 | 16 |
| On tungsten | 51 | 3,377 | ..... | 2,758 | 400–414 | 15–16 |
| | 102 | 2,574 | 316 | 3,448 | 400 | 15.8 |
| | 142 | 2,491 | 316 | 3,516 | 400 | 16.3 |
| | 203 | 2,464 | 316 | 3,654 | 400 | 16.5 |
| SiC-coated | 107 | 2,657 | ..... | 2,413 | 400–414 | 15–16 |
| | 145 | 2,574 | 316 | 2,413 | 400–414 | 15–16 |
| Boron carbide–boron on tungsten | 102 | 2,353 | 316 | 2,689 | 428 | 18.5 |

| | | | | | |
|---|---|---|---|---|---|
| Boron nitride | 7 | 1,910 | 1093 | 1,379 | 89.6 | 4.78 |
| Molybdenum | 25 | 10,214 | 1649 | 2,206 | 359 | 3.58 |
| René 41 | 25 | 8,249 | 982 | 2,000 | 16.5 | 2.06 |
| Silicon carbide, on carbon | 9–11 | . . . . . | . . . . . | 4,137–4,827 | | |
| On tungsten | 102 | 3,183 | 599 | 3,102 | 428 | 13.9 |
| | 142 | 3,128 | 599 | 3,102 | 428 | 13.9 |
| Stainless steel, type 304 | 12–15 | 8,027 | 871 | 2,069 | 186 | 2.36 |
| Titanium boride | . . . . . . . . | . . . . . | 2204 | 103 | 48 | |
| Tungsten | 13 | 19,293 | 3149 | 3,999 | 407 | 2.16 |
| Zirconium oxide | . . . . . . . . | 4,844 | 1927 | 2,069 | 345 | 7.26 |
| **Available in whisker form** | | | | | | |
| Aluminum oxide | 3–10 | 3,958 | 1649 | 20,685 | 427 | 11.02 |
| Beryllium oxide | 10–30 | 2,851 | 1926 | 13,100 | 345 | 12.31 |
| Boron carbide | . . . . . . | 2,519 | 1093 | 13,790 | 483 | 19.53 |
| Chromium | . . . . . . | 7,197 | 538 | 8,895 | 241 | 3.4 |
| Copper | . . . . . . | 8,913 | 260 | 3,275 | 124 | 1.42 |
| Graphite | . . . . . . | 1,661 | ¶ | 19,616 | 703 | 431.8 |
| Iron | . . . . . . | 7,833 | 538 | 13,100 | 199 | 2.59 |
| Nickel | . . . . . . | 8,968 | 538 | 3,861 | 214 | 2.44 |
| Silicon carbide | 1–3 | 3,211 | 599 | 20,685 | 483 | 15.44 |
| Silicon nitride | . . . . . . | 3,183 | 599 | 13,790 | 379 | 12.14 |

†Based on *Mater. Eng.*, August 1975, p. 22.
‡If used in a resin matrix, the maximum service temperature depends on the matrix.  §Softening point.
¶600°F (316°C) in air, 5000°F (2760°C) in inert atmosphere.

**2.7**

**Table 2.2** Forms, Material Types, and Processes[38]

| Function | Material — Type | Material — Form | Advantages | Limitations — Material | Limitations — Processing | Material availability, sources, development |
|---|---|---|---|---|---|---|
| | | | **Filament winding** | | | |
| Direct lay-up of reinforcement to shape; winding is circumferential, helical, polar, or a combination of them | Reinforcements: Gr | Roving, yarns, tapes, mat, filler | Can be wound as prepreg or run with direct impregnation; for fill fibers oriented in wind direction; bulk fiber reinforcement or oriented fill, including chopped; filament winding can be done in gathering or in winding; control of fiber orientation placement and tension; high degree of automated control of tension and pattern; high potential for hands-off fabrication; uses continuous filaments and tapes; | | Surface finish | Commercially available |
| | Kv | | | | Fiber fraction | Effective finishes need to be demonstrated |
| | EGl, SGl | | | | | |
| | Boron | | | | Generally heavier than structure from prepreg lay-up | |
| | Hybrids | All above | high capability for fabrication of complex hybrids, including mixing of fibers; fill fibers can be added at predetermined angles to winding axis; most adaptable to surfaces of revolution; other shapes can be wound with adaptations of equipment and techniques | Development required | Material handling | Filament winding can be done in winding process or with hybridized rovings or tapes |
| | Resins: | Liquids | Allow direct impregnation | | All forms generally have short pot life[b] | Commercially available, some development required for films |
| | Epoxy | Prepreg | Offer better control | Higher cost than liquid | Surface finish | |
| | | Film | | | | |
| | Polyesters | Liquid | Cheap and process quickly | Properties, wetting of Kv | Short gel, fast cures possible | Commercially available; those with lower properties than epoxies may be suitable for secondary applications |
| | Vinyl esters | Liquid | Relatively inexpensive, fast processing; properties approaching those of epoxies | Properties, wetting of Kv; resin development for Gr and Kv; volume fraction control; environmental effects | Short gel, fast cures possible; minimum thickness; control of void content and porosity | Commercially available; resin development required for vinyl ester resin |

| | | Automated broad-goods lay-up | | | |
|---|---|---|---|---|---|
| Direct lay-up of laminates using machine placement of broad-goods prepregs in molds or on fixtures; can be used for ply on ply or ply on film | Gr, Kv, Gl, B, hybrids | Epoxy prepregs, woven fabrics, unidirectionals, cross-plied goods, wide tapers, mat or nonwovens, bias wovens | High rate of material lay-down; possible savings in reduced scrap of prepreg; reduced labor, flow time, and material open time; can be used for preplying material for kitting or lay-up of preforms; can be computer-linked and automated; hands-off lay-up or flat panels and large simple curves; control of reinforcement orientation and location; possible precompaction during lay-up | Formability and conformability to mold surfaces, drape; open time, tack; preplied materials may require large feed-roll radii to prevent wrinkling | Surface finish on part; fiber fraction; mold and mold-form shapes; fiber directions and construction of fabric may hinder ability to conform to compound curves; parts radii critical, may pose problems especially with Kv | Broad goods and unidirectional tapes commercially available as prepregs; preplied and hybrid materials available from most major prepreg suppliers; preplied hybrids or special orientations can be made using broad-goods lay-down machinery; broad goods good for flats and gentle straight-line curves |
| | | Automated tape lay-down | | | |
| Direct lay-up of laminates using machine placement or tape prepregs in molds or on fixtures; can be used for ply on ply or ply on film | Gr, Kv, Gl, B, hybrids | Tape widths to 12 in (304.8 mm) mat or nonwovens; Ep prepregs, woven, unidirectional, cross-plied tapes | Generally same as broad goods but use narrower tapes, especially with angle-cutting capabilities in machine; offers potential for lower waste and lay-up of parts nearer net size | Same as for automated broad-goods lay-up | | Commercially available; more adaptable than broad-goods lay-up because of greater number of degrees of freedom, e.g., prepreg width, ply angles, angle trimming of tapes on machine during lay-down; moderate risk for flat and simple curvatures; high risk for compound complex curves; lower risk than broad goods |

**Table 2.2** Forms, Material Types, and Processes[a38] *(continued)*

| Function | Material | | Advantages | Limitations | | Material availability, sources, development |
|---|---|---|---|---|---|---|
| | Type | Form | | Material | Processing | |
| | | | Self-contained tooling methods[c] | | | |
| Provide heat, pressure, and control for curing composite parts and structures; main objective is elimination of autoclave | Gr, Kv, Gl, B, hybrids | Roving, yarns, tapes, wovens, nonwovens, mixtures | Elimination of autoclave; improved parts-flow and tool-turnaround times; cost savings through less waste and minimization of shop expendables; reduced labor; reduced need for bagging film, etc. | Conformability for shaping; differences in thermal transmission and expansion; control of prepreg thickness critical | Size of tooling, heating requirements, heat transfer; thermal-expansion differences between tool and parts lay-up; differential heating and pressurization; heat-transfer and expansion characteristics of tool and part require careful analysis; better characterization of expansion properties of EMC rubber; control of mating surface and coordination points and surfaces; part and mold volumes critical; additional correlation data required for characterization of effects of process variables (temperature, pressure, properties) | Commercially available, better, more reliable, and predictive methods required for design of tools, especially to overcome pressure and geometry differentials between cure temperatures and room temperature; viscosity profiles of materials required; low risk for structural elements; moderate risk for relatively simple structures and stiffener-panel combinations; high risk for combined, complex, co-cured cobonded structures with multiple controlled surfaces |
| | Epoxy | Liquid | | | | |
| | Polyesters | Film | | | | |
| | Vinyl esters | Prepregs | | | | |
| | Thermoplastics | Molding components for BMC, SMC, XMC; reservoir molding (RIM, RRIM, VARI vacuum-assisted resin injection) | | | | Control of pressures and temperature with EMC or internal pressurization |
| | Urethanes | | | | | |

| Process | Fibers | Reinforcement form | Matrix | Matrix form | Characteristics | Development | Limitations | Availability/risk |
|---|---|---|---|---|---|---|---|---|
| Composite equipment of extrusion; direct conversion of continuous fiber reinforcements into finished profiles; impregnated reinforcements are continuously pulled through heated die, where they are shaped and cured; profile leaves die as finished part requiring only cutting to length | Gr, Kv, Gl, B, hybrids | Roving, yarns, tapes, wovens, nonwovens | Epoxies | Liquid, prepreg, or film | Relatively low-cost production method; continuous processing; uses prepregs or indirect impregnation; produces cured finished sections needing only to be cut to length; parts of infinite length; high degree of automation possible and required for complex reinforcement orientation; fibers in various layers can be oriented as desired by tape addition or overwrapping | Conformability for shaping critical in overwrapped or angle-plied materials; resin development for Gr and Kv; debulking of Gr | Uniform cross sections only; limited ability to postform parts; minimum thickness about 0.094 in (2.39 mm); pultrusion shapes and geometries limited only by limitations of folded surfaces | Commercially available; low risk for relatively simple reinforcement orientations; moderate risk for complex combined geometries; excellent for straight components of constant section and relatively simple geometries; good probability of success for complex-shaped cross sections where section can be modeled as a continuous surface |
| | | | Polyesters | | Preforms possible with epoxies and thermoplastics | | Polyesters and vinyl esters easier to process but are moisture-sensitive and exhibit lower interlaminar shear properties | |
| | | | Vinyl esters, thermoplastics | | Can use thermoset or thermoplastic resin systems as liquid, film, or prepregs | | | , |

| Process | Processes | Characteristics | Development | Limitations | Availability/risk |
|---|---|---|---|---|---|
| Thermoplastic materials formed into products by application of heat and pressure; almost all thermoplastics are in sheet stock or films | Processes include vacuum forming, drape forming, pressure-plug assist, vacuum snapback, trapped-sheet pressure forming, and air-slip forming | Process yields nearly finished parts, needing only trimming; easily automated; not labor-intensive; cost-effective, with cycle times on order of minutes; most reinforcements in short fiber lengths and thermoplastics found as sheet | Formability, properties, solvent resistance | Reinforcement constraining stretch and formability; severely limited by fiber reinforcement and shapes limited by limits of tool geometry and ability to remove parts from tools | Commercially available; generally fiber volume low; varied shapes possible and moderate sizes, limited by materials availability and equipment sizes; development required for thermoforming with continuous reinforcement, high fiber fractions generally limit degree of elongation and hence formability; woven and long-fiber reinforcements tend to constrain elongation at forming temperatures so that billowing or blowing thermoforming may be impractical if not impossible |

2.11

**Table 2.2** Forms, Material Types, and Processes[238] (continued)

| Function | Material | | Advantages | Limitations | | Material availability, sources, development |
| --- | Type | Form | | Material | Processing | |
| --- | --- | --- | --- | --- | --- | --- |
| **Blow molding** | | | | | | |
| Almost all thermoplastics initially in granular or pelletized form for blow molding | | | Automated; film orientation possible; improved directional strength | Properties | Reinforcement lengths and orientation | Fiber volumes low; development required for blow molding with fiber reinforcement |
| **Matched-mold forming or press molding** | | | | | | |
| Analogous to stretch forming, drop-hammer forming, or hydroforming in metals | Thermoplastic resins with most reinforcements; some with fabric; most thermosets | Granulated, chopped squares, sheet stock | Easily automated and not operator-dependent; process yields net size and parts trimmed in mold; unreinforced and reinforced materials practical with continuous fiber reinforcement | Wettability of fibers; properties | Limited by fiber reinforcement but practical where some fiber misorientation allowed | Commercially available; small to moderate size and high complexity possible; some reinforcement fiber misorientation and distortion to be expected; most dependable system for compound complex contours |
| **Stamping[f]** | | | | | | |
| Analogous to metal stamping, coining, extrusion, and heading | Most thermoplastics, some thermosets, most reinforcements in short lengths | Sheet stock, billet, slug, preform | Easily automated, not operator-dependent; process yields net size parts | Formability, flow, properties | Reinforcement-fiber orientation and lengths; fiber wrinkling | Commercially available; small to moderate size and complex shapes possible; melt-flow stamping practical production process with or without reinforcement; development required for solid-phase forming and thermoplastic forging of thermoplastics with fiber orientation |
| **Rotational molding and fusion molding** | | | | | | |
| | Thermoplastics, some thermosets, short-length reinforcements | Pellet, powder, liquid | Film orientation possible; reinforcement can be incorporated into any continuous surface part | Generally fiber reinforcement of short lengths; matrix properties | Complex shapes possible; reinforcement lengths; wetting and physical hindrances; flow or viscosity of matrix material | Fiber volume low; woven reinforcements can be added as details or inserts; limited by matrix wetting of reinforcement fibers |
| **Injection or transfer molding** | | | | | | |
| | Thermoplastics, some thermosets, most reinforcements in short fiber lengths | Pellet, ground stock, some liquids (extruder feed-stock) | High-precision fast process; yields nearly finished parts | Properties; solvent resistance | Fiber orientation; flow characteristics of material; reinforcement lengths generally limited to relatively short fibers | Commercially available; generally random orientation of reinforcement fibers; development analogous to resin injection or impregnation possible |

| Process | Process/Materials | Material form | Advantages | Limitations | Remarks |
|---|---|---|---|---|---|
| Uses two matched dies, a male (core) and a female (cavity); when closed and filled with reinforced plastic composite, the matched-die mold is subjected to heat and pressure to cure the plastic | | | Part-to-part uniformity; high production rate; ability to produce detailed configurations and finished surfaces on both sides of part; uses both thermoplastics and thermosets; not labor-dependent; high capability for automation; inserts can be incorporated and molded in; complex shaped parts with minimal secondary operations other than flash removal; short cure cycle and material or parts flow times | Material flow | Available for molding compounds; some development required for oriented reinforcement systems; higher first cost for tooling than open-molding techniques but practical for long production plans or high production rates |

**Compression molding**

| Process/Materials | Materials | Material form | Advantages | Limitations | Remarks |
|---|---|---|---|---|---|
| Matched-die molding of plastic materials and curing in the mold | Reinforcements: Kv, Gr, Gl, B, hybrids | Preforms, mat, broad goods, cross ply, chopped-fiber nonwovens | Relatively low tooling costs; minimum fiber degradation during molding; reinforcements can be impregnated in molding; hybridization; close tolerances and high mechanical properties; excellent control of tolerances; material flow short, reducing stress in molded parts; short process and cure times with little or no loss of mechanical properties due to deformation of reinforcement; higher initial mold costs but mold maintenance costs lower, overall cost of parts production lower, production rate high, and erosion and damage rates for tooling low | Reinforcement constraints to flow; formability of reinforcement before distortion or fiber damage; fiber crimping or wrinkling; longer cycle time for thick sections; flash lines and control; placement of material in mold may require preforms molded in; holes should not exceed 2.5 times core-pin diameter; mold taper (draft) design changes costly due to mold-rework and mold-modification time; postcure fixturing may be required during mold cool-out | Thin blades and inserts subject to damage; commercially available; development for molding with continuous or woven reinforcements of high-modulus fibers; press capacity |
| | Resins: epoxies, polyesters, vinyl esters, thermoplastics, thermosets | Prepregs, liquids, or films | | Thermal expansion and shrinkage characteristics; reinforcements conformability and matrix-flow characteristics | |
| | Phenolics | Granular | | | |
| | Premixed molding compounds | BMC, SMC, oriented-fiber molding compounds (XMC, TMC), chopped prepregs | | | |

**Table 2.2** Forms, Material Types, and Processes[38] *(continued)*

| Function | Material | | Advantages | Limitations | | Material availability, sources, development |
|---|---|---|---|---|---|---|
| | Type | Form | | Material | Processing | |
| **Injection molding** | | | | | | |
| Heated plastic material forced into cavity of matched-die mold; pressure used to force heated material from plasticator through sprue or runner to mold | Reinforcements; same as compression molding but generally in short fiber lengths; thermosets; thermoplastics | Pellet, granulated; slug or rope; premixed; some liquid resins (extruder feedstocks) | Fast mold cycles; some randomization of fiber orientation, most in plane but some through thickness; automated operation; part uniformity; improved properties; flashless molding; co-injection molding possible with high production rate | Material flow and expansion and shrinkage characteristics; fiber length and orientation limited | Equipment size; mold complexity; some loss of mechanical properties due to fiber damage as molding materials are forced through sprues and gates; depth of molded-in holes made by core pins should not exceed 3–5 times pin diameter; knit lines; moisture and gas entrapment | Moderate to large size and simple to highly complex and intricate shapes; extruder-type injection-molding equipment requires development, as does long-fiber reinforcement; high initial capital costs |
| **Transfer molding** | | | | | | |
| Similar to both compression and injection molding; molding compound put in heated pot and then forced by plunger into heated closed mold, where the compound cures | Same as injection molding | | Fast molding cycles generally analogous to injection molding; can be pressure-impregnated in mold by modified VARI technique (VARI: vacuum-assisted pressure-injection molding); excellent system for limited production; lower capital cost than injection molding; faster than compression molding | Material flow and expansion and shrinkage characteristics | Equipment and mold size capacity; mold complexity; charge preheating required; moisture and gas entrapment | Small to moderate and simple to highly complex and intricate shapes; generally use multiple cavities; commercially available; development required for long-fiber reinforcement |
| **Foam-reservoir molding** | | | | | | |
| Resin-impregnated foam sheet or mat is sandwiched between reinforcement plies, and pressure and heat are applied while resin is squeezed out and into the reinforcement skins | Same as compression molding | Liquid resins in carrier, foam or mat used with nonimpregnated reinforced skin materials | Clean approach to wet lay-up; nonimpregnated reinforcements impregnated during molding cycle; controlled impregnation, wet lay-up system, and hybridization also possible; lowest cost and pressures of all closed-mold systems; lower-cost tooling and reduced labor costs | Handling of resin-impregnated reservoir foam or mat; lower fiber fractions | Mold size; incompatible for co-cured skin structures with integral ribs; void content; control of composite density | Moderate to large sizes and simple to moderately complex shapes; development required in staging and control of resins in reservoir carrier and high-modulus fibers; commercially available |

| | | | | | | |
|---|---|---|---|---|---|---|
| Reinforced plastics (generally thermoplastics) are heated and placed in open matched mold; the mold closes, or strikes the material, dwells for cure or chill, then opens for part removal and reloading | Same as compression molding | Sheet stock, slug, billet, preform | Analogous to forging, coining, and extrusion stamping of metals; fastest of closed-mold processes; well-suited to thermoplastics | Fiber damage, wrinkling, and distortion; flow and formability | Limited hybridization; postcure fixturing during cool-down may be required | Small to moderate size and simple to moderate complexity; commercially available; development required for very-short-cure-time materials |

| | | | | | | |
|---|---|---|---|---|---|---|
| Lay-up of subsections or structural shapes by direct placement of fibers in overlay-type braiding equipment | Reinforcements: Gr, Kv, Gl, hybrids Resins: epoxies, polyesters | Prepregs or dry for impregnations with liquid resin after braiding | Can produce surfaces of revolution or smooth curved surfaces; geometry changes of round to rectangular can be accommodated within limitations of fiber orientation changes; low cost and potentially continuous process; multidirectional reinforcements possible; relatively low-cost tooling; prepreg braiding of tow or roving more economical than postbraiding impregnation; fast method where fiber fraction (volume) and fiber-angle orientation not critical; good potential for fabricating preforms | Parts shape affects braided reinforcement; fiber fractions generally low; fiber angles and ability ro form and turn tight radii in machinery and slip into place on mandrel without fiber damage | Difficulties in debulking; control of resin, voids, resin-rich areas; machine-handling capability of braiding fibers (fiber damage and low fiber-volume fractions); tapered structures require dropping of yarn ends and significant changes in fiber reinforcement angles; number of braiding yarn carriers in braiding machine and amount of yarn that can be spooled | Small to moderate size and simple to moderate complexity; development required for complex primary structures; commercially available but usually must be converted to twisted yarn or binder-stabilized rovings |

**Table 2.2** Forms, Material Types, and Processes[a] *(continued)*

| Function | Material | | Advantages | Limitations | | Material availability, sources, development |
|---|---|---|---|---|---|---|
| | Type | Form | | Material | Processing | |
| | | | **Gore-sections buildup** | | | |
| Lay-up of subsections or details (preforms which are grouped or combined in a larger mold, where they are co-cured and cobonded to produce a large complex structural assembly) | Gr, Kv, Gl, B, hybrids | Epoxy prepregs, woven fabrics, unidirectionals, cross-plied goods, nonwovens, bias wovens; nonimpregnated reinforcements can be used with liquid resins or film systems; preplied, debulked, recompacted prepregs | High rate of material lay-down; reduced scrap of reinforced material; time saving; tooling-cost savings by fabricating many like sections; individual sections laid up outside of main mold to reduce mold loading time; cost saving in repetitive sections or gores to build up assembly lay-ups and material savings by net cutting gore sections | Material accurately cut; ability to make and handle preforms; part shape | Lay-up–joint design; need for additional adhesive; complexity of subsection preforms and detail handling | Available conical or cylindrical parts which can be divided into like sections |

[a]Abbreviations: Kv = Kevlar, EGl = E glass, SGl = S glass, Gl = glass, B = boron, EMC = elastomeric molding tooling compound, Ep = epoxy, BMC = bulk-molding compound, SMC = sheet-molding compound, XMC = directionally reinforced molding compound; TMC = thick molding compound, RIM = reaction injection molding, RRIM = reinforced reaction injection molding.

[b]Cure: elevated temperatures for several hours.     [c]Cure: room to elevated temperatures; short cycle times possible.

[d]Cure: Short cure times, heat and pressure used, temperature generally moderate, radio-frequency-augmented cures possible. Dielectric preheating can be used for thick parts.

[e]Includes melt-flow stamping, solid-phase forming, and thermoplastic forging.     [f]Includes matched-metal-die molding.

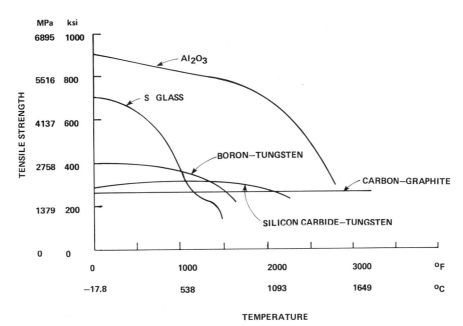

**FIG. 2.1** Temperature dependence of tensile strength of fibrous materials.

the weight of material required to achieve them. For the manufacturer of fiber-reinforced plastic components, therefore, using prepregs reduces the work involved and provides finished components that are consistently stronger, stiffer, more reliable, lighter, and cheaper than equivalent components produced by wet-lay-up techniques.

The most common epoxy resin prepregs are shown in Table 2.3. Boron prepreg tapes generally include a supporting carrier of woven glass-reinforced plastic fabric. Graphite prepregs are the same for the three principal types of fiber: high modulus, high strength, and intermediate strength. Of the various thicknesses of graphite prepregs available cured thicknesses of 0.0056 to 0.0064 in (0.14 to 0.16 mm) per ply are the most common. Length and width dimensions of broad goods depend on the equipment used in prepregging.

Advanced composites are thickness-critical. Fiber or filament volume fraction and mechanical properties are directly related to final ply thickness. Typical uncured and cured lamina characteristics are shown in Table 2.4.

## MATRIXES AND THEIR ROLE

Depending on the application, it is possible to view the role of the matrix in two distinct ways, namely, as the binder that contains the major structural elements (the fibers) and transfers load between them or as the primary phase which is merely reinforced by the secondary, fiber phase. The first approach is the traditional one, since most composites to date have used a relatively soft matrix, a thermosetting plastic of the polyester, phenolic, or epoxy type. The strength of such composites is almost entirely that of the fibers, and for

**Table 2.3**  Prepreg Forms for Epoxy Matrix Composites†

| Prepreg form | Diameter of boron reinforcement | | Graphite reinforcement | |
| --- | --- | --- | --- | --- |
| | 0.004 in (0.1 mm) | 0.0056 in (0.14 mm) | Short staple | Continuous |
| 3-in (76-mm) tape | Continuous roll | Continuous roll | . . . . . . . . . . . . . | Continuous roll |
| Narrow and wide tape | Special order | Special order | . . . . . . . . . . . . . | Special order 0.125–14 in (3.2–356 mm) |
| Sheet | . . . . . . . . . . . . . | . . . . . . . . . . . . . | 12 × 45 in (318 × 1143 mm) | |
| Broad goods (drum-wrapped) | Special order | Special order | . . . . . . . . . . . . . | Special order |

†Based on *Mater. Eng.*, August 1977, p. 50.

efficiency it is desirable to maximize the fiber content. Thus, with some important exceptions, small improvements in the matrix's structural properties are of little value; its adhesion and processing characteristics are paramount.

While the quality of the matrix depends somewhat on the impregnation method, specific resin processing parameters are involved most significantly:

Viscosity and pot life during impregnation

Control of B stage (interim thermosetting cure)

Wettability to filaments

Processing and/or curing temperatures and pressures

Shrinkage during cure

Liberation of volatiles in processing and/or curing

Toxicity

Shelf life for preimpregnation

The lesser structural role played by the binder type of matrix must not obscure the need for it to maintain structural integrity, where interlaminar shear strength, elongation, and notch toughness can be critical. The importance of matrix integrity becomes clear when it is realized that the continuous filaments used in filament-wound structures do not necessarily remain unbroken under normal stress; minute imperfections lead to local failures. These failures do not significantly impair load-carrying ability as long as loads can be efficiently transferred, through the matrix, to other fibers.

The applications which call for significant improvements in matrix properties are those subject to environmental extremes, particularly of temperature. Elevated-temperature applications, such as high-acceleration missiles and reentry vehicles, may require matrix materials with a high degree of heat resistance. At the cryogenic end of the scale, down even to the −423°F (−253°C) of pressure tanks exposed to liquid hydrogen, matrixes must retain their ability to elongate. Advanced resins have increased extensibility and better shear transfer than those now in use; some epoxy types developed have 3% ultimate elongation at −320°F (−196°C), double or triple that of standard epoxies.

Thus, matrix design is rather tightly constrained, not only in its mechanical properties but in its weight. To optimize composite weight without sacrificing structural integrity the matrix portion is usually pared to the minimum giving adequate shear strength and low void content. Filament content of glass composites is typically about two-thirds in volume

**Table 2.4** Typical Uncured and Cured Lamina Characteristics†

| Lamina material | Uncured ply thickness $T_u$ | | Cured ply thickness $T_c$ | | | Bulk factor $T_u/T_c$ | Cured laminate fiber (filament volume fraction) |
|---|---|---|---|---|---|---|---|
| | in/ply | mm/ply | | in/ply | mm/ply | | |
| B–Ep, 0.004 in (0.1 mm), unsupported | 0.0048–0.005 | 0.12–0.13 | max | 0.0047 | 0.12 | 1.15 | 60–65% B, 35–40% Ep |
| | | | nom | 0.0045 | 0.11 | | |
| | | | min | 0.0043 | 0.11 | | |
| Supported | 0.0067–0.0069 | 0.17–0.18 | max | 0.0054 | 0.14 | 1.31 | 52% B, 40% Ep, 8% Gl |
| | | | nom | 0.0052 | 0.13 | | |
| | | | min | 0.0051 | 0.13 | | |
| 0.0056 in (0.14 mm), supported | 0.0088–0.0091 | 0.22–0.23 | max | 0.0069 | 0.18 | 1.31 | 52% B, 40% Ep, 8% Gl |
| | | | nom | 0.0068 | 0.17 | | |
| | | | min | 0.0067 | 0.17 | | |
| Gr–Ep‡ | 0.0081–0.0087 | 0.21–0.22 | max§ | 0.0064 | 0.16 | 1.37 | 60% Gr, 40% Ep |
| | | | nom | 0.0060 | 0.15 | | |
| | | | min | 0.0056 | 0.14 | | |

†Based on *Mater Eng*, August 1977, p. 50.   ‡Oriented-ply lay-ups of all fiber types.
§Most prevalent range; when required, uncured ply thickness can be obtained to allow design as thin as 0.002 in (0.05 mm) or as thick as 0.010 in (0.25 mm) per ply.

and four-fifths in weight; thus, 0.088 lb/in³ (2436 kg/m³) epoxy resin would combine to give a composite density of 0.073 lb/in³ (2021 kg/m³), about one-fourth the density of steel and less than one-half that of titanium.

## Matrix Types

Most structural composite components produced involve resin-matrix materials. Epoxy pre-preg tapes and chopped fiber-filled thermoplastic injection-molding compounds are the most widely used materials. Polyimide, structural thermoplastics (with continuous fibers), polyester, and vinyl ester matrixes are developmental. Fiber-resin prepregs are typically produced by material suppliers. The most frequently used metal-matrix material is aluminum. Titanium, magnesium, and copper alloys are being developed.

### Resin (Organic)

The matrix determines the service operating-temperature and laminating-process parameters. Thanks to widespread fiber-glass–epoxy production experience, nonmetallic matrixes are the most widely used; the 350°F (177°C) stable hot-melt Novolac-epoxy systems predominate. These formulations combine good mechanical properties, thermal stability, and compatibility with existing manufacturing processes. The materials are normally cured for 1 h at 350°F (177°C) under 50 lb/in² (345 MPa) pressure for boron or 75 to 100 lb/in² (517 to 690 MPa) pressure for graphite. As a result of hybridization, several Novolac-epoxy matrixes are frequently cured simultaneously in the same part. The hybrid curing cycles attempt to optimize the response of each component in the final composite at the service-temperature range. These cycles are established from data based on coupon tests of each material molded using the recommended cure cycle for each material. Hybrids will be discussed in greater detail later in the chapter.

Epoxy systems for 250°F (121°C) service have also been qualified. They use much shorter autoclave cycles than the 350°F (177°C) systems, providing production-cost savings. Second-generation condensation-type polyimide systems have been qualified for large-scale space structures with graphite fibers but apparently are not suited for boron composites. Improved thermoplastic polyimide resins have been developed for use with graphite.

### Metallic

The metal matrixes used with boron or Borsic [silicon carbide (SiC)–coated boron] filaments are primarily aluminum alloys; 6061 aluminum is used more frequently than either 2024 or 1100 (pure aluminum). The 2024 alloy provides the highest strength, the 1100 alloy has superior Charpy impact resistance, while the 6061 alloy provides a good combination of strength, toughness, and corrosion resistance. Any of these alloys can be combined with boron or Borsic fiber; however, the bonding temperature with boron is limited to about 930°F (449°C), while with Borsic the bonding temperature can be increased to 1050°F (566°C) if more matrix deformation is desired. Foils of titanium or titanium alloy interleaved between selected plies increase shear strength. Limited developmental programs have been conducted using titanium or molten aluminum with Borsic filaments or aluminum with graphite filaments, but these systems are not in current use for production parts. In the final analysis, the selection of an aluminum matrix material will depend on a trade-off between such properties as tensile strength, shear strength, hardness, impact strength, corrosion resistance, and ease of processing. Bonding of 1100 aluminum is significantly more difficult than with the 2024 or 6061 alloys.[1,2]

Titanium alloys are very amenable to the diffusion-bonding process and can be used as matrix materials for Borsic or SiC fiber, either alone or hybridized with aluminum. Titanium matrix composites provide two advantages over boron-aluminum (B–Al) composites: higher temperature capability [800 to 1000°F (427 to 538°C) vs. 600°F (316°C) for aluminum] and higher transverse properties. Their disadvantage is greater density, 0.13 vs. 0.10 lb/in³ (3598 vs. 2768 kg/m³) for 50 vol% composite material. SiC-coated boron fiber or all-SiC fiber must be used, since the high bonding temperatures would quickly degrade uncoated boron fiber. The properties of various matrix materials are given in Table 2.2.

## Temperature Limitations

### Epoxy

The 350°F (177°C) dry and 260°F (127°C) wet systems are primarily used for military aircraft and high-performance commercial applications. The more economically cured 260°F (127°C) systems are used in helicopter blades and most commercial parts (see Table 2.5).

### Polyimide

Gr–PI materials are being developed for such structures as missile and engine components which must function in temperature ranges beyond the capability of epoxy-resin matrixes. Three classes of polyimide resins are being developed to meet various service temperature regimes. For 260 to 450°F (127 to 232°C) applications the bismaleimide systems cure through a free-radical addition mechanism using autoclave cycles similar to those used for the 350°F (177°C) curing epoxies but higher temperature oven postcures. Currently these systems are being characterized. For the 450 to 600°F (232 to 316°C) service-temperature range, addition polyimides are useful but must be cured at a pressure of 200 lb/in² (1.4 MPa) and a temperature of 600°F (316°C). For service temperatures above 600°F (316°C) condensation-type polyimides are required. Polyimide resins that are normally used with glass fabrics are difficult to process when impregnated on graphite fibers because of volatile entrapment. Some systems have shown excellent thermal stability. Although the material requires curing conditions of 200 lb/in² (1.4 MPa) and 700°F (371°C), engine flaps have been fabricated by press molding.

### Thermoplastic

Thermoplastic matrixes have been used with both continuous and chopped graphite fibers. Recent development programs have shown that conventional thermoplastic matrixes have useful structural properties for lightly loaded structures. Thermoplastic matrixes offer reduced molding costs because the parts require only a few minutes above the matrix softening temperature under pressure. Limited probe data suggest that thermoplastic matrixes may also have better moisture resistance than epoxy systems.

**Table 2.5** General Characteristics of Matrix Systems[3]

| Matrix material | Maximum temperature† | | General characteristics |
|---|---|---|---|
| | °F | °C | |
| Modified epoxy: | | | Thermosetting resin used for low- |
| Continuous service‡ | 350 | 177 | pressure laminating up to 100 |
| Intermittent service | 420 | 216 | lb/in² (690 MPa) requiring a minimum of 350°F (177°C) cure for 350°F service applications |
| Polyimide: | | | Thermosetting resin used for low- |
| Continuous service | 550 | 288 | pressure laminating [up to 100 |
| Intermittent service | 700 | 371 | lb/in² (690 MPa)] requiring 350–600°F (177–316°C) cure plus extended postcure; these resins characterized by high cost, difficult processing, good dielectric properties, and low cured-laminate outgassing |
| Aluminum, continuous service | 600 | 316 | 6061 and 2024 alloys generally require press diffusion bonding under vacuum at about 930°F (499°C) under 3–6 kips/in² (2–41 MPa) pressure to consolidate composite; 713 aluminum braze filler metal requires 1050°F (566°C) under vacuum and about 100 lb/in² (690 MPa) pressure to consolidate (used with Borsic filaments only) |

†Minimum service temperature = $-67°$ $(-55°C)$.

‡Even though modified epoxies are generally considered capable of continuous service up to 350°F (177°C), most experience with composites has involved only intermittent service as high as 350°F and there is some question about degradation of properties after long-time exposure to humidity which limits the service temperature to 275°F (135°C).

## Reinforcements (Filaments)

### Glass

Glass is the most widely used reinforcing material. Glass fiber accounts for almost 90% of the reinforcement in thermosetting resins. Forms of glass-fiber materials are roving (continuous strand), chopped strand, woven fabrics, continuous-strand mat, chopped-strand mat, and milled fibers [0.032 to 0.125 in (0.8 to 3.2 mm long)]. The longer fibers provide the greatest strength; continuous fibers are the strongest. These are all produced in the standard (E-glass) reinforcement grade. The higher-strength forms are also available in S-glass fiber, about one-third stronger than E glass and costing considerably more. Uses are almost entirely in aircraft and military components. The other type of glass fiber is D glass.

All three glasses have very high tensile-strength-to-weight ratios although glass is among the densest of the synthetic inorganic fibers. Glass does not burn and retains good mechanical properties up to approximately 50% of its strength to 700°F (371°C) and 25%

of its strength to 1000°F (538°C). Moisture resistance is excellent, and glass fibers do not swell, stretch, disintegrate, or undergo other chemical change when wet. Except for strong alkalies and hydrofluoric acid glass has excellent corrosion resistance.

Glass is relatively inexpensive. E glass, the most widely used fiber glass, costs between 38 and 43 cents per pound (84 and 95 cents per kilogram), but specialty fibers, such as S glass, cost $7.52 per pound ($16.54 per kilogram) (1980).

**Glass Production:** It is possible to draw many glasses of varying compositions into continuous filaments or as staple fibers. The continuous filament is an individual fiber having an indefinite length (measurable in miles). The staple fiber is an individual fiber 8 to 15 in (203 to 381 mm) long. Most continuous filaments are mechanically drawn from a stream of molten glass at speeds exceeding 2 mi/min (54 m/s). Staple fibers are made by a blowing process in which air, steam, or hot gas is directed at a stream of molten glass. Both continuous and staple fibers can be fabricated into strands, yarns, and cords through conventional twisting, plying, and cabling operations. Staple fibers, however, are seldom used for the reinforcement of structural composites.

In producing glass fibers, a glass batch is compounded and mixed with great care to ensure consistent glass composition. Constantly controlled glass composition is important from the standpoint not only of the chemical and physical properties of the glass fibers but also of the efficiency with which they can be produced.[6]

Production of continuous filament fibers proceeds by feeding the raw glass, in the form of marbles or batch, into an electrically heated fiberizing element referred to as a *bushing*. A typical bushing has 204 or more orifices at the bottom, each generally 0.032 to 0.125 in (0.8 to 3.2 mm) in diameter. Once inside the bushing and in the molten state, the glass flows through the orifices in the form of glass streams. The amount of glass above the orifices (pressure head), viscosity of the melt, the number and size of the orifices, and the rate at which fibers are drawn or pulled from the bushing influence the diameter of the filaments produced.

Just below the bushing is a sizing applicator, which applies an organic sizing material (an aqueous or solvent dispersion) to each filament. The sizing performs a threefold function: (1) as a film former it protects the filaments from abrading each other, (2) it acts as a lubricant, enabling filaments to move relative to each other, and (3) it provides the compatible interface between the inorganic glass filaments and the organic resin matrixes they reinforce. Below the sizing applicator filaments are gathered into a bundle, called a *strand*, by a gathering shoe. Thence the strand passes to a winder, where it is wound onto a forming tube. The strand material so formed is frequently referred to as a *forming cake*. When the cake has been built up to the desired size or weight, it is removed from the winder and dried, usually in an oven. This removes the solvent or water associated with application of the size.

A textile-fiber forming operation must operate efficiently; otherwise sufficient material cannot be produced in the form of usable forming cakes to make the operation economically practical. The process is very sensitive to the processes parameters (the batch, the temperature, the bushing environment, etc.). If any of them is significantly out of control, the bushing and drying apparatus will not form fibers consistently enough to produce the high percentage of full forming cakes required for an efficient operation. Therefore all process parameters must be under control at the same time or the operation will be so inefficient that an inadequate number of full forming cakes will be produced. The glass-fiber

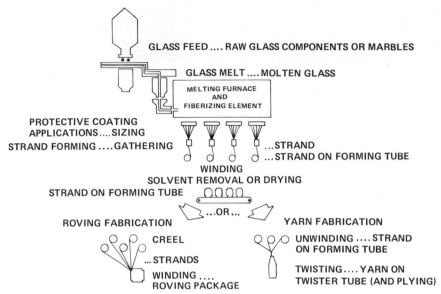

**FIG. 2.2** Glass-fiber manufacturing process.

manufacturing process (Fig. 2.2) therefore has a built-in quality assurance over and above all the testing performed on the material produced to assure material of consistent quality.

**Glass Composition:** The major constituent of most commercial inorganic glasses is silica. The silicon dioxide molecule has a tetrahedral configuration, consisting of a central silicon ion surrounded by four oxygen ions. The three-dimensional network of silica tetrahedra is the basis of the various and unusual properties of glass. By addition of modifying ingredients such as metallic oxides, which may become part of the silica network or disrupt it, the properties of the amorphous glass can be varied and adjusted to various levels of performance.

## E Glass

The first glass developed specifically for production of continuous fibers was a lime-alumina-borosilicate glass designed primarily for electrical applications. Designated E glass, it was found to be adaptable and highly effective in a great variety of processes and products ranging from decorative to structural applications; it has become known as the standard textile glass. Most continuous-filament glass produced today is E glass. E glass may vary in composition as shown in Table 2.6. Changes within the ranges shown in Table 2.6 do not influence the electrical or mechanical properties of E glass.

A property widely used in the glass industry is virgin tensile strength, defined as the tensile strength of glass fibers which have not contacted other gas, solid, or liquid materials and which have been stored at room temperature. It is an excellent indicator of the strength of a given glass composition and an index to the effectiveness of various fiber-production parameters (see Table 2.7). Other properties which are more or less characteristic of the

**Table 2.6**  Composition of E Glass[37]

|  | Range, wt % |
| --- | --- |
| $SiO_2$ | 52–56 |
| $Al_2O_3$ | 12–16 |
| CaO | 16–25 |
| MgO | 0–6 |
| $B_2O_3$ | 8–13 |
| $Na_2O$ and $K_2O$ | 0–3 |
| $TiO_2$ | 0–0.4 |
| $Fe_2O_3$ | 0.05–0.4 |
| $F_2$ | 0–0.5 |

glass industry are softening point, strain point, and annealing point. Each is an arbitrarily defined temperature which helps define glass viscosity.

### S Glass

S glass is a high-tensile-strength glass. Its tensile strength is 33% greater and its modulus of elasticity almost 20% greater than that of E glass. Significant properties of S glass for aerospace applications are its high strength-to-weight ratio, its superior strength retention at elevated temperatures, and its high fatigue limit. It has found wide application in rocket-motor cases (Fig. 2.3), high-performance aircraft parts, and other areas where mechanical performance is the primary requirement. Typical composition is 65% $SiO_2$, 25% $Al_2O_3$, and 10% MgO. Typical properties for S glass are given in Table 2.7.

### D Glass

D glass is an improved dielectric glass developed for high-performance electronic applications. Although the mechanical properties of D glass are exceeded by those of E and S glass, its lower dielectric constant and lower density make it attractive for radome construction. Typical properties of D glass are given in Table 2.7.

### Other Glasses

Other alphabetically named glasses developed for specific uses are not in general demand.

### Glass-Fiber Forms

Continuous-filament glass textile fibers normally have diameters ranging from 0.00010 to 0.00075 in (0.003 to 0.02 mm). Although fibers of greater diameter can and have been made, they have reduced flexibility and begin to assume the properties of the counterpart bulk material. In other words, with increased diameter they perform more as rods rather than as fibers. As commercial reinforcements, glass fibers are produced as roving, chopped strands, mats, fabrics, and woven rovings. A brief description of typical reinforcement forms and their uses is provided in Table 2.8.

**Table 2.7** Typical Properties of E, S, and D Glass[37]

| Property | Temp °F | Temp °C | E glass kips/in² | E glass GPa | S glass kips/in² | S glass GPa | D glass kips/in² | D glass GPa |
|---|---|---|---|---|---|---|---|---|
| Specific gravity† | | | 2.54 | | 2.49 | | 2.16 | |
| **Mechanical†** | | | | | | | | |
| Virgin tensile strength | 72 | 22.2 | 500 | 3.45 | 665 | 4.6 | 350 | 2.41 |
| Yield strength | 1000 | 538 | 120 | 0.83 | 275 | 1.9 | | |
| Ultimate strength | 1000 | 538 | 250 | 1.7 | 350 | 2.41 | | |
| Modulus of elasticity | 72 | 22.2 | 105,000 | 724 | 124,000 | 855 | 75,000 | 517 |
| After heat compaction | 72 | 22.2 | 124,000 | 855 | 135,000 | 931 | | |
| | 1000 | 538 | 118,000 | 814 | 129,000 | 890 | | |
| Elastic elongation | 72 | 22.2 | 4.8% | | 5.4% | | 4.7% | |
| **Thermal‡** | | | | | | | | |
| Coefficient of linear expansion, °F⁻¹ / °C⁻¹ | | | $2.8 \times 10^{-6}$ | $5.04 \times 10^{-6}$ | $1.6 \times 10^{-6}$ | $0.89 \times 10^{-6}$ | $1.7 \times 10^{-6}$ | $3.06 \times 10^{-6}$ |
| Specific heat | 75 | 23.8 | 0.192 | | 0.176 | | 0.175 | |

| | E glass °F | E glass °C | S glass °F | S glass °C | D glass °F | D glass °C |
|---|---|---|---|---|---|---|
| Softening point | 1515 | 824 | 1778 | 970 | 1420 | 771 |
| Strain point | 1140 | 616 | 1400 | 760 | 890 | 477 |
| Annealing point | 1215 | 657 | 1490 | 810 | 970 | 521 |

†Measured on glass fibers.  ‡Measured on bulk glass.

**FIG. 2.3** A filament-winding machine making a rocket-motor case. (*McClean-Anderson.*)

## Roving

*Roving* refers to a group of essentially parallel strands or *ends* of glass fibers which have been gathered into a ribbon and wound onto a cylindrical tube. *Continuous-strand roving* consists of parallel-wound strands available in a variety of number of ends, or yields. *Spun-strand roving* is a bulkier roving in which the continuous-filament strands are looped back and forth upon themselves and held in roving form by a slight twist and the use of a resinous sizing.

For some years roving was designated on the basis of end count such as 12, 20, 30, 60, and 120 ends. With the development of large high-throughput bushings, glass suppliers and users have come to designate roving on the basis of yield. Rovings are primarily rated and selected according to the following characteristics:

*Tensile strength:* Primarily attributed to the glass composition although influenced by the performance of the sizing material

*Strand integrity:* Degree of bonding between filaments in the individual strands (ends)

*Ribbonization:* Measure of the degree to which the strands are held together in the roving bundle

*Catenary:* Degree of sag between the individual strands within the roving bundle

*Wet-out:* Speed and degree of roving wetting by the resin

*Chopping characteristics:* Ease of chopping, retention of strand integrity, and the amount of static electricity produced by the roving when it is processed into chopped strands, chopped-strand mats, or preforms

**Table 2.8** Commercial Forms of Glass-Fiber Reinforcements[37]

| Nominal form | General description | Process | Nominal glass content of typical laminates, % | Typical application |
|---|---|---|---|---|
| Rovings | Continuous strands of glass fibers | Filament winding, continuous panel, preforming (matched die molding), spray-up, pultrusion | 25–80 | Pipe, automobile bodies, rod stock, rocket-motor cases, ordnance |
| Chopped strands | Strands cut to lengths of 0.125–2 in (3.2–50.8 mm) | Premix molding, wet slurry preforming | 15–40 | Electrical and appliance parts, ordnance components |
| Reinforcing mats | Continuous or chopped strands in random matting | Matched-die molding, hand lay-up, centrifugal casting | 20–45 | Translucent sheets, truck and auto body panels |
| Surfacing and over-laying mats | Nonreinforcing random mat | Matched-die molding, hand lay-up, and filament winding | 5–15 | Where smooth surfaces are required (automobile bodies, some housings) |
| Yarns | Twisted strands | Weaving, filament winding | 60–80 | Aircraft, marine, electrical laminates |
| Woven fabrics | Woven cloths from glass fiber yarns | Hand lay-up, vacuum bag, autoclave, high-pressure laminating | 45–65 | Aircraft structures, marine, ordnance hardware, electrical flat sheet and tubing |
| Woven roving | Woven glass fiber strands (coarser and heavier than fabrics) | Hand lay-up | 40–70 | Marine, large containers |
| Nonwoven fabrics | Unidirectional and parallel rovings in sheet form | Hand lay-up, filament winding | 60–80 | Aircraft structures |

Rovings are used to make filament-wound products such as rocket-motor cases, pressure bottles, tanks, and pipe. They are also used in a variety of other processes including preforming, spray-up, and pultrusion.

## Chopped Strands

Continuous or spun roving or strand can be chopped into short lengths, usually 0.125 to 2 in (3.2 to 50.8 mm) long. Chopped strands are available with different sizings for compatibility with most plastics, the amount and type of size having a major influence on the integrity of the strand before and after chopping. Strand of high integrity is termed *hard* and strand which separates more readily is *soft*. Chopped-strand materials are used in premix and wet-slurry molding as well as for reinforcement in thermoplastic molding compounds. Strand, usually hard, is sprayed simultaneously with liquid resin (spray-up process) to build up reinforced plastic parts on a mold. Short chopped strands tend to reduce strand orientation and permit the molding of relatively thin wall sections. In premix molding longer chopped strands may be mixed with shorter ones to improve fill-out in parts having extreme curvature as well as increasing the glass content in areas prone to impact and crazing.

## Milled Fibers

Continuous strands can be hammer-milled into short nodules of glass ranging in length from 0.015 to 0.25 in (0.38 to 6.4 mm). The length of milled fibers refers to the diameter of the screen openings through which the fibers pass during processing. Many sizing materials are available on milled fibers to provide compatibility with polyesters, epoxies, etc. Milled fibers are generally used to provide anticrazing, body, and dimensional stability to potting compounds, adhesives, patching compounds, and putties.

## Reinforcing Mat

Glass-fiber mat is a blanket of chopped strand or of continuous strands laid down as a continuous thin flat sheet. The strands are evenly distributed in a random pattern and are held together by adhesive resinous binders or mechanically bound by *needling*. The reinforcing ability of continuous-strand and chopped-strand mat is essentially the same, but they have different handling and molding characteristics. Continuous-strand mat can be molded to more complicated shapes without tearing. Needled mat, which has some fibers vertically oriented, is softer and more easily draped than nonneedled mat and therefore generally used only where reinforcement conformability is a particular requirement.

Reinforcing mats are distinguished by the binder used to hold them together, which may be of high or low solubility. "Solubility" designates the rate of binder dissolution in the liquid resin matrix. Mats with a high-solubility binder are used in hand lay-up processes or wherever rapid wet-out and the contour matching is important. Mats with a low-solubility binder are used in press molding or wherever the flow of the liquid matrix resin may wash away or disrupt the strands, leaving resin-rich areas.

Mat is used in hand lay-up, press molding, bag molding, autoclave molding, and in various continuous impregnating processes. The isotropic laminates which result from the random reinforcement of mat may have glass contents of from 15 to 50 wt %.

## Surfacing and Overlay Mat

For some applications it is desirable to have a resin-rich surface layer which will be both visually and physically smooth. To obtain such surfaces, very thin surfacing or overlay mats can be used. Surface mat is stiff due to a high binder content, and its use is generally restricted to simple shapes. Overlay mat, with a low binder content, is very soft and drapable and can be used on complex shaped parts. Neither the surfacing mat nor the overlay mat is considered a bona fide reinforcing mat.

## Yarns

A yarn is an assemblage of fibers or strands, generally fewer than 10,000, which is suitable for use in weaving into textile materials. Tow is a large bundle of continuous fibers, generally 10,000 or more, not twisted. The simplest yarn is an individual strand of glass fibers, commonly referred to as a *singles yarn*. Heavier yarns are obtained by combining single strands by twisting and plying, typically by simply twisting two or more single strands together and subsequently plying (twisting two or more of the twisted strands together).

A yarn or strand has an *S twist* if, when held in a vertical position, the spirals conform in slope to the central portion of the letter S; it has a *Z twist* if the spirals conform in slope to the central portion of the letter Z. Strands which are simply twisted will kink, corkscrew, and unravel because of their twist in only one direction. The plying operation normally eliminates this problem by countering the twist in the twisted singles yarn with an opposite twist in the plied yarn. For example singles yarn with an S twist is plied with a Z twist to result in a balanced yarn. The twisting and plying operations vary the yarn strength, diameter, and flexibility and are important steps in producing the variety of fabrics which composite fabricators require.

An exact system for identifying glass-fiber yarns is required because of the wide variety of types available. Yarn nomenclature consists of two basic parts, one alphabetical and one numerical. Typical examples are ECG 150 2/2 and SCD 450 1/2. The first letter indicates the glass composition such as E (electrical glass) or S (high-strength glass). The second letter indicates whether the fibers are continuous (C) or staple (S). The third letter indicates the average diameter of the fibers from which the yarn is made:

Code for Fiber Diameter, $\mu$in

| | | | | | |
|-----|---------|---|---------|---|---------|
| B   | 100–149 | F | 300–349 | L | 550–599 |
| C   | 150–199 | G | 350–399 | M | 600–649 |
| D   | 200–249 | H | 400–449 | N | 650–699 |
| DE  | 230–279 | J | 450–499 | P | 700–749 |
| E   | 250–299 | K | 500–549 |   |         |

Thus, the letters ECG indicate a yarn of E-glass composition produced as a continuous filament of average fiber diameter between 350 and 399 $\mu$in (0.009 and 0.01 mm).

The numbers used in glass-yarn nomenclature identify the basic strand weight and the yarn construction. The strand weight is indicated by the first series of numbers following the letters. It numerically represents approximately one-hundredth of the yards per pound† of basic glass-fiber strand. The second series of numbers such as 1/2, 2/2, 1/0, designates

---

†To convert yards per pound to meters per kilogram, multiply by 2.016.

the yarn construction. It tells how many basic single strands are in the yarn. The first digit indicates the number of basic single strands which are plied together. To find the total number of strands in the yarn, multiply these two numbers together (0 is multiplied as 1).

The approximate yards per pound of fabricated yarn can be computed by dividing the basic strand weight by the number of strands in the yarn. The calculated yards per pound will be approximate because the actual yards per pound is reduced slightly in the twisting and plying operations. An example of glass-yarn nomenclature is ECG 150 1/2, where

E  = electrical glass

C  = continuous filament

G  = G filament diameter 350 to 399 $\mu$in (0.009 to 0.01 mm)

150  = 15,000 yd/lb nominal of basic strand

1/2  = single strands twisted and two of the twisted strands plied together

Since there are two basic strands in the fabricated yarn,

$$\frac{15,000}{2} \approx 7500 \text{ yd/lb (15,120 m/kg)}$$

for fabricated yarn.

## Woven Fabrics

A fabric is a material constructed of interlaced yarns, fibers, or filaments, usually a planar structure. Typical glass-fiber fabrics are manufactured by interlacing warp (lengthwise) yarns and fill (crosswise) yarns on conventional weaving looms. Such fabrics are woven into a variety of styles which permit quite exact control over thickness, weight, and strength. The principal factors which define a given fabric style are fabric count, warp yarn, fill yarn, and weave.

The fabric count refers to the number of warp yarns (ends) per inch ( 25.4 mm) and number of filling yarns (picks) per inch. For example, a fabric count of 57 x 54 means that there are 57 ends per inch running in the warp (lengthwise) direction and 54 picks per inch running in the fill (transverse) direction. Fabric count plus the properties of the warp and fill yarns used to weave fabrics are the principal factors which determine fabric strength.

The weave of a fabric refers to how warp yarns and fill yarns are interlaced. Weave determines the appearance and some of the handling and functional characteristics of a fabric. Among the popular weave patterns are plain, twill, crowfoot satin, long-shaft satin, leno, and unidirectional. *Plain weave* is the oldest and most common textile weave. One warp end is repetitively woven over one fill yarn and under the next. It is the firmest, most stable construction, providing porosity and minimum slippage. Strength is uniform in both directions (see Fig. 2.4). *Twill weaves* have one or more warp ends passing over and under two, three or more fill picks in a regular pattern. Such weaves drape better than a plain weave. In the *crowfoot* and *long-shaft satin weaves* one warp end is woven over several successive fill yarns, then under one fill yarn. A configuration having one warp end passing over four and under one fill yarn is called a *five-harness satin weave*. Similarly, in an eight-harness satin fabric, one warp end passes over seven fill yarns, then under one fill yarn. The satin weave is more pliable than the plain weave. It conforms readily to compound curves and can be woven to a very high density. Satin weaves are less open than other weaves;

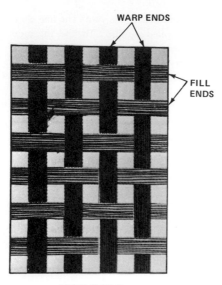

WARP ENDS

FILL ENDS

PLAIN WEAVE

**FIG. 2.4** Plain weave. *(Union Carbide.)*

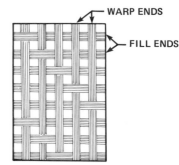

WARP ENDS

FILL ENDS

FIVE-HARNESS SATIN WEAVE

**FIG. 2.5** Satin weave. *(Union Carbide.)*

strength is high in both directions (Fig. 2.5). The *leno weave* has two or more parallel warp ends interlocked. It tends to minimize sleaziness. *Unidirectional weave* involves weaving a great number of larger yarns in one direction with fewer and generally smaller yarns in the other direction. Such weaving can be adapted to any of the basic textile weaves to produce a fabric of maximum strength. Other weaves include basket, semibasket, mock-leno, and high-modulus weaves.

Glass fabrics are used to make a great variety of consumer products, e.g., boats, aircraft, pipe, tanks, electrical laminates, and ballistic armor. A description of some common glass fabrics is given in Table 2.9.

## Nonwoven Fabrics

A nonwoven fabric is a sheet of parallel strands, yarns, or rovings held together by an occasional small transverse strand or by a periodic cross bond with resin. It is also available

**Table 2.9** Some Common Glass Fabrics[37]

| Style | Count | Warp yarn | Fill yarn | Weave | Nominal thickness in | mm |
|-------|-------|-----------|-----------|-------|------|-----|
| 108 | 60 × 47 | 900 1/2 | 900 1/2 | Plain | 0.002 | 0.05 |
| 112 | 40 × 39 | 450 1/2 | 450 1/2 | Plain | 0.003 | 0.08 |
| 116 | 60 × 58 | 450 1/2 | 450 1/2 | Plain | 0.004 | 0.10 |
| 120 | 60 × 58 | 450 1/2 | 450 1/2 | Crowfoot | 0.004 | 0.10 |
| 143 | 49 × 30 | 225 1/2 | 450 1/2 | Crowfoot | 0.009 | 0.23 |
| 181 | 57 × 54 | 225 1/3 | 225 1/3 | Satin | 0.0085 | 0.22 |
| 1581 | 57 × 54 | 150 1/2 | 150 1/2 | Satin | 0.0085 | 0.22 |
| 7581 | 57 × 54 | 75 1/0 | 75 1/0 | Satin | 0.0085 | 0.22 |

in a resin-preimpregnated form, held together by the matrix resin. This unidirectional material permits the application of filament-winding design and analysis techniques to the fabrication of parts that are not easily adaptable to a typical filament-winding process. Nonwoven fabrics are used frequently in both hand lay-up and filament-winding processes.

## Woven Roving

Rovings can be woven into a product called *woven roving* similar to the way yarns are woven into fabrics. Woven rovings are heavier and thicker than fabrics since rovings are heavier than yarns. Woven rovings typically weigh 12 to 40 oz/yd$^2$ (0.041 to 0.136 g/cm$^3$) and have thicknesses of 0.02 to 0.05 in (0.50 to 1.3 mm). They are usually provided in a plain weave although special weaves have been developed. Woven rovings are usually molded by hand lay-up. Typical applications include boats and cargo containers.

## Zero-Twist Fabrics

These are a form of woven roving using finer, untwisted rovings woven with four-shaft twill or crowfoot satin construction to produce a soft, thin fabric readily formable over simple contours. S-glass yarns have been woven this way.

## Other Reinforcement Forms

Glass fabrics are also available in tapes, contoured fabrics, and fluted-core fabrics. Tapes are simple narrow fabrics. Contoured fabrics are fabrics woven into a given symmetrical shape such as a dome or cone. Fluted-core reinforcements are three-dimensional, typically having one or two plies of fabric serving as a core separating two or more plies of fabric which serve as skins. Usually the cross-sectional configuration of the core plies is triangular or rectangular. Such fluted-core fabrics may be integrally woven or the skin fabrics may be sewed to the core fabrics at appropriate intervals.

## Hollow Fibers

It is possible to draw hollow fibers from a bushing in essentially the same way as solid fibers are drawn. Since hollow fibers hold the promise of lighter weight, greater stiffness, and a lower dielectric constant, they have been studied for possible application in hydrospace vehicles and radomes. To date, however, solid fiber predominates in structural and electrical composites.

## Surface Treatments

An important requirement of glass-fiber-reinforced plastics is good adhesion between the glass and the plastic matrix. The adhesion must be strong and not significantly weakened by environmental conditions. If adhesion is weak or weakened by environment, stresses will not be effectively transferred from fiber to fiber and complete reinforcing action of the fibers will not be realized. Good adhesion is obtained by the application of coupling agents which react with, or bond strongly to, the glass-fiber surface and to the plastic matrix. Coupling agents are applied to fibers as a chemical surface treatment in the form of a size or a finish. A *size* is a surface treatment applied as the fibers are formed. A *finish* is a surface treatment applied after fibers have been fabricated into yarns or woven fabric. *Compatible sizes* are used for roving, mat, and chopped strand, and *temporary sizes* are used for producing yarns and fabrics. A compatible size typically contains a coupling agent, a lubricant, and a film former. The coupling agent provides a good glass-to-resin bond, as discussed above; the lubricant protects fibers from glass-against-glass abrasion; and the film former protects fibers and bonds them in a strand to improve subsequent textile operations. Compatible sizes are categorized by their coupling agent, which is generally a silane or a chrome complex.

With mat reinforcements compatible sizes are applied as the fibers are formed, but they are supplemented in mat production by the addition of a small quantity of binder resin which holds the strands together in a sheet that can be handled. Different binder resins are used to control the softening and solubility of the bond upon addition of resin and heat.

Temporary sizes typically contain such ingredients as dextrinized starch, gum, hydrogenated vegetable oil, nonionic emulsifying agent, cationic lubricant, gelatin, and polyvinyl alcohol. They are commonly called *starch-oil sizes* and serve to protect fibers and hold strands together during handling and weaving. The presence of a starch-oil size on fabric is satisfactory for some applications, but it is not compatible with some resins, does not wet out well, and gives poor laminate wet-strength retention. Consequently, it is usually removed and a finish applied. Removal is accomplished by burning or distilling off. The burning-off process, known as *heat cleaning,* involves heating the greige goods or loom-state fabric to 650°F (343°C) or higher, where most of the organic sizing is burned off. Fabrics that undergo this treatment are virtually pure glass and are easily damaged unless protected by a finish. Sometimes starch-oil-sized fabric is heated only enough to burn off part of the sizing, and the remainder is caramelized. This fabric has a tan color and is especially suited for melamine resins.

Finishes contain strong coupling agents and are designed to provide not only a strong glass-to-resin bond but fiber protection, ease of fabric wet-out, and various other special fabric characteristics. A variety of high-temperature, high-strength finishes are available. For example, HTS or 901 is a compatible chemical surface treatment applied during the glass-forming process. It results in a strand of maximum strength, which can be made into roving and yarn and woven into fabric. No heat cleaning or finishing is required. It is epoxy-compatible and is available on S glass. There are many other standard and widely used finishes for glass fabrics, some of which are proprietary to individual industrial weavers.

## Glass Strength

The strength that can be developed by glass filaments in glass composites is surprisingly high, up to 80% of the tensile strength of the virgin filament. Surface defects on the glass filament can reduce the final composite strength significantly. This effect decreases with decreasing filament diameter. Glass filaments and unidirectional composites are relatively poor in compressive strength. Glass filaments also lack the stiffness that can be obtained in other materials, such as boron. To offset the poor compressive composite strength cross-plies are used in laminates and in filament-wound structures.

When specifying the requirements for a high-strength, high-temperature [350°F (177°C)] epoxy-resin-impregnated glass fabric (E glass) for use on thermal-expansion molded parts typically one must show the following:

*Glass-fabric resin content:*   Depending on type of fabric required, this percent can vary from 34 to 40 and 38 to 44 wt %.

*Glass-fabric resin flow:*   This can vary from 6 to 18 and 6 to 24%.

The cured laminate or sandwich must comply with specified mechanical-property requirements (Table 2.10).

# PRINCIPLES FOR GLASS-FIBER-REINFORCED PLASTICS

Four major principles should be recognized in using glass fibers as composite reinforcement. Mechanical properties depend on the combined effect of the amount of glass-fiber reinforcement used and its arrangement in the finished composite. Chemical, electrical, and thermal performance are influenced by the resin system used as the matrix. Materials selection, design, and production requirements determine the proper fabrication process to be used. Finally the cost-performance value achieved in the finished composite depends upon good design and judicious selection of raw materials and processes.

## Principle 1

### Amount of Glass

The strength of the finished object is directly related to the amount of glass in it. Generally speaking, strength increases directly in relation to the amount of glass. A part containing 80 wt % glass and 20 wt % resin by weight is almost 4 times stronger than a part containing the opposite amounts of these two materials.

### Arrangement of Glass

Strength is related to the arrangement of glass in the finished object. Consider three cases: (1) all glass strands laid parallel to each other, (2) half the strands laid at right angles to the other half, and (3) strands arranged in a random manner. In case 1 maximum strength and modulus are obtained in the filament direction. Such a parallel arrangement is used in the design of rocket-motor cases, golf clubs, and fishing rods. In case 2 strength is highest in the two directions of the strands. Although strength is less than with parallel arrangement, it is still considerable. Bidirectional laminates find application in boats, airplane-wing tips, and swimming pools. In case 3 strength is no longer concentrated in one or two directions. Safety helmets, chairs, electrical parts, luggage, and machine housings utilize this strength. This random arrangement results in equal but lower strength in all directions. This condition is called isotropic.

### Arrangement and Glass Content

There is a relationship between how the glass is arranged and the amount of glass that can be loaded in a given object. As in stuffing a shoebox full of objects, the neater the arrangement the more objects can be placed in a given volume. By placing continuous strands next

Table 2.10 Minimum Required Mechanical Properties of Laminate†

| | 75 + 5°F (23.8 + 3°C) | | | | 350 + 5°F (117 + 3°C) | | | |
| | Dry | | Wet | | Dry | | Wet | |
| Property | ksi | MPa | ksi | MPa | ksi | MPa | ksi | MPa |
|---|---|---|---|---|---|---|---|---|
| Compression strength, av | 70 | 483 | 55 | 379 | 40 | 276 | 30 | 207 |
| Individual | 60 | 414 | 50 | 345 | 35 | 241 | 27 | 186 |
| Compression modulus, av | 3200 | 22,064 | 2900 | 19,996 | 2700 | 18,617 | 2200 | 15,169 |
| Flexural strength, av | 75 | 517 | 70 | 483 | 65 | 448 | 60 | 414 |
| Individual | 70 | 483 | 65 | 448 | 60 | 414 | 55 | 379 |
| Flexural modulus, av | 3200 | 22,064 | 2700 | 18,617 | 2700 | 18,617 | 2400 | 16,548 |
| Tensile strength, av | 55 | 379 | 50 | 345 | 45 | 310 | 40 | 276 |
| Individual | 50 | 345 | 45 | 310 | 40 | 276 | 35 | 241 |
| Tensile modulus, av | 3000 | 20,685 | 2500 | 17,238 | 2200 | 15,169 | 2000 | 13,790 |
| Interlaminar shear, av | 2.8 | 19 | .... | .... | 2.5 | 17 | | |
| Individual | 2.5 | 17 | .... | .... | 2.0 | 14 | | |

†Minimum required Barcol hardness = 70.

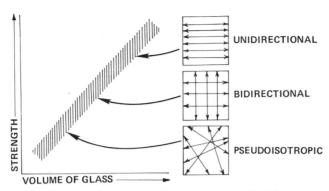

FIG. 2.6   Relation between strength and arrangement of reinforcement. (*Owens-Corning Fiberglas Corp.*)

to each other in a parallel arrangement more glass can be placed in a given volume. Glass loadings range from 45 to 90 wt %. When half the strands are placed at right angles to the other half, glass loadings range from 55 to 75 wt %. A random arrangement gives glass loadings in a range of 15 to 50 wt %. The relationship of amount of glass, strength characteristics, and arrangement of glass is shown in Fig. 2.6. Continuous parallel strands give the highest strength range, bidirectional arrangement gives a middle-strength range, and random arrangement gives the lowest strength range.

## Glass Reinforcements

Glass fibers are sold in forms which permit the designer to use this directionality to maximum advantage. The basic forms of glass are continuous strand, fabric, woven roving, chopped strand, reinforcing mat, and surfacing mat. Continuous strand or roving gives reinforcement in the direction of the lay of the strands or roving. Fabric essentially reinforces the object in two directions. Woven roving gives high strength and is lower in cost than conventional glass fabrics. Chopped strands give a random reinforcement. Reinforcing mats are lower in cost than fabric and give random reinforcement. Surfacing mat gives virtually no reinforcement but gives a smooth decorative surface finish.

## Principle 2

Resins used in glass-fiber-reinforced plastics vary in resistance to corrosion and heat. Formulation of the resin mix also influences corrosion and heat resistance but has a less pronounced effect. Varying ingredients such as filler, pigment, and catalyst system changes the performance of the resin mix. Resin also helps prevent abrasion of the glass fibers by maintaining the position of the fibers and keeping them separated. Polyester resins are used in approximately 85% of all glass-fiber-reinforced plastics because they are economical. Other resins in use are epoxies, phenolics, silicones, melamines, acrylics, and polyesters modified with acrylics. Some thermoplastic resins (nylon, polystyrene, polycarbonate, and fluorocarbons) are reinforced with fibrous glass.

## Principle 3

Processes are not alike in their ability to make use of different arrangements of glass, different amounts of glass, and different resins. A given combination of raw materials required to meet performance criteria in a given application narrows the choice of processes to those which can successfully and economically form the raw material into a completed part. Production flexibility of a process is often the single most important economic factor. If many parts are to be made from one mold, for example, the lowest total cost is achieved by using presses and molds and automated materials handling. Conversely, if only a few parts are required, a process minimizing investment in molds and other equipment would be the logical choice.

## Principle 4

Economical cost and performance result from good design based on judicious selection of both raw materials and process. Proper materials must be combined in a process or processes so that potential performance is realized at an economical cost of manufacturing. Design of the part must take advantage of the material's maximum capabilities.

### Aramid (Kevlar) †

Introduced commercially in the seventies, Kevlar aramid is an aromatic organic compound of carbon, hydrogen, oxygen, and nitrogen. Kevlar aramid fiber is produced by spinning long-chain polyamide polymers using standard textile techniques. The low-density, high-tensile-strength, low-cost fiber produces tough, impact-resistant structures with about half the stiffness of graphite structures. Because the compressive properties of Kevlar laminates are low (due to poor coupling of resin matrixes to the aramid fibers), applications are either secondary structures or tension-critical, filament-wound overwraps of rocket motors. The aromatic polyamide fibers are characterized by their high tensile strength and high modulus compared with other organic fibers. The fiber was originally developed to replace steel in radial tires and has found increasing use in the belts of radial car tires and carcasses of radial truck tires, where it saves weight and increases strength and durability. Kevlar 29 is the low-density, high-strength aramid fiber designed for ballistic protection, slash and cut resistance, ropes, cables, and coated fabrics for inflatables and architectural fabrics. Kevlar 49 aramid fiber is characterized by low density and high tensile strength and modulus (Fig. 2.7). These properties are the key to its successful use as reinforcement for plastic composites in aircraft, aerospace, marine, automotive, and other industrial applications and in sports equipment. Kevlar 49 has been made in continuous-filament yarns and rovings and chopped fiber. The yarns have been converted by weavers into broad goods, woven fairings, and tapes for plastic reinforcement applications. Other fabrics and tapes preimpregnated with thermoset resins have been produced. Chopped fibers can be combined with thermosets or thermoplastics to make chopped-fiber molding compounds. The Kevlar 49 fibers sell (1980) for about $13 per pound ($28.60 per kilogram).

Kevlar 49 aramid is used in high-performance composite applications where light weight, high strength and stiffness, vibration damping and resistance to damage, fatigue,

†Trademark of E. I. du Pont de Nemours & Co.

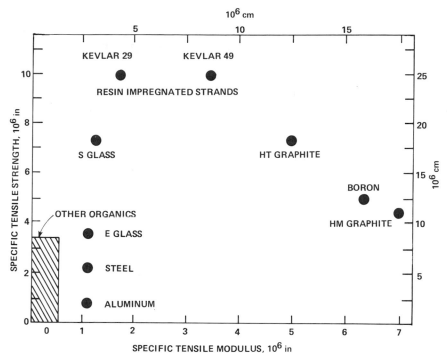

**FIG. 2.7**   Specific tensile strength and specific tensile modulus of reinforcing fibers, i.e., tensile strength and tensile modulus divided by density. *(Du Pont Co.)*

and stress rupture are key properties. Reinforced composites can save up to 40% of the weight of glass-fiber composites at equivalent stiffness. The aramid composites resist shattering upon impact, and the presence of the fiber inhibits propagation of cracks (Fig. 2.7). Depending upon the selection of resin systems, aramid composites have a useful temperature range from $-320$ to $400°F$ ($-196$ to $204°C$).

Kevlar 49 is not a carbonized or graphitized material. Unlike other organic materials, its stress-strain behavior is linear to ultimate failure in tension at 340 kips/in² (2344 MPa) and 1.8% elongation. Toughness of the fiber composites is significantly higher than that of boron or graphite composites. Furthermore, the very low density of the fibers provides a higher specific strength than glass, boron, or graphite reinforcing fibers. The specific modulus is between 4 and 5 times higher than that of glass fiber and equivalent to that of some graphite fibers.

Epoxies reinforced with the new fiber bridge the modulus gap between boron and glass-fiber composites and have a modulus equivalent to that of some graphite composites but with higher tensile strength. Reinforced epoxies have an elongation to failure about twice that of boron or graphite composites and 60% that of S-glass-reinforced epoxies. Usable strength of Kevlar 49 reinforced epoxy is about 4 times that of 7075-T6 aluminum at less than half the density.

The specific ultimate tensile strength of reinforced epoxy composites is higher than that of high-strength inorganic fiber composites; the specific modulus is roughly 3 times that of SGl–Ep and one-third that of boron-epoxy (B–Ep).

**Table 2.11** Compressive-Strength Data

| Material | Compressive strength | | Specific compressive strength | |
|---|---|---|---|---|
| | kips/in$^2$ | MPa | $10^6$ in | $10^3$ m |
| Aligned fiber (60 vol %) in epoxy | 35–50 | 241–345 | 0.7–1 | 18–25 |
| Gr–Ep (some) | 50–110 | 345–758 | 0.9–2.1 | 23–53 |
| Aluminum 7075-T6 | 73 | 503 | 0.73 | 19 |
| Ti–6Al–4V | 155 | 1069 | 0.97 | 25 |

Although the compressive strength of aligned fibers (60 vol %) in epoxy is less than that of the other materials shown in Table 2.11, its specific compressive strength compares favorably.

Impact strength of the composite transverse to fiber alignment is about midway between that of EGl–Ep and Gr–Ep composites. The Charpy impact strength is nearly 7 times higher than for Gr–Ep and 4 times higher than for B–Ep but a little over half that of EGl–Ep.

The aramid composites are nonconductive and exhibit no galvanic reaction in contact with metals. While the tensile behavior is linear and failure occurs at high stress, aramid composites exhibit ductile behavior under compressive and flexural loads and the ultimate strength is lower than that of either glass or graphite composites. Aramid can be combined with glass or graphite in hybrid composites to take advantage of the unique properties of each fiber. Hybrids allow the design engineer to achieve a balance of properties not available from a single reinforcing fiber.

The aircraft industry generally uses Kevlar 49 to reinforce epoxy composites, while the marine industry generally uses orthophthalic and isophthalic polyesters and vinyl ester resin systems. Polyimide, phenolic, and other thermoset and thermoplastic resins also can be reinforced.

Aramid fibers can be processed in essentially the same ways as glass and graphite to produce composites. Vacuum-bag molding of fabric prepregs is most often used in the aircraft industry. High-performance pressure vessels and missile-engine cases are filament-wound with rovings. Rovings can also be used in pultrusion, allowing continuous manufacture of parts of constant cross-sectional area. In the marine industry, fabrics and woven rovings of aramid are used in the wet (hand) lay-up method. Fabrics, chopped fiber, and continuous fibers can be compression-molded.

## Early Development of Carbon (Graphite)

Graphite fibers have been used for electric light filaments since 1880. It is claimed that in 1925 Thomas Edison turned on several electric lights in his laboratory which have been burning ever since. The filaments in those light bulbs were carbon fibers that Edison produced by "baking threads in an oven."

By far the most common reinforcement for plastics in ablative and structural-composite applications has been glass fibers. Although they have outstanding strength characteristics and low density, they are relatively low in elastic stiffness. For this reason about 25 years ago experimental work was carried out on the thermal conversion of various organic precursor materials into carbon and graphite fibers and fabrics. Five years later carbon and graphite cloth were in commercial production for extensive use in phenolic composites as ablative components in missile rocket motors. Shortly thereafter, carbon and graphite felts followed for use as high-temperature insulators. In the next 5 years a significant milestone was reached with the commercial production of carbon yarn by a new continuous process yielding a strong and uniform material. This achievement was remarkable in view of the raw material used and the properties obtained compared with conventional forms of graphite. The rayon precursor used for making these filaments contained only 44% carbon and lost 75 wt % during pyrolysis. These early carbon fibers were 50 times stronger and 3 times stiffer than quality-grade bulk graphites, but remarkable as they were as a form of graphite, they did not compare favorably in properties with other available filamentary materials such as fiber glass.

During this same period, theoretical work indicated that a graphite single crystal should have an elastic modulus of $140 \times 10^6$ lb/in$^2$ (965 GPa). On the basis that the theoretical cohesive strength of a solid is approximately 10% of its modulus, one could predict a tensile strength of $14 \times 10^6$ lb/in$^2$ (96.5 GPa). Experimental work with graphite whiskers grown from boules very nearly confirmed theory by displaying a modulus of $100 \times 10^6$ lb/in$^2$ (690 GPa). The tensile strength of these graphite whiskers was $3 \times 10^6$ lb/in$^2$ (20.7 GPa), again indicating the enormous potential for filamentary graphite.

One line of development used rayon as the precursor, and the other used polyacrylonitrile (PAN). Although the detailed processing conditions for converting cellulose or PAN into carbon and graphite fibers differ in detail, they both consisted fundamentally of a sequence of thermal treatments to convert the precursor into carbon by breaking the organic compound to leave a "carbon polymer." The fibrous carbon formed by the controlled pyrolysis of organic precursor fiber was viscous rayon or acrylonitrile. Processing the material at 1292°F (700°C) gave a partially carbonized fiber comprising about 85% carbon. This fiber had reasonably good strength and handling characteristics but contained residual volatile components that restricted its use to relatively low temperatures. Further pyrolysis of the 85% carbon fiber at 2552 to 3092°F (1400 to 1700°C) resulted in a fiber of 95 to 99% carbon content. When pyrolysis was performed above 3992°F (2200°C), the resulting graphite fiber was thermally stable and soft and had good electrical conductivity.

Carbon fibers produced by the rayon-precursor method have fine-grain, relatively disordered microstructure which remains even after treatment at temperatures up to 5432°F (3000°C). Graphite crystallites with a long-range three-dimensional order do not develop. The individual layers do not grow larger than a few tens of nanometers in diameter and remain highly cross-linked. This structure is intrinsically strong since there is little space for slip on the basal plane.

In both the rayon and PAN processes a high degree of preferred crystal orientation was responsible for the high elastic modulus and tensile strength. Tension was applied to the rayon precursor during the graphitizing step. In the PAN process the fibers were restrained from shrinking during the conversion into graphite, creating the desired crystal alignment.

**Carbon vs. Graphite:**   Although the names "carbon" and "graphite" are used interchangeably when relating to fibers, there is a difference. Typically, PAN-based carbon fibers are 93 to 95% carbon by elemental analysis, whereas graphite fibers are usually 99+%. The basic difference is the temperature at which the fibers are made or heat-treated. PAN-based carbon is produced at about 2400°F (1316°C), while higher-modulus graphite fibers are graphitized at 3450 to 5450°F (1899 to 3010°C). This also applies to carbon and graphite cloths. Unfortunately, with only rare exceptions, none of the carbon fibers are ever converted into classic graphite regardless of the heat treatment.

Controlling the reaction temperature results in different properties in the fiber. Generally, as the modulus increases, ultimate strength and elongation decrease. A high-modulus graphite fiber, for example, exhibits a lower strain to failure than a high-strength carbon fiber. In some aircraft applications, where composite parts are continually stressed and flexed, the fiber with a higher strain to failure (higher elongation) may be chosen.

## Carbon

Carbon fibers are characterized by a combination of light weight, high strength, and high stiffness. Their high modulus and (to a lesser extent) strength depend on the degree of preferred orientation, i.e., the extent to which the carbon-layer planes are oriented parallel to the fiber axis. All carbon fibers are made by pyrolysis of organic precursor fibers in an inert atmosphere. Pyrolysis temperatures can range from 2012 to 5432°F (1000 to 3000°C); higher process temperatures generally lead to higher-modulus fibers. Only three precursor materials, rayon, polyacrylonitrile (PAN), and pitch, have achieved significance in commercial production of carbon fibers.

The first high-strength and high-modulus carbon fibers, discussed above, were based on a rayon precursor. These fibers were obtained by being stretched to several times their original length at temperatures above 5072°F (2800°C). The high cost of this stretching and some uncertainty about the continued supply of suitable rayon precursors have essentially made these fibers obsolete. Nevertheless, considerable quantities of rayon-based carbon cloth, which is not hot-stretched in processing, continue to be used in aerospace applications.

The second generation of carbon fibers is based on a PAN precursor and has achieved market dominance through a combination of relatively low production costs and good physical properties. In their most common form these carbon fibers have a tensile strength ranging from 350 to 450 kips/in$^2$ (2413 to 3102 MPa), a modulus of 28 to 75 × 10$^6$ lb/in$^2$ (0.2 to 0.5 GPa), and a shear strength of 13 to 17 kips/in$^2$ (90 to 117 MPa). This last property controls the transverse strength of composite materials.

PAN-based carbon fibers are offered as yarns containing 1000 to 12,000 filaments and tows containing up to several hundred thousand filaments. The lower-filament-count (1000 to 6000) yarns are also woven into fabrics of various constructions for making composites (see Fig. 2.8). The higher-modulus carbon grades are naturally higher in cost.

The high-modulus fibers are highly graphitic in crystalline structure after being processed from PAN at temperatures in excess of 3600°F (1982°C). Higher-strength fibers, obtained at lower temperatures from rayon, feature a higher carbon crystalline content. There are also carbon and graphite fibers of intermediate strength and modulus (see Table 2.12).

The cross-sectional configuration of the fiber depends primarily on how the precursor was spun. Circular, dog-bone, popcorn, and kidney shapes have been produced.

Graphite fibers are available in five major forms:

1. Continuous fiber for filament winding, braiding, spray-up, and pultrusion; one common continuous yarn consists of 10,000 untwisted filaments, but yarns of 1000 to 160,000 filaments are available, and lengths of 1500 to 3000 ft (457 to 914 m) are typical.
2. Unidirectional prepreg, either in tow or tape, 3 to 36 in (76 to 914 mm) wide for lamination.

**FIG. 2.8**  Single tow of PAN fiber spread to show individual filaments.

3. Chopped fiber for injection or compression molding.
4. Pultruded structural shapes.
5. Woven fabric, either as dry cloth or as a prepreg for lamination; this will be discussed later in the chapter.

The third generation of carbon fibers is based on pitch as a precursor. Ordinary pitch is an isotropic mixture of largely aromatic compounds. Fibers spun from this pitch have little or no preferred orientation and hence low strength and modulus. Pitch is a very inexpensive precursor compared with rayon and PAN. High-strength and high-modulus carbon fibers are obtained from a pitch that has first been converted into a mesophase (liquid crystal). These fibers have a tensile strength of more than 300 kips/in$^2$ (2069 MPa), and a Young's modulus ranging from 55 to 75 $\times$ 10$^6$ lb/in$^2$ (0.38 to 0.52 GPa); they are available as 1000- and 2000-filament yarns and in various cloth forms (Table 2.13). The average filament diameter of continuous yarn is 0.0003 in (0.008 mm). The fiber, like PAN or rayon-based graphite, can be surface-treated for improved interlaminar shear strength. Pitch-based carbon and graphite fibers are expected to see essentially the same applications as the more costly PAN and rayon-derived fibers, e.g., ablative, insulation, and friction materials and in metals and resin matrixes.

**Table 2.12** Fiber Properties[3]

| Category | Density lb/in$^3$ | kg/m$^3$ | Ultimate tensile strength ksi | MPa | Tensile modulus 10$^6$ psi | GPa |
|---|---|---|---|---|---|---|
| Alumina, polycrystalline | | | | | | |
| (Al$_2$O$_3$), FP-1 | 0.1337 | 3702 | 200 | 1379 | 55 | 379 |
| FP-2 | 0.1337 | 3702 | 250 | 1724 | 55 | 379 |
| SiO$_2$-coated | 0.1337 | 3702 | 275 | 1896 | 55 | 379 |
| Aramid, Kevlar 29 | 0.0520 | 1440 | 550 | 3792 | 9 | 62 |
| Kevlar 49 | 0.0534 | 1479 | 550 | 3792 | 19 | 131 |
| Boron, tungsten core, 4-mil | 0.0939 | 2600 | 500 | 3447 | 58–60 | 400–414 |
| 5.6-mil | 0.0896 | 2481 | 510 | 3516 | 58–60 | 400–414 |
| 8-mil | 0.1221 | 3381 | 530 | 3654 | 58–60 | 400–414 |
| Carbon core, 4.2-mil | 0.0759 | 2102 | 475 | 3275 | 53 | 365 |
| Glass, E | 0.0917 | 2539 | 500 | 3447 | 10.5 | 72.4 |
| S | 0.0899 | 2489 | 650 | 4482 | 12.4 | 85.5 |
| Graphite, PAN, low-cost | | | | | | |
| high-strength (LHS): | | | | | | |
| Celion | 0.0632 | 1749 | 419 | 2889 | 34 | 234 |
| A-S | 0.0654 | 1810 | 420 | 2896 | 34 | 234 |
| T-300 | 0.0636 | 1760 | 400 | 2758 | 33 | 228 |
| Type III | 0.0643 | 1780 | 350 | 2413 | 33 | 228 |
| 3T | 0.0650 | 1799 | 300 | 2068 | 30 | 207 |
| Intermediate-modulus | | | | | | |
| (IM): | | | | | | |
| T-400 | 0.0643 | 1780 | 425 | 2930 | 33 | 228 |
| Type II | 0.0618 | 1711 | 360 | 2482 | 40 | 276 |
| HTS | 0.0658 | 1821 | 410 | 2827 | 36 | 248 |
| 4T | 0.0650 | 1799 | 350 | 2413 | 38 | 262 |
| High-modulus (HM): | 0.0683 | 1891 | 350 | 2413 | 53 | 365 |
| T-50 | 0.0603 | 1669 | 300 | 2068 | 57 | 393 |
| Type I | 0.0672 | 1860 | 350 | 2413 | 56 | 386 |
| 5T | 0.0668 | 1849 | 400 | 2758 | 48 | 331 |
| 6T | 0.0686 | 1899 | 420 | 2896 | 58 | 400 |
| Ultrahigh-modulus | | | | | | |
| (UHM), GY-70 | 0.0672 | 1860 | 250 | 1724 | 75 | 517 |
| Pitch, type P | 0.0730 | 2021 | 250 | 1724 | 50 | 345 |
| Silicon carbide, 5.6-mil, | | | | | | |
| carbon core | 0.1109 | 3070 | 550 | 3792 | 50 | 414 |
| Tungsten core | 0.1192 | 3299 | 550 | 3792 | 60 | 414 |

The cost difference can be explained by comparing the processes for producing PAN- and pitch-based fibers. Raw material for the pitch process (the pitch itself) costs (1980) about 10 cents per pound (22 cents per kilogram), while for the PAN process acrylonitrile costs about 30 cents per pound (66 cents per kilogram). The first step in each process is to polymerize the raw material, the pitch into the mesophase and the acrylonitrile into PAN. Both polymerizations are batch processes and involve comparable costs. The next step in each process is to spin the polymers into fibers. Spinning the PAN involves solvent-recovery steps not required in spinning pitch, which add to the process cost. Both types of spun precursor fibers must then be stabilized by heating in an oxygen-containing atmosphere. Because of differences in the ordering of the polymer structures of the precursor fibers as they emerge from spinning, only the PAN fibers must be held under tension during stabilization. This further raises the PAN process cost.

**Table 2.13** Comparison of Properties of Pitch and PAN Fibers†

| Nominal fiber modulus | | Fiber strength | | Tensile modulus | | Composite strength‡ | | Composite modulus‡ | | Interlaminar shear |
|---|---|---|---|---|---|---|---|---|---|---|
| 10⁶ psi | GPa | ksi | MPa | 10⁶ psi | GPa | ksi | MPa | 10⁶ psi | GPa | |
| | | | | | PAN fibers | | | | | | |
| 30 | 2.1 | 400 | 2758 | 33 | 228 | 220L 7T | 1517L 48T | 20L 7T | 138L 48T | 14 |
| 50 | 3.5 | 350 | 2413 | 55 | 379 | 175L 5T | 1207L 35T | 32L 5T | 221L 35T | 10 |
| 70 | 4.8 | 300 | 2069 | 75 | 517 | 110L 4T | 758L 28T | 44L 4T | 303L 28T | 7 |
| | | | | | Pitch fibers | | | | | | |
| 50 | 3.5 | 300 | 2069 | 55 | 379 | 150L | 1034L | 32L | 221L | 4.5 |
| 75 | 5.2 | 300 | 2069 | 75 | 517 | 150L | 1034L | 44L | 303L | 3.0 |

†Based on data from Union Carbide Corp.   ‡L = longitudinal; T = transverse.

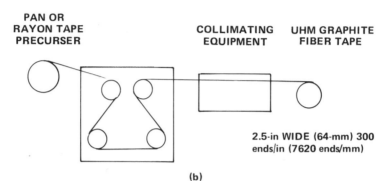

**PAN OR RAYON TOW**
**1,000–10,000 ENDS**

**LHS GRAPHITE FIBER**

(a)

**PAN OR**
**RAYON TAPE**
**PRECURSER**

**COLLIMATING**
**EQUIPMENT**

**UHM GRAPHITE**
**FIBER TAPE**

**2.5-in WIDE (64-mm) 300**
**ends/in (7620 ends/mm)**

(b)

**FIG. 2.9** Schematic diagram of fabrication of PAN tapes: (*a*) furnace operating at about 3092°F (1700°C) with trace of oxygen in atmosphere; combine 10,000 ends tow with resin to form 3-in (76.2-mm) tape; (*b*) furnace operating at about 5072°F (2800°C) with a trace of oxygen in atmosphere; spread to 3-in (76.2-mm) width and combine with resin to form 3-in (76.2-mm) tape.[37]

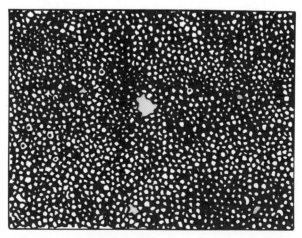

**FIG. 2.10** High-strength Gr–Ep.

Both types of fibers finally are carbonized by heating to temperatures exceeding 1832°F (1000°C). The carbon yield from mesophase pitch is about 75 to 80%, while that from PAN is only about 50%. The greater carbon yield characteristic of the pitch permits higher process speeds because less waste gas evolves and less structural rearrangement occurs during carbonization (Fig. 2.9). The fiber tension and temperature determine the final properties of the graphite fiber. Typically, increasing stiffness reduces both strength and strain to failure. Low-cost high-strength (LHS) graphite fiber is the most common type because it offers the best balance of properties and cost (Fig. 2.10). Lower-cost derivatives of the LHS fiber are used for commercial products. The PAN-based intermediate- and high-modulus fibers (Fig. 2.11), are not widely used in aerospace applications at present because of their low

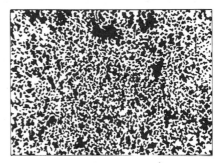

FIG. 2.11   High-modulus Gr–Ep.[3]

strain to failure; they are used, however, in developmental graphite-aluminum (Gr–Al) composites. Ultrahigh-modulus graphite fiber is used primarily for structures in optical systems having low coefficients of thermal expansion.

**Discontinuous Carbon Fibers:**   Carbon fibers can be long and continuous or short and fragmented; they can be directionally or randomly oriented; and they can be dispersed in a thermoplastic or a thermosetting material. Each fiber form has its limitations: In general, short fibers cost the least, and fabrication costs are lowest; but properties of the resulting composite are also lower than those obtainable with longer or continuous fibers.

*Milled Fibers:*   The shortest carbon fibers used for reinforcement are finely ground, or milled, fibers. Commercial milled carbon fibers range in length from 0.0012 to 0.12 in (0.03 to 3 mm), averaging about 0.012 in (0.3 mm). The mean length-to-diameter ratio $L/D$ of these fibers is 30. Incorporating 30% (by weight) of these fibers in a nylon 6/6 matrix substantially improves many properties. More detailed information on matrixes and properties of laminates can be found later in this chapter and in Chap. 3. As shown in Table 2.14, with only a 14% increase in specific gravity of the systems, tensile strength increases by 43%, tensile modulus by 217%, flexural strength by 58%, flexural modulus

**Table 2.14**   Typical Properties of Reinforced Nylon 6/6†[37]

| Reinforcement | sp gr | Elonga-tion, % | Tensile strength ksi | Tensile strength MPa | Flexural strength ksi | Flexural strength MPa | Compres-sive strength ksi | Compres-sive strength MPa | Tensile modulus $10^6$ psi | Tensile modulus GPa | Flexural modulus $10^6$ psi | Flexural modulus GPa |
|---|---|---|---|---|---|---|---|---|---|---|---|---|
| None | 1.14 | 60 | 11 | 76 | 16 | 110 | 9 | 62 | 0.40 | 2.76 | 0.41 | 2.83 |
| 30% milled carbon‡ | 1.30 | 2.7 | 15.8 | 109 | 25.3 | 174 | 14 | 97 | 1.27 | 8.76 | 1.49 | 10.27 |
| 30% 0.25-in (6.4-mm) carbon | 1.28 | 2.4 | 35 | 241 | 51 | 352 | .. | ... | 1.46 | 10.07 | 2.9 | 20 |
| 25% milled carbon‡ + 20% 0.25-in (6.4-mm) glass | 1.42 | 3.8 | 16.5 | 114 | 34 | 234 | 17 | 117 | 1.71 | 11.79 | 1.92 | 13 |

†Based on injection-molded test specimens.      ‡Milled pitch-base fibers.

by 263%, and compressive strength by 55%. The large drop in ductility is indicated by the greatly reduced elongation. The reinforced system has less reaction to thermal gradients (lower thermal expansion coefficient) and is electrically conductive. Table 2.14 also lists properties of hybrid composites (discussed later in the chapter) containing milled carbon fibers and 0.25-in (6.4-mm) chopped glass fibers in a nylon 6/6 matrix. At a cost of about 14% less and a slight weight increase (9%) over the compound containing only milled carbon fibers, this hybrid offers still higher properties, particularly in the flexural mode.

*Chopped Fibers:* Higher values of the flexural modulus can be achieved using 0.25-in (6.4-mm) chopped carbon fibers to reinforce the matrix material. Because of the way these fibers are manufactured, their length distribution is very uniform, with an $L/D$ ratio of approximately 800. Typical values obtained by adding 30% of these fibers to a nylon 6/6 resin are also shown in Table 2.14.

The longer carbon-fiber reinforcement increases both strength and modulus significantly while maintaining good electrical conductivity. Such increases are obtained at a cost, however; a molding compound reinforced with 0.25-in (6.4-mm) fibers costs (1982) about $12 to $14 per pound ($26.40 to $30.80 per kilogram), while one containing milled carbon fibers sells for $6 to $7 per pound ($13.20 to $15.40 per kilogram).

*Longer Chopped Fibers:* Chopped reinforcing fibers 0.5 to 2 in (12.7 to 50.8 mm) long are commonly added to a thermosetting polyester resin (usually 25 to 50% fibers) in manufacturing sheet molding compound (SMC), which is normally processed by compression molding. Glass-fiber reinforcement is adequate for many applications, even where modest load-bearing requirements must be met, but adding carbon fiber to such compounds enhances the performance of the molded part, especially where deflection resistance is important (Table 2.13). Neither injection molding nor compression molding need be altered to accommodate the addition of carbon fibers.

The processing techniques employed in manufacturing SMCs give reinforcing fibers a preferred orientation. Properties at 90° (1.6 rad) to the machine direction are usually 15 to 20% lower than those shown in Table 2.15. Thus, the anisotropic nature of this form of composite material requires an appropriate adjustment in design calculations. Sometimes the charge pattern can be modified to achieve equal properties in both planes.

Progress has been made in controlling the orientation of discontinuous carbon fibers in the manufacture of SMCs. Fibers over 2 in (50.8 mm) long can be placed in a preferred direction. Even allowing for some disorientation from flow while the mold is being closed, such composites offer the potential of greatly increased resistance to load deformation in a specified direction. Structural properties approaching those of die-cast metals can be expected.

**Table 2.15** Typical Properties of Sheet Molding Compounds†

| Fiber reinforcement | sp gr | Elongation, % | Tensile strength | | Tensile modulus | |
|---|---|---|---|---|---|---|
| | | | kips | MPa | $10^6$ psi | GPa |
| 33% glass | 1.67 | 1.4 | 15 | 103 | 1.7 | 11.7 |
| 35% carbon | 1.66 | 0.6 | 15 | 103 | 3.9 | 26.9 |
| 24% glass, 26% carbon | 1.65 | 0.5 | 17 | 117 | 3.5 | 24 |

†Data generated from flat plates from an SMC formulation using an isophthalic polyester matrix resin. Molding conditions: 1 kip/in² (7 MPa) at 300°F (149°C) for 3 min. Reinforcing fibers ranged from 0.5 to 2 in (12.7 to 50.8 mm) long.

**Continuous Carbon Fibers:** Where the ultimate in performance or weight reduction is required, continuous carbon fibers are usually the preferred reinforcement. Continuous fibers are also specified where thermal expansion must be kept to very low levels or where matching the expansion characteristics of an adjoining part made from another material is necessary.[7]

Continuous fibers are available in a variety of forms, including yarns or tows containing from 400 to 160,000 individual filaments, unidirectional impregnated tapes up to 48 in (1.22 m) wide, multiple layers of tape with individual layers or plies at selected fiber orientation, and fabrics of many weights and weaves.

The individual carbon filaments, which are the basic elements of the reinforcement, are usually 0.0003 in (0.008 mm) or less in effective diameter. Tensile strengths can exceed 400 kips/in$^2$ (2758 MPa); moduli, 70 $\times$ 10$^6$ lb/in$^2$ (0.48 GPa). Commercial fibers are available in three modulus ranges (Table 2.16), which satisfy most design needs.

**Table 2.16** Modulus Ranges of Commercial Fibers

| 10$^6$ psi | GPa |
|---|---|
| 30–35 | 0.21–0.24 |
| 50–55 | 0.35–0.38 |
| 70–75 | 0.48–0.52 |

Today's manufacturing technology can produce efficiencies approaching 100% in property translation. The degree to which the strength and modulus of a composite approach 100% of the corresponding filament properties (multiplied by the filament's volumetric fraction) is governed by the uniformity of filament properties, filament alignment within the composite, and the integrity of the fiber-matrix interface.

Choice of the form of the continuous fiber often depends upon the fabrication process selected. Filament-winding processes usually dictate the use of lower-filament-count yarns to minimize the catenary effects (and resultant looseness of windings) common with high-filament-count yarns. Autoclave, vacuum-bag, and compression molding of relatively flat or simple curvature parts can use the oriented unidirectional tapes, laid up in situ or preplied.

*Unidirectional Tape:* Properties of a composite transverse to the fiber direction depend largely upon the matrix material but in any case are very low compared with the longitudinal properties. Consequently, in the design of most structures subjected to both longitudinal and transverse loadings the fibers must be oriented in specific directions to withstand these loads. Chapter 3 will show the lay-up patterns required for combining layers of unidirectional tape to achieve desired directional properties.

Design charts for the $0°/\pm45°/90°$ ($0/\pm0.79/1.6$ rad) family of carbon fiber–epoxy (C–Ep) composites are given for several types of fibers in Ref. 3.

*Woven Fabrics:* Although woven fabrics are more expensive than unidirectional tapes, significant cost savings are often realized in the molding operation because labor requirements are reduced. Complex part shapes or processes requiring careful positioning of the reinforcement can benefit from the use of the more easily handled woven forms of carbon fiber.[8]

Some fabrics are essentially unidirectional. Most of the carbon fibers in these fabrics or even all are oriented in one direction and are held in position by nonstructural tie yarns. Satin-weave fabrics, particularly the commonly used eight-harness satin, retain most of the fiber characteristics in the composite and can easily be draped over complex mold shapes. Plain-weave fabrics are less flexible and are suitable for flat or simple contoured parts, at a slight sacrifice in fiber-property translation (Fig. 2.12).

PAN carbon fabrics are usually made by weaving carbonized yarn. Weaving costs increase with increasing fiber modulus because of the mechanical difficulty in working a yarn with low extensibility. Pitch-derived fibers, on the other hand (because of the ease of developing a high modulus in these materials), can be woven at an intermediate processing stage, then converted into the high-modulus product while in fabric form. Although the strength properties of pitch-based carbon fabrics do not yet approach those of PAN-based products, the inherent lower cost potential for the pitch materials and the continually improving properties make this product attractive. Typical composite properties of a variety of presently available carbon fabric–epoxy composites are given in Table 2.17.

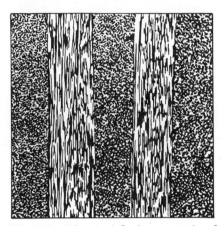

**FIG. 2.12** Bidirectional five-layer composite of PAN graphite fiber in epoxy-resin matrix.[8]

**Carbon and Graphite Yarns:** Graphite fibers are produced commercially in the form of yarn (compared with a monofilament) for reasons of economy of manufacture and for fabrication of composite articles. A yarn is used as the reinforcement when maximum composite strength is desired, as in a filament-wound or unidirectional-tape laminate. The strength of individual filaments in a yarn is not significantly greater than that of similar filaments in a cloth when both are measured in short-gage lengths.

Carbon yarns have higher surface area, higher electrical resistivity, and lower carbon assay and density than graphite yarns; the carbon yarns have a tendency to pick up more moisture, are wet with resin more easily, and provide lower-thermal-conductivity yarn-resin composites. Thus, the graphite yarns have lower surface area, higher density, higher carbon assay with inherent higher purity, and lower electrical resistivity than the carbon yarns; and the graphite yarns have less tendency to absorb moisture, are more oxidation-resistant, and have more lubricity than the carbon yarns.

**Carbon and Graphite Cloths:** The unique characteristics of carbon and graphite cloths make them versatile materials for use in high-temperature laminates. With excellent thermal stability and strength at temperatures to 4500°F (2482°C), coupled with low density, they offer ideal materials for ablative components. Single filaments from carbon or graphite cloth have tensile strengths of 50 to 100 kips/in$^2$ (345 to 690 MPa). The average breaking strength of the cloth is 7 to 24 kips/in$^2$ (48 to 165 MPa), depending on the processing. Carbon and graphite cloths have been made in a variety of weave constructions, ranging from a very light lace to a multiple thickness. There is no technical reason why almost any woven construction of a selected raw material cannot be processed to carbon or graphite textile form.[9]

**Table 2.17** Typical Properties of Carbon Fabric–Epoxy Composites†

| Fiber | Tensile strength | | Tensile modulus | | Flexural strength | | Flexural modulus | | Shear strength | |
|---|---|---|---|---|---|---|---|---|---|---|
| | ksi | MPa | $10^6$ psi | GPa | $10^6$ psi | GPa | $10^6$ psi | GPa | ksi | MPa |
| PAN: | | | | | | | | | | |
| 8-harness satin: | | | | | | | | | | |
| Warp | 90 | 621 | 10 | 689 | 120 | 827 | 9.5 | 66 | 9.0 | 62 |
| Fill | 90 | 621 | 10 | 689 | | | | | | |
| Plain weave: | | | | | | | | | | |
| Warp | 80 | 552 | 9.5 | 66 | 110 | 758 | 9.5 | 66 | 8.5 | 59 |
| Fill | 80 | 552 | 9.5 | 66 | | | | | | |
| Pitch: | | | | | | | | | | |
| 8-harness satin (low modulus): | | | | | | | | | | |
| Warp | 55 | 379 | 6.0 | 41 | 70 | 483 | 5.9 | 41 | 4.5 | 31 |
| Fill | 55 | 379 | 5.8 | 40 | | | | | | |
| 5-harness satin (high modulus): | | | | | | | | | | |
| Warp | 50 | 345 | 14 | 96.5 | 60 | 414 | 13.5 | 93 | 3.5 | 24 |
| Fill | 48 | 331 | 13 | 90 | | | | | | |

†Nominal carbon content = 60 wt %. Table based on data from Union Carbide Corp.

## Carbon and Graphite Matrixes

*Resin:* Generally, resin matrixes are used with carbon-graphite (C–G) filaments (Table 2.17). The most popular resin is epoxy. Low viscosity and low flow rates are required during impregnation to prevent misalignment of the wispy filaments. Each of the three major types of epoxies (conventional, epoxy Novalacs, and cycloaliphatic epoxies) has been used to prepare matrix materials for carbon and graphite composite structures. Epoxy compositions are selected to meet specific temperature requirements, including working life. The principal disadvantage is their upper temperature limit of 450°F (232°C), above which polyimides, aluminum, titanium, and carbon matrixes are required.

*Polyethersulfone:* A new reinforcement for carbon fiber is polyethersulfone (PES), a thermoplastic matrix prepreg[10,11] that combines outstanding high-temperature characteristics with dimensional stability and surface resistivity. It is expected to be suitable for structural parts for aircraft and in aerospace applications, where heat resistance and high strength-to-weight ratios are vital, and in rotating machinery parts, where dissipation of static charges is required. This new material can be processed on conventional injection-molding equipment similar to that used for glass-reinforced compounds. Some of the material properties include tensile strength, 26 kips/in² (179 MPa), flexural modulus, 2.1 X 10⁶ lb/in² (0.014 GPa), and flexural strength, 37 kips/in² (255 MPa).

*Fluorocarbon:* Parts made from fluorocarbon resins reinforced with graphite fibers provide outstanding toughness, compressive creep resistance, and chemical inertness at high temperatures and pressures. The fluorocarbon resin parts are chemically resistant to a wide range of substances, including hydrogen halides, halogens, organic reagents and solvents, inorganic salt solutions, strong acids, boiling water, and steam. The parts are also resistant to high and low temperatures and have good stiffness. In addition, high tensile and impact strength and good wear properties result from the graphite-fiber–fluorocarbon combination. Uses include valve seats, compressor seals, and bearing applications (Fig. 2.13).

*Phenolic:* Phenolic-impregnated graphite fabrics have been used for many years in missiles and other applications requiring exceptional heat resistance. The material is strong at high temperatures, erosion-resistant, and impervious to gases and liquids. Reinforced pyrolyzed plastic (RPP) composites are a new family of high-temperature materials featuring high strength-to-weight ratios in densities of only 112.4 lb/ft³ (1.8 g/cm³). The materials are intended for long-term use in high-temperature, highly oxidixing, high-heat-flux environments. They are made by reinforcing phenolic with graphite fibers or fabric, fabricating or machining the part to the shape wanted, e.g., a rocket nozzle, then carbonizing the composite part under high heat [up to 4600°F (2538°C)] into a porous skeleton. This skeleton is then coated and infiltrated by slurry or chemical-vapor deposition with refractory materials to improve thermal and/or ablative characteristics and resistance to handling and impact. Performance of the structure depends greatly on the coating used. Typi-

**FIG. 2.13** Seals of graphite-fiber-reinforced fluorocarbon resins.[8]

**Table 2.18**  Mechanical Properties of Polyimide[12]

| Temperature | | Flexural strength | | Flexural modulus | | Short-beam shear strength | |
|---|---|---|---|---|---|---|---|
| °F | °C | ksi | MPa | $10^6$ psi | GPa | ksi | MPa |
| 68 | 20 | 208 | 1434 | 16.6 | 114.5 | 15.8 | 109 |
| 450 | 232 | 103 | 710 | 15.0 | 103.4 | 6.4 | 44 |

cal coatings include pyrolitic graphite, zirconium carbide [good for relatively mild combustion environments to about 0.3 kips/in² (2 MPa)], and silicon carbide (higher pressures and more severe environments).

  *Polyimide:*  Graphite-polyimide materials are generally characterized by poor handling characteristics, erratic processing response, highly variable molded-product quality, and high shrinkage that can cause microcracking and/or delamination of complex contoured parts. Within the family of polyimide resins, however, the ease of processing varies widely with the type of resin selected; in general the more heat-resistant the resin the greater the fabrication and processing difficulties. Some polymides are capable of withstanding temperatures of up to 700°F (371°C), so that structural weight savings for some applications can raise their design operating temperatures from 350°F (177°C) to at least 500°F (260°C). Typical properties are shown in Table 2.18.

  Typical of applications and their cure cycles for graphite-polyimide (Gr–PI) are Figs. 2.14 to 2.17. Current and future studies looking to increase the higher-temperature capa-

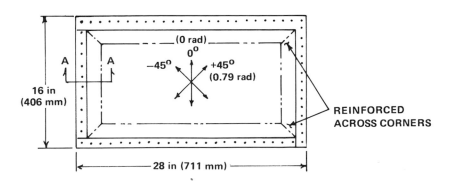

SECTION A–A

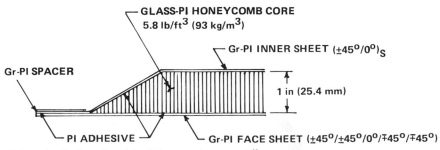

**FIG. 2.14**  Configuration of Gr–PI honeycomb-core panel.[13]

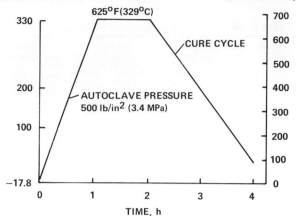

FIG. 2.15   Cure cycles for Gr–PI panel components.[13]

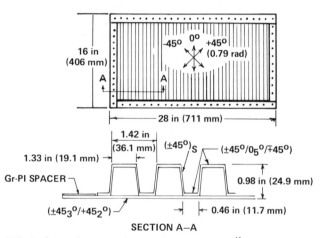

SECTION A–A

FIG. 2.16   Configuration of Gr–PI skin-stringer panel.[13]

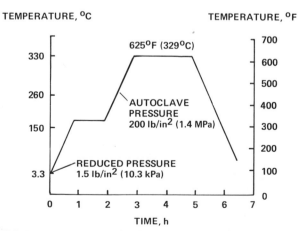

FIG. 2.17   Cure cycle for Gr–PI skin-stringer panels.[13]

bilities of graphite-fiber-reinforced epoxy composites are developing imide-amine hardeners and aromatic amine hardeners.[13-15]

## Graphite

Graphite-fiber-reinforced graphite composites can be used to temperatures in excess of 6332°F (3500°C). No compatibility problems exist because the graphite fiber or filament is in a graphite matrix. This composite system is good in reducing environments; in air or oxidizing atmospheres special protective coatings are sometimes needed.

The graphite matrix is produced by the pyrolytic decomposition of polymeric systems in which the graphite fiber or filaments are originally embedded. Polymeric binders are selected to provide minimum volatile outgassing and high carbon residues on pyrolysis to produce matrixes of maximum density. While many of matrix starting materials are considered proprietary, usable polymer systems include phenolic, furfuryl ester, and epoxy resins.

Graphite-graphite composites, even those fabricated with low-modulus materials, are up to 20 times stronger than conventional carbon and graphite materials. At a density of approximately 0.05 lb/in$^3$ (1384 kg/m$^3$) they also are about 30% lighter than conventional carbons. They provide a very high strength-to-weight ratio at temperatures to 5972°F (3300°C) and exhibit superior thermal stability. They also are resistant to physical shock. Both thermal conductivity and strength increase with rising temperature, providing excellent composite dimensional stability and resistance to thermal shock. Most Gr–Gr composites are more than 99% carbon (carbon content about 99.5 to 99.9%). This high purity provides good chemical inertness and corrosion resistance. Gr–Gr composites are not wetted by molten metals, which makes them ideally suited to metallurgical applications where high strength, light weight, erosion resistance, and good thermal conductivity are important. They are relatively ductile and do not exhibit the typical brittle fracture characteristics of bulk graphite. The principal failure mode of these composite materials is by interlaminar shear, although three-dimensional weaving and other special fabrication methods circumvent this problem to some extent. Final strength of the Gr–Gr composite, of course, is directly related to the strength of the reinforcing fiber or filament and a wide range of composite strengths is available, although in most cases the high-temperature characteristic of the composite is the most important factor in component design.

Typical properties are tensile strength, 8.2 kips/in$^2$ (56.5 MPa), flexural strength, 11 kips/in$^2$ (76 MPa), compressive strength, 40 kips/in$^2$ (276 MPa), and modulus, 2.5 × 10$^6$ lb/in$^2$ (17.2 GPa).

**Graphite-Aluminum:**  Graphite-fiber-reinforced aluminum (Gr–Al) alloys have been fabricated into plate, cube, and cylinder shapes in an effort to provide useful forms of the composite for evaluation in applications. The problem of voids and some fiber misalignment have been solved, and results indicate that fabrication of shapes is feasible by the processes examined.[16]

One of the basic materials used for fabrication was Gr–Al wire, 0.062 in (15.8 mm) in diameter, consisting of eight strands of surface-treated graphite yarn (11,000 individual fibers) impregnated with one of several aluminum alloys. Some of the specific alloys included Al–13Si (Al3), Al–10Mg (220), and Al–1Mg–0.6Si (6061). The composite normally consists of unidirectional or 0 ± 90° (0 ± 1.6 rad) cross-ply configuration made by hot-pressing aligned Gr–Al wire segments between face sheets of pure aluminum or titanium or without face sheets. Table 2.19 presents typical tensile properties of several test

**Table 2.19**  Tensile Properties of 35 vol % PAN Gr–Al Alloy Panels[17]

| Aluminum alloy | Filament orientation† | Face sheets | Longitudinal | | | | Transverse | | | |
| | | | Strength | | Modulus | | Strength | | Modulus | |
| | | | ksi | MPa | $10^6$ psi | GPa | ksi | MPa | $10^6$ psi | GPa |
| 220 | UD | None | 70 | 483 | .. | .. | | | | |
| 6061 | CP | None | 32 | 221 | .. | .. | 15 | 103 | | |
| 6061 | CP | Al | 29 | 200 | .. | .. | 16 | 110 | | |
| Al–5 wt % Mg | UD | Ti | 71 | 490 | .. | .. | 27 | 186 | | |
| 6061 | UD | Ti | 65 | 448 | 18 | 124 | 22 | 152 | 4 | 28 |
| 6061 | UD | Ti‡ | 68 | 469 | .. | .. | 20 | 138 | | |
| 6061 | CP | Ti‡ | 52 | 359 | 11 | 76 | 24 | 165 | 7 | 48 |

†UD = unidirectional; CP = cross ply, two layers at 0° (0 rad), one at 90° (1.6 rad).
‡Also contained Ti sheets between cross-ply layers.

specimens. No values are given for tranverse properties of the unidirectional panel without face sheets because the transverse strength is only 1 kip/in² (7 MPa) or less. The improved transverse strengths in the specimens with cross plying and use of aluminum face sheets demonstrate their advantage. Additional improvements in transverse strength are possible by using titanium face sheets instead of aluminum. New horizons for Gr–Al are seen in the availability of ultrahigh-modulus graphite fibers and a liquid-metal infiltration process which can produce composite "wires" with high specific strength and high specific modulus.

In the infiltration process, multifilament graphite fiber rows are intimately coated with a fine layer of Ti–B by the reduction of $TiCl_4$ and $BCl_3$ with zinc vapor. The Ti–B coating activates the surface of the fibers and promotes wetting and infiltration by the molten aluminum. This provides intimate bonding of the fibers to the aluminum. Another technique for reinforcing Gr–Al composite is to use nickel-plated graphite fibers. Electroplating of nickel on fibers was initially conducted as a batch operation with nickel sulfate–boric acid and nickel sulfamate plating baths. In this method 3 in (76.2 mm) of single-ply yarn was coated in 2 to 5 min. Since the operation was tedious, a continuous process for electrocladding the fibers has been developed. Figure 2.18 shows the equipment designed to provide reasonably uniform nickel coatings on all filaments in the yarn. The two plies of the yarn are separated and stored on the two spools $A$ (Fig. 2.18). The spools rotate independently and in two directions: one rotary motion removes twist in the yarn introduced mainly during the separation of the two plies, and the second motion permits free feeding of the untwisted single-ply yarns to the plating bath. The furnace $B$ heats the yarn to 1562°F (850°C) under an argon atmosphere to remove the polyvinyl alcohol finish coating remaining on the filaments. Electrical contact is made with the yarn through the roller at $C$. The larger rollers are made of copper and the smaller rollers are of soft sponge rubber. Satisfactory plating results are obtained with a current of approximately 1.5 A through the two-ply yarn. The yarn is washed with hot water in the U-tube vessel at $D$ to remove salts, and the plated yarn is stored on clutch-monitored spools at $E$.

Although Watt's, nickel sulfate–boric acid, and sulfamate plating baths provide comparable coating homogeneity around all filaments (Fig. 2.19), the best composite properties and bath lifetime are achieved with a Watt's plating solution. With this equipment two plies of yarn are coated simultaneously and continuously at a rate of 2 in/min (50.8 mm/min), providing plated yarn containing 50 vol % fiber.

A later technique applies an electroless nickel coating to graphite-fiber yarn, which is

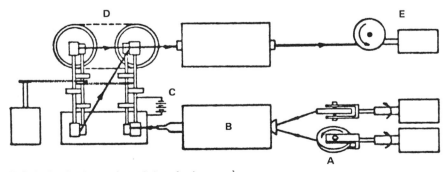

**FIG. 2.18** Continuous electroplating of carbon yarn.[3]

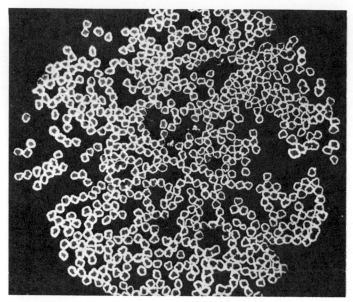

**FIG. 2.19**   Nickel coating (white areas) around carbon filaments. One ply of yarn is shown.[3]

aligned between aluminum sheets in a stacked array. The array is heated to obtain diffusion bonding, and the material becomes one piece. The graphite fibers remain intact throughout the process and the undesirable aluminum-carbon reaction at the fiber interface is prevented by the nickel coating. Since high-strength graphite fiber is less expensive than other common reinforcing materials, future manufacturing costs are likely to decrease. Unlike other fibrous materials (particularly boron), graphite yields a composite material that can be bent or otherwise formed without breaking the fibers. Other techniques for fabricating Gr–Al composites include pultrusion,[17] coating the fibers with titanium,[18] and impregnating the graphite fibers either by drawing them through a molten metal bath or by infiltration.

**Graphite-Magnesium:**   Composites of magnesium alloy matrix AZ91C or AZ61A, magnesium alloy surface foil, and 10 vol % titanium coating have been made. Typical longitudinal tensile strength for a unidirectional sample of this composite was 25.7 kips/in$^2$ (177 MPa) and a modulus of 11.6 to 13 $\times$ 10$^6$ lb/in$^2$ (80 to 90 GPa). Comparable values for monolithic magnesium are 9.4 kips/in$^2$ (65 MPa) and 6.5 $\times$ 10$^6$ lb/in$^2$ (44.8 GPa).

**Graphite-Lead:**   Work has been performed with graphite fibers to effect a great increase in the strength of lead and its alloys; this could enhance lead's potential in the chemical, battery, building, and bearing industries. Research was directed toward improved composites for possible battery and bearing applications. Systems have been produced using PAN fiber and PAN graphite yarn. Lead containing 1.5 to 3 wt % magnesium (to aid wetting) reinforced with 39 vol % PAN fiber has been made in wire form. Failure stress has averaged 110 kips/in$^2$ (758 MPa) and modulus of elasticity 23 $\times$ 10$^6$ psi (15.9 GPa).

Using hot compacting to convert the wire into bar form resulted in a drop in ultimate tensile stress to 45 kips/in$^2$ (310 MPa), but the modulus remained at 23 $\times$ 10$^6$ lb/in$^2$ (15.9 GPa). One of the problems to be overcome before the material becomes commercially available is to convert the wire into bar or sheet form by hot pressing. Efforts have included methods to liquid-phase-bond using diffusion agents, including indium, 60Sn–40Pb, zinc, and cadmium. Wires made using a Pb–10Sn–Gr system containing 42 vol % fiber have provided remarkably high ultimate strengths, testing as high as 160 kips/in$^2$ (1103 MPa).

A major criterion for battery anodes is obviously electrical resistivity. Theoretical calculations for a lead 40 vol % fiber composite indicate that a resistivity of 36 $\mu\Omega \cdot$ cm should be attained. Actual conductivity tests on composite wire with 37 vol % fiber showed an average value of 44.5 $\mu\Omega \cdot$ cm, which was much higher than expected (Fig. 2.20). Lower values, on the order of 35 to 36 $\mu\Omega \cdot$ cm, were obtained with the Pb–10Sn 42 vol % graphite fiber composite. Work continues in developing material with much lower fiber content (20 to 30 vol %) to gain lower resistivity.

Mechanical properties are very promising. Where pure lead has a strength of 2 kips/in$^2$ (14 MPa) and a typical lead-base bearing alloy has 10.5 kips/in$^2$ (72 MPa), the composites show values of 50 to 160 kips/in$^2$ (345 to 1103 MPa). The modulus of elasticity

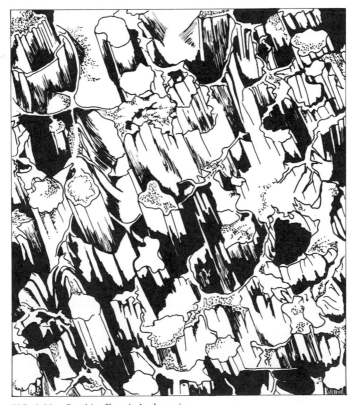

**FIG. 2.20** Graphite fibers in lead matrix.

of the composites is also 4 to 8 times higher than that of the bearing alloy. Another important property, fiber density, is low; thus a 40 vol % loading has great impact on the total density of the composite. As a result, strength and stiffness-to-weight ratios make these lead composites similar to the low-alloy medium-strength steels.

**Graphite-Copper:**  Copper-graphite (Cu–Gr) composite materials were formerly fabricated by electroplating copper onto graphite fibers and then hot pressing the material, but rule-of-mixture studies indicated to researchers that better properties could be attained. By using a liquid-infiltration method similar to the process used to fabricate Gr–Al materials wires were made by passing fibers of graphite through a molten-copper bath. The wire, about 50 $\mu$m in diameter, was compacted into larger sizes by hot pressing. Development test samples measure 1 in (25.4 mm) wide and 6 in (152.4 mm), long and strength and modulus values have been obtained for composites containing 35 to 39 vol % graphite. The significance of the composite lies in whether the Cu–Gr will retain its room-temperature properties at temperatures near the melting point of copper. The degradation of pure copper's strength properties even at moderately elevated temperatures requires going to special cooling systems or using a metal alloy like beryllium copper. Of course, one seldom gets something for nothing in materials engineering. The graphite increases strength, but the composite loses electrical conductivity compared with pure copper. On the other hand, this comparison is rather meaningless under temperature conditions where pure copper cannot be used at all. At higher temperatures beryllium copper is actually 6 times as strong as pure copper compared with 3 to 4 times for the new Cu–Gr. This difference, however, is more than compensated for in electrical conductivity: Cu–Gr has about 65% the conductivity of pure copper, while that of beryllium copper is only 25% that of the pure metal.

**Graphite-Cobalt:**  Composites consisting of PAN-carbon filaments in a cobalt matrix have been fabricated by swaging and hot isostatic compaction of electroless cobalt-coated carbon filaments. This work was developmental; the room-temperature tensile strength of composites containing up to 65 vol % carbon filaments [63.4 kips/in$^2$ (437.1 MPa)] indicates it was about 50% below the expected value. Tensile strengths at elevated temperatures were only 25% of the expected values. More development work is necessary since the reduction in strength was attributed to fiber breakup during consolidation resulting in a reduced length-to-diameter ratio.


# FIBER-REINFORCED METAL MATRIX

## Introduction

Metal-matrix composites (MMCs) are of interest today because they offer the opportunity to tailor a material with a combination of properties unavailable in any single material, e.g., combining the very high tensile strength and modulus of elasticity of various types of fibers with the low density of a metal such as aluminum, titanium, or magnesium to obtain a composite material with a higher strength-to-density or modulus-to-density ratio than any single known alloy. In addition, because the reinforcing agent in the composite has a relatively high melting point, barring degradation of the fiber by chemical interaction with the matrix the strength properties of the composite can be retained at relatively high temperatures. When very high service temperatures are contemplated, high-melting-point metals with significant resistance to oxidation, sulfidation, and thermal shock can be used for the matrix; the resultant composite often has a creep strength significantly higher than that of the matrix alone. Fiber-reinforced MMCs are also of interest because of the improvement in fatigue resistance they offer when certain types of fibers are used.

"Fiber" is used here as a generic term to denote four types of reinforcing agents: wires, filaments, whiskers, and indigenous phase particles in a eutectic alloy. (Phase particles improve the mechanical properties of a matrix by dispersion strengthening.) The most common types of wires used to reinforce MMCs are beryllium, molybdenum, steel, and tungsten. Filaments used as reinforcing agents for composites are themselves composites. They consist of a very fine wire substrate [usually about 0.005 in (0.13 mm) in diameter] coated with an inorganic material such as boron, silicon carbide, boron carbide, or silicon boride. Whiskers, on the other hand, are single crystals formed directly from the vapor phase; one axis of the whisker is considerably longer than the other two. Normally, whiskers are significantly less than 1 in (25.4 mm) in length. Aluminum oxide, boron carbide, and silicon carbide whiskers are the types most commonly used for the reinforcement of MMCs. Rod- and lamella-shaped phase particles, produced during the controlled unidirectional solidification of eutectic alloys, are sometimes called whiskers, but the term is properly reserved for the vapor-deposited material just described. Examples of rod- and lamella-shaped phase particles are $Ta_2C$ and NiBe, respectively.

Wires and filaments may be used in composites either as continuous lengths extending from one end of the composite to the other or as discontinuous (chopped) lengths. Whiskers and phase particles are relatively short and therefore discontinuous within the composite. Each of the various types of fibers may be random or aligned in one, two, or three directions in the composite.[19]

## General Composite-Preparation Methods

The methods of preparing fiber-reinforced MMC can be classified as diffusion, liquid, or deposition processes (Table 2.20), although there is some overlap between categories. For instance, when the matrix is formed by a deposition process, the composite is usually densified by a diffusion process.

**Table 2.20**  General Processes Used for the Preparation of Fiber-Reinforced Metal-Matrix Composites[3]

| Diffusion processes | Liquid processes | Deposition processes |
|---|---|---|
| Pressing and sintering of powdered matrix and bare or coated fibers | Infiltration of liquid matrix metal between fibers (matrix flows due to gravity or pressure differential) | Electrodeposition of matrix around prearranged fibers (generally followed by pressing) |
| Pressing (often followed by sintering) of coated fibers | Pressing a mixture of powdered matrix and fibers at a temperature above the solidus of the matrix | Plasma spraying of matrix around prearranged fibers (generally followed by pressing) |
| Pressing of bare fibers between thin foils of matrix metal (occasionally powdered matrix is used in conjunction with the foils) | Unidirectional solidification of eutectic alloy | Vacuum deposition of matrix around fibers to form thin-film composite |

## Theory of Fiber Reinforcement of Metals

To avoid misconceptions, it is well to point out that the mechanism by which metals are "strengthened" by fibers is different from the method whereby they are strengthened by dispersed particles (as in thoria-dispersed nickel, sintered aluminum powder, and the like). In dispersion strengthening, the particles strengthen by blocking the movement of dislocations in the matrix; in fiber reinforcement, on the other hand, the fibers themselves carry most of the applied load and are not generally considered as barriers to dislocation motion. Thus, the fibers do not strengthen the matrix but unite with it to form a strong composite body.

### *Continuous Fibers*

Continuous fibers are those which extend the length of the composite specimen; they may be either wires or filaments. In composites reinforced with continuous fibers, fibers and matrix are generally considered to be loaded simultaneously, e.g., by the grips of a tensile machine. On this premise the primary purpose of the matrix is to act as a "glue" to hold the fibers together in a bundle; little or no load is transferred between the matrix and the fibers.

For a composite consisting of uniaxially oriented continuous fibers which are well bonded to the matrix, tensile deformation in the fiber direction will cause the strain to be the same in the fibers as in the matrix. In the most common case (strong, brittle fibers in a ductile matrix) the fibers behave elastically to failure while the matrix first deforms elastically and then plastically. As long as the composite contains more than a certain minimum volume percent of fibers $V_{min}$ (which depends on the matrix-fiber combination being considered) and the fibers are not degraded by chemical reaction with the matrix, the composite will fail at a total strain equal to the strain of the fibers at their ultimate tensile strength.

Figure 2.21 indicates the theoretical effect of the volume percent fibers $V_f$ on the ultimate tensile strength of a composite containing uniaxially aligned continuous fibers. Here $\sigma_u$ is the ultimate tensile strength of the matrix, which naturally includes a contribution due to work hardening of the matrix; $\sigma'_m$ is the tensile stress in the matrix when the fibers are strained to their ultimate tensile strain and includes less of a contribution from work hardening; and $\sigma_f$ is the ultimate tensile strength of the fiber.[20] It can be seen that the addition of strong, brittle fibers up to a content $V_{min}$ causes a decrease in tensile strength. The ultimate tensile strength of composites with this fiber content depends primarily on the strength of the work-hardened matrix; the brittle fibers fail at a strain which is less than the strain at which the composite fails and are assumed not to strengthen the matrix after they fail. Above $V_{min}$ the ultimate strength of the composite $\sigma_c$ is given by

$$\sigma_c = \sigma_f \frac{V_f}{100} + \sigma'_m l - \frac{V_f}{100}$$

where $V_f$ is the volume percent fibers in the composite. This equation is often referred to as the *rule of mixtures*. Note from Fig. 2.21 that a fiber content of $V_{crit}$ or more must be attained before the ultimate tensile strength of the composite exceeds that of the fiber-free work-hardened matrix.

The situation is somewhat different for ductile continuous fibers (such as metallic wires) (Fig. 2.21). Since ductile fibers are considered not to fracture much before the ultimate

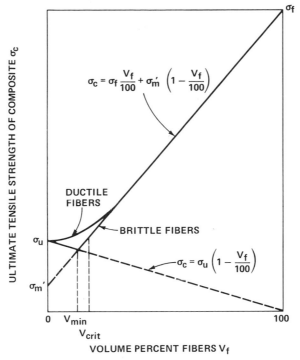

**FIG. 2.21** Theoretical variation of composite strength $\sigma_c$ with fiber content $V_f$ for reinforcement with continuous fibers.[20]

strain of the composite, they theoretically exhibit a strengthening effect even at low fiber contents.

## Discontinuous Fibers

Since discontinuous fibers are not long enough to reach the length of the composite specimen, fiber ends are present in the composite. This type of fiber may be either whiskers, indigenous phase particles in a eutectic alloy, or pieces of cut wire or cut filaments. In composites where the fibers are uniaxially aligned and well bonded to the matrix but discontinuous the matrix must not only bind the fibers together but also transfer stress from fiber to fiber. Stress transfer is accomplished by shear stresses at the fiber-matrix interface. The stress varies along the length of discontinuous fibers, unlike that for continuous fibers, where the stress is considered to be constant.

If the length of the discontinuous fibers in a composite is greater than the critical fiber length $l_c$, the strain of the fibers is believed to be essentially the same as that in the matrix. $l_c$ depends on the ultimate tensile strength and diameter of the fiber and the shear strength of the matrix. For the most effective utilization of a given volume percent of discontinuous fibers the length of the fibers should be greater than $l_c$. A composite consisting of uniaxially aligned discontinuous fibers will obey the rule of mixtures if the fiber lengths are greater than $l_c$ and the volume percent of fibers is greater than a certain minimum value.

## Key Considerations

In using fiber-reinforced MMC it is important to be aware of problems that may arise, problems that vary with the particular matrix-fiber combination being considered. Reaction between fibers and matrix at elevated temperatures, either as the composite is being prepared or under service conditions is one of the most serious drawbacks in the use of some filaments in MMC. Certain coatings, however, can reduce the reactivity of the bare filament and the matrix material. For example, while unprotected boron filaments degrade after a few minutes' exposure to air at elevated temperatures, SiC-coated boron filaments show no loss of strength after 2000 h exposure in air at 1000°F (538°C). Even more important is the fact that SiC also protects boron filaments in molten matrixes. Tests indicate that the coated filament's mechanical properties do not degrade during limited exposure to the molten aluminum required in composite fabrication. The coated filament is made with the same techniques used for producing ordinary boron filaments.

Aluminum coatings on beryllium wire eliminate wire-to-wire contact in beryllium-aluminum (Be–Al) composites. Whether coating is accomplished by lacquer bonding of aluminum powders or free evaporation of aluminum in vacuum, the coating becomes the composite matrix upon consolidation. Lacquer bonding results in Be–Al composites with fiber loadings up to 41 vol %; evaporation-coated wire produces composites with fiber loadings of 92 vol %. Both types of coated wires made fiber alignment easier in hot-pressing dies. Improved fiber alignment reduced scatter in the tensile-strength data and resulted in better ductility in composites with more than 25 vol % of reinforcement.

Coating 0.005-in (0.13-mm) beryllium wire with unalloyed titanium increases its tensile strength at room temperature to 165 kips/$in^2$ (1138 MPa), but strength deteriorates considerably at 900°F (482°C). Thus, Be–Ti composites fabricated by hot-pressing coated wire are sensitive to pressing time and temperature. For example, Be–Ti composites hot-pressed for 15 min at 900°F (482°C) and 100 kips/$in^2$ (689.5 MPa) had a tensile strength of about 120 kips/$in^2$ (827 MPa), but the same composite pressed at 8 kips/$in^2$ (55 MPa) for 4 h at 900°F (482°C) had less than 100 kips/$in^2$ (689.5 MPa) tensile strength. This can probably be explained by recrystallization of the beryllium wire or reaction with the titanium cladding. In either case, it is clear that Be–Ti composites need high pressure and hot pressing.

## Boron

Boron fiber used as a reinforcement for polymeric and metallic materials is available in many forms, several diameters, and on substrates of tungsten or carbon. Since vapor deposition of boron on a carbon monofilament substrate is still relatively new, when one is talking about the material as a reinforcement, "boron" refers to boron deposited on tungsten.

**Boron-Tungsten:** The most common method for producing continuous boron filament is a chemical-vapor plating process in which the reduction of boron trichloride by hydrogen gas takes place on a moving incandescent tungsten filament. A boron production line basically consists of an assembly of plating chambers (Fig. 2.22). The tungsten filament, 0.0005 in (0.01 mm) in diameter, is drawn continuously through the length of the reactor tube. At both ends of the tube it passes through liquid mercury (held in place by capillary forces) serving as electrodes connecting the filament to a power supply. Current from the

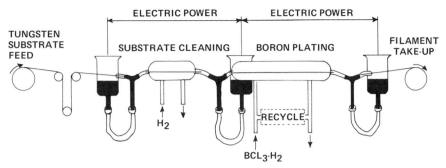

**FIG. 2.22** Schematic diagram of boron deposition process.[35]

power supply resistively heats the section of the filament in the tube to about 2200°F (1204°C). A gaseous mixture of hydrogen and boron trichloride flows through the reactor and reacts chemically with the heated filament. A coating of boron is deposited on the filament substrate, and hydrogen chloride gas is the by-product. While the filament passes through the plating chamber, the diameter is enlarged by chemical deposition. The boron filament is finally taken up on the spool at the right in Fig. 2.22. The hot tungsten substrate is given an outgassing treatment before boron is deposited onto it. The filament is heated to about 2200°F (1204°C) in a hydrogen atmosphere to remove any surface contamination (such as die lubricant) resulting from the drawing operation and any oxide layer on the surface.

Since in any given pass through the reactor only about 5 to 6% of the hydrogen–boron trichloride mixture reacts with the tungsten filament, a chemical recovery system is essential. The hydrogen chloride gets scrubbed out, unreacted hydrogen is vented to the atmosphere, and unreacted boron trichloride is condensed and recovered, mixed with fresh hydrogen, and recirculated through the reactor.

B–W is available in continuous lengths with diameters of 0.002, 0.004, 0.0056, and 0.008 in (0.05, 0.10, 0.14, and 0.20 mm). In each case the diameter of the tungsten substrate is 0.0005 in (0.01 mm). The specific modulus and the density of B–W increases with the final diameter. The modulus of this material is exceeded only by that of silicon carbide deposited on tungsten, but an obvious disadvantage of B–W is its density, which is notably higher than that of graphite.

Most commonly specified is the 0.002-in (0.05-mm) material; approximately 90% of the annual boron production is this grade, which generally goes into a B–Ep composite. The rest is 0.0056-in (0.14-mm) material, and for all practical purposes, only a few pounds of 0.008-in (0.20-mm) B–W is produced, for use in developing a jet-engine blade. Development work continues on diameters of 0.010, 0.012, 0.015 and 0.020 in (0.25, 0.30, 0.38 and 0.51 mm) deposited on the 0.0005-in (0.01-mm) substrate.

The 0.004-in (0.10-mm) B–W, with 400 kips/in² (2758 MPa) tensile strength, costs (1979) between $250 and $270 per pound ($550 and $594 per kilogram). The 0.008-in (0.20-mm) grade costs approximately $100 less. Production of 1 lb (0.45 kg) of 0.004-in (0.10-mm) fiber consumes about $40 worth of boron trichloride and about $50 worth of tungsten. The manufacturing and subsequent prepregging steps approximately double the value of the material, yielding the costs above.

Boron-fiber composites cost more than carbon-fiber composites and have superior mechanical properties. The 0.004-, 0.0056-, and 0.008-in (0.10-, 0.14-, and 0.20- mm) boron fibers have tensile strengths of 400, 510, and 530 kips/in² (2758, 3516, and 3654 MPa), respectively. All three fibers have a modulus of approximately $58 \times 10^6$ lb/in² (400 GPa) tensile strength and $33 \times 10^6$ lb/in² (228 GPa) modulus for a PAN carbon fiber.

One boron-fiber manufacturer can produce 35,000 lb (15,900 kg) of the 0.004-in (0.10-mm) fiber per year, which is enough fiber for more than 50,000 lb (22,500 kg) of B–Ep prepreg per year.

The 0.0056-in diameter (0.14-mm) fiber has become a standard of sorts for MMC. For a given mass of boron fibers in a given mass of aluminum the total number of fibers is less when the diameters are larger. The geometry of the arrangement then allows for more plastic deformation of the material upon impact, a desirable property for jet-engine fan blades, where the problem of foreign-object damage is paramount.

Four fibers, adjacent but not touching, can be visualized making up the corners of a square within an aluminum matrix composite viewed in cross section. For a given boron-to-aluminum mass ratio, the larger the diameter of the boron fibers the greater the mass of aluminum contained in the square. It is the aluminum matrix that lends the composite whatever measure of plastic deformability, or give, it has.

## Matrixes

**Boron-Epoxy:**   The oldest of the advanced composites, B–Ep is still popular with aerospace designers. Stronger and stiffer than carbon composites, it provides the rigidity of steel at about one-fourth the weight (see Fig. 2.23).

An interesting application is a compressor disk nearly one-fourth lighter than the all-metal version from which it was derived. The disk incorporates a boron filament in an epoxy-resin matrix system. By winding the filament into two grooves 0.50 in (12.7 mm) wide on the outer rim of the Ti–6Al–4V disk the design engineer was able to eliminate the bore and web sections, giving a 23% weight reduction. The diameter of the boron

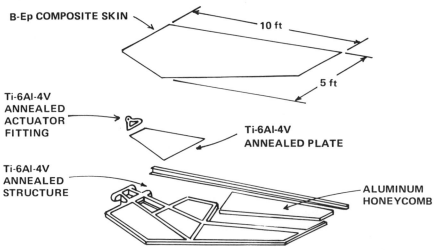

**FIG. 2.23**   This aircraft part is one of the largest B–Ep parts fabricated (5 ft = 1.52 m, 10 ft = 3.05 m).

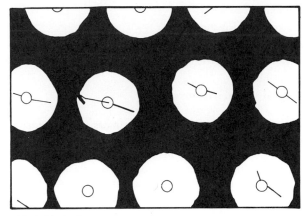

FIG. 2.24   B–Ep.

filament was 0.0036 in (0.09 mm), tensile strength 400 kips/in² (2758 MPa), and modulus of elasticity $63 \times 10^6$ lb/in² (434 GPa). Approximately 16 mi (25.7 km) of filament was wound into each groove, until the composite was about 0.54 in (13.7 mm) deep. When boron fiber, with its thin core of tungsten, is combined with epoxy, the resulting prepreg sells (1980) for approximately $200 per pound ($440 per kilogram) in tonnage orders (Fig. 2.24). Compressive strengths of 400 kips/in² (2758 MPa) are attainable in B–Ep compared with 200 kips/in² (1379 MPa) for C–Ep composites.

Although B–Ep has been predominantly used by industry, B–PI composites show excellent high-temperature capabilities up to 500°F (260°C) (Table 2.21). Polyimides, however, have one significant drawback: they cure with a release of volatiles. This can cause significant voids in the final composite, seriously reducing its strength. Development work over the past few years has been able to prepare B–PI composites with a void content of less than 1%, which is significantly lower than the 6 to 10% void content found previously.

Early developmental applications for boron-polymer composites reported in the literature include:[21]

Floor beams for supersonic aircraft
Helicopter rotor and tail-rotor blades
Landing-gear doors
Actuated wing panels
Wing box and fence components

**Table 2.21**   Properties of B–PI Composites†

| sp gr | Boron, vol % | Resin, vol % | Interlaminar shear | | Axial compression | | Axial tension | |
|---|---|---|---|---|---|---|---|---|
| | | | ksi | MPa | ksi | MPa | ksi | MPa |
| 1.88 | 54.3 | 17.5 | 8.5 | 58.6 | 216 | 1489 | 175 | 1207 |
| 1.86 | 53.1 | 21.9 | 10.1 | 69.6 | . . . | . . . | 180 | 1241 |
| 1.87 | 52.6 | 20.5 | 8.0 | 55.2 | 192 | 1324 | | |

†Based on *Ind. Res.*, October 1969, p. 72.

Flaps, rudders, and fire-access doors

Aircraft tail, air doors, and deflectors

Engine turbine compressor, fan and disk assembly

Reentry-vehicle cylindrical midsection

Aircraft fuselage assembly

Aircraft horizontal and vertical stabilizers

**Boron-Aluminum:**    The B–Al metal-matrix system has many potential aerospace applications, including jet-engine fan blades and tubular struts in the midfuselage section of the space shuttle. The B–Al composite has the weight of aluminum and the strength and stiffness of steel. The densities of aluminum and B–Al are about the same, and the modulus of B–Al along the direction of the fibers is about $32 \times 10^6$ lb/in$^2$ (221 GPa) compared with about $30 \times 10^6$ lb/in$^2$ (207 GPa) along any direction for steel.

A limitation inherent in common boron fibers makes the process currently used to fabricate B–Al parts expensive. Typically, a simply shaped part is fabricated by laying up plies of B–Al, each ply consisting of a layer of boron fibers on a layer of aluminum foil. To make a ply, a single layer of fibers one fiber thick and with all fibers parallel is laid on a sheet of aluminum foil, which is generally about 0.002-in (0.05-mm) thick. The foil is usually 2024 or 6061 aluminum alloy. The fiber layer is bonded temporarily to the foil by an acrylic adhesive, or binder. Plies of B–Al are cut to appropriate planform shapes and stacked on top of each other to form a part lay-up. The process is similar to that used with B–Ep or C–Ep in aircraft parts currently in production. The lay-up then is put into an airtight bag and heated to 700°F (371°C) while a vacuum is being drawn, removing the volatile acrylic binder.

Next the lay-up is put under a transverse pressure of 4 to 5 kips/in$^2$ (28 to 35 MPa) between matched dies in a press while simultaneously being heated to 975°F (524°C). This causes consolidation of the aluminum-alloy matrix material by diffusion welding, a solid-state joining process. The matrix does not actually melt under these conditions, but the layers of aluminum foil flow together around the entrapped fibers and join wherever they touch. The result is a solid aluminum part with the fibers embedded in it, alignment of the fibers being determined by the orientation of the plies in the lay-up (Fig. 2.25). The

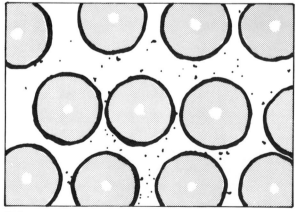

**FIG. 2.25**    B–Al.

**Table 2.22** Reactivity of Boron with Metals[2]

| Metal | Little or no reaction | | | Definite reaction | | |
|-------|-----------|-----|---------|-----------|-----|---------|
| | Temperature | | | Temperature | | |
| | °F | °C | Time, h† | °F | °C | Time, h† |
| Ni | .... | .... | .... | 1112 | 600 | 100 |
| | .... | .... | .... | 1292 | 700 | 100 |
| | .... | .... | .... | 1652 | 900 | 1 |
| Fe | 1112 | 600 | ‡ | 1292 | 700 | 50 |
| | | | | 1652 | 900 | 1 |
| Co | 1112 | 600 | ‡ | 1292 | 700 | 50 |
| | | | | 1652 | 900 | 1 |
| Al | 1112 | 600 | 1 | 1292§ | 700§ | 0.2 |
| Mg | 1112 | 600 | 100 | | | |
| | 1292§ | 700§ | 0.2 | | | |
| Be | 1832 | 1000 | 24 | | | |
| Ti | 1112 | 600 | 100 | | | |
| Cr | .... | .... | .... | 1652 | 900 | 1 |
| Ag | 1652 | 900 | 2 | | | |

†Time required to consolidate metal powders with fibers by hot pressing.
‡Reaction time is so short that as soon as fibers are hot-pressed, the reaction is finished.
§Molten.

process is expensive primarily because the high pressures involved require expensive tooling, such as matched die sets.

B–Al materials are available in three forms: (1) foils with boron filaments attached by fugitive organic resins, (2) foils with filaments attached with plasma-sprayed aluminum, and (3) fully or partially diffusion-welded foil-filament sandwiches. Because the demand is not high, no standardized foil-fiber form has yet been established.[22]

Researchers are continually trying to reduce production costs of B–Al parts, and their goal is to develop a fabrication method that can be carried out using higher temperatures and lower pressures. The efforts are aimed at modification of the boron fiber.

Consolidation is carried out at 975°F (524°C) in the current process because at temperatures above 1000°F (538°C) the boron starts to react with the aluminum, degrading the mechanical properties (strength and stiffness) of the fibers and thus of the composite itself (Table 2.22).

*Coating Compatibility:* The chemical and physical properties of the surface of the boron filament have been investigated because of its obvious involvement in bonding to the matrix material in a composite. Boron filament composites were postulated by analogy to fiber-glass technology, where silane finishes on the fiber were used to improve the bonding of the fiber to the resin. It was speculated that suitable finishes applied to boron filament might improve its bonding to the matrix resin. A good adhesive bond between the reinforcing filament and the matrix resin in a composite probably depends in part on adequate wetting of the filament by the uncured resin. Contact-angle measurement of a number of liquids indicated that the critical surface tension for wetting boron filament is 0.3 to 0.32 mN/cm. This means that a resin liquid having a surface tension below this value should completely wet the surface of a boron filament.

The potential of B–W as a reinforcement for aluminum is obvious, but the problem is the reaction between boron and aluminum already mentioned. It is overcome by vapor-depositing 0.0001 in (0.003 mm) of SiC on B–W. The coating is a diffusion barrier that

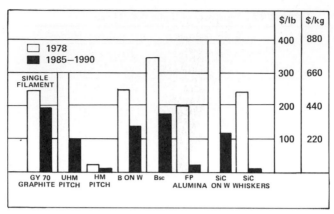

**FIG. 2.26**  Projected costs for fibers.[39]

prevents interaction between boron and metals at elevated temperatures. This fiber, known as Borsic, is marketed in 0.0042 and 0.0057 in (0.107 and 0.145 mm) and costs (1979) $350 per pound ($770 per kilogram), as shown in Fig. 2.26.

A 36-in blade fan rotor Borsic composite of B–Al has been tested for 500 h in a jet engine successfully. It provided reduced rotating mass, lighter weight (by 40%), higher efficiency, higher blade-tip speeds, and elimination of part-span shrouds needed to prevent blade flutter. Typical physical properties of Borsic are shown in Table 2.23.

For exposure to temperatures greater than 1000°F (538°C) recent experimental work has shown that boron coated with boron carbide ($B_4C$) in a titanium matrix can be processed at relatively low consolidation pressures and hence at lower cost. This barrier coating on boron has also been used in aluminum matrixes. Figure 2.27 shows a metallographic cross section taken from a hot-pressed monolayer tape. The 6061 aluminum-alloy powder was plasma-sprayed onto the $B–B_4C$ fibers, which were processed at 1112°F (600°C) at 426.6 lb/in² pressure (2.94 MPa). The unidirectional composite with 50% filament volume achieved tensile strengths of 26 kips/in² (179 MPa) and a modulus of elasticity of $32 \times 10^6$ lb/in² (221 GPa). Concurrent with this work on MMC is a program to incorporate boron coated with $B_4C$ into organic matrixes (epoxy and polyimides).

**Boron-Titanium:**  Limited work has been done with Borsic fibers in titanium matrixes. Titanium matrix materials have included Ti–3Al–2.5V and Ti–13V–10Mo–5Zr–2.5Al. The test work was performed in vacuum to temperatures of 800 to 1800°F (427 to 982°C) for up to 240 h. Strength was degraded significantly by the longer exposure times at temperatures of 1400°F (760°C) and greater. For all exposure times the strength after exposures

**Table 2.23**  Typical Physical Properties of Borsic (47 vol % Filament)[2]

| | Tensile strength | | | | Modulus of elasticity | |
|---|---|---|---|---|---|---|
| | Longitudinal | | Transverse | | | |
| Orientation | ksi | MPa | ksi | MPa | $10^6$ psi | GPa |
| Unidirectional | 175 | 1207 | 1.5 | 10 | 32 | 0.22 |
| Cross-plied, 0/90° (0/1.6 rad) | 7.5 | 52 | 7.5 | 52 | 20 | 0.14 |

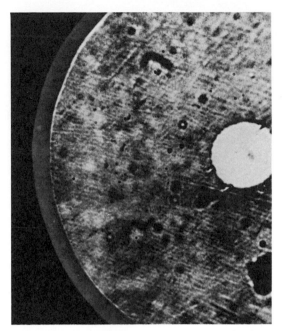

**FIG. 2.27** Metallographic cross section showing the three concentric layers of $WB_2$, boron, and $B_4C$.

at 1000 and 1200°F (538 and 649°C) was greater than after 800°F (427°C) exposures. Test results suggested a two-stage chemical reaction: (1) the simultaneous interdiffusion of the silicon, carbon, and titanium, resulting in the depletion of the SiC coating and formation of titanium silicides and (2) significant formation of TiB and a higher rate of formation of $TiSi_2$. The strength of the composite was degraded before the formation of any identifiable boride compounds.

**Boron-Carbon:** Tungsten is an expensive and dense substrate for vapor deposition of boron; substituting a carbon monofilament for the tungsten is both practical and economical. Manufacturers have successfully vapor-deposited boron onto a carbon monofilament, which differs from the carbon fiber used in C–Ep composites. The carbon monofilament has a diameter of 0.0013 in (0.03 mm), and the mechanical properties of boron fiber made from it are somewhat less than those of boron on tungsten, though still superior to those of regular carbon fiber. Because the carbon monofilament substrate has a larger diameter than the tungsten filament substrate, a boron-on-carbon fiber contains less boron than a boron-on-tungsten fiber for a given final fiber diameter.

A 0.0042-in (0.107-mm) boron-on-carbon fiber has been rated at a tensile strength of 475 kips/in² (3275 MPa) and a modulus of $53 \times 10^6$ lb/in² (365.4 GPa). A 0.0056-in (0.14-mm) fiber has the same tensile strength and $55 \times 10^6$ lb/in² (379 GPa) modulus. The 0.0042-in (0.107-mm) fiber can now be directly substituted for the 0.004-in (0.10 mm) boron-on-tungsten fiber in a B–Ep prepreg tape.

The physical properties of the carbon monofilament substrate give it a production advantage in the vapor-deposition process. Since reactors can run 40% faster making boron

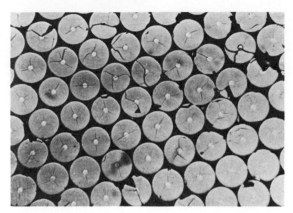

**FIG. 2.28** Cross section of boron-reinforced magnesium composite made by liquid-metal infiltration.[7]

on carbon, fewer reactors are needed to make a given amount of fiber in a given amount of time. The carbon monofilament is made from petroleum pitch and graphitized to give it the conductivity needed to carry the resistive heating current required in vapor deposition.

**Boron-Magnesium:**   MMCs have been produced by continuous-casting methods. The process has made preform shapes (rods, tubes, I beams, Z stiffeners, D shapes, tapes) of boron-reinforced magnesium. In lieu of hot-pressure-bonding fabrication techniques the continuous-casting process was adopted to reduce cost and attain consistency of the filament incorporated into the MMC (Fig. 2.28). The process is adaptable to any system in which the filament is stable in the matrix and cast cross-sections are uniform. The magnesium-boron (Mg–B) (combination) is a model MMC system exhibiting an excellent interfacial bond and outstanding load-redistribution characteristics. Tests have shown flexural strengths of 162 kips/in$^2$ (1117 MPa) at 25% filament volume and 426 kips/in$^2$ (2937 MPa) at 75%. Density of 0.087 lb/in$^3$ (2408 kg/m$^3$) has been attained and ultimate tensile strength of 189 kips/in$^2$ (1303 MPa) [compared with 140 kips/in$^2$ (965 MPa) for steel] and flexural modulus of 35 $\times$ 10$^6$ lb/in$^2$ (241.3 GPa).

### Silicon Carbide

Silicon carbide is more oxidation-resistant than boron and less susceptible to attack by molten aluminum and retains more of its strength and stiffness at elevated temperatures. SiC filaments were initially produced by two general methods: (1) reacting silicon halides, e.g., silicon tetrachloride, with organic gases, e.g., hydrocarbons, on a hot metallic filament and (2) thermal decomposition of organosilanes on a hot metallic filament. In the first method the SiC bond is formed during the deposition; in the second, which is used today, this bond is already formed in the precursor molecule. Like boron fiber, SiC fiber is made by chemical vapor plating; both tungsten and carbon monofilament substrates can be used to make a fiber.

The boron and SiC fiber reactors are similar. In the SiC process, the silane compounds take the place of BCl$_3$. Manufacturers are capable of converting each stage of the process from boron to SiC, depending on the future demands of industry. Typically a 0.0005-in

(0.01-mm) tungsten filament is exposed to 2192°F (1200°C) in the presence of hydrogen and the thermal decomposition of the silanes, e.g., ethyltrichlorosilane. The plating rates for SiC are comparable to those for boron.[23,24]

A fluidized-bed technique to produce SiC fibers appears to be more economical than the hot-wire vapor-deposition technique. Seed fibers 0.001 in (0.03 mm) diameter by 1 in (25.4 mm) long are fluidized (subjected to an upward flow of gas containing the reactants) in a heated reactor. The fibers are made at 2200 to 2300°F (1204 to 1260°C) with an inlet composition of 79% argon, 18% hydrogen, 1.8% methyltrichlorosilane, and 0.8% methane. The strength of the fibers [<250 kips/in$^2$ (1724 MPa)] is not sensitive to moderate variations in the above concentrations, but fiber strength is drastically reduced when small amounts of impurities are present in the inlet gas.

**Silicon Carbide–Tungsten:** SiC on 0.0005-in (0.01-mm) tungsten is commercially available in 0.004- and 0.0056-in (0.10- and 0.14-mm) diameters. Even though SiC–W is useful up to very high temperatures, there is the problem of a reaction between the SiC and the tungsten at 1700°F (927°C). The relatively high density of SiC–W is also a disadvantage, but the fiber's superior thermal stability above 660°F (349°C) makes it important for high-temperature applications. SiC filaments are intended primarily for reinforcement of alloys used in the high-temperature section of jet engines, current effort being concentrated on reinforcement of Ti–6Al–4V. Sections of the engine include compressor blades, which must withstand temperatures of up to 900°F (482°C) under great rotational stress and also drive shafts and engine core shafts.

Since SiC does not suffer chemical degradation in contact with molten aluminum, it is compatible with the higher-temperature aluminum-matrix composite-processing techniques. Since no special coatings are necessary, processing costs are reduced.

The mechanical properties of SiC–W fibers at room temperature are similar to those of boron, 500 to 600 kips/in$^2$ (3447 to 4137 MPa) tensile strength and a modulus of 60 × 10$^6$ lb/in$^2$ (413.7 GPa). The tensile strength of SiC has been measured at elevated temperatures up to about 2500°F (1371°C). Test results showed an approximate 30% loss in strength at this temperature. Boron filaments lost all their strength at about 1200°F (649°C). Figure 2.29 compares the room-temperature tensile strength of boron filament, SiC filament, and SiC-coated boron filament after heat soaking in air at 1000°F (538°C) for times up to 2000 h. SiC- and SiC-coated boron showed essentially no loss in strength after 2000 h, whereas boron showed a 50% loss after about 400 h at 600°F (316°C).

**Silicon Carbide-Carbon:** SiC on carbon is relatively new and still essentially a developmental material. SiC–C fibers, 0.0056 in (0.14 mm) diameter, promise better stability at higher temperatures for lower production cost than the tungsten-substrate fiber. SiC vapor deposited on a carbon monofilament avoids the high-temperature internal reaction associated with SiC–W. Also, SiC–C is less dense than SiC–W, and the developmental fiber shows tensile strengths approaching 700 kips/in$^2$ (4827 MPa) and modulus of 60 × 10$^6$ lb/in$^2$ (413.7 GPa). Current typical average SiC–C filaments show a 100-h 2000°F (1093°C) stress rupture of over 200 kips/in$^2$ (1379 MPa), making this material attractive for high-temperature superalloy or ceramic reinforcement.

Other government-sponsored research programs are examining and developing a mixture of silicon nitride–SiC fibers, which can be obtained by preconditioning and pyrolysis of organosilicon polymeric-fiber precursors using techniques similar to those for the conversion of PAN and other organic fibers into graphite.

Carbosilane and carbosilazane polymers have yielded inorganic fibers in the laboratory

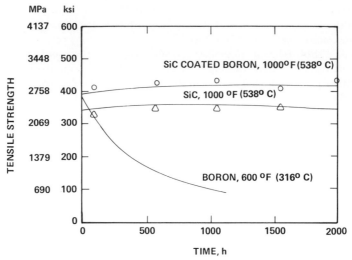

**FIG. 2.29**  Room-temperature tensile strength after heat soaking in air.

with attractive strength and modulus values, but retention of these properties in practical fiber quantities and lengths has not been demonstrated. Further work on these materials will show whether they have the potential to compete with graphite fibers in properties and cost effectiveness. The new fibers will be similar to graphite fibers but will not be electrically conductive.

**Silicon Carbide Whiskers:**   SiC whiskers (as opposed to continuous filaments), initially investigated in 1961, have received renewed impetus. Interest lagged because of high cost and processing difficulties. Whisker processing has recently moved forward more rapidly. Potentially, high-performance whiskers could sell (1981) at less than $10 per pound ($22 per kilogram). One development includes the manufacture of SiC whiskers from rice hulls.

The whiskers themselves are tiny, typically 8 to 20 $\mu$in (20 to 51 nm) in diameter and about 0.0012 in (0.03 mm) long. Whisker-reinforced or short-fiber-reinforced metals offer an intermediate alternative between homogeneous metals and continuous-filament-reinforced composites. The attraction of whisker-reinforced metals is that they can be formed with relative ease. Such processes as extrusion, rolling, swaging, and press-brake operations are possible. Damage from deformation processing is not likely to degrade mechanical properties. Other advantages include:

The ease with which the whiskers can be incorporated into a metal matrix by powder metallurgy. Many complex components can be hot-pressed isostatically to final shape with a minimum of machining, and hence the component price is low (Fig. 2.30).

The potential of higher-strength composites than can be formed from fibers, since the whiskers are expected to have strengths of over 1000 kips/in² (6895 MPa).

The highest transverse thermal conductivity of any potential reinforcing fiber. This quality, along with thermal expansion, is the most important design parameter in many space structures, where thermal distortion must be minimized.

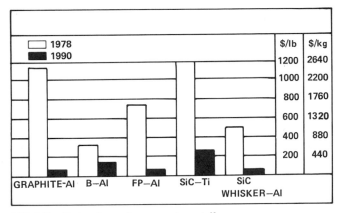

**FIG. 2.30** Projected costs for structural panels.[39]

Metals reinforced with SiC whiskers are the only composites evaluated that can be formed by normal metal-working processes.

The machinability of aluminum reinforced with SiC whiskers is considerably better than that of aluminum reinforced by continuous fibers of SiC, boron, or alumina.

Extruded and rolled materials having tensile strengths up to 88 kips/in² (607 MPa) have been obtained with a 7075 aluminum alloy. For space applications whiskers markedly reduce the coefficient of thermal conductivity. The strength and stiffness of the composites are remarkably isotropic even after extrusion.

### Matrixes

**Silicon Carbide–Epoxy:**   Composites consisting of 0.004-in-diameter (0.10-mm) SiC filaments in an epoxy resin matrix have been fabricated and tested. Unidirectional composites containing 62 vol % SiC had an average flexural strength of 260 kips/in² (1793 MPa) and an average flexural modulus of 33.4 × 10⁶ lb/in² (230.3 GPa). The filaments had an average tensile strength of 36 kips/in² (248 MPa) and an average modulus of elasticity of 61 × 10⁶ lb/in² (420.6 GPa).

**Silicon Carbide–Graphite:**   Custom-machined components can be made from a SiC–Gr composite, a recently developed SiC-processed material that imparts many of the characteristics of ceramics to ordinary graphite. A vapor-phase SiC is formed at and below the surface of machined graphite without altering the configuration of the part or greatly reducing the inherent properties of the graphite. The SiC–Gr composite increases hardness, wear, and abrasion resistance and inhibits attack by many acids and alkalies. The composite is capable of withstanding temperatures up to 3272°F (1800°C) in reducing or inert atmospheres and can be used for precision fixtures in semiconductor manufacturing. Its thermal and chemical resistance suggests use in pump seals, metallurgical applications, and similar areas for crucibles, casting molds, pouring nozzles, and bearings. Tests also show that its surface hardness is far superior to that of carbon graphite.

**Table 2.24** Composite Fabrication Methods[25]

| Process | Fiber |
|---|---|
| Liquid-metal infiltration | SiC yarn |
| Powder metal with liquid-phase sintering under pressure | SiC yarn |
| Tape fabrication, then low-pressure bond using preheated press platens in a vacuum chamber to reduce the bonding time from several hours to less than 30 min | SiC monofilament |
| Tape fabrication, then high-pressure bond using the method described above | SiC monofilament |

**Silicon Carbide–Aluminum:** Fabrication methods to produce a usable MMC are being developed with both private and government funds. One company has investigated making SiC–Al composites using (1) tows of SiC yarn of relatively small diameter [0.012 to 0.040 in (0.3 to 1.02 mm)] and (2) SiC monofilaments of 0.140 in (3.6 mm) diameter.[25] Some of the composite fabrication methods evaluated are shown in Table 2.24. Among the aluminum alloys investigated were 2024 and 6061. Of the processes in Table 2.24 the last two appeared to be most successful in terms of achieving composite mechanical properties attractive for engineering applications. The highest tensile-strength levels were obtained by the high-pressure-bond method, presumably because less degradation occurred at the fiber-matrix interface at the lower pressing temperatures used.

The SiC monofilament composites exhibited the highest axial strength values and compared favorably with those for B–Al, although the transverse tensile strength did not. The SiC monofilament composites also showed better impact properties than both alumina and SiC yarn composites (Fig. 2.31).

SiC–Al composites have been successfully produced in the 2000, 6000, and 7000 series aluminum alloys, and currently the government is sponsoring development of the 5000 marine alloy system. SiC–Al shows promise for ship superstructures and decking, underwater applications, and torpedo and mine structural applications.

**Silicon Carbide–Titanium:** Titanium-alloy matrix composites reinforced with SiC filaments have potential application as turbine engine fan and compressor blade materials operating at temperatures above 800°F (427°C). Recently completed work shows that the new titanium composites are more economical than B–Ti and Bsc–Ti.[26] Care must be exercised to avoid any significant filament breakage when incorporating SiC filaments into titanium-alloy matrixes by the hot-press diffusion-welding method. SiC filaments are subject to surface compressive damage, which can cause serious strength losses. SiC is more sensitive to surface damage than boron or Borsic filaments.

Room-temperature strengths of up to 170 kips/in$^2$ (1172 MPa) have been obtained with 40 to 5 vol % Sic–Ti–6Al–4V [0° (0-rad) filament orientation]. The composite modulus was 38 × 10$^6$ lb/in$^2$ (262 GPa). Off-axis strengths are higher than those for B–Al and Bsc–Ti at angles to the filament direction greater than about 5 and 15° (0.09 and 0.26 rad), respectively. Transverse failures occur by filament-matrix debonding and not by filament splitting, and secondary creep rates are very low; however, work continues to improve materials and optimize the processing of SiC fibers to provide the required mechanical properties for high-temperature applications. An example of this work is a government program which evaluated an SiC–Ti composite fabricated by hot-pressing alternating layers of collimated SiC fiber and unalloyed titanium foil. Test results showed that SiC–A–70Ti composites had a higher room-temperature modulus of elasticity [35 × 10$^6$ lb/in$^2$ (241.3 GPa)] than either stainless steel [30 × 10$^6$ lb/in$^2$ (206.9 GPa)] or

900 μ

400 μm

FIG. 2.31  Fracture surface of a Charpy impact specimen, 50% fiber content Si–C-reinforced 6061 aluminum.[25]

titanium [18 × 10⁶ lb/in² (124.1 GPa)]. This advantage persisted[27] up to and including test temperatures of 1110°F (599°C).

Notched miniature Izod impact strength of SiC–A–70Ti composites, in the as-fabricated condition, averaged 60 ft·lb/in² (126 kJ/m²). This was increased to 90 ft·lb/in² (190 kJ/m²) by changing the matrix to A-40 unalloyed titanium, which was lower in oxygen content and higher in ductility. The impact strength of the A-40 titanium composites was equal to that of Ti–6Al–4V alloy, 95 ft·lb/in² (200 kJ/m²). Finally the average room-temperature tensile strength of the SiC–A–70 titanium composite was 95 kips/in² (655 MPa), which is nearly identical to that of the unreinforced A-70 titanium, 97 kips/in² (669 MPa). At temperatures above 800°F (427°C) the composite was superior in tensile strength to unreinforced unalloyed A-70 titanium and had lower tensile strength than the Ti–6Al–4V (Fig. 2.32).

**Silicon Carbide–Silicon:**  A newly developed composite material can operate at temperatures as high as 2500°F (1371°C), some 400°F (204°C) above the limits of parts fabricated from the most heat-resistant structural metals and alloys. This new lightweight composite consists of SiC filaments separated by a silicon filler. The SiC provides the com-

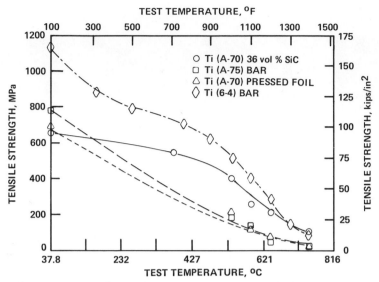

**FIG. 2.32**  Room- and elevated-temperature tensile strengths of 36 vol % SiC–A-70 titanium composites, hot-pressed A-70 foil, and A-75 and Ti–6Al–4V bar stock.[27]

posite with its strength at high temperatures, and the silicon filler imparts toughness, ability to bend under stress, and other metallic characteristics.

The material is produced by a proprietary process that involves reacting molten silicon directly with carbon fibers, converting the carbon fibers into SiC. The spaces between the resulting SiC filaments are then filled with silicon. Since SiC cannot react with silicon, reactions between reinforcing filaments and the matrix do not occur.

A major advantage of the new material is the speed and relatively low cost with which complex parts can be produced. This fabrication process, considerably faster and less complicated than that required to produce parts from most high-temperature ceramics, shows great potential for high-volume applications. In the production of a part, the composite is precision-cast in a mold into a shape that experiences little or no shrinkage. With a minimum of machining, the part is then finish-machined with diamond cutting tools, producing the final shaped part. Another benefit of the new composite is that it is fabricated at approximately 2700°F (1482°C), while processing temperatures for high-temperature ceramics run to 3500°F (1927°C) and higher.

Although the SiC–Si composite is still developmental, parts produced from it are being evaluated in gas turbines and aircraft engines. Other potential applications are in coal gasifiers and other high-temperature machinery.

### Aluminum Oxide (Alumina)

A polycrystalline alumina ($Al_2O_3$) fiber, called FP, has been developed for use in MMCs, but also shows promise for resin-matrix composite applications, especially teamed with Kevlar in a hybrid. Fiber FP is produced in the form of round filaments, each with a diameter of approximately 0.0008 in (0.02 mm). The material has greater than 99%

purity, and its melting point is 3713°F (2045°C), which makes it attractive for use with high-temperature MMC processing techniques (Fig. 2.33).

The alumina filaments are given a coating of silica (silicon dioxide, or $SiO_2$) and combined into a continuous yarn comprising 210 filaments that makes up the usable product. For uncoated filaments the tensile strength is 200 kips/in$^2$ (1380 MPa), and the modulus is $55 \times 10^6$ lb/in$^2$ (379.2 GPa).

Thanks to a mechanism currently not explainable by the developer of FP fibers, the silica coating results in an increase in the tensile strength of the filaments to 275 kips/in$^2$ (1896 MPa) even though the coating is approximately 10 $\mu$in (0.25 $\mu$m) thick and the modulus does not change. The increase in tensile strength is of practical significance only in resin-matrix applications. During MMC (reinforced-casting) processing the silica coating on the filaments generally reacts with the molten metal, leaving the fibers bare in the matrix. For that reason all values for the physical properties of uncoated fibers are the same as for the metal matrix. For fibers that are to be used in a resin matrix, the values quoted are those for coated fibers, because the coating is not affected by the processing. For reinforced-casting applications, the main function of the silica coating is to improve the wetting of the fibers by the molten matrix metal during processing. This improves the matrix-to-fiber bond. There is a difference in the coating application, depending on whether the fibers are destined for use in resin-matrix or reinforced-casting applications.

Fiber FP has been demonstrated as a reinforcement in magnesium, aluminum, lead,

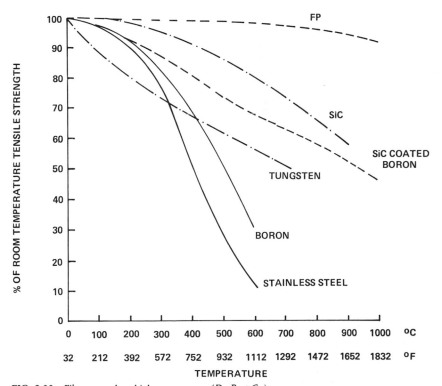

**FIG. 2.33**  Fiber strength at high temperature. *(Du Pont Co.)*

copper, and zinc, with emphasis to date on aluminum and magnesium materials. One advantage of the alumina fiber is its compatibility with vacuum infiltration casting, a manufacturing technique that could improve the economics of producing many MMC components. Since the process involves exposing the reinforcement fibers to molten metal, they must be able to withstand high temperatures. Because of its high melting point and other

properties, fiber FP has been used successfully. In the process, fiber FP yarn (Fig. 2.34) is assembled into tapes consisting of parallel yarn fibers held together by a temporary organic binder, which is later burned away during processing. The flat tapes are laid up in a casting mold, the layers being arranged to give fiber orientations and densities that will provide the desired properties in the finished component. The mold is sealed and then heated to burn off the organic binder, which is drawn out of the cavity in the form of a gas. More details on the process will be given in Chap. 4.

A vacuum is drawn in the mold cavity, and the molten matrix metal is injected. The molten metal flows between the enclosed filaments and surrounds them before it is cooled. The cooling process is controlled to assure the metallurgical integrity of the finished part. The result is a reinforced casting in which the reinforcing fibers are fixed within the matrix in designer-specified densities and orientations. The key to success with the vacuum-infiltration process is how well the molten matrix metal adheres to the fibers and flows along them. Uniform infiltration is necessary because gaps in the casting or weakness at the matrix-to-fiber interfaces decrease mechanical properties. Even though FP fiber is relatively new, it is available in several different forms.

**FIG. 2.34**   FP yarn. *(Du Pont Co.)*

*Tape:*   Fiber FP is made in tape form with 10 to 15% of a fugitive acrylic binder. The tapes are flexible and easy to handle and are typically used in making preforms for reinforced castings. Tapes up to 36 in (914 mm) long and 22 in (559 mm) wide are available in thicknesses of 0.01 to 0.15 in (0.25 to 4 mm). The fugitive binder is easily burned off in air above 932°F (500°C).

*Plates and Prepreg:*   Unidirectional plates of FP aluminum (FP–Al) and FP magnesium (FP–Mg) are available in fiber volume fractions of 35 and 55%. FP lead is also available in plate form with 50 vol %. The FP Kevlar resin prepreg is available in strips 3 to 36 in (76 to 914 mm) wide with maximum length to 120 in (3048 mm). Any resin which can be applied from solution (epoxies, polyimides, and phenolics) can be used for prepregging.

*Fabrics:*   Fiber FP has been successfully woven into a variety of fabrics. The brittle ceramic fiber is first overwrapped with a low-denier rayon yarn to improve handleability on the loom. With this technique, open weave, 8- and 12-harness satin fabrics have been made. They are used to make angle-ply lay-ups in both resin-matrix composites and reinforced castings.

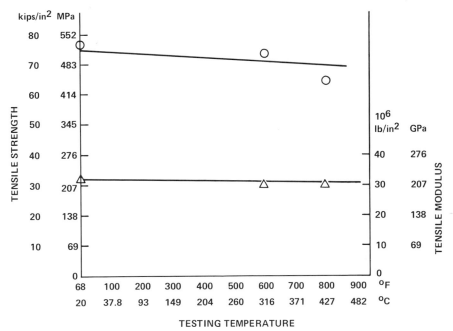

FIG. 2.35    Tensile properties of FP–Mg composites. (*Du Pont Co.*)

## Matrixes

**FP Magnesium and Aluminum:**
Magnesium and its alloys adhere well to
fiber FP. Magnesium matrix composite
samples containing 70 vol % fibers have
been produced. Alloys of magnesium which
have successfully been vacuum-infiltrated
include ZK60A and ZE41A (Mg–4.25Zn–
1.25 rare earths–0.50Zr). Typical properties of FP–Mg are shown in Fig. 2.35 and
a typical cross section in Fig. 2.36.

To achieve the wetting action needed
with other matrix materials it is necessary
to modify the matrix alloy by adding small
amounts of "active" metals (lithium, calcium, or magnesium). Particular success has
been found with Al–Li alloys. Recent tests
have shown that fiber FP is not degraded by
molten magnesium or Al–Li alloy during
infiltration at 1382°F (750°C); the effect of
infiltration on the strength of fiber FP is
shown in Table 2.25.

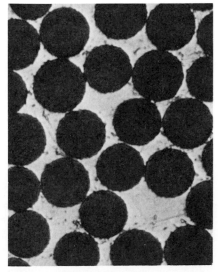

FIG. 2.36    FP–Mg composite. (*Du Pont Co.*)

**Table 2.25**  Effect of Infiltration on Strength of Fiber FP

| | Unidirectional fiber strength | | | |
| | Before infiltration | | After infiltration | |
| Fiber | ksi | MPa | ksi | MPa |
|---|---|---|---|---|
| FP–Mg | 217 | 1496 | 204 | 1406 |
| FP–Al–4 wt % Li | 210 | 1448 | 206 | 1420 |

**FP-Ep and Kevlar:**  The use of fiber FP with Kevlar in hybrid resin-matrix composites offers promise in several applications:

Radar-transparent structures

Circuit boards

Sporting goods

Ballistic armor

Antenna supports

Although the alumina fiber does not have the tensile strength of other reinforcement fibers, its compression properties in a matrix exceed those of other fibers. The compressive strength along the axis of the fibers of a unidirectional fiber FP–Ep composite is 330 to 350 kips/in$^2$ (2275 to 2413 MPa). The corresponding value for a Kv–Ep composite is 40 kips/in$^2$ (276 MPa), and that for a typical C–Ep composite is about 200 kips/in$^2$ (1380 MPa). Therefore, combined in a hybrid composite, fiber FP and Kevlar tend to complement each other. Fiber FP contributes to compression strength, where Kevlar is weak, and Kevlar contributes to tensile strength, where fiber FP is weak.

Since both fibers are electrically nonconductive, the hybrid can be used where a radar-transparent material with good structural properties is needed. The ability of the matrix material to withstand heat is the thermal limiting factor for such a hybrid, because both fibers can take a temperature of 1832°F (1000°C) with no change in strength or modulus.

**Others:**  Composites of fiber FP in lead have been produced containing 45 vol % fiber FP with unidirectional tensile strengths of 65 to 75 kips/in$^2$ (448 to 517 MPa) and modulus of 26 to 28 × 10$^6$ lb/in$^2$ (179.3 to 193.1 GPa). Potential applications include large storage-battery plates, chemical reaction vessels, radiation shielding, and reinforced projectiles.

For applications requiring higher temperature capabilities than those of epoxy matrixes, polyimide matrixes are being investigated. Prepregs have been prepared by impregnating fiber FP with a polyimide resin; subsequently these prepregs were made into laminates by hot pressing at 800°F (427°C) and 2 kips/in$^2$ (14 MPa). Figure 2.37 shows the effect of temperature on the flexural strength of FP–PI, indicating the high-temperature strength of this combination.

Limited evaluation of the use of alumina fiber to reinforce glass and ceramics such as silica has been conducted. The prime interest is in radomes and nose cones because of their resistance to erosion and laser damage. Figure 2.38 shows a 50% fiber FP in fused silica. The material was made by impregnating the fiber with a slurry of fine glass particles, drying, and sintering. Initial tests of flexural strength were 55 kips/in$^2$ (379 MPa) and flexural-modulus values were 30 × 10$^6$ kips/in$^2$ (206.9 GPa).

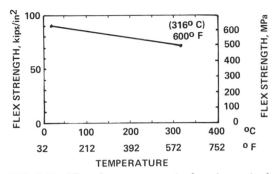

FIG. 2.37   Effect of temperature on the flexural strength of
FP–PI laminates (55 vol % FP). *(Du Pont Co.)*

Other matrix materials reinforced with fiber FP include copper and zinc; future research
will include titanium and superalloys.

**Sapphire-Titanium:**   Continuous single-crystal sapphire (alumina filaments) 0.010 in
(0.25 mm) in diameter have unusual physical properties: high tensile strength [over 300
kps/in$^2$ (2069 MPa)] and modulus of elasticity of 65 to 70 $\times$ 10$^6$ lb/in$^2$ (448.2 to 482.7
GPa). The filaments combine the chemical inertness, temperature resistance, strength, and
stiffness that are especially needed for use in metal composites at elevated temperatures and
in highly corrosive environments.

An unusual method for producing high-strength single-crystal fibers in lieu of a crystal-
growing machine is the floating-zone fiber-drawing process. The fibers are produced directly
from a molten ceramic without using a crucible. To form the controlled-purity single-
crystal ceramic fibers a laser is focused on the tip of a feed rod of alumina and the power
increased until the tip of the feed rod begins
to melt. A single-crystal seed of the same
material is brought into contact with the
molten tip of the feed rod to initiate single-
crystal growth onto the seed from the molten tip. The seed crystal is drawn away at a
predetermined rate to achieve the desired
dimensions for the solidified single-crystal
filament. This system is capable of making
single-crystal fibers of any ceramic material
with a melting point up to 7257°F
(4014°C). The process has also been successfully used to produce titanium boride
(TiB$_2$) and titanium carbide (TiC) fibers.

Sapphire continuous filaments have been
incorporated into titanium-alloy matrixes
by hot-press diffusion-welding methods
without any significant filament breakage.
Sapphire is more sensitive to surface damage than boron or Borsic filaments.[26]

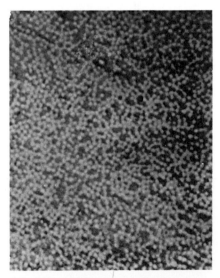

FIG. 2.38   FP–Gl composite. *(Du Pont Co.)*

The highest unidirectional strength of a sapphire–Beta III† (Ti–11Mo–6Zr–5Sn) titanium alloy composite is approximately 143 kips/in² (986 MPa) for a 45 to 50 vol % reinforcement, and this system has a modulus of $40 \times 10^6$ lb/in² (275.8 GPa). Compared with SiC–Ti-reinforced matrixes, the sapphire–Ti has very high failure strains in unidirectional material. Transverse strengths are also low[27] [55 kips/in² (379 MPa)].

## Miscellaneous Filaments

Several other materials have been produced as filaments by a variety of methods. Although they were never produced in volume (for one or more reasons), their usefulness should be noted.

**Titanium Diboride:** $TiB_2$ has a higher modulus than boron or SiC, but its density is also high, so that its stiffness-to-density ratio is somewhat less than that of the other materials. It has an advantage over them in terms of chemical resistance to gaseous fluorine-containing species at elevated temperatures, which might make it applicable as a reinforcement for composites used as ablative liners for rocket nozzles employing advanced propellant systems. It may also be more resistant chemically to a titanium matrix at elevated temperatures. $TiB_2$ involves the deposition on an incandescent tungsten substrate from a mixture of $BCl_3$, $H_2$, $TiCl_4$, and argon at a temperature of 2282 to 2372°F (1250 to 1300°C). Filaments with diameters of 0.003 and 0.004 in (0.08 and 0.10 mm) have been produced by this chemical vapor-deposition method. The filament is heated resistively like boron and also by radio-frequency coupling. Representative values of the tensile strength are 250 to 300 kips/in² (1724 to 2068 MPa). The tensile modulus is $71 \times 10^6$ lb/in² (489.5 GPa).

**Boron Carbide:** High-strength BC yarn has been produced by a process in which a precursor carbon fiber is chemically converted by reaction with a boron halide compound and hydrogen to yield continuous-filament BC yarn with high strength and high modulus. By controlling the degree of conversion 50 to 100% fibers are obtained with various properties. The structure of the new fiber is highly crystalline and more resistant to self-abrasion than conventional boron fibers. BC fibers also exhibit exceptional thermal stability and show no loss in properties when heated to 3992°F (2200°C) for several hours.

The translation of the BC fiber properties to composites has been observed since composites with BC fibers have been fabricated using resin, metal, and ceramic matrixes. SiC reinforced with BC fibers demonstrates that the BC fibers provide high strength and resistance to shock at temperatures up to 2012°F (1100°C) in air and could represent a major advancement for turbine blade materials. A composite of BC fibers in a BC matrix could be a promising lightweight armor material.

BC composites have also been made from BC filament tape impregnated with a thermosetting resin. The plastic is converted into a carbon matrix by heating at elevated temperatures. The surface of the composite is a thin film of SiC. In air, no oxidative degradation is observed at 2500°F (1371°C). The low specific gravity of the composite offers a higher strength-to-weight ratio than metal alloys or silicon-based ceramics. In comparison, the best metal alloys start to lose mechanical properties about 1700°F (927°C). The fiber-

†Trade name.

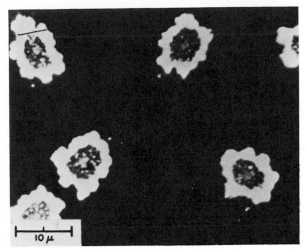

**FIG. 2.39**   BC-fiber composites. *(The Carborundum Co.)*

reinforced structure avoids the inherent brittleness of other ceramics. This represents one of the first systems to combine the outstanding thermal resistance of ceramics with the high directional strength of reinforced composites (Fig. 2.39).

**Boron Nitride:**   BN fibers in a silicone resin matrix show promise as a material for high-temperature electric-wire insulation. The high thermal conductivity of BN (equal to that of stainless steel) dissipates heat, while its high dielectric properties ensure good insulation. The heat-dissipation characteristics of BN fibers help maintain the structural integrity of the composite by reducing the danger of heat degradation of the matrix. Wires insulated with BN–silicone heated to 1500°F (816°C) and then tested at room temperature can take extremely high (1000$g$) impact loads without cracking.

**Tantalum Carbide:**   Wires used for reinforcing cobalt-base MMC and other applications can be given a protective coating of TaC by a new chemical vapor-deposition process. The carbide has a melting point near 7257°F (4014°C) and exceptional hardness characteristics; it is applied as a tightly bonded layer to a continuously moving wire. In the coating process, a resistively heated filament of tungsten (or other metal) is drawn through a chamber containing $CH_4$ and $TaCl_5$ to form a surface carbide layer (Fig. 2.40). The $TaCl_5$ is formed in situ by reacting tantalum chips with either HCl or $Cl_2$ in an attached chlorinating chamber. The continuous wire filament is coated at a rate of 5.5 in/min (2.33 mm/s).

## Hybrids

Hybrid composites, which combine two or more different fibers in a common matrix, greatly expand the range of properties that can be achieved with advanced composites. They may cost less than materials reinforced only with graphite or boron.

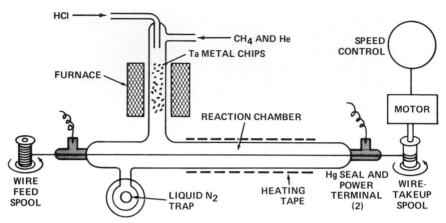

**FIG. 2.40**  Apparatus for chemical vapor deposition of TaC.[36]

**Uniqueness of Hybrids:**  Generally, "hybrid" applies to advanced composites and refers to use of various combinations of continuous graphite, boron, Kevlar, or glass filaments in either thermoset or thermoplastic matrixes. Combining continuous and chopped fibers in a common matrix also qualifies the material as a hybrid composite. Hybrids have unique features that can be used to meet diverse and competing design requirements in a more cost-effective way than either advanced or conventional composites. Some of the specific advantages of hybrids over conventional composites are balanced strength and stiffness, balanced bending and membrane mechanical properties, balanced thermal-distortion stability, reduced weight and/or cost, improved fatigue resistance, reduced notch sensitivity, improved fracture toughness and/or crack-arresting properties, and improved impact resistance.

**Types:**  The four basic types of hybrid composites are interply, intraply, interply-intraply, and superhybrid. The first of these is subdivided into interspersed or core and shell variations. *Interply hybrids* consist of plies from two or more different unidirectional composites (UDC) stacked in a specific sequence. *Intraply hybrids* consist of two or more different fibers mixed in the same ply. *Interply-intraply hybrids* consist of plies of interply and intraply hybrids stacked in a specific sequence. *Superhybrids* consist of resin-matrix composite plies stacked in a specific sequence.[28,29] The interply and intraply hybrids generally have the same matrix, and the laminate is fabricated by the co-curing procedure outlined by the prepreg producer. If the plies of these hybrids are made from different matrixes, the hybrid must be fabricated by a curing procedure compatible with both systems. Superhybrids are fabricated by adhesively bonding metal foils, the MMC, the resin-matrix UDC, and resin-fiber prepreg, with an adhesive that has the same curing cycle as the prepreg tape.

**Currently Applicable Fibers:**  Various types of graphite, boron, glass, and Kevlar fibers are used in hybrids, as are cloth and fabric woven from them. Typical stress-strain diagrams of some of these fibers are shown in Fig. 2.41, from which one can see that fibers are available with the following range of mechanical properties: tensile strength of from 2500 to 5000 kips/in$^2$ (17,238 to 34,475 MPa); fracture strain ranging from 0.4 to 4.0%, and tensile modulus of from 10 to 60 $\times$ 10$^6$ lb/in$^2$ (68.95 to 413.7 GPa). Essentially, three different types of graphite fibers are commercially available for hybrid use (Fig. 2.41). Graphite fibers offer high stiffness and strength and low density, but impact resistance is low. The leading commercial aromatic polyamide fiber is Kevlar, which combines high strength, high modulus, and high impact resistance but suffers from fairly low compressive strength. Another drawback is the difficulty of producing chopped fibers on standard equipment. Of the two types of glass fiber currently used, S glass is more rigid and stronger than E glass. Their main limitation is their density compared with that of graphite and Kevlar fibers.

**Currently Applicable Resins:**   Both thermoset and thermoplastic resins are being used for hybrid composites. The epoxies remain the chief thermoset, but thermoset polyester may emerge as a strong competitor, especially for automotive applications. The epoxies are available with a wide range of properties (Fig. 2.42), but intermediate-modulus epoxies are used in most hybrids. Epoxies have good to moderate elevated-temperature properties, but there may be a problem with moisture absorption when they are exposed to temperatures near 350°F (177°C). Polyimides find use as matrixes for hybrids where extended operation at temperatures near 500°F (260°C) and higher are required.

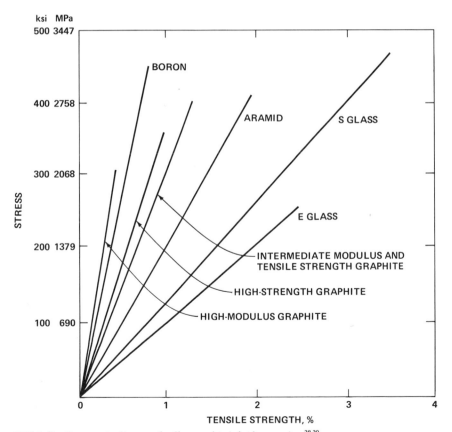

**FIG. 2.41**   Stress-strain diagrams for fibers used in hybrid composites.[28,29]

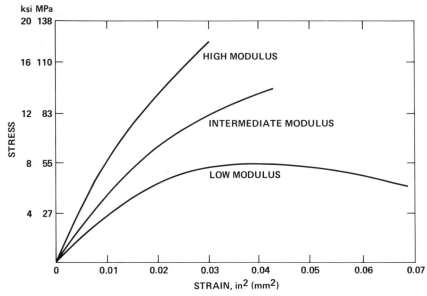

**FIG. 2.42** Stress-strain diagrams for three epoxy matrix materials.[28,29]

Thermoset polyester can be formulated to be brittle and hard, tough and resilient, or soft and flexible. These resins, reinforced with chopped glass, are already used in a wide variety of products ranging from truck cabs to pleasure boats. Thermoplastic resins as matrixes for advanced and hybrid composites are relatively new. Nylon, polysulfone (PS), polyphenylene sulfide (PPS), and polybutylene terephthalate (PBT), better known as *thermoplastic polyester,* are likely matrix systems for hybrids used in automotive and aerospace applications. Polysulfone, though costly, is a rigid, strong plastic that is easily extruded, molded, and thermoformed. It can tolerate long-term exposure to temperatures from 300 to 340°F (149 to 171°C). Polyphenylene sulfide has a melting point of 550°F (288°C), outstanding chemical resistance, thermal stability, and nonflammability. It is not attacked by any known solvents below 375 to 400°F (191 to 204°C), and its mechanical properties are excellent up to 250°F (121°C). High-performance thermoplastic polyesters are extremely easy to process and can be molded in very fast cycles. Reinforced with glass, they are comparable and in some cases superior to some thermosets in electrical, mechanical, dimensional, and creep properties at temperatures of or near 300°F (149°C).

**Fiber-Resin Combinations:** The potential number of fiber-resin combinations for hybrids is vast. With just two resins (epoxy or polyester) and three types of fibers (graphite, Kevlar, or glass) an almost unlimited number of different hybrid composites can be produced, depending on the fiber content and orientation of each fiber in the matrix.[20]

With two or even three different fibers in a common matrix, it is possible to make the most effective use of each fiber. For example, graphite fibers have high tensile strength and high modulus but poor impact resistance, whereas Kevlar fibers have good impact properties but low modulus compared with graphite. When they are combined in a single matrix, the result is a hybrid to which each fiber contributes its best properties (Table 2.26).

**Table 2.26**  Properties of Fiber-Resin Combinations

| Property | Graphite | Kevlar |
|---|:---:|:---:|
| Low density | + | + + |
| Tensile modulus | + + | + |
| Compressive strength | + + | − − |
| Flexural yield strength | + + | − − |
| Flexural ultimate strength | + + | + |
| Fail-safe | − | + |
| Fracture toughness | − − | + + |
| Impact resistance | − − | + + |
| Coefficient of thermal expansion | Compatible | |

Therefore, a hybrid of graphite and Kevlar is a natural combination. A hybrid containing 50% graphite and 50% Kevlar reportedly shows flexural strength about 3 times that of straight Kevlar. Another reason these two fibers are so suitable for hybridization is that their coefficients of thermal expansion are similar; in fact, they are slightly negative. This minimizes internal thermal stresses.

There are also hybrids in which graphite and Kevlar are used to improve the stiffness of glass-fiber-reinforced composites. Unidirectional Kv–Gl hybrids are stiffer than glass alone and may have higher compressive strengths than Kevlar alone; however, the compressive strength of the hybrid is lower.

Another potential combination with several appealing features is graphite and glass in polyester. Table 2.27 shows the results when the fiber content of a polyester matrix is varied from 100% glass and no graphite to 25% glass and 75% graphite. Substituting increasing amounts of graphite results in notable increases of most mechanical properties. Varying the fiber content, as shown in Table 2.27, also is a good way to control the price of the composite. If slightly lower mechanical properties are acceptable, users will pay less for the composite because less graphite and more glass is used. The combination of graphite and glass is an excellent mix. Glass is less expensive but has only one-third to one-sixth the modulus of carbon fibers, depending on the grade of carbon fiber selected. Carbon fibers contribute greater fatigue performance, electrical conductivity, or torsional rigidity.

For the mass-production industries evaluating large-scale application, cost is a critical factor. Hybrids give designers the opportunity to achieve weight saving, enhanced performance, and simpler designs with minimal use of higher-performance reinforcements. Hybrids will become even more attractive as the price of carbon fiber and other high-performance reinforcements drops as markets expand. An important factor that will deter-

**Table 2.27**  Properties of Carbon-Glass-Polyester Composites†

| Carbon-glass ratio | Tensile strength | | Modulus of elasticity in tension | | Flexural strength | | Flexural modulus | | Interlaminar shear strength | | Density | |
|---|---|---|---|---|---|---|---|---|---|---|---|---|
| | ksi | MPa | $10^6$ psi | GPa | ksi | MPa | $10^6$ psi | GPa | ksi | MPa | lb/in$^3$ | kg/m$^3$ |
| 0:100 | 87.7 | 605 | 5.81 | 40 | 137 | 945 | 5.14 | 35.4 | 9.5 | 66 | 0.069 | 1660 |
| 25:75 | 93 | 641 | 9.27 | 63.9 | 154 | 1062 | 9.2 | 63.4 | 10.8 | 74 | 0.067 | 1885 |
| 50:50 | 100 | 690 | 13.0 | 89.6 | 177 | 1220 | 11.4 | 78.6 | 11.0 | 75.8 | 0.065 | 1799 |
| 75:25 | 117 | 807 | 17.9 | 123 | 183 | 1262 | 16.3 | 112 | 12.0 | 83 | 0.060 | 1660 |

†All fiber content by volume; matrix resin 48% thermoset polyester plus 52% continuous unidirectionally oriented fiber by volume, equivalent to 30% resin and 70% glass by weight. Properties apply to longitudinal fiber direction. Based on *Mater. Eng.*, August 1978, p. 37.

**Table 2.28**   Interply Hybrids†

| SGl–Gr–SGl | Gr–Gr–Gr | Gr–Kv |
|---|---|---|
| Gr–B–Gr | SGl–Gl cloth–SGl | Gr–Gl–Gl |
| B–Gr–Gr | Gl cloth–Gl cloth | Kv–Gl–Kv |
| SGl–B–SGl | SGl–Kv–SGl | |
| KV–Gr–Kv | Gr–Gl cloth | |

†Based on *Mater. Eng.*, August 1978, p. 37.

mine the future success of hybrids is rapid processing. The slower, precise tape-lay-up operations used by the aerospace industries are not suitable for high-speed automotive and other mass-production operations. To adapt hybrids to mass-production use, machines and automated equipment must be developed to handle filament winding, sheet molding, and pultrusion.

Many other combinations of fibers and matrix resins are possible (Tables 2.28 and 2.29).

**Hybrid Forms:**   Hybrids are available in several different forms. For filament winding or tape lay-up, the secondary reinforcement may be introduced as intermixed fibers in the same layer or as separate layers oriented to derive maximum benefits from each material. Available commercial fabrics contain a mix of fibers in the wrap and/or fill or carbon in one direction and the secondary fiber in the other (Fig. 2.43).

One of the newest hybrids combines continuous reinforcements, either all carbon or mixed fibers, with a chopped secondary fiber, usually glass. This type of hybrid can be molded very much like SMC, and the continuous reinforcement needs to be positioned only in areas that require stiffening or strengthening. A typical hybrid consists of 30% continuous graphite fiber and 45% chopped glass. Longitudinal and transverse properties compare well with those of the closest all-glass system. Hybrid systems are currently being classified according to applications:

As sandwich structures with glass-reinforced SMC cores overlaid with carbon-fiber unidirectional tapes or woven cloth

As laminated structures with unidirectional plies, each ply reinforced with a different fiber and oriented in different directions

As laminates made of reinforcing cloth woven with different fibers, usually one fiber for the warp and the other for the filling

**Hybrid Processes:**   Hybrids can be fabricated into structures by most of the processes commonly applied to glass-fiber composites. Compression molding, transfer molding, pultrusion, and filament-winding systems can easily be modified to accommodate hybrid materials. Injection-molding compounds with thermoplastic resin matrixes can be processed with chopped hybrid reinforcements (see Chap. 8). Carbon fibers in these compounds add stiffness, fatigue strength, lower coefficient of friction, longer wear life, and electrical conductivity. The fibers are usually short: 0.25 in (6.4 mm) or less. Another process receiving serious consideration for hybrid formulations is reaction-injection molding (RIM), a fast process for molding large, lightweight parts for cars; eventually the system may be applied to trunk lids and doors, where stiffer materials are needed.

Of all these processes, the major interest for volume applications, such as automotive, centers on compression molding. Carbon fibers in chopped or continuous form are com-

**Table 2.29** Properties of Various Hybrid Composites†

| Material‡ | Configuration,§ deg | Tensile failure stress | | Compressive failure stress | | Modulus of elasticity | | | | | | | |
|---|---|---|---|---|---|---|---|---|---|---|---|---|---|
| | | | | | | Tension | | | | Compression | | | |
| | | | | | | Longitudinal | | Transverse | | Longitudinal | | Transverse | |
| | | ksi | MPa | ksi | MPa | $10^6$ psi | GPa | $10^6$ psi | GPa | $10^6$ psi | GPa | $10^6$ psi | GPa |
| Gl-Gr | $0_4/\pm45_2$ | 141.5 | 976 | 96.5 | 655 | 6.9 | 47.5 | 3.1 | 21 | 5.7 | 39 | 3.3 | 23 |
| | $0_4/\pm45$ | 109.0 | 752 | 57.9 | 399 | 2.8 | 19.3 | 1.1 | 7.58 | 5.3 | 37 | 2.7 | 19 |
| Gr-B | $0_4/\pm45_2$ | 157.5 | 1086 | 98.7 | 681 | 21.4 | 148 | 4.4 | 30 | 17.0 | 117 | 2.6 | 18 |
| B-Gr-Gr | $0_3/\pm45/90$ | 124.2 | 856 | 95.0 | 655 | 22.0 | 152 | 8.4 | 58 | 3.5 | 24 | 2.2 | 15 |
| Gl-B | $0_5/\pm45$ | 241.5 | 1665 | 75.1 | 518 | 7.2 | 49.6 | 4.3 | 29.6 | 7.1 | 49 | 3.4 | 23 |
| Kv-Gr-Kv | $0_2/\pm45_2/90$ | 72.0 | 496 | 25.5 | 176 | 7.0 | 48 | 4.0 | 27.6 | 5.7 | 39 | 3.0 | 20.6 |
| Gr-Gr | $0_4/\pm45_3$ | 92.0 | 634 | 89.1 | 614 | 10.8 | 74 | 3.6 | 24.8 | 10.4 | 72 | 3.1 | 21.4 |
| Gr-B | $0_5/\pm45$ | 116.0 | 800 | 90.7 | 625 | 10.8 | 74 | 3.6 | 24.8 | 12.8 | 88 | 2.5 | 17 |

†Based on *Mater. Eng.*, August 1978, p. 37.    ‡Gl is S glass.

§All configurations are symmetrical about the last ply in the sequence. Subscripts denote the number of plies. $0° = 0$ rad, $45 ° = 0.79$ rad, and $90° = 1.6$ rad.

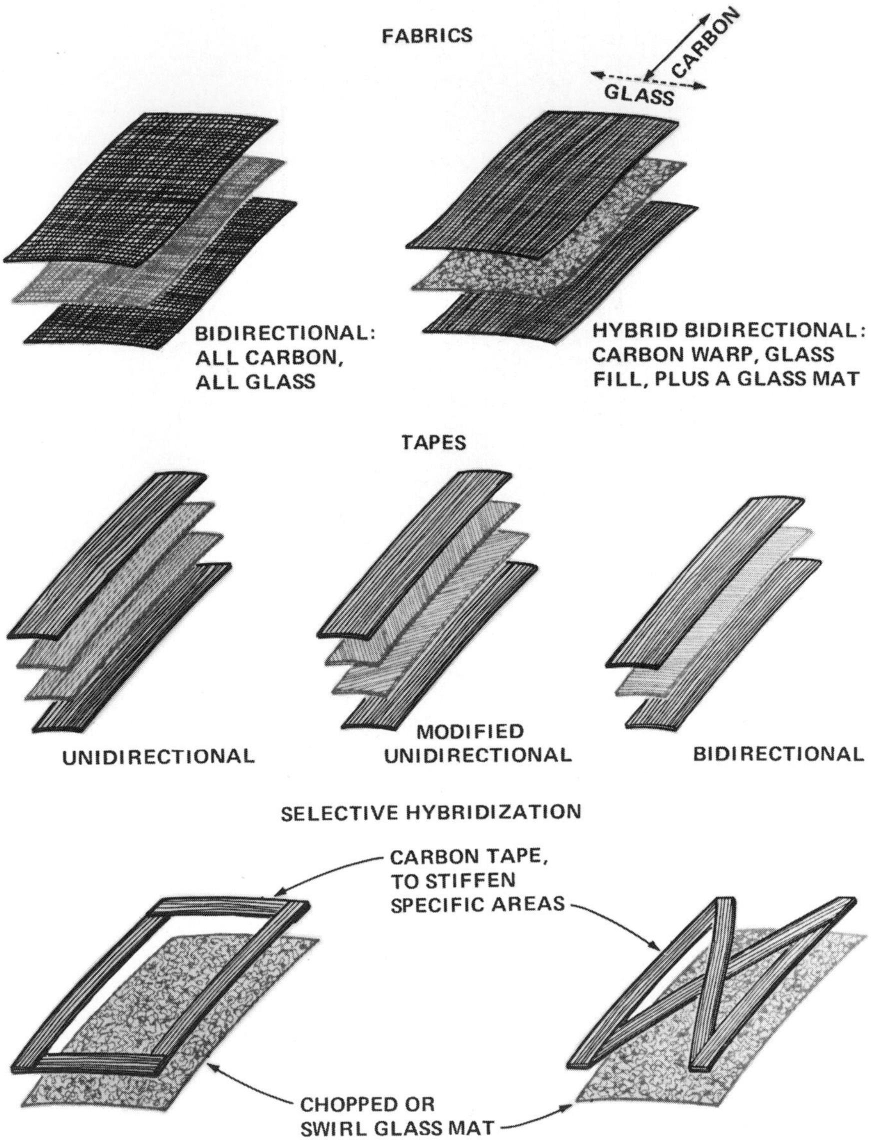

**FIG. 2.43** Several possibilities in tapes, fibers, and selective combinations; carbon-fiber components in black, glass (or other fiber components) in gray. *(Union Carbide Corp.)*

bined with chopped or continuous glass fibers to produce SMC that is readily processed by existing technology. The carbon reinforcement (preferably continuous) adds stiffness, dimensional stability, and sometimes electrical conductivity or electromagnetic shielding. The glass, randomly oriented in most systems, imparts transverse tensile strength, flow control in molding, and lower costs. The newer forms of SMC with all continuous fibers offer strength and flexural properties and can be formulated to produce such structural parts as bumper parts, door intrusion beams, chassis members, and structural supports for cars and trucks.

Engineers have also evolved vacuum-bag, autoclave molding, and filament-winding methods using various orientations of glass, Kevlar, and carbon fibers. Finally the co-curing process may be unique to hybrids since each composite within the hybrid is optimized with respect to a fiber-matrix combination. The optimum cure cycle is a compromise between the cure cycles of the individual composites. Co-curing appears to improve mechanical properties.

**Hybrid Applications:** The special features of hybrids can be used advantageously in countless products ranging from automobiles and aerospace hardware to sporting goods and textile machinery. Several typical hybrid resin-matrix composites and the performances and cost benefits derived from hybridization are listed in Table 2.30.

*Aircraft:* One of the largest hybrid composite structures built to date is the horizontal stabilizer of an advanced supersonic bomber. The structure is a hybrid of aluminum and Gr–B–Ep 30 ft (9.14 m) long with an area of 480 ft$^2$ (44.6 m$^2$). Weight savings over conventional aluminum and titanium construction amount to 15 to 18%. Not only is the structure lighter but it uses only 108 parts compared with 270 in the all-metal version. Fewer parts mean fewer holes to drill, countersink, and inspect and fewer fasteners to join. All these factors help reduce the expense of production worker-hours.

*Automotive:* The automobile industry promises to be the biggest user of hybrids. Experimental parts are already being evaluated, and a hybrid driveshaft may be the first such part to reach production. In this application, graphite is needed primarily for stiffness,

**Table 2.30**  Resin-Matrix Composite Hybrid Combinations[3]

| Fiber combination | Application | Benefit |
|---|---|---|
| B–Gr–Ep | Highly loaded fastened joints; uniaxial boron strips | Increased bearing strength to eliminate splice plates; high stiffness; reduced hole penalty |
| Gr–Gl–Ep, Kv–Gl–Ep | Secondary structure | Reduced net section thickness compared with all glass; lower costs than all-advanced composite |
| Gr–Kv–Ep, Gr–Gl–Ep | Improving damage tolerance of primary structure | Surface plies or crack-stopping strips |
| Kv–quartz–FP fiber–Ep (or PI) | Radomes | Improved electrical and structural properties; polyimide usually permits higher power output |
| Low-cost high-strength ultrahigh-modulus graphite | Optical-support structure | Coefficient of expansion matches that of glass components; low-cost high-strength graphite reduces material cost |
| Kv–FP fiber–Ep | | Tension-compression properties balanced |

so that the fibers are oriented along the axis of the shaft. The glass fibers carry torsional loads. Continuous graphite and glass in a polyester matrix is the hybrid being tried for auto bumpers. Other parts which have demonstrated the effective use of Gr–Ep and Gl–Ep hybrids include engine brackets and leaf springs.

*Sporting Goods:* Sporting goods where hybrids are making inroads include golf clubs, tennis rackets, skis, and fishing rods. In a tennis racket or golf club, for example, torsional rigidity can be provided by high-modulus carbon fibers [$50 \times 10^6$ lb/in$^2$ (344.8 GPa) or above] and tensile strength and toughness improved by adding glass or Kevlar fibers. Gr–Kv–Ep and Gr–B–Ep hybrids are being used in golf clubs and tennis rackets. The Gr–Kv–Ep tennis racket is 2.5 to 50 times stronger in torsion than a racket built using conventional laminating techniques. The Gr–B–Ep golf shaft is more rigid than an all-graphite shaft.

## Metal-Metal Composites

Future composites of different metals or metals and ceramics will be able to take advantage of the desirable properties of the constituents while minimizing their undesirable properties. Metallic composites are composed of a metal matrix reinforced with metallic fibers. There is usually sufficient reaction between the metal fiber and the metal matrix to produce a satisfactory bond. In fact, there is often too much reaction, and keeping it to a minimum is a major fabrication problem. The reaction normally results in the formation of brittle intermetallic compounds, resulting in poor bond strength and absence of mechanical interaction between composite components.

### Match Expansions: Mismatch Moduli

Matching thermal coefficients of expansion is important in selecting components for metal-metal composites. Mismatch problems are less severe in metal-than ceramic-reinforced composites. Mismatch causes internal residual stresses, which can be high where service temperature ranges are broad. The effect is particularly detrimental for continuous filaments. Mismatch stresses fracture filaments and reduce reliability. Whereas thermal-expansion differences must be kept to a minimum in metal-metal composites, the opposite is true of modulus. Here a mismatch is required; otherwise loads are not transferred but are carried by the fiber, negating the advantages of the composite configuration. This is the problem with stainless-steel-reinforced aluminum composites since the modulus of both materials is nearly the same. Wires used to reinforce MMC include tungsten, molybdenum, beryllium, and some superalloys.

### Pros and Cons of Reinforcement Wires

Refractory wires have many advantages as reinforcements and a few disadvantages. Their poor oxidation resistance can be solved by embedding them in an oxidation-resistant matrix, but more critical is the fact that they are embrittled at fairly low temperatures by some elements in nickel-base alloys. This is particularly unfortunate because nickel-base alloys are most attractive for high-temperature applications. The dilemma can be resolved by barrier coatings to restrict interdiffusion between the fiber and matrix.

Tungsten wires have been incorporated in cobalt-base alloy matrixes with some success. For example, 33 vol % of tungsten wires 1 in (25.4 mm) long and 0.005 in (0.13 mm) in diameter in a cobalt-base alloy produced a composite 2.3 times stronger than the matrix material alone at 1600°F (871°C). Other reinforced cobalt-base alloys containing 18 to 30.5% continuous 0.010-in (0.25-mm) tungsten wire is 9 to 17 times stronger than the matrix alone at 2000°F (1093°C). Tungsten-alloy filaments have also been produced in order to create a high-strength oxidation-resistant composite. Filaments of W–3Re in Cb–Ti–Cr–Al and Cb–Ti–Cr–Al–V matrixes have been developed for use in gas turbines at temperatures exceeding 2000°F (1093°C). The composites have good oxidation resistance but must be coated for fail-safe protection. W–Hf–C and W–Re–Hf–C wires show promise as potential fiber reinforcement in the 2000 to 2200°F (1093 to 1204°C) range and are excellent candidates for use in high-temperature turbine blades as reinforcement in superalloys.[31] These composites have high low- and high-cycle fatigue strength. Superalloys reinforced with tungsten offer a 400°F (204°C) increase in blade temperature—up to 10 times the 1000-h rupture strength at 2000°F (1093°C) and more than 3 times the tensile strength of the strongest unreinforced superalloys. The superstrength tungsten fibers (W–Re–Hf–C) can impart 4 times the strength of conventional superalloys.

TZM† molybdenum wires have been incorporated in a number of superalloy matrixes and in tungsten. The Mo–W composite has been used for rocket-nozzle applications because of its good thermal-shock resistance. Nozzles made from the composite have performed well in test firings where temperatures ran to 6500°F (3593°C), pressures to 1 kip/in² (6.89 MPa), and firing durations over 60 s. The composite's good thermal-stock resistance was attributed to increased crack path length. Molybdenum contents of 35% proved optimum. Incorporating TZM molybdenum wire in nickel-base matrixes has produced good results.

Another promising wire is beryllium. Although cost, brittleness, and toxicity have limited its broader use, there is hope that beryllium fibers in a Ti–6Al–4V matrix will result in a structural, cost-effective material.

The major effect of beryllium fibers on the properties of Ti–6Al–4V is to increase its specific modulus by a simultaneous increase in elastic modulus and a decrease in density. A 50 vol % Be–Ti composite shows a modulus improvement of from $16 \times 10^6$ lb/in² (110.3 GPa) for Ti–6Al–4V to 26 to $28 \times 10^6$ lb/in² (179.3 to 193.1 GPa) for the composite. Density drops from 0.16 to 0.12 lb/in³ (4429 to 3322 kg/m³). Since the beryllium is not the high-strength component of the Be–Ti system, the composite shows some decrease in ultimate and yield strength. Decrease in strength level depends on the form and strength of the beryllium used, fine-wire composites showing the least strength loss.

One production method consists of drilling a solid Ti–6Al–4V billet with a hole pattern calculated to produce a composite with the desired volume percent. Beryllium rods are inserted and the assembly is canned in mild steel, evacuated, sealed, and hot-extruded to the final specified shape (Fig. 2.44). This method can be used to produce sheet or round shapes.[32] A shape for forging or rolling can also be produced in Be–Ti sheets; the beryllium flattens during extrusion.[33]

Well-bonded composite sheet material has been made in 0.30-in (7.6-mm) thickness by hot rolling extruded flat material. The beryllium fibers in 0.030-in (0.76-mm) Be–Ti sheet are about 0.030 to 0.040 in (0.76 to 1 mm) thick and 0.4 in (10 mm) wide. The Be–Ti interface appears to be well bonded; the intermetallic compound $TiBe_2$ probably forms. Mechanical properties of well-bonded coextruded and rolled sheet material of 37.9 vol % are 115.9 kips/in² (799 MPa) ultimate tensile strength, 72.5 kips/in² (500 MPa)

†Trade name.

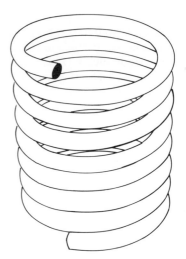

**FIG. 2.44** Ti–6Al–4V extrusion billet showing drilling pattern and beryllium rod before extrusion.[33]

**FIG. 2.45** Potential application of a Be–Ti composite coil spring.[33]

yield strength, and 0.87% elongation at failure (Table 2.31). Applications include that shown in Fig. 2.45 and third-stage compressor blades for a jet engine.

The fibrous nature of Be–Ti composites gives them unusual resistance to crack propagation, resulting in good fracture toughness and impact resistance. As a result of their high beryllium content, Be–Ti composites have untypically high specific-heat values. Thermal conductivity is higher than that of monolithic materials such as steel and titanium. These physical properties suggest applications where a lightweight structurally rugged heat sink is needed.

Beryllium wires have also been produced in aluminum matrixes by several methods. By vacuum hot-pressing of 0.003-in (0.08-mm) aluminum foil and 0.0043-in (0.11-mm) beryllium wire a composite having an overall wire content of 43 vol % has been produced. With the filaments in a unidirectional orientation ultimate tensile strengths of 48.2 kips/in² (332 MPa), 0.2% offset yield strength of 40.7 kips/in² (281 MPa), elongation of 6.6%, and a modulus of 2.9 × 10⁶ lb/in² (20 GPa) have been achieved. Another method (Fig. 2.46) involves coating beryllium wire with a slurry of aluminum powder and a clean-burning resin in several organic solvents. The wet wire is wound onto a mandrel to form a mat, which is then dried, removed from the mandrel, and hot-pressed. Elastic moduli measured after binder removal and compaction show reasonably good agreement with values predicted by the rule of mixtures.

Steels and nickel-base alloys have also been made into wires and subsequently introduced into metal matrixes. Yarns of Chromel-R (74Ni–20Cr–3Al–3Fe) have been manufactured. Yarns are considered more economical and exhibit mechanical properties equivalent to yarns formed by bundling singly drawn fibers. Chromel-R 0.002-in (0.05-mm) diameter fibers have been successfully reduced to 0.005 in (0.01 mm).

In superconducting magnets copper is used to prevent the columbium-tin from losing its superconductivity with sudden current changes. In producing superconducting magnets from Cb–Sn it is necessary to produce miles of wire containing filaments of the Cb–Sn

**Table 2.31** Mechanical Properties of Be–Ti Composites and Ti–6Al–4V[33]

| Density | | Modulus | | | | Shear strength | | Temp | | Ultimate tensile strength | | Elongation, % |
|---|---|---|---|---|---|---|---|---|---|---|---|---|
| | | Tension | | Torsion | | | | | | | | |
| lb/in³ | kg/m³ | 10⁶ psi | GPa | 10⁶ psi | GPa | ksi | MPa | °F | °C | ksi | MPa | |
| Be–Ti, 60 vol % | | | | | | | | | | | | |
| 0.103 | 2851 | 31.2 | 215 | 14.3 | 98.6 | 67 | 462 | 68 | 20 | 123 | 848 | 2.7 |
| | | | | | | | | 300 | 149 | 136† | 938† | 15 |
| | | | | | | | | 600 | 316 | 110 | 758 | 13 |
| Be–Ti, 40 vol % | | | | | | | | | | | | |
| 0.123 | 3405 | 26.3 | 181 | 11.3 | 77.9 | 80 | 552 | 68 | 20 | 163 | 1124 | 5.3 |
| | | | | | | | | 300 | 149 | 138 | 952 | 14 |
| | | | | | | | | 600 | 316 | 123 | 848 | 13 |
| Ti–6Al–4V | | | | | | | | | | | | |
| 0.161 | 4456 | 16 | 110 | 6.2 | 42.7 | 79 | 545 | 68 | 20 | 160 | 1103 | 12 |
| | | | | | | | | 300 | 149 | 134 | 924 | 12 |
| | | | | | | | | 600 | 316 | 118 | 814 | 12 |

†Superior to that at room temperature because of greater elongation.

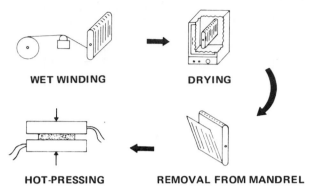

WET WINDING          DRYING

HOT-PRESSING      REMOVAL FROM MANDREL

**FIG. 2.46** Process for fabricating fiber-reinforced metal composite with beryllium wire in aluminum slurry with resin.

embedded in a copper matrix. These filaments must be very long and thin and homogeneously spaced. The composite wire must contain about 70% copper and 30% Cb–Sn. In the process (Fig. 2.47) an arc is struck whereby copper and columbium are melted; the alloy drips off in a puddle and solidifies in uniform dispersion of randomly oriented columbium filaments, called *dendrites,* in a copper matrix. Drawing the alloy through a series of dies of decreasing size aligns these filaments along the wire axis and reduces them in size. The cast Cu–Cb alloys are so ductile that they can be drawn from a diameter of 0.5 to 0.004 in (12.7 to 0.1 mm) without even heating the material. The tin is introduced as shown in Fig. 2.47, and the alloy is extruded into a hollow cylindrical shape; tin is poured in, and then the wire is drawn.

Wire produced in this way is called *in situ wire* because the columbium is introduced right on the spot by casting. Recent experiments with in situ wires show that they carry electric currents as high or higher than the commercial bronze wires. In addition, their ability to be coiled without breakage of the Cb–Sn is slightly better than that of commercial wire. This new wire is so flexible that it can be bent around one's finger without affecting its performance. The wire is marketed in diameters ranging from 0.015 to 0.078 in (0.4 mm to 2.0 mm), capable of carrying currents ranging from 90 to 1500 A, respectively, in a field of 50,000 G (5 T) (Fig. 2.48).

## Alternatives to MMC

Outstanding high-temperature properties can be achieved with directionally solidified eutectics, natural composites consisting of high-strength fibers or plates in a ductile matrix. Eutectics (natural composites in that fiber reinforcement of the matrix is governed by physical laws) are thermodynamically in equilibrium even at the melting point. Thus, they offer a unique solution to the problem of fiber-matrix incompatibility. An objection to the use of eutectics for turbine blades might seem to be their relatively low melting point, but to avoid the oxidation problem inherent in refractory metals and their high cost, an iron, nickel, or cobalt-base matrix is desirable. Since the eutectic must melt at a lower temperature than the matrix, its melting point would be about 2500°F (1371°C) at best—not much better than that of superalloys. Fortunately, however, the trend in high-pressure turbine blade design is to higher stresses and lower allowable temperatures. This is the result

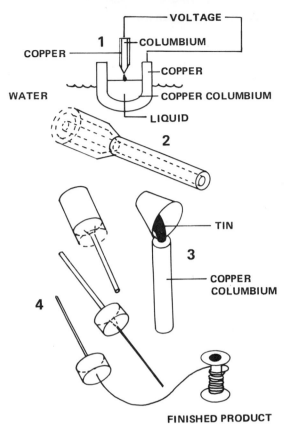

FIG. 2.47 In situ process. Step 1: Cu–Cb ingot produced by consumable arc casting; step 2: extrusion to hollow cylinder; step 3: filling with tin; step 4: drawing to wire, requiring no anneals.

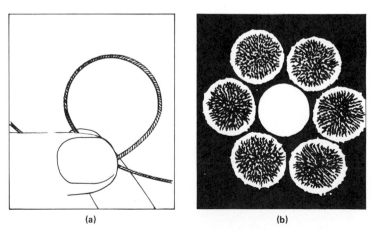

(a)  (b)

FIG. 2.48 (a) Flexible Cu–Sn superconductor and (b) cross section of cable, ×270. Cable shows how six copper wires, each containing hundreds of Cb–Sn filaments, are wrapped around a stainless-steel wire. (General Electric Co.)

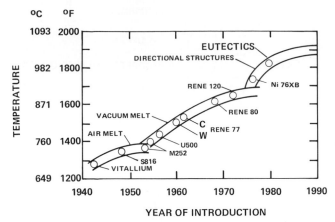

**FIG. 2.49** Progress in turbine-blade materials and the promise of DS eutectics; strength at 30 kips/in² (207 MPa) after 5000 h.[34]

of higher centrifugal stresses associated with design trends to higher turbine tip speeds and superimposed thermal stresses arising from complex air-cooling techniques. The conditions shown in Fig. 2.49 represent limiting stresses in current turbine blades for current alloys, that is, 30 kips/in² (207 MPa) for 5000 h and 1600 to 1700°F (871 to 927°C). Thus, there is room for substantial temperature improvement without concern over the melting point limit of eutectics.[34]

## Directionally Solidified (DS) Eutectics

The fibrous or composite morphology of eutectic systems is achieved by directionally solidifying a homogeneous melt of the eutectic under appropriate solidification rates and temperature gradients. Both lamellar and rodlike structures have been produced and studied. First-generation materials are NiTaC-13 and $\gamma\gamma'\delta$, both of which are blade alloys. Second-generation materials are $\gamma\gamma'\alpha$ and NiTaC3-116A for blades and $\gamma\beta$ and COTAC-74 for vanes. The compositions and properties of these DS eutectics are given in Tables 2.32, 2.33.

**Table 2.32** Composition of DS Eutectics†

|    | $\gamma\gamma'\delta$ | NiTaC-13 | $\gamma\gamma'\alpha$ | NiTaC3-116A | $\gamma\beta$ | COTAC-74 |
|----|------|------|------|------|------|------|
| Ni | bal | bal | bal | bal | bal | bal |
| Co | — | 3.3 | — | 3.7 | 10.0 | 20.0 |
| Cr | 6.0 | 4.4 | — | 1.9 | — | 10.0 |
| Al | 2.5 | 5.4 | 6.0 | 6.5 | 11.0 | 4.0 |
| Cb | 20.1 | — | — | — | — | 4.9 |
| Ta | — | 8.1 | — | 8.2 | — | — |
| W | — | 3.1 | — | — | 8.0 | 10.0 |
| Re | — | 6.2 | — | 6.3 | — | — |
| V | — | 5.6 | — | 4.2 | — | — |
| Mo | — | — | 32.0 | — | — | — |
| C | 0.06 | 0.48 | — | 0.24 | — | 0.6 |

†$\alpha$ = alpha, $\beta$ = beta, $\gamma$ = gamma, $\delta$ = delta. Table based on *Mater. Eng.*, December 1977, p. 22.

**Table 2.33** Properties of DS Eutectics†

| Eutectic† | Density | | mp | | Potential service temp‡ | | Strength§ | | Transverse tensile elongation,¶ % | Resistance to oxidation and heat corrosion |
|---|---|---|---|---|---|---|---|---|---|---|
| | lb/in³ | kg/m³ | °F | °C | °F | °C | ksi | MPa | | |
| γγ'δ | 0.310 | 8580 | 2320 | 1271 | 1832 | 1000 | 25 | 172 | 0.5 | Poor |
| NiTaC-13 | 0.315 | 8719 | 2640 | 1349 | 1841 | 1005 | ... | ... | 3 | Poor |
| γγ'α | 0.307 | 8498 | 2390 | 1310 | 1859 | 1015 | 31.9 | 220 | 5 | Good |
| NiTaC3-116A | 0.310 | 8580 | ..... | ..... | 1886 | 1030 | ... | ... | ... | Good |
| γβ | 0.29 | 8027 | 2500 | 1371 | .... | .... | 24.6 | 170 | <1 | Excellent |
| COTAC-74 | 0.310 | 8580 | 2425 | 1329 | .... | .... | 49.3 | 340 | 2 | Good |

†Based on *Mater. Eng.*, December 1977, p. 22.
‡Base-line service temperature is 1760°F (960°C) for DS MAR-M-200 + HF.
§Shear-rupture strength at 1400°F (760°C) for 100 h.
¶At 1400°F (760°C).

## CONCLUSION

Developments in high-modulus fibers are occurring rapidly as the state of the art advances, but it is almost certain that improvements in existing filamentary materials will lead to a certain amount of jockeying for position. At present no filament for advanced composites can be said to have the future to itself. Moreover, it is almost as certain that wholly new filaments will become added starters in the race. The insatiable demand from aerospace, missile, and deep-submergence fields will push the development of existing high-modulus filaments and spawn new ones. It is likely, too, that most if not all the filaments will eventually find employment somewhere in the spectrum of needed properties and performance. Problems being overcome include:

Adequate techniques for joining

Nondestructive testing

Optimized deposition parameters

Improved reproducibility of filament material properties from batch to batch and even within a batch

Tailor-made materials with selected combinations of properties by combining fibers

## REFERENCES

1. Morin, D.: Boron-Carbide-Coated Boron Filament as Reinforcement in Aluminum Alloy Matrices, *J. Less Common Met.*, 47(1):207–213 (June 1976).

2. Prewo, K. M., and K. G. Kreider: High-Strength Boron and Borsic Fiber Reinforced Aluminum Composites, *J. Compos. Mater.*, 6:338–357 (July 1972).

3. "Advanced Composite Design Guide," 3d ed., Air Force Materials Laboratory, Wright-Patterson AFB, Ohio, 1973.

4. Stuckey, J. M., and W. G. Scheck: Development of Graphite-Polyimide Composites, *SAMPE Tech. Conf., Huntsville, Ala., Oct. 5–7, 1971*, pp. 747–754.

5. Mace, M.: Graphite-Polyimide Processing, *SAMPE Int. Conf., Cannes, Jan. 13, 1981.*

6. Coggin, C. H.: New-Processes for Glass Fibers, *Reinf. Plast. Conf., El Segundo, Calif., Dec. 1980.*

7. *Mat. Eng.*, January 1980, pp. 26–66.

8. Dwyer, John J., Jr.: Composites, *Am. Mach.*, July 13, 1979, pp. 87–96.

9. Mayfield, J.: Carbon Fiber Replacements Developing, *Aviat. Week Space Technol.*, Jan. 22, 1979, pp. 49–52.

10. Hill, S. G., E. E. House, and J. T. Hoggatt: Advanced Thermoplastic Composite Development, Boeing Co., Seattle, Wash., Contr. N00019-17-C-0561, for Naval Air Systems Command, Washington, May 1979.

11. Hoggatt, J. T., S. G. Hill, and E. E. House: Environmental Exposure on Thermoplastic Adhesives, Boeing Aerospace Co., D180-24744-1, Seattle, Wash., December 1978.

12. Curley, R. C., P. W. Harruff, H. S. Parechanian, and P. R. Scherer: Manufacturing Methods for Reentry Vehicle—Advanced Composite Structure, III: Graphite Polyimide Substructure, *McDonnell Douglas Astronautics Co. Final Rep.* F33615-76-C-5013, *June 1978–August 1979*, Huntington Beach, Calif. AFML-TR-79-4094.

13. Bales, T. T., E. L. Hoffman, L. Payne, and L. F. Reardon: Fabrication Development and Evaluation of Advanced Titanium and Composite Structural Panels, *NASA Tech. Pap.* 1616, March 1980.

14. Winters, W. E., and P. J. Cavano: PMR-II Polyimide Composites, *10th Nat. SAMPE Meet., Kiamesha Lake, N.Y., Oct. 17–19, 1978,* pp. 661–671.

15. Scola, D. A.: Imide Modified Epoxy Matrix Resins, *UTRC 7th Mon. Rep.,* Contr. NAS 3-22032, NASA Lewis Research Center, Cleveland, Ohio, Mar. 20, 1980.

16. Pepper, R. T., and F. Bucherati: Development of Aluminum Graphite Shapes, *Fiber Materials Final Rep.* AMMRC CTR 74-7, Contr. DAAG 46-73-C-0221, Biddeford, Me, January 1974.

17. Gigerenzer, H., and G. C. Strempek: Fabrication of Graphite-Aluminum Composites via Pultrusion, *Fiber Materials Final Rep.* AMMRC-TR-78-16, Contr. DAAG-46-77-C-0036, Biddeford, Me., March 1978.

18. Armstrong, H. H., A. M. Ellison, D. H. Kintis, H. Shimizu, and D. E. Kizer: Development of Graphite/Metal Advanced Composites for Spacecraft Applications, *Lockheed Missiles and Space Co., Interim Rep.* LMSC-D671326, Contr. F33615-78-C-5235, Sunnyvale, Calif. Apr. 1, 1979.

19. "Fiber Composite Materials," American Society for Metals, Metals Park, Ohio, 1965.

20. Kelly, A., and G. J. Davies: The Principles of the Fibre Reinforcement of Metals, *Metall. Rev.* 10(37):1–77 (1965).

21. Engler, E., and R. Hadcock: Development of Boron-Epoxy Tubular Struts for a One-Third Scale Shuttle Booster Thrust Structure, *SAMPE Tech. Conf. Huntsville, Ala., Oct. 5–7, 1971,* pp. 83–98.

22. Dolowy, J. F., Jr.: Diffusion Bonded B–Al Sheet and Plate, *Amercom, Inc., Interim Tech. Rep.* IR-364-i(II), Contr. F33615-72-C-1329, Air Force Materials Laboratory, Wright-Patterson AFB, Ohio, October 1972.

23. Turpin, M., and A. Robert: Structure of SiC Ribbons and Continuous Fibers Grown by Chemical Vapor Deposition, *Proc. Br. Ceram. Soc., 22:*337–353 (1973).

24. DeBolt, H., and V. Krukonis: Improvement of Manufacturing Methods for the Production of Low-Cost Silicon Carbide Filament, *AVCO Corp. Final Rep. April 1972–March 1973,* AFML TR-73-140, Contr. F33615-72-C-1177, September 1973.

25. Prewo, K. M.: Fabrication and Evaluation of Low-Cost Alumina Fiber Reinforced Metal Matrices, *United Technologies Res. Cent. Final Rep.* R79-912245-6, Contr. N00014-76-C-0035, E. Hartford, Conn., November 1978.

26. Brentnall, W., and I. Toth: High Temperature Titanium Composites, TRW, Inc., Cleveland, 1974.

27. Jech, R. W., and R. A. Signorelli: Evaluation of Silicon Carbide Fiber/Titanium Composites, *NASA Lewis Res. Cent. Rep.* TM-79232, Cleveland, Ohio, July 1979, N79-31349.

28. Chamis, C. C., and R. F. Lark: Hybrid Composites: State of the Art Review; Analysis, Design, Application and Fabrication, *NASA Lewis Res. Cent., Rep.* TM X-73545, Cleveland, Ohio, 1977.

29. Chamis, C. C., R. F., Lark, and T. L. Sullivan: Super-Hybrid Composites: An Emerging Structural Materials, *NASA Lewis Res. Cent., Rep.* TM X-71836, Cleveland, Ohio, 1977.

30. Zweben, C., and J. C. Norman: Hybrid Composites for Aerospace Applications, *SAMPE Q.,* July 1976.

31. Petrasek, D. W., High Temperature Strength of Refractory Metal Wires and Consideration for Composite Applications, *NASA Rep.* TN D-6881.

32. London, G., W. Taylor, and M. Herman: Co-extruded Beryllium-Titanium Alloy Composites for Light Weight Structures, Kawecki Berylco and Detroit Diesel Allison Division of General Motors, 1972.

33. Goodwin, V. L.: Beryllium/Titanium Composites, Brush Wellman, Inc., Cleveland, 1972.

34. Jahnke, L. P., H. J. Brands, and G. D. Oxx, Jr.: Directionally Solidified Eutectics in Gas Turbine Design, *AGARD, NATO, Toulouse, Sept. 22, 1972.*

35. Fiber-Reinforced Metal-Matrix Composites: Government-Sponsored Research, 1964–1966, *Def. Mater. Inf. Cent. Rep.* 241, Sept. 1, 1967.

36. *Tech. Note* NTN-77/1020, Rep. WVT-TR-74037, September 1977, AD-784502.

37. Lubin, G.: "Handbook of Advanced Composites," 2d ed., Van Nostrand Reinhold, New York, 1982.
38. Monteleone, R., B. Kay, and J. Ray: Airframe Preliminary Design for an Advanced Composite Design Airframe Program (ACAP), Sikorsky Aircraft, Division of United Technologies, Stratford, Conn., Q. Rep. 1–3, July 1981, and October 1981, January 1982, DAAK51-81-C-0017, Army Technology Laboratory, Ft. Eustis, Va.
39. *Iron Age*, July 9, 1979, p. 52.

## BIBLIOGRAPHY

Agarwal, B. D., and L. J. Broutman: "Analysis and Performance of Fiber Composites," Society of Plastics Engineers, Brookfield Center, Conn., 1980.
Ahmad, I., et al.: Metal Matrix Composites for High Temperature Applications, *Watervliet Arsenal* WVT-7266, AD-756867, 1972.
Amateau, M. F., W. C. Harrigan, and E. G. Kendall: Metal Matrix–Graphite Composites, *Aerospace Corp. Summ. Rep.* ATR-74 (8162)-4, April 1974.
Bulloch, D. F., and A. L. Dobyns: Development of Fabrication Techniques for Borsic Aluminum Aircraft Structures, *Boeing Co. Final Rep., June 1970,* AFML-TR-72-25, Contr. F33615-70-C-1541, Air Force Materials Laboratory, Wright-Patterson AFB, Ohio, January 1972.
Bunsell, A. R., and B. Harris: Hybrid Carbon/Glass Fibre Composites, *1st Int. Conf. Composite Mater. Metall. Soc. AIME, Geneva and Boston, Apr. 1975,* vol. 1, no. 2, pp. 174–190.
Carlson, C. E. K.: Diffusion Bonded Boron Aluminum Spar Shell Fan Blade, United Technologies-Hamilton FR 6/77-5/78, June 1980, N8025382, NASA CR 159571.
Carlson, R. G.: Metal Matrix Composite Bonding Characteristics and Impact Properties, General Electric Co., FR 3/77-2/80, F49620-77C0067, April 1980, R80AEG345, AFOSR TR80-045.
Champion, A. F., and H. K. Street: Fabrication of Boron-Reinforced Magnesium Composites by Diffusion Bonding of Plasma-Sprayed Monolayer Tapes, *Sandia Lab. Summ. Rep.* SC-DR-72-677, September 1972.
Crane, R. L., and V. J. Krukonis: Strength and Fracture Properties of Silicon Carbide Filament, *Am. Ceram. Soc. Bull,* 54:184–188 (February 1975).
Didchenko, R.: Graphite Fibers from Pitch, *Union Carbide Corp. Final Tech. Rep. April 1975–May 1976,* AFML-TR-73-147-Part 4, Contr. F33615-75-C-5109, September 1976.
Ezekiel, H. M.: Formation of Very High Modulus Graphite Fibers from a Commercial Polyacrylonitrile Yarn, *Air Force Mater. Lab. Final Rep. December 1966–May 1968,* AFML TR-72-210, November 1972.
Fitzer, E., and B. Terwiesch: Carbon-Carbon Composites Unidirectionally Reinforced with Carbon and Graphite Fibers, *Carbon,* 10(4):383–390 (August 1972).
Forsyth, R. B.: New Low Cost "Thornel" Fibers from Pitch Precursor, *19th Natl. SAMPE Symp., Bueno Park, Calif., Apr. 23–25, 1974.*
Frost, L. W.: Silicone Modified Resins for Graphite Fiber Laminates, *Westinghouse Electric Corp. Final Rep.* 9/79, N8022407, NASA CR159750.
Gell, M., and M. J. Donachie: Directionally Solidified Eutectics for Use in Advanced Gas Turbine Engines, *"SAMPE: Materials on the Move,"* 6th Natl. SAMPE Tech. Conf. Dayton, Ohio, Oct. 8–10, 1974.
Gigerenzer, H., et al.: Hot Drawing of Fiber (Filament) Reinforced Metal-Matrix Composites, *Proc. 1978 Int. Conf. Composite Mater., Toronto, Apr. 16–20, 1978.*
Holister, G. S., and C. Thomas: "Fiber Reinforced Materials," Elsevier, New York, 1966.
Hoover, W. R.: Graphite/Aluminum: An Evaluation of State-of-the-Art Material, *J. Compos. Mater.* 11:17–29 (January 1977).
Imprescia, R. J., L. S. Levinson, R. D. Reiswig, T. C. Wallace, and J. M. Williams: Carbide-Coated Fibers in Graphite-Aluminum Composites, *Los Alamos Sci. Lab. Prog. Rep.* 4, *January–June 1975,* NASA CR-2711, July 1976.

Joseph, E.: Boron-Aluminum Titanium Hybrid Composites, *Air Force Mater. Lab. Final Rep.,* AFML TR-74-236, May 1975.

Kendall, E. G., M. F. Amateau, D. M. Goddard, and W. H. Harrigan: The Development of Lead-Glass Composites, *Aerospace Corp. Summ. Rep.* ATR-74(8162)-2, April 1974.

Klein, M. J., and A. G. Metcalfe: Tungsten Fiber Reinforced Oxidation Resistant Columbium Alloys, *International Harvester Co., Solar Div. 3d Q. Prog. Rep.,* Contr. N00019-72-C-0230, NASC, October 1972.

Lin, R. Y., J. Economy, and H. N. Murty: Exploratory Development on Formation of High-Strength, High-Modulus Boron Nitride Continuous Filament Yarns, *Carborundum Co. Final Tech. Rep. March 1971–February 1972,* AFML TR-72-72, Contr. F33615-71-C-1377, June 1972.

London, G. J., W. Taylor, and M. Herman: Co-extruded Beryllium-Titanium Alloy Composites for Lightweight Structures, *SAMPE Tech. Conf., Huntsville, Ala., Oct. 5–7, 1971,* pp. 69–80.

Lovell, D. R.: Hybrid Laminates of Glass/Carbon Fibres, 1, *Reinf. Plast.,* 22(7):216–221 (July 1978).

————: Hybrid Laminates of Glass/Carbon Fibres, 2, *Reinf. Plast.,* 22(8):252–256 (August 1978).

May, G. J.: The Fatigue Behavior of an Aligned Ni–Al–Cr–C Eutectic Alloy, *Metall. Trans.,* 6A(5):115–117 (May 1975).

McGann, T. W.: Chemical Composition and Processing Specifications for Air Force Advanced Composite Matrix Materials, *Rockwell International–Air Force Mater. Lab. Final Rep.* TR-79-4180, *September 1977–August 1979,* Contr. F33615-77-C-5243.

McMahon, P. E., and F. S. Campbell: High-Strength, High-Modulus Graphite Fiber, *Celanese Research Co.–Air Force Mater. Lab. Tech. Rep.* TR-66-161, pt. VII, Contr. F33615-71-C-1537, November 1972.

Morley, J. G.: Fibre Reinforcement of Metals and Alloys, *Int. Met. Rev.* 21:153–170 (September 1976).

Pepper, R. T., et al.: Review of the Processing and Properties of PAN Graphite/Aluminum Composites, T. R. Shives and W. A. Willard (eds.), pp. 93–107 in *"Engineering Design,"* NBS Spec. Publ. 487, 1977.

Phillips, L. N., and D. J. Murphy: Carbon Fibre Reinforced Plastics: The New Technology, *RAE Tech. Mem.* MAT 273, April 1977.

Rabinovitch, M., et al.: Applications of DS Composites in Aircraft Gas Turbines, pp. 289–317 in "Advances in Composite Materials," Applied Science Publishers, London, 1978.

Reece, O. Y.: Molybdenum Wire Reinforced FS-85 Columbium, *SAMPE Tech. Conf., Huntsville, Ala., Oct. 5–7, 1971,* pp. 81–83.

Renton, W. J.: "Hybrid and Select Metal Matrix Composites: A State of the Art Review," American Institute of Aeronautics and Astronautics, New York, 1977.

Salkind, M. J.: Fibre Composite Structures, *Proc. Int. Conf. Compos. Mater., Geneva, April 1975,* vol. 1, no. 2, pp. 5–30, Metallurgical Society of AIME, New York, 1976.

Sheldon, D. Q., and D. Lewis: Fabrication and Property of a Ceramic Fiber–Ceramic Matrix Composite, *J. Am. Ceram. Soc.,* 59(7–8):372–374 (July–August 1976).

Shibata, N., A. Nishimura, and T. Norita: Graphite Fiber's Fabric Design and Composite Properties, *SAMPE Q.,* 7(4):25–33 (July 1976).

Sullivan, P. G., and L. Raymond: Graphite Fiber Morphology: Effect on Behavior in Graphite-Aluminum Composites, *SAMPE 10th Natl. Meet., Kiamesha Lake, N.Y., Oct. 17–19, 1978,* pp. 466–474.

Summerscales, J., and D. Short: Carbon Fibre and Glass Fiber Hybrid Reinforced Plastics, *Composites,* 9(3):157–166 (July 1978).

Thompson, E. R., and F. D. Lemkey: Directionally Solidified Eutectic Superalloys, "Composite Materials," vol. 4, "Metallic Matrix Composites," pp. 102–156, Academic, New York, 1974.

Tsai, S. W., and H. T. Hahn: "Introduction to Composite Materials," Technomic, Westport, Conn., 1980.

Walrath, D. E., and D. F. Adams: Development of a Graphite/Aluminum/Epoxy Hybrid Composite, *J. Compos. Mater.,* 10:44–54 (January 1976).

A Gr-Ep fitting.   *(Hercules, Inc.)*

# Composite Designs and Joint Criteria, Properties, and Quality Assurance

Composite structures open up new dimensions of design freedom not available with isotropic materials. Modulus, strength, and thickness can be varied freely within a single component; parts can be "softened" locally so that loads are transferred from highly stressed areas; and large, one-piece structures can be designed without the need for a multitude of fasteners and parts. Design considerations for composites are entirely different from those for homogeneous, isotropic materials having well-defined elastic and plastic stress-strain behavior. For structural parts, all failure modes must be investigated, and particular attention must be paid to interlaminar tensile and shear stresses because strengths in these modes depend principally on the matrix resin, not on the reinforcing fiber.

Components made from carbon-resin or C–Gl resin combinations, often called *advanced composites,* are usually of laminated construction. They can be essentially isotropic, quasi-isotropic, or very anisotropic, depending on material form, lay-up configuration, and fabrication method. Most parts are designed to be anisotropic to exploit fully the highly directional properties of the fibers.

Carbon, glass, or Kevlar fiber composites are becoming well known as the ultimate stiffness and/or weight performers. These materials offer new dimensions in design freedom, but along with the advantages come special problems for the designer, who is suddenly confronted with a new set of rules. Elastic modulus, for instance, becomes a design variable, along with strength, coefficient of thermal expansion, and material cost.

# MATERIALS

## Properties

### Stiffness

A major advantage of using various forms of reinforcing fibers in designing structural parts is that these composite systems can be tailored to suit specific needs. Stiffness, for example, can be varied significantly in different areas of a composite part by selecting the type and form of fiber, by judicious orientation, and by controlling local concentrations of fibers.

### Ductility

Like most rigid materials, carbon-reinforced plastics are relatively brittle because carbon fibers are brittle. This characteristic requires particularly close attention to cutouts and areas involving fasteners, where stress concentrations can cause failure. Carbon-fiber composites have no yield behavior, and resistance to impact is low.

### Conductivity

Incorporating carbon fibers into plastic resins makes the compounds conductive, both electrically and thermally. Electrical conductivity provides the benefits of static drain and radio-frequency suppression in such applications as electronic-equipment enclosures, automotive ignition-system devices, and small-appliance housings. This property is also an advantage in providing electrostatic paintability. Thermal conductivity provides the benefits of heat dissipation in such components as gears, bearings, brake pads, and other friction-related products and in cryogenic processing equipment.

### Thermal Expansion

The coefficient of thermal expansion of a composite depends not only on the orientation of the fibers but also on the thermal behavior of the matrix material. Thus, care must be exercised to ensure that volumetric expansion of the composite in the (usually) unrestrained direction does not exceed a tolerable limit. If such expansion is troublesome, changing to randomly oriented chopped fibers may be an acceptable solution. Since matrix materials expand more than carbon fibers, thermally induced stresses are easily set up in a composite. Because matrix resins have larger elastic strain regions than the carbon fibers and are relatively ductile, they absorb the thermally induced strains. As long as the fiber-matrix interfacial bond strength is not exceeded, the properties of the composite will not be affected.

Service-temperature limitations of a composite part depend entirely on the matrix system. For high-temperature epoxies, the service-temperature range can cover several hundred degrees, to a maximum of about 350°F (177°C). The design guidelines described later in this chapter apply principally to high-performance carbon-fiber structures, but they also can be used to design with the mixed-fiber hybrids and the lower-cost glass composites. Since the applications of these materials are structural, if stress levels beyond 30 kips/in² (207 MPa) and elastic moduli in the range of 5 to 30 $\times$ 10⁶ lb/in² (34.5 to 207 GPa) are expected, some of the reinforcing fiber must be in continuous form (see Table 3.1).

## Forms

Designing a part requires, first, an understanding of the available material forms and fabrication methods. Although structural composites can take many forms, they all consist of long fibers in a resinous matrix, typically epoxy or polyester. Short, random fibers may also be used with the long fibers, usually as a core filler or as an impact-absorbing surface layer.

The material form is usually associated with a specific fabrication process. Forms and processes can be grouped into those involving mixing dry fiber and resin at the point of part formation (such as filament winding, pultrusion, or open-mold spray-up) and those requiring an intermediate prepreg or sheet-molding form, such as compression or autoclave molding.

### Conventional Prepreg

Thin sheets of uni- or bidirectional (fabric) fiber, impregnated with B-staged resin, can be laid up in plies and cured into typical aerospace composites. These multiple-ply laminates can be cured into complex shapes using vacuum-bag, autoclave, rubber-tooling, or compression-molding techniques. Each ply is about 0.006 in (0.15 mm) thick, and fiber direction, type (high-strength or high-modulus carbon, epoxy, S glass, or Kevlar), and layering sequence are variable. Depending upon the loading and stiffness requirements in the part, any number of plies can be laid up to develop properties in various directions.

## Design Considerations

In the past direct substitution of composites rather than new part design was usually the design engineer's approach. For example, turbine-engine applications of composites have generally been pursued as a means of lowering the weight and cost of existing components. The composite part design was therefore forced to be compatible with the neighboring hardware, which often was not being considered for composites. As a result, the composite components invariably contained redundancies and extra features which would not be required in original engine designs. Furthermore, the full advantage of the higher specific stiffness and specific strength of structural composites was not taken advantage of. The higher costs and weight that resulted from interchangeability requirements made the composite components less attractive to engine designers. New engine designs incorporating composites have projected substantial (20 to 40%) cost and weight reductions for components throughout the engine.

Composites have not been extensively used in gas-turbine engines because the flight-critical nature of the engine system has made the designer extremely conservative with respect to material system changes. Since new advanced supersonic fighter engines are designed to yield the maximum performance at the lowest weight, the application of com-

**Table 3.1** Recommended Design Values for Carbon Reinforcements†

| Form | Design value | | | | Common processing method | Remarks |
|------|------|------|------|------|------|------|
| | Tensile strength | | Modulus | | | |
| | ksi | MPa | $10^6$ psi | GPa | | |
| Chopped, 0.25 in (6.4 mm) | 30 | 207 | 2.5 | 17 | Injection or compression molding | Highest production rates; lowest cost; properties vary with part complexity |
| 0.5–2 in (12.7–50.8 mm) | 32 | 221 | 3.0 | 21 | Injection or compression molding | Moderate production rates; brittle; properties vary with part complexity; matrix controls fatigue properties |
| Bidirectional cloth | 85 | 586 | 10.5 | 72 | Compression molding or autoclave | High cost (weaving operation; good for complex parts; good fatigue properties) |
| Tape | 190 | 1310 | | | Compression molding, autoclave, or filament winding | Highest performance; difficult to use in complex parts; high production costs; good fatigue properties |
| | 140‡ | 970 | 21§ | 145 | | |
| Yarn | 190 | 1310 | | | Filament winding | Low production costs; avoids prepregging; suitable for limited shapes; good fatigue properties |
| | 140‡ | 970 | 18§ | 124 | | |

†Based on data from Union Carbide Corp.    ‡Compressive strength.    §Intermediate-modulus fiber.

posites to these systems will undoubtedly be more actively pursued. Commercial jet-turbine engine applications of flight-critical composite components will be based on the successful use of composites in the military and stationary turbine engine systems.

The total temperature range of materials in an operational jet engine is from $-65$ to approximately 2200°F ($-54$ to 1204°C). With one exception, most composite materials can be used to a maximum temperature of approximately 600°F (316°C). This generally limits their use in aircraft engines to the nose cone, the fan, some of the cases, external accessories and plumbing, and possibly part of the low-pressure compressor. The one exception is Bsc–Ti, which is usable to 1000°F (538°C). This composite material can be used farther back in the engine, where inlet temperatures may exceed 600°F (316°C).

Structural composites are particularly attractive for fan blades, which are normally stiffness-limited. In this case, the high stiffness and low density of composites enable the designer to eliminate part-span shrouds or bumpers, which reduces weight and cost and increases fan or compressor efficiency. Even more attractive is the possibility of operating composite fan stages at rotational speeds up to 50% higher than now possible with metal parts. This can eliminate entire fan stages (and the turbine stages that drive them) because of increased work per stage at the higher speeds, significantly reducing weight and cost.

The high temperatures in engines generally restrict the types of composite matrixes to metal (aluminum or titanium) or polyimide resin. Epoxy resin-matrix composites are usually restricted to the cooler parts of subsonic engines. New composite systems such as the fiber-reinforced superalloys being developed may provide additional applications in the compressor and turbine sections.

Table 3.2 lists the turbine engine components that can effectively make use of composite materials, the temperature range, controlling parameters, weight payoff, and degree of technical difficulty in making the components from composites. The nose cone or spinner is not included, since it is usually made of a conventional composite material, such as Gl–Ep. Table 3.2 does not show applications of new composite materials because they are not yet commercially available and their viability has not yet been fully demonstrated.

Suitability of a given composite material for the components listed in Table 3.2 means only that it can provide the specific stiffness and/or specific strength required for the component. For example, not all the composite materials listed as suitable would be applicable

**Table 3.2**  Composite Candidates for Engine Components[1]

| Component | Max temp range | | Limiting parameter | Weight payoff | Degree of technical difficulty | Suitable composite materials† |
|---|---|---|---|---|---|---|
| | °F | °C | | | | |
| Fan: | | | | | | |
| Blade | 300–600 | 149–316 | Torsional stiffness | High | Moderate | 1–6 |
| Vane | 300–600 | 149–316 | Torsional stiffness | High | Low | 1–6 |
| Disk | 200–750 | 93–399 | Stiffness or strength | Moderate | Moderate | 3–6 |
| Low compressor: | | | Torsional stiffness | Low | Moderate | 1, 2, 5, 6 |
| Blade | 350–800 | 177–427 | Torsional stiffness | Low | Low | 1, 2, 5, 6 |
| Vane | 350–750 | 177–399 | Stiffness | Low | Moderate | 3–6 |
| Disk | 400–850 | 204–454 | Containment, | | | |
| Cases, ducts | 275+ | 135+ | stiffness, buckling | Moderate | High | 3, 5 |
| Accessories | Ambient+ | 20+ | | | Low | 3, 4 |
| Plumbing | Ambient+ | 20+ | | | Low | 3, 4 |

†1 = Bsc–Al or B–Al, 2 = Bsc–Ti, 3 = Gr–Ep, 4 = B–Ep, 5 = Gr–PI, 6 = B–PI.

for all fan blades because of specific special requirements. The application of composites to fan components requires not only high specific strength and stiffness but also high resistance to foreign-object damage (FOD). The table shows both organic and MMC as being suitable for fan blades, but MMCs are generally favored for the high-performance high-speed fan-blade designs because of the greater FOD resistance of MMC.

One engine manufacturer's use of reinforced composites in its 40,000-lb-thrust (18,000-kg) engine (carbon fiber in the fan blades and casing and glass fiber in the intermediate-pressure compressor) is saving more than 700 lb (318 kg). Another engine manufacturer has run a fan equipped entirely with boron-filament–aluminum-matrix blades. Compared with titanium alloys carbon-reinforced composites have about 4 times the strength-to-density ratio; for boron composites, it is a factor of 2.5. Modulus and/or density advantages of these composites, at factors ranging from 3 to 5.5, also pay off in stiffness-critical components; boron fan blades, for example, may not need part-span shrouds, thereby eliminating their drag effect and increasing blade efficiency.

B–Al has been examined as a substitute in turbofan engine blades (Fig. 3.1), which are

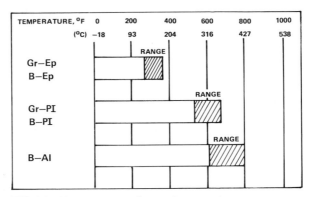

FIG. 3.1   Use temperatures of composite systems.

now either stainless steel or titanium. If the impact resistance of a polymer composite is low, the "bird problem" rules them out unless they are covered with monolithic metal at the edges. B–Al may solve this FOD problem. B–Al took a quantum leap over titanium when it passed the federal tests for bird ingestion.

Design engineers have considered applications of carbon and boron extending past the fan to most of the remaining compressor stages, using a titanium matrix, and in the turbine, using a nickel or cobalt matrix. High-temperature alloys are keeping pace via new methods for casting directionally solidified blades with grain boundaries parallel to the main stress axis. The estimated result of these materials advances is a 25% saving in engine weight.

For airframe applications, too, there is a definite move toward composites. The weight-saving stiffness of boron and graphite composites makes them particularly competitive. The first applications naturally tend to be in relatively small, secondary structures such as control surfaces, spoilers and nacelles, rather than in sections critical to flight safety in fittings, where the advantage gained is not worth the cost involved. But composites also have great potential for the major elements of wing, fuselage, and tail.

The ascendancy of composites follows from the increasing emphasis on fatigue and structural life as a principal problem area, as opposed to the industry's well-established capability in static-stress design. Using boron reinforcement in an aluminum matrix, for example, might lead to weight reductions for fatigue design of 65 to 80% and for static design of perhaps 25 to 50%. Demonstration parts using boron and epoxy [with tensile strength about 400 kips/in$^2$ (2758 MPa)] are now realizing up to 25% weight savings in direct material substitution.

The trend in composites design has paralleled material and polymer-chemistry high-temperature development. Until recently the application of polymer-matrix composite materials has been limited to structural components whose use-temperature requirements could be met by epoxy resins. Although high-temperature-resistant polymer-matrix composites provided an opportunity to design structures with nearly a twofold increase in use temperature, the chemistry and severe processing requirements of the first (condensation-type) high-temperature polymers made it impractical and difficult to fabricate high-quality structural components. In contrast, fiber-reinforced epoxy resins can easily be processed using a variety of techniques at relatively low temperatures and pressures.

Persistent experimentation, however, eventually led to the development of a class of polyimides known as polymerization-of-monomer-reactants (PMR) polyimides. The reinforcing fibers are impregnated with a solution containing a mixture of monomers dissolved in a low-boiling alkyl alcohol solvent. The monomers are essentially unreactive at room temperature but react at elevated temperatures to form a thermooxidatively stable polyimide matrix. These highly processable addition polyimides can be processed by compression- or autoclave-molding techniques and are now making it possible to realize much of the potential of high-temperature polymer-matrix composites.

Research has identified monomer-reactant combinations for two PMR polyimides differing in chemical composition. Because of their excellent processing characteristics and commercial availability, polyimide materials are used in a number of diverse structural components. The inner cowl for an experimental turbofan engine was fabricated with a maximum diameter of approximately 35.4 in (900 mm) and has undergone ground engine tests. Compressor-blade skins for spar-shell blades having an 11.8-in (300-mm) chord and a 59-in (1500-mm) span and an oil-tank bracket using a chopped-graphite fiber molding compound have been fabricated.

## Environmental Considerations

Composite hardware has been designed to compensate for such environmental effects as lightning strike, electromagnetic pulse, nuclear blast, and impact damage. Integrally molded aluminum-foil, wire-mesh, and flame-spray coatings have qualified as lightning-strike protection systems. Aluminum mesh and flame-spray coatings provide adequate shielding for electronic components located behind graphite-epoxy access doors. Gloss insignia white polyurethane paint provides protection against high-intensity light generated by nuclear blast. S-glass crack stoppers within graphite-epoxy plies and stitching are being evaluated as a means of reducing impact damage.

## Material Substitution

Although most structural-composite production hardware today still involves designs based on material substitution, as design technology matures, use of composites in actual airframe, automotive, and commercial structures will increase. As reliable and reproducible manufacturing techniques are implemented and critical flaws are detectable by available nondestructive inspection, the unique properties of composite materials will be more fully realized. Chapters 4 and 5 will discuss manufacturing processes; nondestructive inspection is discussed later in this chapter.

# MECHANICAL BEHAVIOR OF COMPOSITE MATERIALS

Composite materials have many characteristics that differ from those of more conventional engineering materials. Although some characteristics are merely modifications of conventional behavior, others are totally new and require new analytical and experimental procedures. Most common engineering materials are *homogeneous and isotropic*. Bodies with temperature-dependent isotropic material properties are not homogeneous when subjected to a temperature gradient but still are isotropic. In contrast, composite materials are often both *inhomogeneous* (or heterogeneous) and *nonisotropic* (orthotropic or, more generally, anisotropic).

Some composite materials have very simple forms of inhomogeneity. For example, laminated safety glass has three layers, each of which is homogeneous and isotropic; thus, the inhomogeneity of the composite is a step function in the direction perpendicular to the plane of the glass. Some particulate composites are inhomogeneous yet isotropic although some are anisotropic. Other composite materials are typically more complex.

Because of the inherent heterogeneous nature of composite materials, they are conveniently studied from two points of view, micromechanics and macromechanics. In *micromechanics* the interaction of the constituent materials is examined on a microscopic scale. In *macromechanics* the material is presumed homogeneous and the effects of the constituent materials are detected only as averaged apparent properties of the composite. The difference between the two approaches is the geometrical scale of reference. While micromechanics deals with the constituent properties of the monolayer, macromechanics describes the combined characteristics of a number of monolayers. In order to have a meaningful and sequential approach to the design of composite materials micro and macro methods cannot be totally divorced since the macro properties are totally dependent on the micro properties.

## Micromechanics Technology

Fibers have been added to plastics for decades to improve strength, but no one knew exactly how they worked. The science of micromechanics was created 25 years ago to help explain the complex transfer of stresses and strains between a plastic matrix and the reinforcing fibers. Only recently has this technology been sharpened to the point where it can be used to make more efficient use of the fibers. Although the scientific explanation of how fibers behave in composite materials under load is steeped in engineering jargon and arm-long equations, the basic (highly simplified) principles can be summarized by first illustrating the state of stress in a sample of unreinforced material under load and then demonstrating how most of these stresses can be transferred to reinforcements.

In Fig. 3.2 a sample of material is being pulled from both ends in simple tension, as indicated by the arrows. To understand what takes place throughout the material when the load is applied, an imaginary, very small segment of material, called an *element,* can be

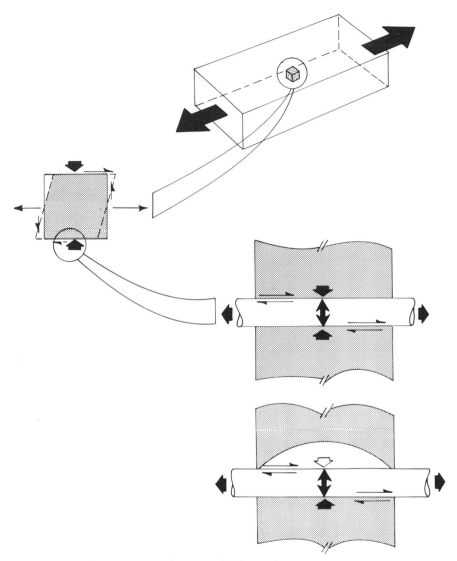

**FIG. 3.2** State of stress and stresses being transferred to reinforcement. *(From Mater. Eng., May 1978, p. 29.)*

selected from anywhere within the sample for examination. Although on a macroscopic scale the sample is in simple tension, internally the material must endure a variety of stresses, including in some cases, compressive stresses, as indicated by in-pointing arrows, and shear stress along each face, shown as half arrows. Under this combined state of stress the element will deform and initiate failure of the entire sample when the strength of the material is exceeded.

When they are properly mixed with the matrix material, fibers form strong, shear-resistant adhesive bonds with the resin. Tensile stresses, formerly applied directly to the unreinforced material, are now transferred to the fiber through this interfacial bond. Theoretically, a glass fiber can endure tensile loads of 500 kips/in$^2$ (3448 MPa) although the strength of a reinforced plastic is primarily limited by the shear strength of the bond. Two factors that reduce the effectiveness of a reinforcement are poor interfacial bond shear strength and voids in the resin. A void shown in the bottom illustration of Fig. 3.2 reduces bonding surface area and creates an imbalance of stresses, transferred between the resin and fiber. Local instability can cause the fiber to shift or buckle, creating a discontinuity in the material that can lead to failure. Generally, a void along 1% of the fiber surface reduces interfacial shear strength by 7%.

The concepts of both macromechanics and micromechanics are used in tailoring a composite material to meet a particular structural requirement with little waste of material capability. The ability to tailor a composite material to its job is one of the most significant advantages of a composite over an ordinary material. Tailoring a composite material yields only the stiffness and strength required in a given direction. In contrast, an isotropic material, by definition, must have excess strength and stiffness in any direction other than that of the largest requirement.

## Laminae

A lamina is a flat (or curved, as in a shell) arrangement of unidirectional fibers or woven fibers in a matrix. Two typical laminae are shown in Fig. 3.3; their principal material axes

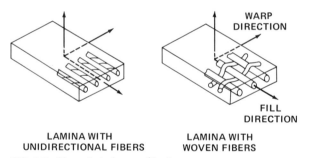

WARP
DIRECTION

FILL
DIRECTION

LAMINA WITH
UNIDIRECTIONAL FIBERS

LAMINA WITH
WOVEN FIBERS

**FIG. 3.3**   Two principal types of laminae.

are parallel and perpendicular to the fiber directions. The fibers, or filaments, the main reinforcing or load-carrying agent, are typically strong and stiff. The matrix may be organic, ceramic, or metallic. Its function is to support and protect the fibers and to provide a means of distributing load among the fibers and transmitting it between them. The latter function is especially important if a fiber breaks. In Fig. 3.4 the load from one side of a broken fiber is transferred to the matrix and subsequently to the other side of the broken fiber as well as to adjacent fibers. The mechanism for the load transfer is the shearing stress developed in the matrix; the shearing stress resists the tendency of the broken fiber to pull out. This load-transfer mechanism is the means by which whisker-reinforced composites carry any load at all above the inherent matrix strength.

The constituents of the lamina, i.e., fiber and matrix, exhibit different types of stress-strain behavior. Fibers generally show linear elastic behavior, although reinforcing steel bars in concrete are more nearly elastic–perfectly plastic. Aluminum and some composites

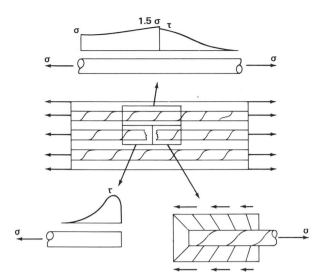

**FIG. 3.4**  Effect of broken fiber on matrix and fiber stresses.

exhibit elastoplastic behavior, which is really nonlinear elastic behavior if there is no unloading. Common resinous matrix materials are viscoelastic if not viscoplastic. The various stress-strain relations are sometimes referred to as *constitutive relations* since they describe the mechanical constitution of the material.

Fiber-reinforced composites such as B–Ep and Gr–Ep are usually treated as linear elastic materials since the fibers provide most of the strength and stiffness. Refinement of that approximation requires consideration of some form of plasticity, viscoelasticity, or both (viscoplasticity). Work remains to be done to implement those idealizations of composite-material behavior in structural applications.

### Laminates

A laminate is a stack of laminae having various orientations of principal material directions (Fig. 3.5). Note that the fiber orientation of the layers in Fig. 3.5 is not symmetric about the middle surface of the laminate. The layers of a laminate are usually bound together by the same matrix material that is used in the laminae. Laminates can be composed of plates of different materials or, in the present context, layers of fiber-reinforced laminae. A laminated circular cylindrical shell can be constructed by winding resin-coated fibers on a mandrel first with one orientation to the shell axis, then another, until the desired thickness is built up.

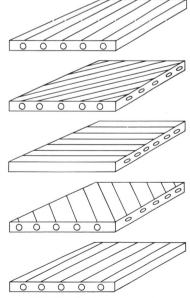

**FIG. 3.5**  Laminate construction.

A major purpose of lamination is to tailor the directional dependence of strength and stiffness of a material to match the loading environment of the structural element. Laminates are uniquely suited to this objective since the principal material directions of each layer can be oriented according to need. For example, six layers of a ten-layer laminate could be oriented in one direction and the other four at 90° (1.6 rad) to that direction; the resulting laminate then has a strength and extensional stiffness roughly 50% higher in one direction than the other. The ratio of the extensional stiffnesses in the two directions is approximately 6:4, but the ratio of bending stiffnesses is unclear since the order of lamination is not specified in the example. Moreover, if the laminae are not arranged symmetrically about the middle surface of the laminate, stiffnesses exist that describe coupling between bending and extension.

A potential problem facing the engineer in the construction of laminates is the introduction of shearing stresses between layers. Shearing stresses arise from the tendency of each layer to deform independently of its neighbors because all may have different properties (at least from the standpoint of orientation of principal material directions). Such shearing stresses are largest at the edges of a laminate, where they may cause delamination; the transverse normal stress can also cause delamination. Composite design engineers and stress analysts must take all these points into consideration in the conceptualization and design of laminated fiber-reinforced composite materials.

## Forms of Constituent Materials

The fibers and matrix material can be obtained commercially in a variety of forms, both individually and as laminae. Fibers are available individually or as roving. The fibers may be unidirectional or interwoven. Fibers are often saturated with the resinous material, e.g., polyester resin, which is subsequently used as a matrix material. This process is referred to as *preimpregnation,* and preimpregnated fibers are called *prepregs.* For example, unidirectional fibers in an epoxy matrix are available in tape form (prepreg tape) where the fibers run in the lengthwise direction of the tape. The fibers are held in position not only by the matrix but by a removable backing that prevents the tape from sticking together in the roll. Prepreg cloth or mats are available in which the fibers are interwoven and then preimpregnated with resin. Other variations exist.

## Laminate Analysis

Laminate analysis is a mathematical means of determining elastic properties of the complete laminate from the angles of fiber orientation and elastic moduli of the individual layers. Conversely, the method is also used to determine stress and strains in each layer produced by stretching and bending forces on the complete laminate.

The elastic properties of a laminate ($E_x$, $E_y$, $G_{xy}$, $\gamma_{xy}$) are determined in three steps:

1. The elastic properties of each layer are determined from the moduli $E_{11}$, $E_{12}$, and $G_{12}$ and Poisson ratio $\gamma_{12}$ of the layer material. These properties are determined from the Hooke's-law relationship between stress and strain in the principal material directions (parallel and perpendicular to the fiber direction).
2. The elastic properties of each layer in the principal laminate directions (parallel and perpendicular to the axis of the laminate structure) are determined from the material-direction properties and the angle between the laminate and material directions.

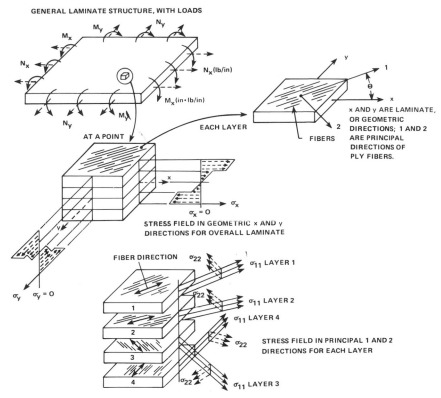

**FIG. 3.6** Elastic properties of a laminate.[46]

3. A summation of the laminate-direction elastic properties of all layers, taking into account their relative positions with respect to the midplane of the complete laminate, yields the elastic properties of the complete laminate (Fig. 3.6).

Laminate analysis is also used to determine stresses and strains in each layer due to forces on the laminate. Layer strains are determined from laminate strains and rotations, taking into account the layer position relative to the middle surface of the complete laminate. Stresses in each layer are then determined from layer strains and the Hooke's-law relationship. Thus, the stresses and strains at a point in the structure can be calculated in every direction, that is, $\sigma_x$, $\sigma_y$, $\tau_{xy}$ and $\sigma_{11}$, $\sigma_{22}$, $\tau_{12}$ in every ply. A designer must be able to calculate the principal stresses ($\sigma_{11}$, $\sigma_{22}$, $\tau_{12}$) because they are the material unidirectional and transverse design allowables that are published in the literature.

## Design Properties

Unidirectional and multidirectional properties of high-strength and high-modulus C–Ep composites are shown in Table 3.3, along with the corresponding properties of lower-cost fiber composites and metals. The longitudinal or 0° (0-rad) values most frequently cited

## Table 3.3  Comparative Properties of Composites and Metals[46]

FIBER DIRECTION

ALTERNATING PLIES, +45°

| | | Unidirectional properties | | | | | | ±45° properties | | | |
| | | Modulus | | | Ultimate strength | | | Modulus | | Ultimate strength | |
| Material† | Density | Axial $E_{11}$ | Trans $E_{22}$ | Shear $G_{12}$ | Axial $\sigma_{11}$ | Trans $\sigma_{22}$ | Shear $\tau_{12}$ | $E_x = E_y$ | $G_{xy}$ | $\tau_{xy}$ | $\sigma_x = \sigma_y$ |
|---|---|---|---|---|---|---|---|---|---|---|---|
| | $lb/in^3$ | $10^6\ lb/in^2$ | | | $kips/in^2$ | | | $10^6\ lb/in^2$ | | $kips/in^2$ | |
| C–Ep: | | | | | | | | | | | |
| High strength | 0.057 | 20 | 1.0 | 0.65 | 220 | 6 | 14 | 2.5 | 4.5 | 20 | 50 |
| High modulus | 0.058 | 29 | 1.0 | 0.70 | 175 | 5 | 10 | 2.5 | 6.5 | 18 | 42 |
| Ultrahigh modulus | 0.061 | 44 | 1.0 | 0.95 | 110 | 4 | 7 | 3.0 | 11.5 | 14 | 30 |
| Kv 49–Ep | 0.050 | 12.5 | 0.8 | 0.3 | 220 | 4 | 6 | 1.1 | 3.0 | 30 | 32 |
| EGl–Ep | 0.072 | 6 | 1.5 | 0.3 | 180 | 6 | 10 | 1.5 | 3.0 | 25 | 40 |
| Chopped-glass SMC† | 0.068 | 2.5 | 2.5 | 1.0 | 30 | 30 | 20 | 2.5 | 1.0 | 20 | 30 |
| | 0.072 | 3.5 | 3.5 | 1.5 | 50 | 50 | 40 | 3.5 | 1.5 | 40 | 50 |
| Steel | 0.294 | 29.5 | 29.5 | 11.5 | 60 | 60 | 35 | 29.5 | 11.5 | 35 | 60 |
| Aluminum | 0.098 | 10.5 | 10.5 | 3.8 | 42 | 42 | 28 | 10.5 | 3.8 | 28 | 42 |
| | $kg/m^3$ | GPa | | | MPa | | | GPa | | MPa | |
| C–Ep: | | | | | | | | | | | |
| High strength | 1578 | 138 | 6.90 | 4.48 | 1517 | 41 | 97 | 17.2 | 31.0 | 138 | 345 |
| High modulus | 1605 | 200 | 6.90 | 4.83 | 1207 | 34 | 69 | 17.2 | 44.8 | 124 | 290 |
| Ultrahigh modulus | 1688 | 303 | 6.90 | 6.55 | 758 | 28 | 48 | 20.7 | 79.3 | 97 | 207 |
| Kv 49–Ep | 1384 | 86.2 | 5.52 | 2.07 | 1517 | 28 | 41 | 7.58 | 20.7 | 207 | 221 |
| EGl–Ep | 1993 | 41.4 | 10.3 | 2.07 | 1241 | 41 | 69 | 10.3 | 20.7 | 172 | 276 |
| Chopped-glass SMC† | 1882 | 17.2 | 17.2 | 6.90 | 207 | 207 | 138 | 17.2 | 6.90 | 138 | 207 |
| | 1993 | 24.1 | 24.1 | 10.3 | 345 | 345 | 276 | 24.1 | 10.3 | 276 | 345 |
| Steel | 8138 | 203 | 203 | 79.3 | 414 | 414 | 241 | 203 | 79.3 | 241 | 414 |
| Aluminum | 2713 | 72.4 | 72.4 | 26.2 | 290 | 290 | 193 | 72.4 | 26.2 | 193 | 290 |

†Fiber content is 65 vol % except for steel and aluminum, where it does not apply. For chopped-glass SMC it is 30 vol % for the first row in the table and 65 vol % for the second row.

in the literature are not very useful in design, however, because they are ultimate values of carefully molded panels. The real world includes many property-reducing factors such as stress concentrations, fiber wash, resin-rich areas, and voids. Although the laboratory values give preliminary design direction, substantial reduction in mechanical properties can be expected, depending on part complexity and molding conditions.

In many structural designs, either stress level or deformation is the material-limiting factor. With conventional materials, a material is chosen, the modulus is set, and deformation and stress level are controlled by cross-sectional design. For composites, on the other hand, the scenario is usually one of iterative optimization; changing the angularity of certain plies changes the structural stiffness and in turn affects the load pattern, which again requires a change in angulation.

Predicting modulus and stiffness in carbon-fiber composites is straightforward and accurate. Predicting ultimate strength is more difficult because of the multiplicity of failure modes involved. Tensile modulus is the slope of a material's stress-strain curve. Stiffness is that modulus multiplied by the section moment or area (Fig. 3.7). An effective graphic design method used by many engineers starts by assuming that a laminate consisting of plies at only 0°, ±45°, and 90° (0, ±0.79, and 1.6 rad) effectively resists all loads and is reasonably close to optimum from the design, weight, and fabrication standpoints. Mod-

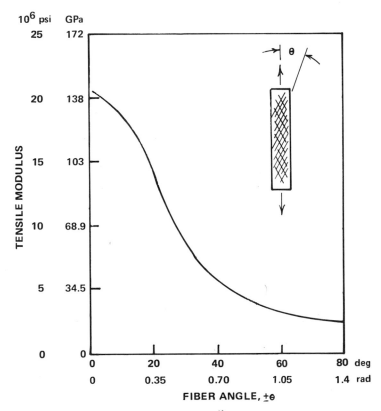

**FIG. 3.7**  Tensile modulus of C–Ep composites.[46]

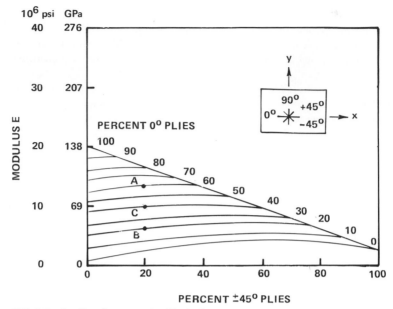

**FIG. 3.8**  Families of curves to simplify the first stages in designing C–Ep composites. The curves are used to determine modulus values for 0°/±45°/90° (0/0.79/1.6 rad) composite laminates. The elastic modulus for composites is selected from the curves. The chart is based on room-temperature values of multiple-ply composites containing 65 vol % high-strength carbon fiber in an epoxy matrix.[46]

ulus and strength levels for the 0°/±45°/90° family of high-strength C–Ep composites are shown in Figs. 3.8 and 3.9 for composites containing 65 vol % carbon fibers in an epoxy matrix.[1]

Allowable tensile strength (Fig. 3.9) includes a safety factor of about 35%. The design factor, or safety margin, may change, of course, for different applications. For instance, an aircraft drone would be designed with a much lower safety factor than a commercial transport. Changing the safety margin would have an accordionlike effect on the general shape of the design curves, moving point $D$ up and down along the axis. To use the curves in Figs. 3.8 and 3.9 in designing a C–Ep composite:

1. Determine the modulus level $E_x$ needed.
2. From Fig. 3.8 select a percentage of plies at 0, ±45, and 90° (0, 0.79, and 1.6 rad) to match that level.
3. From Fig. 3.9 determine the allowable longitudinal tensile strength $\sigma_x$ based on the stacking choice.
4. If the strength value is too low or too high, select a new stacking arrangement and repeat the procedure.

For example, suppose a material is needed which is stronger than steel in the $x$ direction, say $\sigma_x = 70$ kips/in$^2$ (483 MPa), and as stiff as aluminum in the $y$ direction [$E_y = 10$

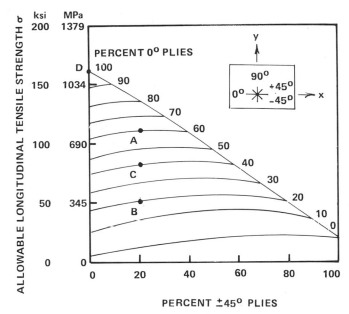

**FIG. 3.9** Curves of strength values for $0°/\pm45°/90°$ composite laminates. Allowable tensile strength (point *D*) for 100% $0°$ fibers in a composite is 160 kips/in² (1103 MPa). This corresponds to the ultimate strength of 220 kips/in² (1517 MPa) in Table 3.3. The chart is based on room-temperature values of multiple-ply composites containing 65 vol % high-strength carbon fiber in an epoxy matrix.[46]

$\times$ 10⁶ lb/in² (69 GPa)]. Pick a material combination, i.e., percent plies at 0, $\pm45$, and 90° and locate the choice in Figs. 3.8 and 3.9. An arbitrary choice might be

60% at 0° (0 rad)
20% at 90° (1.6 rad)
20% at $\pm45°$ (0.79 rad)

This combination is shown as point *A*, where $\sigma_x$ = 105 kips/in² (724 MPa) and $E_x$ = 13 $\times$ 10⁶ lb/in² (90 GPa). On interchanging the x and y coordinates and the 0 and 90° (0- and 1.6-rad) percentages (point *B*) we see that $\sigma_y$ is about 55 kips/in² (380 MPa) and $E_y$ = 6 $\times$ 10⁶ lb/in² (414 GPa). Since $\sigma_x$ is higher than necessary and $E_y$ is too low, the first adjustment would take some of the 0° material and move it to 90°; for example,

40% at 0° (0 rad)
40% at 90° (1.6 rad)
20% at $\pm45°$ (0.97 rad)

which is represented by point *C*, having x properties of $\sigma_x$ = 85 kips/in² (586 MPa) and $E_x$ = 10 $\times$ 10⁶ lb/in² (69 GPa). Reversing x and y (40% at 0°, 40% at 90°, and 20% at $\pm45°$) again gives point *C*, with $\sigma_y$ = 85 kips/in² (586 MPa) and $E_y$ = 10 $\times$ 10⁶ lb/in² (69 GPa). The initial requirements have been matched, and further fine tuning would be only marginally beneficial.

Layering sequences other than the $0°/\pm45°/90°$ arrangement can also satisfy a given set of requirements. For example, to get quasi-isotropic properties, a $0°/\pm30°/\pm60°/90°$ (0/0.52/1.04/1.6 rad) ply stacking could be used. But the beauty of the $0°/\pm45°/90°$ system is that (1) it is mathematically simple; the property plots do not work for four-angle combinations; (2) it is the best combination for biaxial and shear loads; (3) fabrication of $\pm45°$ (0.79-rad) layers is simplified by using 0° or 90° fabric laid up on a 45° bias. In addition to the modulus and tensile-strength plots in Figs. 3.8 and 3.9, Ref. 1 also provides curves for shear strength, compression strength, shear modulus, and coefficient of thermal expansion, all as functions of ply stacking at $0°/\pm45°/90°$. Property curves[2] have been developed for a family of $0°/90°$ or $\pm\theta°$ (fiber angle) carbon-face-sheeted SMC materials.

## Design and Environmental Effects

Moisture and temperature are two key environmental concerns in designing composite structures. Moisture attacks composites by migrating along fiber-resin boundaries and destroying the adhesive interface. In glass, the moisture actually attacks the fiber as well. The result is a gradual softening of the structure. High temperatures also soften and weaken a composite structure. The worst case is high temperature with high humidity. A typical environmental check of a composite material would be immersion in boiling water for 24 to 72 h, followed by mechanical test (flexure) at high temperature. From a design standpoint, the best ways to combat these effects are to:

Define the worst-case environmental condition accurately

Choose the matrix (polyester, epoxy, or polyimide) based on the highest anticipated temperature

Base the design on an ultimate material strength measured at the anticipated service temperature rather than at room temperature

Typically, structural composites retain one-half of their ambient tensile strength when tested at a recommended maximum operating temperature. With current technology, coatings are not an acceptable means of protecting against moisture and temperature.

### Symmetry Effects on Composites

Plies of a laminate must be stacked symmetrically and overall balance must be maintained to avoid coupling or bending, stretching, and torsion. Balance implies that for every +45° (0.79-rad) ply there exists a −45° ply in the laminate. Symmetry requires a mirror image of ply stacking about the midplane. Figure 3.10 shows how simple loads result in unusual deformations because of coupling action. With balance and symmetry these effects disappear.

### Failure Modes of Composites

Tensional strength of a C–Ep structural composite is always related to fiber direction. A simple tensile test shows strikingly different failure behavior in composites having different

ALL PLIES AT 0° (FIBER ANGLE).
AXIAL LOAD RESULTS IN
STRETCHING-SHEARING BEHAVIOR.

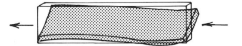

TWO PLIES AT ±0 (ANY ANGLE).
OPPOSING SHEAR DEFORMATIONS
IN THE PLUS AND MINUS PLIES
RESULT IN STRETCHING-
TORSION INTERACTION.

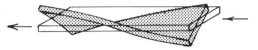

A 0°/90° (0 = rad/1.6 rad) STACKING.
THIS ARRANGEMENT BENDS
UNDER PURE TENSION BECAUSE
THE MODULUS-WEIGHTED
CENTROID IS NOT COINCIDENT
WITH THE GEOMETRIC CENTROID,
RESULTING IN AN OFFSET LOAD
PATH.

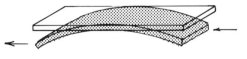

ANOTHER 0°/90° (0 = rad/1.6 = rad)
STACKING. BECAUSE OF DIFFERENT
THERMAL EXPANSION CHAR-
ACTERISTICS IN EACH LAYER, THIS
STACKING DEFORMS INTO
A "SADDLE" WHEN HEATED.

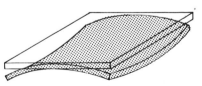

**FIG. 3.10**   Symmetry effects on deflection of composites.[46]

fiber orientation. In a multidirectional composite, single plies can fail without overall structural failure. Recognition of the various failure modes and knowledge of how composites
fail are prerequisites to correcting a problem (Fig. 3.11).

### Design in Joining Composites

Although high stiffnesses and strengths can be attained for composite laminates, these characteristics are quite different from those of ordinary materials to which we generally want

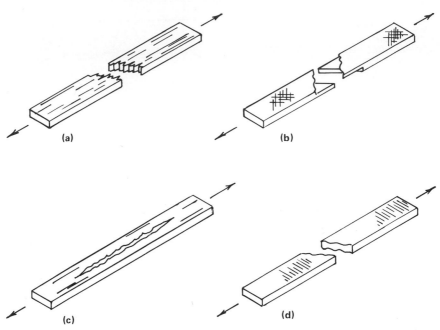

**FIG. 3.11** Failure modes on deflection of composites: (*a*) fiber tensile failure for all fibers in direction of load 0° (0 rad); resin shear failure (*b*) for fibers at ±45° (0.79 rad) and (*c*) through the thickness between fibers at 0°, usually caused by poor fiber-to-resin adhesion; (*d*) resin tensile failure between fibers at 90° (1.6 rad) to load.[46]

to fasten composite laminates. The full strength and stiffness characteristics of the laminate often cannot be transferred through the joint without a significant weight penalty. Thus, joints and other fastening devices are critical to the successful use of composite materials. The specific design of a joint is much too complex to cover in detail in this manual. The published state of the art of laminate joint design is summarized in Refs. 1 and 3.

## Bonded vs. Bolted Joints

Advanced composites introduce design problems different from those common to metal and molded-plastic construction, particularly where components must be fastened together. Because most types of composite joints, whether they are adhesively bonded or mechanically fastened, involve some cutting or machining of the strength-providing fibers, joint configurations require careful planning to minimize the possibility of failure. The planning must consider weakness in in-plane shear, transverse tension, interlaminar shear, and bearing strength relative to the primary asset of a lamina, namely, the strength and stiffness in the fiber direction.

**Bolted Joints:**   Because structural composites are made with thermosetting resins, they cannot be joined by welding methods as thermoplastics and metals can. The choice for composites lies between mechanical methods and adhesive bonding. Each technique produces a joint with significant differences both in production and in function; each has its advantages and limitations. Mechanical joints are generally not as adversely affected by thermal cycling or humidity as bonded joints are. They permit disassembly without destroying the substrate, and they are readily inspected for joint quality. Mechanical joints also require little or no surface preparation of the substrate and do not require white-glove and clean-room production conditions.

On the less favorable side, mechanical methods add weight and bulk to a joint, and since they require holes machined in the substrate members to accommodate bolts or screws, the members are weakened. In addition, the stress concentrations produced by mechanical fasteners can cause joint failure (Fig. 3.12). Slight dimensional inaccuracies in bolt patterns in materials that deform plastically, e.g., aluminum or steel, seldom cause problems. The material around the holes can yield to distribute the load to adjacent bolts.

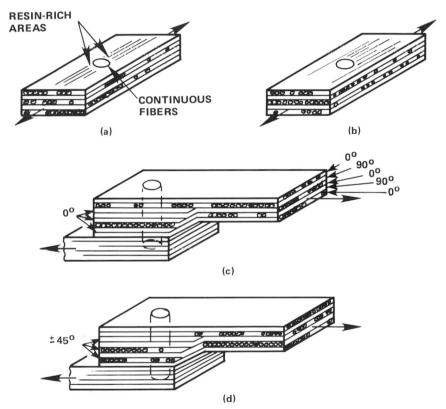

**FIG. 3.12** When holes are needed, bonded joints in advanced-composite structures are preferred where possible, but when strength requirements exceed those available from adhesives alone, bolts are used, often in conjunction with an adhesive. Because of their low strain capacity and inhomogeneity, composites require special design treatment when they are to be joined by mechanical fasteners. (*a*) Avoid molded-in holes, where continuous fiber is forced around the plug that forms the hole, leaving resin-rich areas that are relatively weak. (*b*) Drilled holes are better. The cross fibers at 45 or 90° to the load maintain transverse integrity and keep cracks from forming. (*c*) Do not reinforce a hole or cutout with carbon fibers aligned in the same direction as the load; such reinforcement merely stiffens the joint and makes it more prone to failure. (*d*) A better treatment is to soften the joint by adding staggered plies of ±45° fabric. Glass fiber is better than carbon fabric here because glass has a higher strain capacity. Metal shims are also acceptable.[3]

This is not true of the advanced composites, however, which have no yield point; if one bolt is tighter in its hole than others, the bearing stress at that hole remains higher (Fig. 3.13).

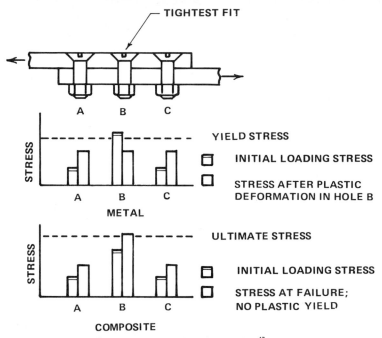

FIG. 3.13    Effect of bolt patterns in metals and composites.[47]

The brittleness of a C–Ep structural composite could cause ultimate failure of the material around a hole and the sudden distribution of its entire load to the other holes. One way to avoid such problems is to soften the laminate in the bolt area by using staggered fabric plies. Another way is to coat the bolts with uncured epoxy resin before attachment. The resin fills the clearances between bolts and holes and, when cured, helps distribute the load evenly.

Advantages of adhesive-bonded joints include light weight, distribution of the load over a larger area than in mechanical joints, and elimination of the need for drilled holes that weaken the structural members. Drawbacks to bonded joints are difficult inspection for bond integrity, the possibility of degradation in service from temperature and humidity cycling, and their relative permanence, which does not permit disassembly without destruction of the joined members. In addition, bonded joints require a rigorous cleaning and preparing of the adhering surfaces.

The principal failure modes of bolted joints are (1) bearing failure of the material, as in the elongated bolt hole of Fig. 3.14a, (2) tension failure of the material in the reduced cross section through the bolt hole in Fig. 3.14b; (3) shear-out or cleavage failure of the material (actually transverse tension failure of the material), as in Fig. 3.14c and d, and (4) bolt failures (mainly shear failures). Combinations of these failures also occur.

The bearing strength of a joint can be increased by using metal inserts, as in the shim

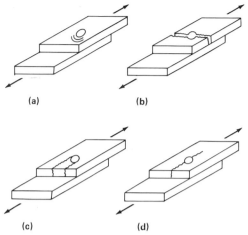

**FIG. 3.14** Bolted-joint failures: (*a*) bearing, (*b*) net tension, (*c*) shear out, and (*d*) cleavage.[3]

joint of Fig. 3.15*d*, or by thickening a section of the composite laminate, as in the reinforced edge joint in Fig. 3.15*c*. Net tension failures can be avoided or delayed by increased joint flexibility to spread the load transfer over several lines of bolts. Since composite materials are generally more brittle than conventional metals, loads are not easily redistributed around a stress concentration such as a bolt-hole. Simultaneously, shear-lag effects due to discontinuous fibers lead to difficult design problems around bolt-holes. A possible solution is to put a strip of ductile composite such as SGl–Ep several times the bolt diameter in line with the bolt rows.

**Bonded Joints:** Performance and safety requirements of various applications for advanced composites differ widely, and the design approaches differ accordingly. For exam-

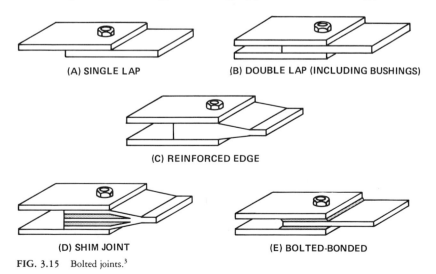

**FIG. 3.15** Bolted joints.[3]

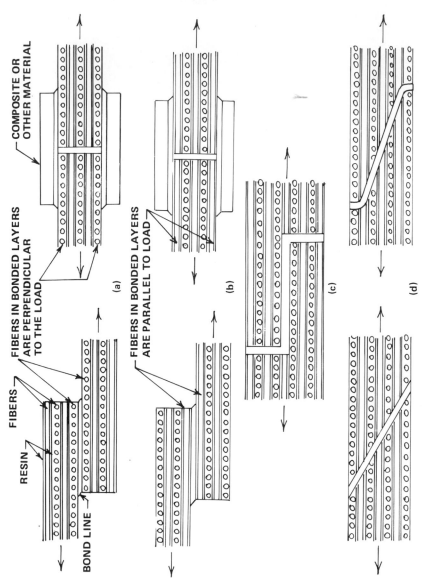

**FIG. 3.16** Multilayer composites must not be stressed in a manner likely to cause delamination. Proper orientation of the surface fibers in a bonded joint, with the fibers in the direction of the load, must be included in design specifications. Otherwise machining of bond-line areas can result in a fiber orientation that drastically weakens the joint. Components illustrated have 0 and 90° fiber orientation. (*a*) Poor design: since the surface fibers are normal to the load, the joint makes no use of fiber strength, only that of the adhesive and the resin. (*b*) Good design: all fibers at the bond lines are oriented in the same direction as the load. (*c*) Poor design: fibers along the length of the joint are normal to the load. The vertical portions of the bond line form butt joints, which are very weak. The bond line should be moved up or down one layer. (*d*) Acceptable design: scarf and landed scarf are acceptable joint configurations because they provide a compromise between fiber orientation and load direction.[3]

ple, the importance of a failure in a carbon-fiber composite tennis-racket frame in no way approaches that of a failure in a composite helicopter rotor blade. Thus, the consequences of component failure determine the type of design data needed and the nature of testing required (Fig. 3.16). The special difference that sets orthotropic materials, such as the advanced composites, apart from isotropic materials is that damage from drilling or machining makes them susceptible to interlaminar shear, delamination, and peeling. For this reason, bonded joints are usually preferred over mechanically fastened joints for assembling advanced composite components to each other or to metal parts. More specifically, lap joints and strap joints are usually best because they require no machining of the adherends (Figs. 3.17 and 3.18). These joints impose a slight weight penalty and are unacceptable if a smooth, uninterrupted surface is absolutely necessary.[3]

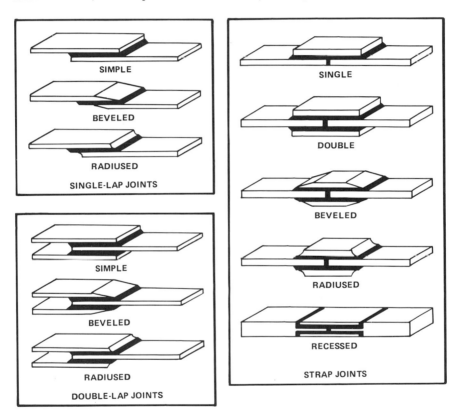

FIG. 3.17   Bonded-joint configurations: lap and strap joints.[3]

Design techniques for bonded composite joints are essentially extensions of those for isotropic materials, adjusted for the way isotropic characteristics alter failure modes. Combining joint stress analyses with data on material properties and failure criteria can provide a fairly accurate prediction of the strength and deflection characteristics of the joint. Values of shear and tensile properties of the composite and the cured adhesive film are required and should reflect the pertinent environmental conditions and any special lay-up and cure characteristics of both composite and adhesive.

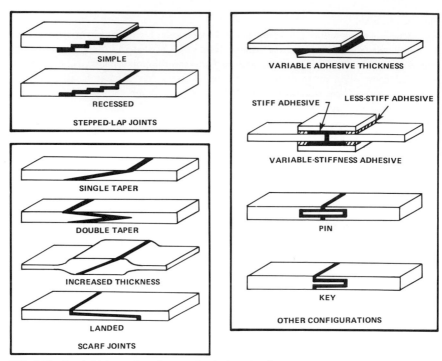

**FIG. 3.18** Bonded-joint configurations: scarf and other joints.[3]

With so many variables involved it is unlikely that such material-characterization data will be readily available in handbook form in the near future. Eventually, all new data may some day be reported in a standardized format and placed in a central data bank, accessible for use by engineers conducting computer-aided design projects.

**Bonded-Bolted Joints:**   Bonded-bolted joints have better performance than either bonded or bolted joints. The bonding results in reduction of the normal tendency of a bolted joint to shear out. The bolting decreases the likelihood of a bonded joint's debonding in an interfacial shear mode. The usual mode of failure for a bonded-bolted joint is tension failure through a section including a fastener, interlaminar shear failure in the composite, or a combination of both. Bonded-bolted joints have good load distribution and are generally designed so that the bolts take all the load if the bond breaks. The bond provides a change in failure mode and a sizable margin against fatigue failure (Fig. 3.15e).

### Filament-Wound Structures

Because of their composite makeup, filament-wound structures present special strengthening problems around openings and holes. Design and materials reinforcement methods used to create effective structures are summarized below. In filament winding there are various winding patterns (Fig. 3.19), and the key problem is how to put openings like holes in a filament-wound structure without degrading strength and stiffness. When filaments are cut by a hole (Fig. 3.20), their primary tension load is transferred into the resin

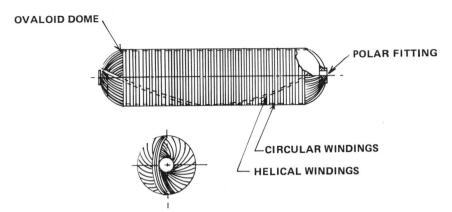

FIG. 3.19   Typical filament-wound structure, showing various winding patterns. *(From Mater. Des. Eng., February 1965, p. 109.)*

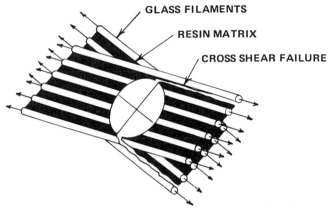

FIG. 3.20   Cutaway view of filament failure when cut by a hole. *(From Mater. Des. Eng., February 1965, p. 109.)*

matrix, which, in turn, transfers the load to uncut filaments at the sides of the opening; the filaments cannot resist the shear loads, and failure occurs along the maximum shear plane. The simplest solution to the problem of premature failure would be to use filament and binder materials having high shear strength, but such materials often do not exist. This leaves four solutions.

**Design Solution:**   Conducting a thorough stress analysis of a filament-wound structure may make it possible to establish winding configurations that permit openings to be placed where the applied stresses are low. Whenever possible, unreinforced openings should be located in areas where overlapping helical windings, e.g., in the dome of a pressure vessel, increase the wall thickness, resin content, and helix angle. These areas have a higher cross-shear strength. One of the most desirable patterns for cutting unreinforced openings is the diamond filament shape produced by helical winding. Reinforcement, in the form of extra helical layers, can be added around the periphery of the openings by using an open-pattern helical winding (Fig. 3.21).

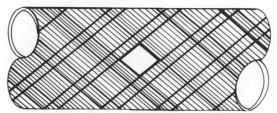

**FIG. 3.21** Open helical pattern reinforces openings. *(From Mater. Des. Eng., February 1965, p. 109.)*

Naturally, the size and shape of openings are important considerations. Since small openings require fewer cut filaments, unreinforced openings should be kept as small as possible to reduce cross-shear stresses. The openings should be shaped so that they enclose the largest area with the fewest filaments cut. All filament-wound structures contain one or two naturally reinforced polar openings. These openings should be used to their maximum extent to eliminate the need for cutting unnecessary openings.

**Wound-in Holes:** Wound-in holes should also be considered as a way of avoiding cutting filaments. They can be made during fabrication by displacing the glass (or carbon) filaments around the opening using a cone installed on the mandrel. This method is best suited for small-diameter openings because when the diameter is large, the winding paths are considerably displaced from their stable patterns and as the hole diameter increases, there is an increasing tendency for gaps to form in the windings around the periphery of the opening.

**Partial Reinforcement:** The cross-shear strength of filament-wound materials can best be improved with partial reinforcements restricted to the cross-shear planes at the periphery of the hole. This can be done by adding filaments normal to the cross-shear planes. Filaments can be added in three ways:

*Addition of filament-wound mat:* Filament-wound mat or glass (or carbon) cloth placed normal to the cross-shear plane (Fig. 3.22) can considerably increase the cross-shear strength of the material with a minimum increase in weight. The mat is fabricated from helical

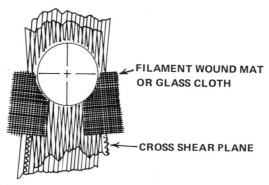

**FIG. 3.22** Filament-wound mat used for partial reinforcement of an opening. *(From Mater. Des. Eng., February 1965, p. 110.)*

windings, wound over a cylindrical mandrel. When the winding is complete, it is split longitudinally, removed from the mandrel, and laid out flat on a cutting board. Mats of any size, shape, or filament orientation are then cut out of this sheet.

*Addition of intermediate-angle windings:* Filaments can be added nearly normal to the cross-shear planes by adding layers wound at a higher or intermediate helix angle, selected so that the windings will cover the high shear areas (Fig. 3.23).

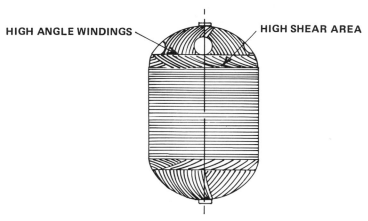

HIGH ANGLE WINDINGS ⟍    ⟋ HIGH SHEAR AREA

**FIG. 3.23** Higher helix winds add cross-shear strength under holes. *(From Mater. Des. Eng., February 1965, p. 110.)*

*Addition of level windings:* When the configuration permits, e.g., on a tube or the cylindrical portion of a pressure vessel, the intermediate angle may be increased to 90° (1.6 rad) and the windings applied locally over the cross-shear planes (Fig. 3.24).

**Total Reinforcement:** The most efficient technique for reinforcing openings in a filament-wound structure is to intersperse glass or carbon-fiber washer-type reinforcements (wafers) in the windings over the area where an opening is to be cut. By framing the opening these reinforcements transfer the tension load in the cut filaments around the opening to the cut filaments on the opposite side; the reinforcement is subjected to internal movements and radial-shear and hoop-tension loads. Several factors are vital to the success of total reinforcement. A good bond must exist between the primary structure and the reinforcement if the load is to be transferred from one to the other. Best load-

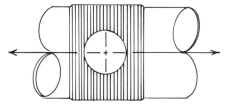

**FIG. 3.24** Level winds applied locally at an opening for reinforcement. *(From Mater. Des. Eng., February 1965, p. 110.)*

transfer results when the stiffness of the reinforcement closely matches the stiffness of the windings. Furthermore, the reinforcements must be partially cured to just the correct degree of stiffness before installation. If a reinforcement becomes too stiff, it is difficult to form it to the required contour. If it is not stiff enough, it will be distorted by the pressure of subsequent windings.

*Types of Reinforcements:* Reinforcements can be classified according to the form in which the glass, carbon, etc., fiber material is used, e.g., cloth, tape, or filament-wound reinforcements. *Cloth reinforcements* (Fig. 3.25), which are relatively inexpensive to fabri-

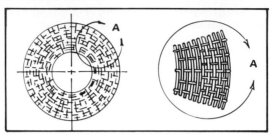

**FIG. 3.25**   Spiral woven cloth. *(From Mater. Des. Eng., February 1965, p.111.)*

cate, are best suited to structures requiring only moderate port reinforcement. Under excessive loads, the woven structure of the cloth reinforcements stretches, causing adhesive-bond failure or cross-shear failure. *Tape reinforcements* can be used in a wide variety of configurations. Most of these consist of strips of preimpregnated tape laid up in equal intervals around, and tangent to, the circumference of the port (Fig. 3.26). Because they are oriented

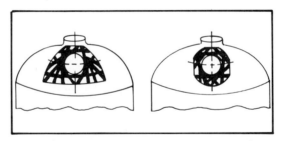

**FIG. 3.26**   Prepreg tape. *(From Mater. Des. Eng., February 1965, p. 111.)*

tangent to the circumference, the tapes resist axial forces in the reinforcement. The tapes can be overlapped to provide an ample cross section to resist bending and shear forces. They also can be oriented in the reinforcement to provide extra cross-shear strength under the hole. Tape reinforcements are fabricated on a form having the same contour as the structure to be fabricated, so that when they are installed the reinforcement closely follows the contour of the winding mandrel. There are four general types of *filament-wound* reinforcements.

▪   Spiral-wound wafers (Fig. 3.27) are fabricated using special winding machines. The roving material is wound on a mandrel consisting of two rigid side plates separated by a spacing disk. The reinforcements are at least equal in strength to tape reinforcements. Furthermore, their fabrication is fully mechanized, so that they have good reproducibility and relatively low cost. Spiral-wound wafers effectively reinforce against internal bending, radial shear, and hoop-tension loads because the filaments are wound normal to any radius of the opening. However, flexural shear is resisted only by the resin bond. Although spiral-wound

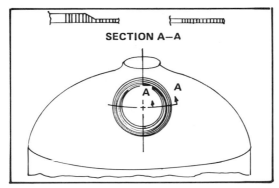

FIG. 3.27 Spiral-wound wafer. *(From Mater. Des. Eng., February 1965, p. 111.)*

wafers provide no reinforcement to the cross-shear area under a hole, the cross shear offers no problem when enough reinforcements are used around the opening. Tension loads in a reinforcement may be considerably higher near the opening. In such cases, the shape of the reinforcement can be tailored to the loads simply by tapering its thickness (Fig. 3.27).

• Spiral-wound teardrop reinforcements (Fig. 3.28) are fabricated like the spiral-wound type except that they have a teardrop shape. They are usually bonded to a piece of

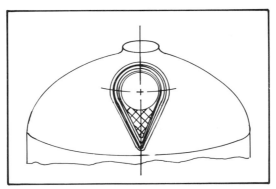

FIG. 3.28 Teardrop wafer. *(From Mater. Des. Eng., February 1965, p. 111.)*

filament-wound mat of the same shape before being wound into a structure. The teardrop shape reduces distortion under load and increases the flexural shear strength of the reinforcement. They are installed with the pointed end below the port, where the wafer and mat combine to reinforce the cross-shear planes.

• Spiral-wound, flattened, hoop-type reinforcements can be used to reinforce two holes with a single wafer (Fig. 3.29). The reinforcements are fabricated like the spiral-wound type except for their flattened hoop shape. The reinforcements are not practical for more than one pair of holes because (Fig. 3.29) failure can occur where the windings of the primary structure bridge over the crossed reinforcements.

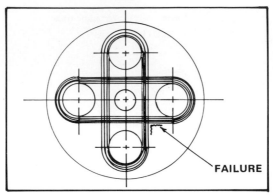

**FIG. 3.29** Flattened-hoop wafer. *(From Mater. Des. Eng., February 1965, p. 111.)*

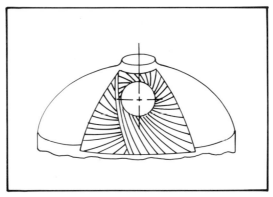

**FIG. 3.30** Filament reinforcement wound to hole contour. *(From Mater. Des. Eng., February 1965, p. 111.)*

■ Port contour reinforcements (Fig. 3.30) are supplementary windings made over a mandrel having a dome of the same contour as the area around the port area of the chamber to be fabricated. When the reinforcement windings have been partially cured, the reinforcement is cut from the mandrel and wound into the structure. Although this method approximates the way polar openings in filament-wound vessels are naturally reinforced, it requires additional mandrels and cutting time and material waste is high.

**Testing Filament Winding:** A major problem in developing design data for any engineering material is correlating the results of standard tests with the performance of the material in its final shape. This is even more critical with filament winding, since the load-bearing characteristics of the material are built in as the part is formed (Fig. 3.31). Because the filaments are the primary load-bearing members and are designed to be stressed only in tension, much of the work in developing meaningful tests for filament-wound reinforced plastics has been aimed at deriving practical tensile-strength values in composite structures. Since the tensile strengths of the filaments are relatively well known, the primary purpose of testing is to determine how well the resin allows the filaments to carry the load efficiently.

FIG. 3.31  Filament winding. *(McClean Anderson.)*

Tensile specimens cut from cured filament-wound shapes are usually curved because of the nature of the structure. Curved specimens will not provide tensile failures if pulled in tension. Also, failures of standard ASTM tensile specimens reinforced with unidirectional filaments are in shear rather than tension. Consequently, special test specimens for use with ASTM standard test methods have been developed.[4] For final proof testing of structures, of course, service or simulated service testing should be performed. The most meaningful data are provided by hydrostatic burst-testing an actual or scaled component or bursting a cylinder with floating seals.

## JOINTS, JOINT CONFIGURATIONS, AND VERIFICATION

The following examples show how the criteria and theoretical and analytical data previously presented are applied by design engineers with the assistance of specialists in stress, weight, and material to test joints and actual parts.

- In the two configurations in Fig. 3.32 the honeycomb panel has the more structur-

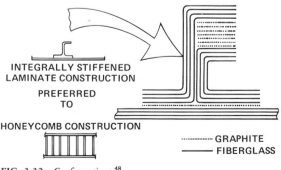

INTEGRALLY STIFFENED
LAMINATE CONSTRUCTION

PREFERRED
TO

HONEYCOMB CONSTRUCTION

·············· GRAPHITE
———— FIBERGLASS

FIG. 3.32  Configurations.[48]

ally efficient geometric shape. Hence the stiffened panel can be weight-competitive only if proper consideration is given to the fact that honeycomb face sheets must exceed a certain minimum thickness in order to survive the normal wear and tear of in-service conditions. For exterior aircraft surfaces, the minimum face sheet thickness varies from one aircraft model to another, but a range of 0.016 to 0.025 in (0.40 to 0.64 mm) is good for reference purposes since it has been used as the minimum gage on several production aircraft programs. Thus, in applications, e.g., external fairings, where these minimum gages must be adhered to, an integrally stiffened laminate fairing of Gr–Ep and Gl–Ep can be weight-competitive with commonly used metal or fiberglass-faced honeycomb fairing. Figure 3.33 shows the wing-to-fuselage fairings which were redesigned from existing honeycomb construction to use integrally stiffened laminate. The integrally stiffened laminate design (Fig.

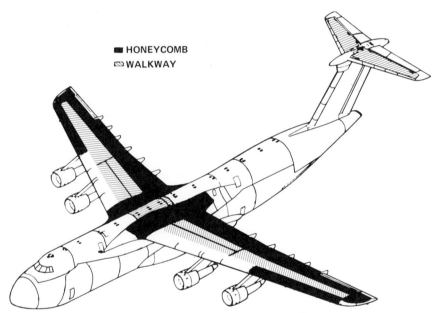

**FIG. 3.33** Wing-to-fuselage aft fairings for C-5A (1 ft = 305 mm).[48]

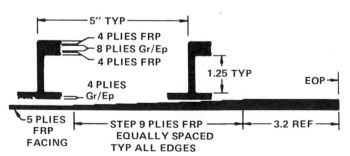

**FIG. 3.34** Integrally stiffened composite design (1 in = 25.4 mm).[48]

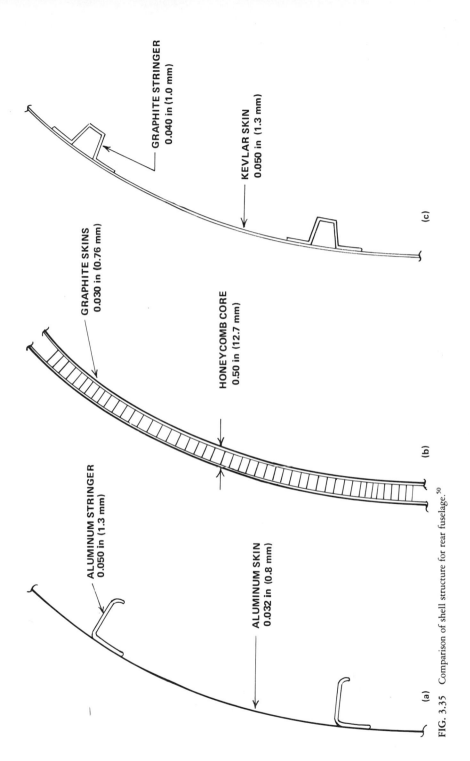

GRAPHITE STRINGER
0.040 in (1.0 mm)

KEVLAR SKIN
0.050 in (1.3 mm)

(c)

GRAPHITE SKINS
0.030 in (0.76 mm)

HONEYCOMB CORE
0.50 in (12.7 mm)

(b)

ALUMINUM STRINGER
0.050 in (1.3 mm)

ALUMINUM SKIN
0.032 in (0.8 mm)

(a)

FIG. 3.35   Comparison of shell structure for rear fuselage.[50]

3.34) had J-section stiffeners of Gr–Ep and Gl–Ep spaced on 5-in (127-mm) centers; panel depth was 1.5 in (38.1 mm) and met four important goals:

1. Fairings would not weigh more than the honeycomb fairings they replaced.
2. Life-cycle costs are reduced since fairings last longer.
3. Fewer and less costly repairs are needed.
4. Only visual inspection is required, not the more costly and time-consuming x-ray and ultrasonic inspection used on honeycomb panels.

■  The results of a recent study of an all-composite rear fuselage for an advanced helicopter showed that Kevlar skins with graphite stiffening elements are lighter and more durable than competing designs (Fig. 3.35 and Table 3.4). The increased durability results

**Table 3.4**  Load Capability of Shell Structures in Fig. 3.35[50]

| Structure | Axial load | | Shear | | Weight | |
|---|---|---|---|---|---|---|
| | lb | kg | lb/in | kg/cm | lb/ft² | kg/m² |
| (a) | 5160 | 2241 | 518 | 92.7 | 0.70 | 3.38 |
| (b) | 5200 | 2359 | 900 | 161 | 0.67 | 3.23 |
| (c) | 6450 | 2926 | 600 | 107 | 0.50 | 2.41 |

from the use of relatively thick, dent-resistant monolithic Kevlar skins. A skin-skeleton concept (Fig. 3.36) is a principal feature in achieving low cost in this design. In skin-skeleton construction the skin is molded as a one-piece component. A mating skeleton integrates frames, longerons, and stringers into a one-piece molding. The two units are then bonded together to form a complete structure.

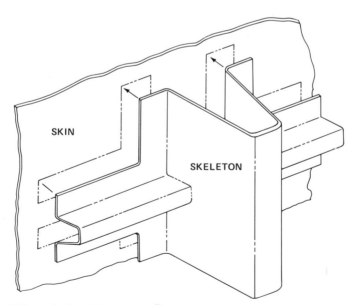

**FIG. 3.36**  Skin-skeleton concept.[50]

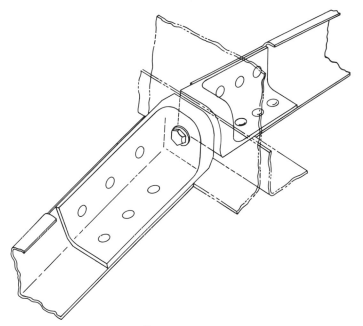

**FIG. 3.37**   Metal stringer joint.[50]

■  The base-line joint in Fig. 3.37 was selected for development in a study of stringer end fittings. Several composite arrangements (Fig. 3.38) were studied and the most promising chosen for development. This arrangement consisted of an all-composite, integrally reinforced angle with an aluminum radius block to distribute bolt tension loads. The joint was designed to carry 6200 lb (2812 kg) tension and 5000 lb (2267 kg) compression. The plies in the joint are oriented so that the tension load is carried into the skin plies in hoop tension of the fibers round the radius block. The compression load is carried partially by transverse shear of the thick section and by fiber end load components at the corner. Test specimens constructed from woven and unidirectional graphite materials for the joint reinforcement and woven Kevlar for the hat stiffener were tested statically in compression and tension; compression failure occurred at 6700 lb (3039 kg) (134% ultimate-load test) due to premature buckling of the unsupported skin, which caused disbonding and collapse of the stiffener. In tension, failure occurred at 8500 lb (3856 kg) (137% ultimate-load test) due to failure of the bolt in bending.

■  Boron-reinforced structures are conventional metallic structural elements (beams, stiffeners, tubes, rods, etc.) selectively reinforced with boron (Fig. 3.39). The reinforcement is an aligned set of boron filaments in a tough, rigid heat-resistant epoxy resin. The boron reinforcement is selectively located to carry the primary loads while preserving the ease and familiarity of metallic attachment schemes such as riveting, welding, and bonding. Boron-reinforced structural elements are 25 to 50% lighter than equivalent all-metal structural elements. Boron-reinforced beams offer the designer new techniques to meet the old requirement of minimum weight and height. The boron reinforcement is selectively located to carry the primary loads, whereas the metal provides the convenience of attachment while

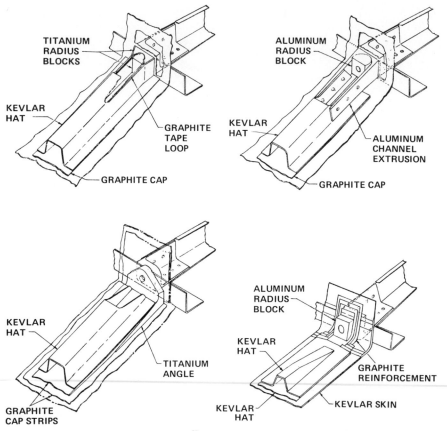

**FIG. 3.38** Composite stringer arrangement.[50]

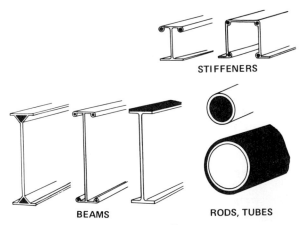

**FIG. 3.39** Boron structural elements.[49]

carrying secondary loads. Figure 3.40 shows a boron-reinforced beam which is typical of a large aircraft floor beam. The design requirements were:

| | |
|---|---|
| Bending moment | 250,000 lb·in (28.3 kJ) |
| Shear | 7500 lb (3401 kg) |
| Stiffness | 23 × 10⁷ lb·in (26 MJ) |
| Length | 20 ft (6 m) |

The charts in Fig. 3.40 reflect the following design savings:

$$\text{Weight at minimum weight} = \begin{cases} 27\% & \text{for equivalent stiffness} \\ 23\% & \text{for equivalent strength} \end{cases}$$

$$\text{Height at same weight} = \begin{cases} 28\% & \text{for equivalent stiffness} \\ 45\% & \text{for equivalent strength} \end{cases}$$

Sturdy stiffeners are often used to increase the compressive-load-carrying ability of a panel structure. As a result of exceptional strength and stiffness, boron-reinforced panel stiffeners readily achieve design objectives of minimum weight and depth. Figure 3.41 compares the weight and dimensions of a boron-reinforced and an aluminum panel.

▪ A new method of joining dissimilar laminated composites without mechanical fasteners consists of a bicomposite transition joint, i.e., a joint made by interleaving the plies of one composite with the plies of another (Fig. 3.42). The interleaving forms a transition area between the composites. Voids in the area are filled with epoxy resin to form a strong, smooth transition between the two materials. There are several advantages:

1. Lower-cost laminates can be used in parts of the structure where more costly components are not necessary.
2. More flexible fibers can be used where a structure has intricate contours.
3. Properties can be tailored to the structure; e.g., a low-thermal-conductivity composite can be used where heat insulation is required and a high-thermal-conductivity fiber where heat is desired.
4. Virtually any combination of composite material can be joined.

▪ Gr–PI prepregs can be manufactured into honeycomb-cored sandwich panels. The panels consist of a honeycomb Gr–PI core covered with thin face sheets of the same material. Panels are useful up to 500°F (260°C), and at elevated temperatures these panels exhibit better properties than aluminum (Table 3.5).

▪ In order to verify composite-stiffened-designed structures NASA sponsored a program to optimize the design and fabrication test envelope to produce a panel that will fail at a prescribed load and in a prescribed manner. Designs were optimized for load vs. weight by changes in fiber pattern for the following stiffener shapes:[5]

Solid blade (Fig. 3.43)
Hat (Figs. 3.44 and 3.45)
Honeycomb blade

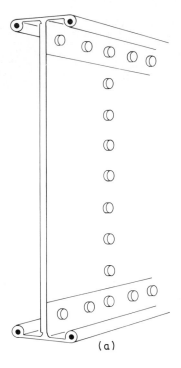

**(a)**

FIG. 3.40 (*a*) Boron-reinforced beam; (*b*) and (*c*), resulting design savings.[49]

J-stiffened panel (Fig. 3.46)

Sine-wave T-stiffened panel (Fig. 3.47)

■ Interconnection of primary structural components fabricated from advanced composites is important for the efficient and reliable use of these materials in aircraft structures. Mechanical fastening using bolts and rivets is required to provide for reasonable component repair and replacement. The lack of plastic redistribution in composites means that the analysis of mechanical fastening must be more precise than for metals. These methods consist of finite-element analysis for accurate prediction of the bolt-load direction (in a complex joint) and the optimization of ply lay-ups through laminate analysis to reduce the stress concentrations. In addition, the design must ensure that interface of the bolt and composite will prevent premature fracture initiated by micro buckling of the outer ply layer. Composite joints require more detailed analysis and design than metal joints, but their advantages justify their use. One major advantage is improved efficiency in fatigue. In normal practice the fatigue strength of metals is less than 10% of the static strength as a result of corrosion, fretting, and the need to reduce stress levels for large fatigue scatter. In composite joints, e.g., the main-rotor damper lug in a commercial helicopter (Fig. 3.48), the fatigue strength, including all factors, resulted in a 40% utilization of static strength.

■ A bolted shear attachment of the main-rotor transmission to a graphite composite beam and frame structure simulating the overhead roof of a military helicopter was designed and tested. This joint was the most complex and highly loaded attachment of the airframe. The results of tests and predicted fracture strength show (Fig. 3.49) that the average margin of safety was 12%; no test point fell below the predicted strength.

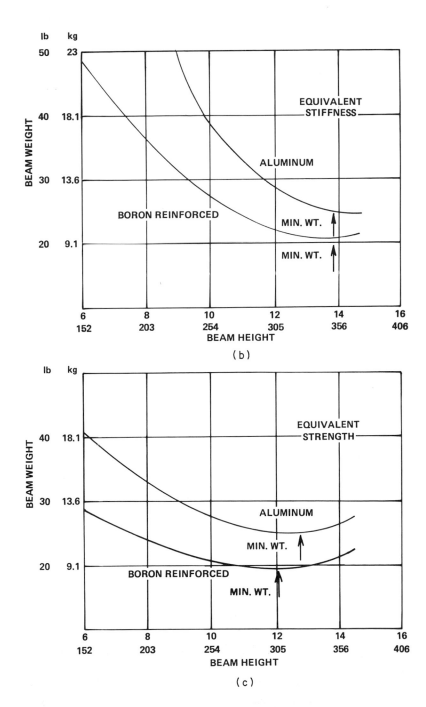

(b)

(c)

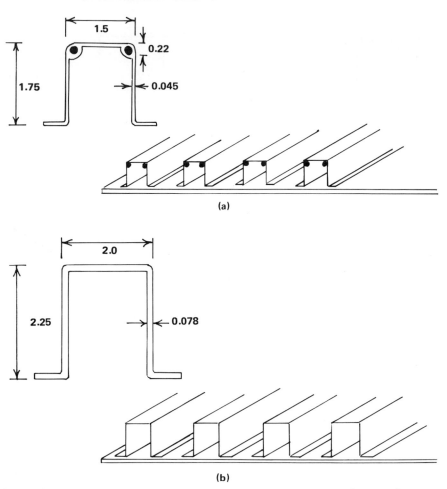

(a)

(b)

FIG. 3.41 (*a*) Boron-reinforced compression panel. Area of aluminum = 0.34 in² (219 mm²); area of reinforcement = 0.076 in² (49 mm²); skin thickness = 0.068 in (1.7 mm); panel weight = 28.6 lb (13 kg). (*b*) Conventional 2024-T3 aluminum panel. Area = 0.58 in² (374 mm²), skin thickness = 0.060 in (1.5 mm); panel weight = 39 lb (17.7 kg); depth = 2.31 in (58.7 mm). For both panels length = 48 in (1219 mm), width = 36 in (914 mm), stiffener pitch = 4 in (102 mm) on center; nonbuckling skin.

■ Four types of structural elements representing typical primary aircraft structure have been developed to confirm the validity of theories (Fig. 3.50). The loaded hole element represents (1) typical mechanical-tension attachment to metal structure, (2) the tension and compression elements, solid tension and compression covers, respectively, and (3) typical substructure of the I-beam element.[6] The loaded hole element (Fig. 3.51) represents a typical root splice joint of a primary composite structure. The tapered Gr–Ep composite cover is mechanically attached to the alumi-

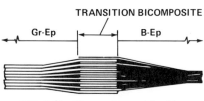

FIG. 3.42   Bicomposite transition joint.

**Table 3.5** Comparison of Honeycomb Materials†

| Material | Density | | Temp | | Shear modulus | | | | Compression modulus | |
|---|---|---|---|---|---|---|---|---|---|---|
| | | | | | Longitudinal | | Transverse | | | |
| | lb/ft³ | g/cm³ | °F | °C | ksi | MPa | ksi | MPa | ksi | MPa |
| Gr–PI | 1.9 | 0.030 | 68 | 20 | 32.4 | 223 | 12.5 | 86.2 | 19.6 | 135.1 |
| | | | 450 | 232 | 27.8 | 191.7 | 13.5 | 92.4 | 19.6 | 135.1 |
| Aluminum | 2 | 0.032 | 68 | 20 | 27 | 186.2 | 13 | 89.6 | 45 | 310 |
| | | | 450 | 232 | 17 | 117.2 | 8.2 | 56.5 | 28.3 | 195.1 |
| Gl–PI | 2.5 | 0.040 | 68 | 20 | 14 | 97 | 6.0 | 41 | 20 | 138 |
| | | | 450 | 232 | 11.5 | 79.3 | 4.9 | 33.8 | 16.4 | 113 |

†Data from Hercules, Inc.

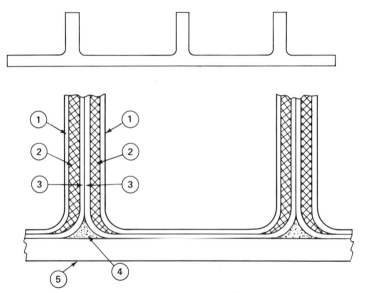

FIG. 3.43  Solid blade stiffened panels: 1 = mainly ±45° (0.79 rad), 2 = 0°, 3 = outer skin, 4 = fiber-glass B-stage layer.[5]

num splice plate with fasteners. Figure 3.52 shows a compression cover section of a primary empennage structure. The double-tapered constant midsection of the composite cover simulates the variable thickness of an actual cover hardware component. Related substructure and tension cover components that are not considered critical test areas are simulated by metallic details; Fig. 3.53 depicts the shear beam element.

■  A preliminary design concept[7] using advanced composites as the structural materials for the primary and secondary structural systems of an advanced launch vehicle is shown in Figs. 3.54 and 3.55.

## Computer-Aided Design Systems

With the development of computers a number of different systems have been built, all basically consisting of an interactive design tool for generation of component geometries on

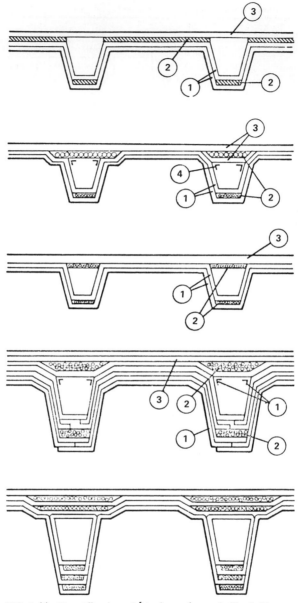

**FIG. 3.44**  Hat-stiffened panels[5] (code numbers as in Fig. 3.43).

a cathode-ray-tube display and permanent storage of the data in a computer. Geometries are defined exactly the same as on the drawing board; coordinates, lengths, angles, etc., are input to the computer using a lightpen and keyboard. The dimensions are stored digitally with great precision, and the computer can be interrogated to determine resultant locations and distances without any calculation by the operator. Geometry is created in two or three dimensions, and hard copies are generated at any desired scale on paper or stable Mylar.

**FIG. 3.45** Skin face of hat-stiffened panel in conjunction with beading for ballistics.[50]

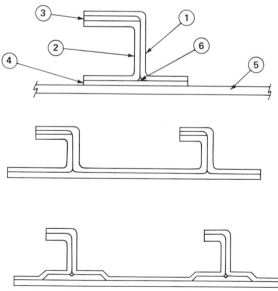

**FIG. 3.46** J-stiffened panels: 1,2 = mainly ±45°, 3,4 = all 0°, 5 = outer skin 0 ± 45°, 90°, 6 = all 0° roll.[5]

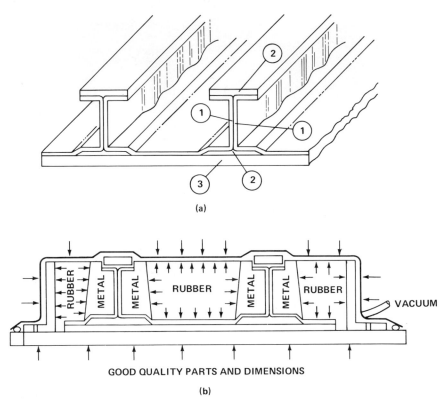

**(a)**

GOOD QUALITY PARTS AND DIMENSIONS

**(b)**

**FIG. 3.47** (*a*) Sine-wave T-stiffened panel: 1 = mainly 45°, 2 = 0°, 3 = 0°/±45°,90°. (*b*) Tooling techniques.[5]

The systems are especially well suited to the design of composite structure, where the designer must engineer the material system as well as the component form. To accomplish this, the fiber orientation within a given ply as well as the location and shape of the ply must be defined. The laminar nature of tape and woven composite materials provides the designer with an opportunity to distribute material efficiently within a component but requires more data to be generated. In particular, since laminate cross sections are very time-consuming to define manually, the ease of drafting with the computer-aided design system allows design iteration with much less effort.

Another major advantage of computer-stored data is the ability to take a "copy" of an existing drawing and use it as the basis for further more detailed design or to extract portions of the data to generate an interface with another component. This data transfer is accomplished error-free and at negligible cost in time since the person drafting does not regenerate any information.

Some of the greatest benefits of computer-aided design systems come in the area of manufacturing. The geometries of individual parts are given an identity in the data base, and to obtain detailed information for tooling purposes manufacturing operations can access the drawing and select the specific part by name or number. This eliminates the need for the stable, highly accurate dimensionless Mylar master drawings used solely to provide

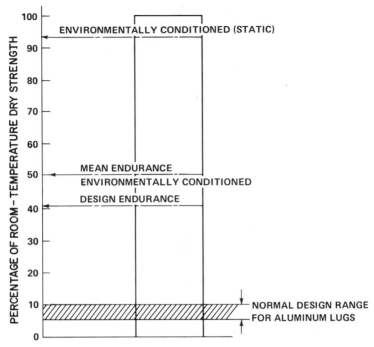

**FIG. 3.48** Composite main-rotor damper-lug specimens showing large design fatigue strength.[50]

information for parts manufacture. Manufacturing selects the required geometry, modifies as required, and creates a new tool file of data for manufacturing use. Information is transferred with no loss of accuracy.

Flat patterns of composite plies are generated with the help of automatic routines (identified by number) and positioned in a "nest" to obtain efficient use of raw material. The nested pattern information can then be used in numerically controlled fabric-cutting equipment, automatic dimensioning, and automatic nesting.

## Applications of Composite Substitution for Metal

### Blades

Incorporation of MMCs into the fan section of a gas-turbine engine has the potential of improving the engine because of the large size and weight of the fan. Significant improvements in cost, weight, efficiency (fuel consumption), and life-cycle costs can be realized. Use of B–Al composites requires low-cost fabrication and improved FOD resistance while maintaining the design properties of the blade. The major conclusion from this program found that B–Al is a viable material for blade application in terms of both properties and cost.[8]

B–Al was also considered for blade use in a typical Supersonic Cruise Aircraft Research

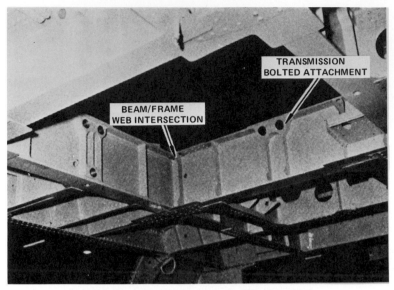

THE TRANSMISSION BOLTED JOINT ATTACHMENT IS
THE MOST DIFFICULT IN THE AIRFRAME

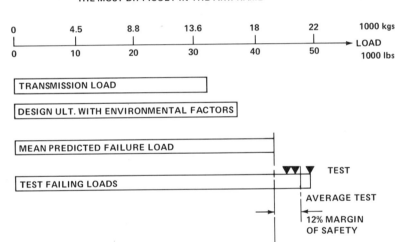

**FIG. 3.49** Transmission joint and test results. *(Sikorsky Aircraft, Division of United Technologies.)*

(SCAR) fan. The trends in testing the blade shown in Fig. 3.56 were that as the impact strength increased, tensile and shear strengths decreased. Of the six-ply orientations investigated, the highest combination of impact and tensile strengths were derived from the ± 15° (0.26-rad) orientation. Using of B–Al resulted in a shroudless fan-blade design which satisfied aeromechanic stability design requirements.[9]

Another program dealing with improving fan-blade impact improvement was directed toward fiber composite materials and foreign-object impact-damage resistance. Figure 3.57 shows a trimetric view of the general arrangement of the plies in the blade.[10]

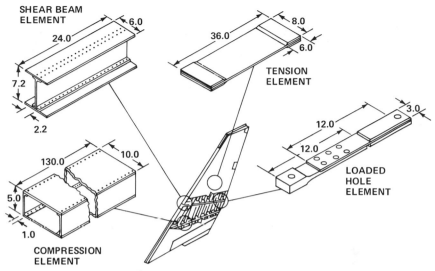

**FIG. 3.50**  Structural elements.[6]

The successful field testing of Bsc–6061 aluminum-alloy composite gas-turbine engine blades was the first of an advanced composite material applied to an engine primary structural component, i.e., the third-stage fan.[11] The principal conclusion was that Bsc–6061 alloy third-stage fan blades, designed, fabricated, and inspected using the procedures developed in Ref. 11 would be acceptable for flight service subject to periodic inspection. The same material was successfully used and tested as a first-stage composite fan blade in another engine application. Various other blade-development programs (design, fabrica-

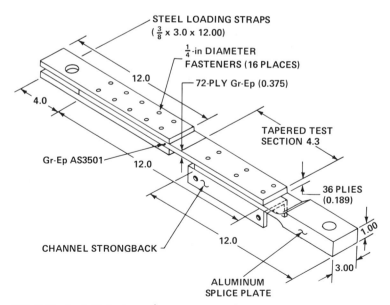

**FIG. 3.51**  Loaded hole element.[6]

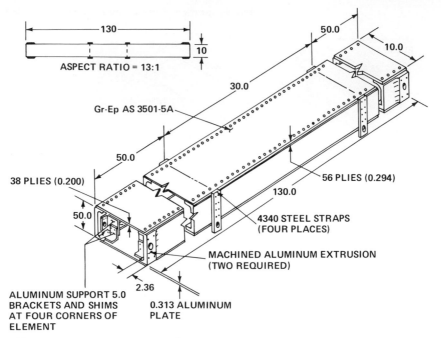

FIG. 3.52    Compression cover element.[6]

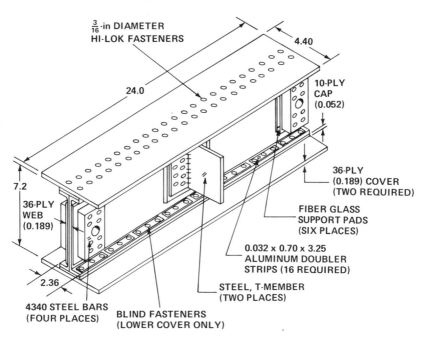

WEBS, CAP AND COVER DETAILS
MADE FROM Gr-Ep AS3501-5A MATERIAL

FIG. 3.53    Shear beam element.[6]

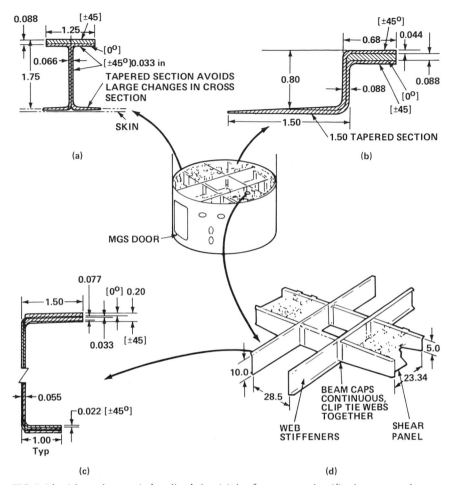

**FIG. 3.54** Advanced-composite base-line design: (*a*) ring-frame cross section, (*b*) stringer cross section, (*c*) beam cross section, (*d*) internal beam arrangement; 0°/45° (0/0.79 rad).[7]

tion, testing) have covered the following materials: Bsc–Al (Ref. 12), B–SiC–Ti–6Al–4V (Refs. 13 and 14), and Gr–Ep and B–Ep (Ref. 15). Other components of gas-turbine engines, vanes, and nozzle flaps have also been considered for composite applications. A composite nozzle flap has been designed and tested for an advanced engine. The performance of this same engine has also been successfully evaluated with first-stage fan vanes (Fig. 3.58).[16,17]

## Aircraft, Cars and Trucks, Auto Components, Space Vehicles

**Cars and Trucks and Auto Components:** Several programs under government and private sponsorship have focused on improving the fuel economy of passenger and freight road-transportation vehicles. More expensive materials are being considered than in the past

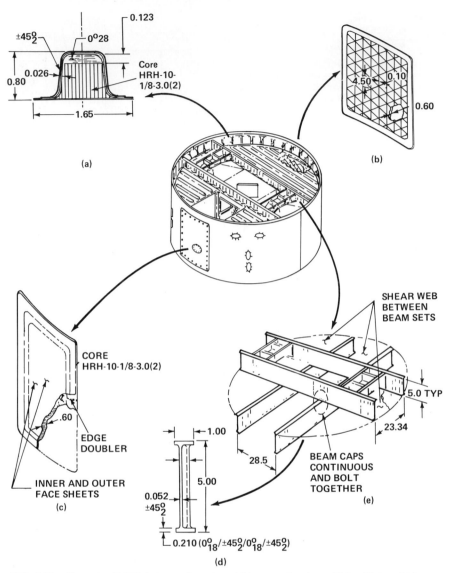

**FIG. 3.55** Alternative 0°/45° design configurations: (*a*) honeycomb stringer, (*b*) isogrid door, (*c*) honeycomb door, (*d*) sine-wave beam, (*e*) sine-wave over-and-under arrangement.[7]

because the cost-effectiveness calculations indicate that in many cases the higher initial cost will be more than offset by decreased fuel costs resulting from the reduced weight. Furthermore, the primary materials under consideration for structural components, Gr–Ep and thermoplastics, offer additional advantages, e.g., reduced corrosion and improved fatigue properties, which further reduce the life-cycle costs for the transportation vehicle. Components that have been designed and tested include driveshafts, crashworthy door beams, bumpers, and leaf springs for automobiles. Truck springs and frame rails have been studied

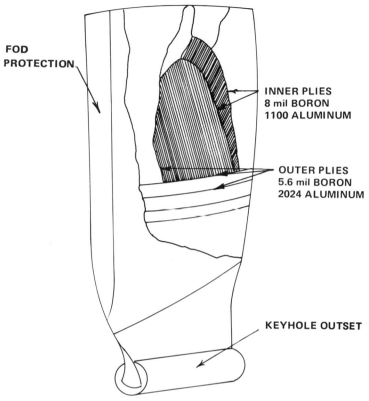

**FOD PROTECTION**

**INNER PLIES
8 mil BORON
1100 ALUMINUM**

**OUTER PLIES
5.6 mil BORON
2024 ALUMINUM**

**KEYHOLE OUTSET**

**FIG. 3.56**  Hybrid B–Al blade design.[9]

by various groups. A government report deals with the design and test of a Gr–Ep composite used to reinforce a standard steel truck frame.[18]

The Gr–Ep-reinforced truck-frame rail is typical of an application for which maximum use is made of the unique properties of the composite. Since the idea behind reinforcing a frame rail is to increase stiffness and provide for greater gross-vehicle-weight capacity, a composite-reinforced rail was designed to place high-stiffness graphite in the areas where it could be used in a cost-effective manner. The standard steel-reinforced rail section and the graphite-composite-reinforced rail section are shown in Fig. 3-59. A 90% reduction in reinforced weight was achieved using composite, which results in a 35% reduction in reinforced-section weight. To demonstrate the capability of the composite-reinforced rail, one-third scale specimens were fabricated and tested.

1. Fatigue failure of the composite or bond does not occur within 0.5 million cycles at a stress level of 34 kips/in$^2$ (234 MPa).
2. Application of graphite-composite caps can increase the ideal buckling-load capacity of a steel channel by at least 50% even after 0.5 million fatigue cycles.

A driveshaft for a light truck (Fig. 3.60) is an example of potential optimal-use material and angulation. First, a filament-winding technique lays down E glass on an extractable

LOCAL TRANSVERSE
STRENGTHENING PLIES

BORON TORSIONAL
STIFFENING PLIES

Gr-K CENTRIFUGAL
AND FLEXURAL
LOAD PLIES

NICKEL PLATE
LEADING EDGE
FOD PROTECTION

S-GLASS FLEX
ROOT PLIES

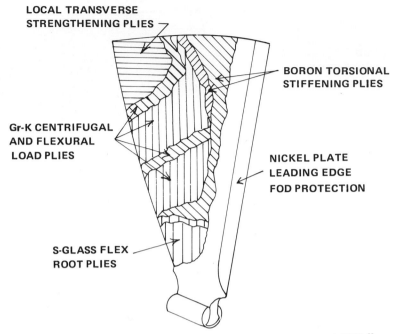

**FIG. 3.57**  Composite blade for quiet, clean short-haul experimental engine (QCSEE).[10]

mandrel at $\pm 45°$ (0.79 rad) to take the torque loads. Carbon is then placed longitudinally at approximately 5 to 10° (0.09 to 0.18 rad) on the tube to develop sufficient axial stiffness to ensure that the resonant frequency of the tube will be above engine speed. Finally, the outer glass hoop wrap [80 to 90° (1.4 to 1.6 rad)] cinches in the inner plies and provides resistance to torsional buckling and to FOD. Although the tube weighs almost 8 lb (3.6 kg), only 1 lb (0.45 kg) of carbon fiber is required. With such composites, a high critical frequency can be designed into the shaft without resorting to a two-piece structure. Current steel assemblies consist of two separate tubes with a center support bearing. Prototype C–Ep driveshafts have been under test since 1979 and should reach production readiness by the mid 1980s.

Facing the same fuel constraints and a federally mandated fuel economy standard that adds a few miles every 2 or 3 years, the auto industry has not been able to apply the new composite technology to their product as readily as the aerospace industry. Part of the problem is price, and another factor is fabrication. Carbon-fiber composites can cost (1984) upwards of $18 per pound ($39.60 per kilogram), while steel can be obtained for 20 cents per pound (44 cents per kilogram) and aluminum for approximately 85 cents per pound ($1.87 per kilogram). Because of hand operations and long hours, composite-materials fabrication currently does not lend itself to automation of the automotive assembly line. Nevertheless, the weight and strength factors of these new materials are too important for the fuel-efficiency-pressed automakers to ignore. A graphite-fiber-reinforced experimental car (Fig. 3.61) containing 160-odd graphite-fiber parts demonstrated the possibilities (not the practicability) for graphite-fiber-reinforced composite in cars. This vehicle weighed 2504 lb (1136 kg), about 1250 lb (567 kg) less than its metal-body counterpart now on the road (Table 3.6).

**Aircraft:** Imaginative application of technology invariably leads to better products and better ways to build them. One of the most significant product improvements is the introduction of a fiber-glass-composite main-rotor blade for a commercial helicopter. The fiber-glass blade (Fig. 3.62), now certified by the FAA and flying in production helicopters, shows better service life than bonded metal blades. The fiber-glass blade is virtually immune to corrosion and much more resistant to fatigue-crack propagation. It has already been approved for a 7200-h service life, compared with 3600 h for a metal blade. Tests continue for ultimate approval for unlimited fatigue life.

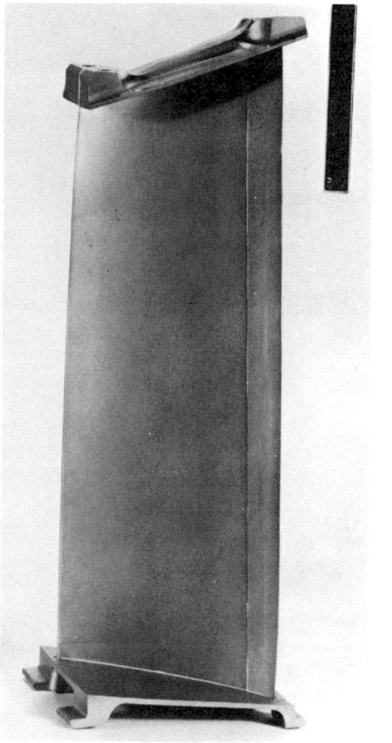

Numerous design, fabrication, and demonstration programs involving structural and nonstructural composite structures have been conducted throughout the aerospace industry over the past dozen years. Most programs were government-sponsored and related to commercial and military aircraft. Some of the more significant follow:

FIG. 3.58 Finished B–Al vane assembly.[17] (*Hamilton Standard, Division of United Technologies.*)

1. Design, manufacture, and ground test of a wing torque box (Fig. 3.63) using B–Ep and Gr–Ep composites.[20]

2. A trailing-edge wing flap and rudder fabricated and successfully test-flown and demonstrated near term production applicability.[21]

3. B–Ep leading-edge composite slats,[22] B–Ep wing boxes, which have been flying for 7 years,[23] and spoilers. One spoiler program involved selection of flight spoilers of the 737 airplane as components to advance the use of Gr–Ep materials for structural applications. In this program spoilers were developed with the aluminum skins replaced by Gr–Ep and the aluminum end ribs replaced by fiber glass. Figure 3.64 shows a schematic of a 737 spoiler with Gr–Ep skins.[24]

4. A Navy-sponsored program for designing, developing, fabricating, and testing spoilers for the S-3A aircraft. The spoilers, with sandwich construction and Gr–Ep skins, have been monitored for 5 years of in-service testing and experience. In addition to exhibiting excellent static strength and fatigue resistance, the graphite spoilers are approximately 40% lower in cost and 30% lower in weight than the metal spoiler.

5. Composite horizontal stabilizers for F-111, F-14A, F-15, B-1, and A-4 aircraft fabricated using B–Ep and Gr–Ep materials. The weight reduction (Fig. 3.65) shows significant savings over the originally designed metal stabilizer.[25,26]

6. Composite rudders of Gr–PI, B–PI, and B–Ep for the F-4 aircraft (Fig. 3.66).[27]

7. Kevlar fairings on the L-1011 airplane, including the center engine and wing-body panels. These fairings are being exposed to commercial airline service for 5-year eval-

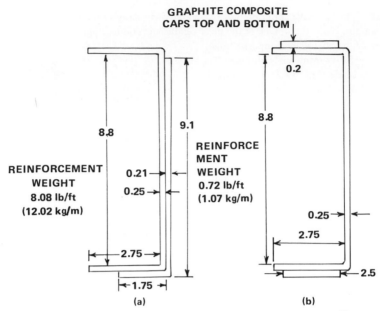

GRAPHITE COMPOSITE
CAPS TOP AND BOTTOM

8.8

9.1

0.2

8.8

REINFORCEMENT
WEIGHT
8.08 lb/ft
(12.02 kg/m)

0.21

0.25

REINFORCE
MENT
WEIGHT
0.72 lb/ft
(1.07 kg/m)

0.25

2.75

2.75

1.75

2.5

(a)

(b)

**FIG. 3.59** Rail sections reinforced with (*a*) steel and (*b*) graphite composite.[18] (*Lewis Research Center.*)

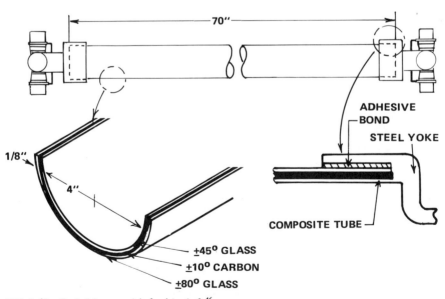

70"

ADHESIVE
BOND

STEEL YOKE

1/8"

4"

COMPOSITE TUBE

±45° GLASS

±10° CARBON

±80° GLASS

**FIG. 3.60** Optimizing materials for driveshaft.[46]

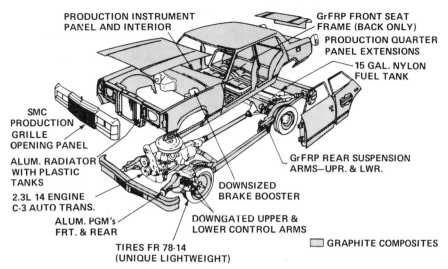

PRODUCTION INSTRUMENT
PANEL AND INTERIOR

GrFRP FRONT SEAT
FRAME (BACK ONLY)

PRODUCTION QUARTER
PANEL EXTENSIONS

15 GAL. NYLON
FUEL TANK

SMC
PRODUCTION
GRILLE
OPENING PANEL

ALUM. RADIATOR
WITH PLASTIC
TANKS

2.3L 14 ENGINE
C-3 AUTO TRANS.

ALUM. PGM's
FRT. & REAR

GrFRP REAR SUSPENSION
ARMS—UPR. & LWR.

DOWNSIZED
BRAKE BOOSTER

DOWNGATED UPPER &
LOWER CONTROL ARMS

TIRES FR 78-14
(UNIQUE LIGHTWEIGHT)

▨ GRAPHITE COMPOSITES

**FIG. 3.61**   Lightweight vehicle.[19]

uations which will provide a degree of confidence in the material which cannot be obtained by any simulated environmental testing on test coupons or part segments.[28]

8. Gr–Ep speed brake for A-7 aircraft. The existing A-7 metal speed brake (Fig. 3.67) is constructed from 7075-T6 aluminum sheet metal and machined forgings. The main-actuator attach fitting is a large aluminum machined forging with separate stainless-steel lugs. The complete speed brake (including chines but not actuator, linkage, or attach hardware) contains about 300 detail parts and weighs 123 lb (56 kg). In design-ing the speed brake using composites, a general design approach and detailed design criteria were formulated. This approach has been adopted by many companies in an effort to maximize effectiveness in obtaining the program goals:

   *a.* The composite must be a functional replacement of the existing metal speed brake, requiring no aircraft modifications, flight restrictions, or decrease in performance.

**Table 3.6**   Weight Summary for Graphite-Fiber-Reinforced Plastic Car[19]

Weight of composite is subtracted from weight of steel to show savings in weight

| Part | lb | kg | Part | lb | kg | Part | lb | kg |
|---|---|---|---|---|---|---|---|---|
| Body | 461.0 | 209.1 | Hood | 49.0 | 22.2 | Wheels | 92.0 | 41.7 |
| | 208.0 | 94.3 | | 16.7 | 7.6 | | 49.3 | 22.4 |
| | 253.0 | 114.8 | | 32.3 | 14.6 | | 42.7 | 19.3 |
| Frame | 282.8 | 128.3 | Deck lid | 42.8 | 19.4 | Doors | 155.6 | 70.6 |
| | 207.2 | 94.0 | | 13.9 | 6.3 | | 61.1 | 27.7 |
| | 75.6 | 34.3 | | 28.9 | 13.1 | | 94.5 | 42.9 |
| Front end | 96.0 | 43.5 | Bumpers | 123.1 | 55.8 | Misc† | 69.3 | 31.4 |
| | 29.3 | 13.3 | | 44.4 | 20.1 | | 35.8 | 16.2 |
| | 66.7 | 30.2 | | 78.7 | 35.7 | | 33.5 | 15.2 |

†Bracketry, seat frame, etc.

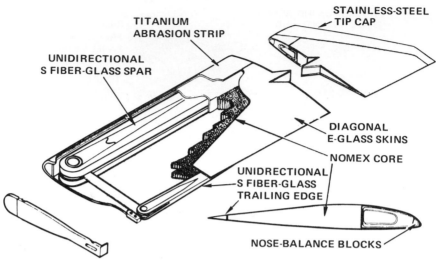

**FIG. 3.62** Breakout of fiber-glass composite main-rotor blade.[36]

   *b.* The composite should use structural composite materials to the maximum extent, so that a maximum amount of design and fabrication experience can be obtained. Metal should be used only in applications where the use of composites would be unfeasible or inefficient.

   *c.* It should be designed for maximum structural efficiency (for minimum weight) within the limits of available manufacturing capabilities. The composite was designed and fabricated primarily from Gr–Ep prepreg material. The total predicted

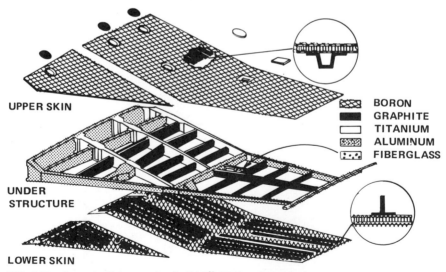

**FIG. 3.63** Composite flight-test wing for F-15.[20] *(McDonnell-Douglas.)*

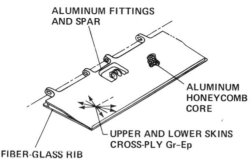

FIG. 3.64 Construction of Gr–Ep spoiler; length 52 in (1.32 m), width 22 in (0.56 m), surface area 15.8 ft² (1.47 m²), weight 12 lb (5.4 kg), weight saving 15%.[1]

weight of the graphite speed brake was 74 lb (34 kg), representing a 40% weight saving.[29]

9. Composite applications for missiles, aircraft, and helicopters (Chap. 8).

   a. Advanced fighter, hybrid supersonic composite inlet (Fig. 3.68).

   b. L-1011 composite aileron which is 26% lighter and has 50% fewer parts than the metal aileron. Two full-scale ailerons (Fig. 3.69) have been designed, fabricated, and tested. The aileron is a multirib configuration with single-piece upper and lower covers mechanically fastened to the substructure. The basic materials were Gr–Ep unidirectional tape, bidirectional Gr–Ep fabric, and a syntactic epoxy core. The aileron covers are thin sandwich plates with three-ply Gr–Ep tape face sheets around a syntactic core. This construction enables a large reduction to be made in the number of ribs needed to stabilize the skin. The composite aileron had 10 ribs compared with 18 for the metal aileron. Where concentrated loads were introduced into the cover, Gr–Ep tape doublers substituted locally for the syntactic epoxy. The intermediate ribs were fabricated of Gr–Ep fabric, as were the three main ribs that bear the hinge and actuator loads. Main-rib caps were reinforced with Gr–Ep tape. Forged aluminum bathtub fittings transfer concentrated

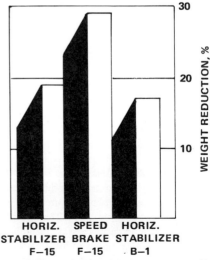

FIG. 3.65 Composite vs. metals for stabilizers.[25]

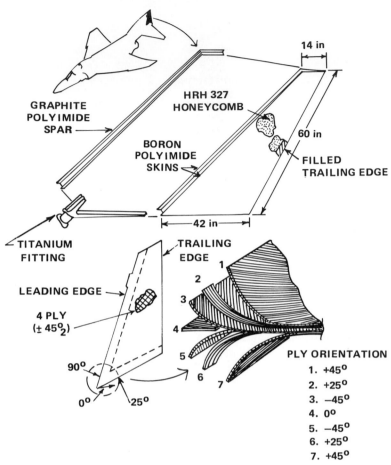

**FIG. 3.66** F-4 polyimide rudder scale-up component; $0°/+25°/-45°/90°$ ($0/0.44/0.79/1.6$ rad).[27]

loads into the ribs. The front spar was Gr–Ep tape laid up in an approximately quasi-isotropic orientation. The aileron assembly was completed by addition of the aluminum leading-edge shroud. Gl–Ep fairings, aluminum hinge-actuator fittings, and a Kv–Ep trailing edge (Fig. 3.70). Certification by the FAA and initiation of commercial flight service of 10 shipsets (the complement for one plane) of L-1011 Gr–Ep inboard ailerons occurred in November 1981. In flight tests the ailerons proved to be flutter-free and exhibited excellent damping characteristics. One test flight included taking the L-1011 to maximum dive speed of Mach 0.95. Aerodynamic characteristics of the composite structures were identical to those of heavier aluminum components currently being used. Weight saving amounts to 65 lb (29.4 kg) per pair (Fig. 3.71). Three years of testing has been required to establish stress and fatigue characteristics, including thorough ground testing of various samples from small coupons to full-scale subcomponents, subassemblies, and complete test articles. Observance and reporting on the performance of the ailerons will continue for another 5 years.

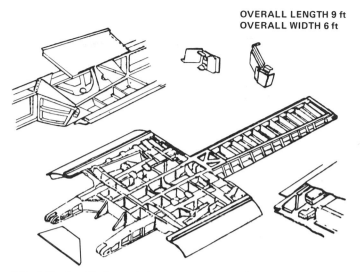

**FIG.** 3.67    Speed-brake metal design; 1 ft = 305 mm.[1]

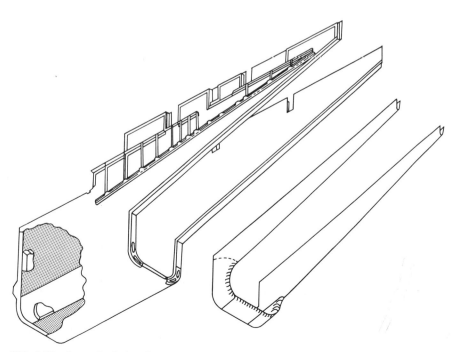

**FIG.** 3.68    Composite design of supersonic inlet.

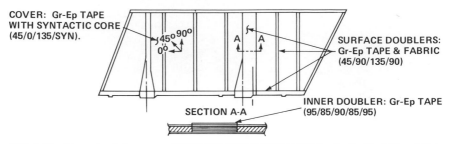

COVER: Gr-Ep TAPE
WITH SYNTACTIC CORE
(45/0/135/SYN).

45° 90°
0°

A

A

SURFACE DOUBLERS:
Gr-Ep TAPE & FABRIC
(45/90/135/90)

INNER DOUBLER: Gr-Ep TAPE
(95/85/90/85/95)

SECTION A-A

**FIG. 3.69**    Schematic of aileron cover 0°/45°/90°/135° (0/0.79/1.6/2.4 rad). *(Lockheed Aircraft Co.)*

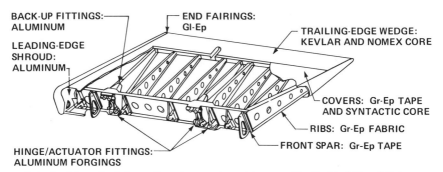

BACK-UP FITTINGS:
ALUMINUM

END FAIRINGS:
Gl-Ep

TRAILING-EDGE WEDGE:
KEVLAR AND NOMEX CORE

LEADING-EDGE
SHROUD:
ALUMINUM

COVERS: Gr-Ep TAPE
AND SYNTACTIC CORE

RIBS: Gr-Ep FABRIC

FRONT SPAR: Gr-Ep TAPE

HINGE/ACTUATOR FITTINGS:
ALUMINUM FORGINGS

**FIG. 3.70**    Schematic of aileron ribs and caps and remaining assembly. *(Lockheed Aircraft Co.)*

**Space Vehicles:**    The anisotropy of graphite fibers with respect to thermal expansion makes them an excellent material for constructing components that must have maximum dimensional stability under conditions of changing temperature. One such structure is the metering truss for NASA's Large Space Telescope (LST). The truss (Fig. 3.72) is required to keep the focal distance between the primary and secondary mirrors of the LST reasonably constant in order to maintain optical characteristics. The LST will be placed in orbit by the Space Shuttle, orbiting at an altitude of 400 mi (643.7 km). The goal of the LST is to be able to look 10 times deeper into space than the largest telescope on earth.[30] Gr–Ep was selected since structural-element tests showed it to have acceptable strength, stiffness, and outgassing characteristics for this application. The truss was designed to have essentially zero coefficient of thermal expansion in the circumferential direction. The graphite fabric and fibers in the rings and struts were given zero coefficient of thermal expansion by shape and ply orientation and the number of plies.

S-glass and Kevlar properties were evaluated for use in a design involving composite material as a means of reducing costs of a disposable tank for liquid hydrogen in the Space Shuttle. After consideration of the mechanical properties of the two composite materials used for the candidate designs in Table 3.7, S-glass was selected.[31]

**Engine Nacelles:**    A design, fabrication, and test verification program has established the effective use of composite turbofan engine frames compared with metal frames.[32] The engine fan frame is representative of a class of structures for advanced turbofans in 1980–1990. The TF34 engine has a sophisticated multimaterial metal frame already optimized for low weight at 186 lb (84.4 kg). The TF34 composite frame (Fig. 3.73) weighs 142 lb (64.4 kg), a 44-lb (20-kg) (24%) weight savings per engine. The designer, materials engineer, production planner, and production supervisor worked closely in reducing the overall

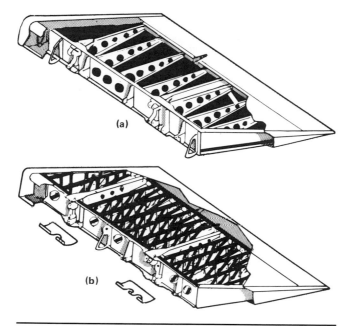

| Type | Weight | | Composite material, % | Number of: | | |
| | lb | kg | | Ribs | Parts | Fasteners |
|---|---|---|---|---|---|---|
| Aluminum | 139.9 | 63.5 | 5.8 | 18 | 398 | 5253 |
| Composite | 107.4 | 48.7 | 61.9 | 10 | 205 | 2574 |

**FIG. 3.71** Comparison of (*a*), aluminum and (*b*) composite aileron for L-1011.[54]

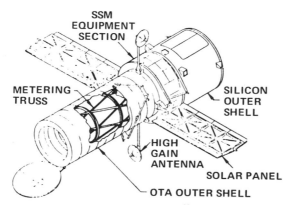

**FIG. 3.72** Gr–Ep for truss construction.[30]

**Table 3.7**  Properties of Uniaxial Filament-Wound Composite Materials for Use in Parametric Study of Filament-Overwrapped Tanks[31]

| Filament | sp gr | Ultimate strength | | Elastic modulus | |
|---|---|---|---|---|---|
| | | ksi | MPa | $10^6$ psi | GPa |
| SGl–Ep | 2.4 | 665 | 4585 | 12.4 | 855 |
| Kv–Ep | 1.4 | 388 | 2675 | 18.5 | 127.6 |

| | | | Composite | | | | | | |

| | Filament fraction, vol % | sp gr | Temp | | Longitudinal modulus | | Longitudinal tensile ultimate strength | | Longitudinal tensile operating stress† | |
|---|---|---|---|---|---|---|---|---|---|---|
| | | | °F | °C | $10^6$ psi | GPa | ksi | MPa | ksi | MPa |
| SGl–Ep | 67 | 2.0 | 350 | 177 | 8.3 | 57 | 174 | 1200 | 104 | 717 |
| | | | 75 | 23.8 | 8.3 | 57 | 220 | 1517 | 132 | 910 |
| | | | −320 | −196 | 9.1 | 63 | 275 | 1896 | 165 | 1138 |
| Kv–Ep | 65 | 1.4 | 350 | 177 | 10.6 | 73 | 144‡ | 993‡ | 96§ | 662§ |
| | | | 75 | 23.8 | 12.2 | 84 | 180 | 1241 | 120§ | 827§ |
| | | | −320 | −196 | 13.9 | 96 | 180 | 1241 | 120§ | 827§ |

†All operating stresses are based on cyclic loading from zero stress to full operating stress, a conservative basis.
‡Estimated value.    §Assumed value based on 1.5 safety factor.

frame costs by a direct reduction in material costs and selecting materials that could be processed more efficiently, including Gr–Ep tape, K–Ep or Gl–Ep fabric, Gl–Ep (compression-molding compound), and 30% fiber-glass-content polyurethane (injection-molding compound).

A composite nose cowl that recently completed 5 years of flight service with no serious defects had the following design features:

Composite used only in outer barrel (Fig. 3.74)

No stiffeners

Dielectric shielding and lightning protection

Resistance to galvanic corrosion and sonic fatigue

Riveted and bolted construction

Noise suppression

## PROPERTIES

The properties of metallic and nonmetallic fibers and laminated materials have already been discussed. Just as important are the special and peculiar properties that depend on particular plastic composites. These properties are usually peculiar to resin-composites and are discussed below. These necessary properties are developed in uncured prepreg materials; every production batch of material must meet stringent criteria. The user or designer must know how the material was cured and processed, how the fibers were made, etc.

### Variation of Properties with Form

Many composite properties can be influenced by reinforcement form (tape, mat, or fabric), composition, i.e., percent of glass, and effect of fiber finish on laminate performance (Table

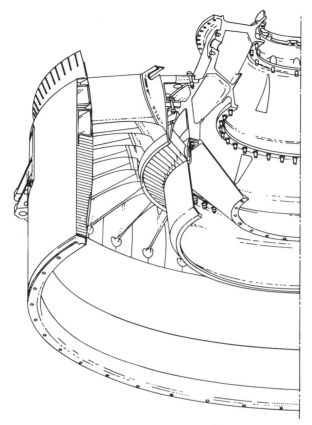

**FIG. 3.73**  TF34 composite fan-frame trimetric.[32]

3.8). To illustrate the effects of the direction and number of lay-ups of material on the properties of composites vs. metal, three typical lay-ups of B–Ep tape 3 in (76 mm) wide (Table 3.9) were chosen (along with the unidirectional laminate) for comparison with those of Gl–Ep, Ti–6Al–4V and the 7075-T6 and 2024-T42 aluminum alloys. Lay-ups *a* and *b* are typical of those used for wing or stabilizer covers. Lay-up *c* gives maximum shear strength and shear modulus and would only be used for shear webs.

Examination of the specific room-temperature tension, compression, and shear strengths in Fig. 3.75 and specific tension and shear moduli in Fig. 3.76 shows that the B–Ep material is by far the most efficient, provided the optimum lay-up is chosen for the specific loading condition.

In another example of tailoring composite properties by lay-up methods (Fig. 3.77) composite structures of graphite and polyimides are laminated to have very low coefficients of thermal expansion. Such structures are light and strong and have many uses where expansion or contraction with temperature change is not desirable. The low thermal expansion is achieved through a special lay-up of the laminate. The unidirectional fiber-reinforced layers are oriented at angles with each other (Fig. 3.77); depending on the materials used, coefficients of expansion for this lay-up are approximately $10^{-7}$ over a temperature range of 75 to 300°F (23.8 to 149°C).

**FIG. 3.74** Composite nose cowl on commercial airline engine being examined after first year of service.[51]

**Table 3.8** Mechanical Properties of Polyester Laminates with Various Finishes on 181 Glass Fabric[53]

| Type | Glass content, wt % | Elongation, % | sp gr | Tensile strength | |
|---|---|---|---|---|---|
| | | | | ksi | MPa |
| Filament-wound Ep | 60–90 | 1.6–2.8 | 1.7–2.2 | 80–250 | 552–1724 |
| Fabric Ep | 50–65 | 1.6–2.0 | 1.55–1.80 | 24–60 | 165–414 |
| Mat polyester | 25–50 | 1.0–1.5 | 1.8–2.0 | 10–24 | 69–165 |
| Premix polyester | 10–45 | 0.3–0.5 | 1.8–2.0 | 5–10 | 35–69 |

| | Tensile modulus | | Compressive strength | | Flexural strength | | Flexural modulus | |
|---|---|---|---|---|---|---|---|---|
| | $10^6$ psi | GPa | ksi | MPa | ksi | MPa | $10^6$ psi | GPa |
| Filament-wound Ep | 4.0–9.0 | 28–62 | 45–70 | 310–483 | 100–270 | 690–1862 | 5.0–7.0 | 35–48 |
| Fabric Ep | 3.0–3.5 | 21–24 | 50–70 | 345–483 | 50–100 | 345–690 | 2.0–4.0 | 14–28 |
| Mat polyester | 0.8–1.8 | 5.5–12 | 18–30 | 124–207 | 25–45 | 172–310 | 1.3–1.8 | 9–12 |
| Premix polyester | 1.6–2.0 | 11–14 | 20–26 | 138–179 | 6–26 | 41–179 | 1.5–2.5 | 10–17 |

**Table 3.9** Three Typical Lay-Ups of 3-in (76-mm) B–Ep Tape[53]

| Lay-up | Layers, % | | |
|---|---|---|---|
| | At 0° (0 rad) | At ±45° (0.79 rad) | At 90° (1.6 rad) |
| *a* | 60 | 30 | 10 |
| *b* | 40 | 50 | 10 |
| *c* | | 100 | |

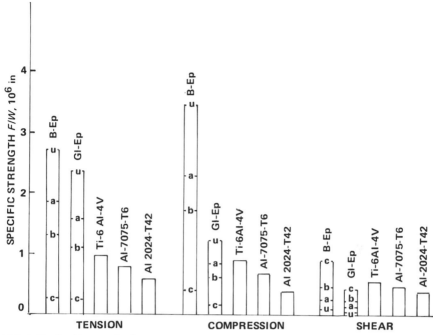

**FIG. 3.75** Specific strength of composite materials and metals at room temperature (all glass is S glass). Typical composite lay-ups: $a$ = 60% at 0°, 30% at ±45°, 10% at 90°; $b$ = 40% at 0°, 50% at ±45°, 10% at 90°; $c$ = 100% at ±45°; $u$ = unidirectional, 100% at 0°.[53]

## Composite Properties

The excellent properties developed in the variety of fibers in use today carry over to the reinforcements of plastic composites. The fibers can be combined with most conventional thermosetting and thermoplastic resins, obtaining excellent properties with no special finishes. Processes used to combine fiber glass with matrix systems are adaptable to graphite with little or no modification. Many composites are fabricated by laminating and curing prepreg (Fig. 3.78). Laminates in which the fibers are all parallel are *unidirectional* or 0° (0 rad). Laminates with alternate layers of 0, 90, +45, and −45° (0, 1.6, and 0.79 rad) plies are *quasi-isotropic;* i.e., the properties are equal in all directions in the plane of the laminate. In real structures the lay-up will be such that properties between these two extremes will be obtained.

## *Fatigue*

A high percentage of the original static and ultimate strength of both unidirectional and cross-plied graphite-composite laminates is retained after 10 million cycles. Figure 3.79 illustrates the constant-amplitude fatigue for higher-strength, lower-cost Gr–Ep in tension-tension cycling. The percentage of strength retention exceeds that of aluminum after 0.5 million cycles.

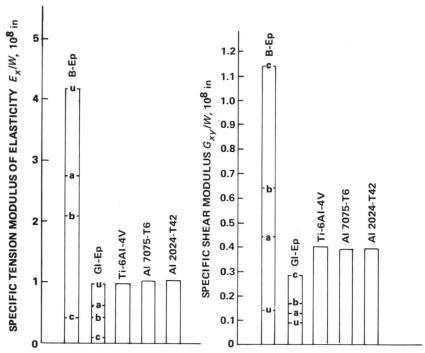

**FIG. 3.76** Specific tension and shear elastic moduli of composite materials and metals at room temperature; lay-ups coded as in Fig. 3.75.[53]

## Creep

Graphite fibers exhibit excellent resistance to creep, while resins do not; thus, the fiber orientation in the composite is most important in determining creep behavior. When load is applied in the fiber direction, the total creep strain is very small and a negligible permanent strain remains after unloading.

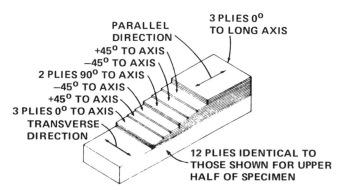

**FIG. 3.77** Low-expansion Gr–PI composite lay-up; 0°/45°/90°. (*From Mater. Eng., December 1974, p. 49.*)

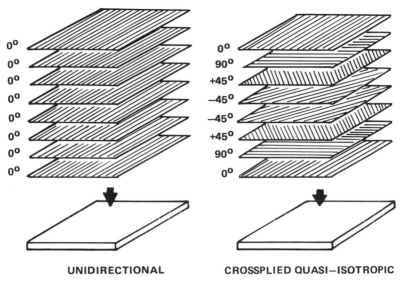

| | | | |
|---|---|---|---|
| 0° | | 0° | |
| 0° | | 90° | |
| 0° | | +45° | |
| 0° | | −45° | |
| 0° | | −45° | |
| 0° | | +45° | |
| 0° | | 90° | |
| 0° | | 0° | |

**UNIDIRECTIONAL**          **CROSSPLIED QUASI–ISOTROPIC**

**FIG. 3.78**  Laminates of composites tailored to meet specific requirements with savings in time and material; 0°/45°/90°. *(Hercules, Inc.)*

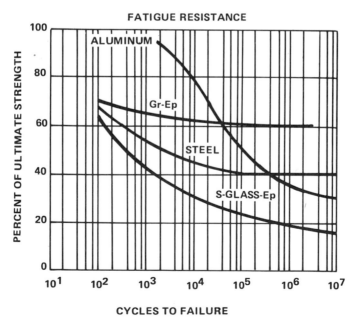

**FIG. 3.79**  Comparison of fatigue strength of graphite, steel, fiber glass, and aluminum (note the greater strength of graphite after 10 million cycles). *(Hercules, Inc.)*

### Sonic Fatigue

Properly designed graphite composite has been used and tested satisfactorily. Experimentation with flat panels has resulted in no damage after different time exposures to aircraft-engine frequencies. Some impact-damaged panels have shown no propagation of damage after 4 h exposure while others have demonstrated that maximum acoustic stress on composites is 25% lower than on aluminum.

### Impact Strength

Boron with 1100 aluminum composite, with impact strengths nearly 5 times greater than titanium alloys, also provides a significantly higher impact strength than composites with 2024, 5052, and 6061 aluminum matrixes. The more ductile matrixes allowed additional energy absorption through matrix shear deformation and multiple fiber breakage. Matrix shear with multiple-fiber breakage provides the highest impact-energy absorption.[33]

### Strength and Stiffness at Elevated Temperature

Many composite compositions have reasonably good to excellent properties at elevated temperatures. Development work continues with polyimides and various resin and metal matrixes to increase their strengths at temperatures above 700°F (371 °C). Comparisons of properties between the elevated-temperature properties of an MMC consisting of 21 vol % SiC whiskers in 2024 aluminum alloy and 2024 aluminum alloy control specimens are shown in Figs. 3.80 to 3.82. Figure 3.80 shows the elastic modulus to be about 50%

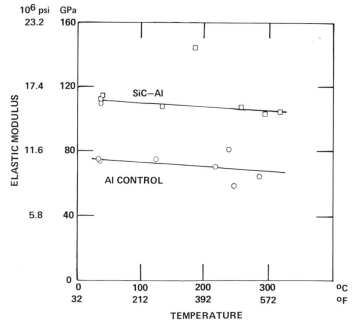

**FIG. 3.80**   Elastic modulus to 572°F (300°C) of SiC–Al and 2024 aluminum.[34]

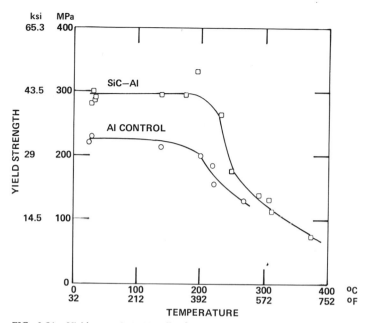

**FIG. 3.81**  Yield strength (0.2% offset) vs. temperature for SiC–Al and 2024 aluminum.[34]

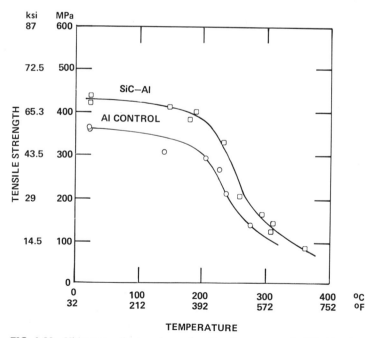

**FIG. 3.82**  Ultimate tensile strength as a function of temperature for SiC–Al and 2024 aluminum.[34]

greater for the composite, at least at temperatures to 572°F (300°C). The yield and tensile strengths are 30 and 15% greater,[34] respectively, at temperatures to about 392°F (200°C) (Figs. 3.81 and 3.82).

**FIG. 3.83** A brazed titanium-clad Bsc–Al skin-stringer panel for YF-12 aircraft.[35] *(Courtesy of Thomas Bales.)*

## Compression

The compressive properties of composites are critical in many structures. Figure 3.83 shows a titanium-clad Bsc–Al skin-stringer panel before testing. In the tests:

1. Titanium-clad Bsc–Al unidirectional sheet materials had longitudinal and transverse tensile strengths 0.9 and 2.5 times that of unclad Bsc–Al unidirectional sheet materials.
2. After local skin buckling, titanium-clad Bsc–Al panels withstood from 15 to 40% more compressive load before failure, whereas unclad Bsc–Al panels exhibited buckling and failure simultaneously.
3. Specific buckling strengths of titanium-clad Bsc–Al skin-stringer panels were 1.5 to 1.7 times those for similar unclad Bsc–Al panels.
4. Strengths of Borsic fibers were unaffected by the stringer fabrication and brazing processes. Compression tests of the hat-stiffened panels and metallurgical investigation of the joints after panel failure indicated no failures in the braze joints. Titanium cladding provided an effective diffusion barrier between the braze alloy and the 6061 aluminum-alloy matrix.

These results have led to flight testing a full-scale wing panel to verify the initial promising results.[35]

## Properties of Some Molding Compounds

Many low-cost manufacturing processes, e.g., injection molding, extrusion, and compression molding, use short- or chopped-fiber reinforcement. In general, the use of chopped or discontinuous graphite fiber in a polymer-matrix molding compound will produce a unique material that has the relative properties depicted in Fig. 3.84 for nylon 6/6 composites. Thus a 40 wt % graphite-reinforced injection molding compound offers the specific strength and modulus of aluminum, the low thermal expansion of Invar, the thermal and electrical conductivity of steel, a heat-distortion temperature of nearly 500°F (260°C), a

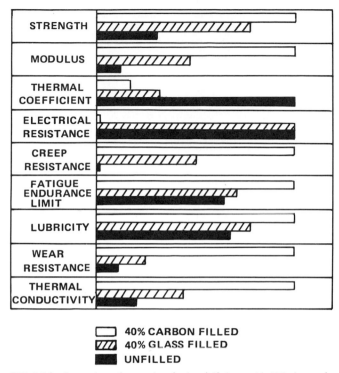

□ 40% CARBON FILLED
▨ 40% GLASS FILLED
■ UNFILLED

**FIG. 3.84** Comparison of properties of nylon 6/6 alone, with 40% glass, and with 40% carbon.[40]

friction coefficient nearly as low as that of Teflon, and unlubricated-wear properties approaching those of lubricated steel. More details of these processes are discussed in Chap. 4.

## Physical Properties

The physical properties of uncured preimpregnated materials should be noted since the acceptance of all materials must conform to rigid requirements. This also includes processing such as autoclave and press molding. These properties include:

Volatile content, weight percent
Resin solids content, weight percent
Resin flow, weight percent
Tack
Drape
Gel time
Filament content, weight per ply

This involves qualification inspection and quality-conformance inspection as well as mechanical-property tests like flexure modulus and strength, shelf life, and out time. All these tests are normally conducted by the supplier or the user to accept or reject materials. Detailed processing and manufacturing specifications are used throughout the industry.[36]

## Unusual Properties

In order for composites to replace metals in many applications new types of tests and new ways of treating composites have been developed.

### Fuel, Sealant, Salt Water, etc.

Graphite-reinforced polysulfone (Gr–PS) is a potential material for light, low-cost primary aircraft fuselage structures. Gr–PS laminates can be made, cured, and postformed any time to a desired contour or shape. Properties compare with those of Gr–Ep composites, but chemistry and shelf-life problems associated with the latter are eliminated. More simplified handling, postforming by simple application of heat and pressure, and recycling possibilities of Gr–PS offer potential cost savings in aircraft. To date, only experimental parts such as target drones have been produced; tests (Table 3.10) have all been met by Gr–PS composites.

**Table 3.10**  Requirements for Gr–PS Composites†

| Property | Requirement | Qualification |
|---|---|---|
| Jet-fuel (JP-4) compatibility | Continuous exposure at 77°F (25°C) | No effect after 28-day exposure |
| Hydraulic-fluid exposure MIL-H-5606 | Casual exposure | No effect after 28-day exposure |
| Saltwater exposure | Withstanding salt spray and saltwater immersion | No effect after 200-day exposure in salt spray or 60-day exposure in salt water |
| Sealant compatibility | Process and chemical compatibility | No effect, no softening or deterioration of composite resulting from application of polysulfide tank sealant |
| Paint coating | Process and chemical compatibility per MIL-F-18204 | Primed and painted successfully per military specification |
| Fire resistance | Nonburning | Qualified per ASTM D635 |
| Synthetic lubricant | Casual exposure | No effect after 28-day exposure |
| Weather exposure | Continuous exposure | No effect per ASTM D1499 |

†Based on *Mater. Eng.*, December 1977, p. 35.

### Cryogenic Conditions

The search for composite materials for cryogenic-fluid tanks in spacecraft is a search for materials that combine the safety of aluminum with the high strength-to-weight ratio of titanium. Aluminum performs well in contact with liquid fuels and oxidizers, but a strong enough container is too heavy. On the other hand, titanium provides low weight but is hazardous, particularly in contact with liquid oxygen. Two lightweight composite systems have been developed to meet these problems, wire-reinforced aluminum and glass-reinforced resin.

**Metal-Metal Composites:**  Stainless-steel-wire-reinforced aluminum composites offer good mechanical properties at cryogenic temperatures plus light weight. They are made by diffusion-bonding NS 355 stainless-steel wire in a 2024 aluminum-alloy matrix. Mechanical properties of stainless-steel–aluminum composites, good at room temperature, improve at lower temperatures. For example, NS 355 stainless-steel–2024 aluminum composite has an impact value of 12 ft·lb (16.3 J) at room temperature (20°C). At −320°F (−196°C) it rises to 18 ft·lb (24.4 J). Tensile strength is 40% higher at −423°F (−253°C) than at room temperature. Elongation at −423°F, though not as high as at room temperature, is higher than predicted by the rule of mixtures, indicating a favorable matrix-reinforcement interaction.

Although stainless-steel wire in aluminum offers a good strength-to-weight ratio, other reinforcements look better where high modulus or stiffness is needed. For example, boron or SiC filaments can be incorporated with aluminum to produce very stiff structures with the same techniques described above. However, since almost all research with these materials has been aimed at high temperatures, development work is required to determine their cryogenic capabilities.[37]

**Glass-Fiber–Plastic Composites:**  Glass-fiber–resin matrix composites also show promise in cryogenic structural applications. High strength-to-weight ratios, low thermal expansion, and good thermal insulation make them prime candidates for metal-lined cryogenic-fluid containers. Recent research indicates that at low temperatures they retain their integrity and improve in mechanical properties. The composite systems examined used an epoxy resin system with filament-wound roving and glass fabric of high-strength S glass.[38]

## Fire Resistance

Composites containing graphite fibers commonly have epoxies as matrix materials, but conventional epoxies have a low char yield and do not resist fire well; exposed to heat, they release the graphite fibers and lose structural strength. A class of imide-amine curing agents developed for such epoxies has increased char yield and thus fire resistance by 200 to 300%. The development is a major breakthrough in composites for aerospace and automotive applications and for other structures that require integrity when exposed to heat. The method consists of adding boron to the resin matrix by dispersion in the impregnation mixture applied to the graphite fiber. Addition of as little as 3% boron powder by weight effectively maintains the structural integrity of the composite and prevents release of the graphite fibers.

Char formed during the burning of resin-matrix composites is a carbonaceous material that oxidizes rapidly in air at temperatures above 1200°F (649°C). Elemental boron powder added to the composite as a filler material stabilizes the char as it is formed and protects it from oxidation. Resin-matrix–graphite-fiber composites modified by the addition of boron powder can be exposed to high thermal heat fluxes in air for prolonged periods of time and maintain their structural integrity. The char formed by the thermally decomposed resin is protected from oxidation in air and remains as a functional matrix material that cements the fibers together.[39,40]

## Corrosion and Moisture

Although many composite structures have been committed to production, some engineers feel that more data are needed about the performance of composites in hostile environments. Resin-matrix composites have some tendency to absorb moisture and are susceptible to lightning damage. Moisture absorbed in sufficient quantity by matrix materials lowers matrix-dependent properties. Full moisture saturation, which represents an extreme condition, is unlikely to occur in actual service. Experimental results showed that moisture levels below 0.6% can be tolerated without notable strength losses. So far, no moisture-absorption problems have been experienced with aircraft flying B–Ep structures. Some composite rudders that have been flying for over 4 years are coated with an epoxy primer and finished with polyurethane paint.

Resin-matrix composites generally behave like unfilled plastics when exposed to detrimental environments; i.e., the resin and not the reinforcement will be degraded. In the case of Gr–Ep, B–Ep, Gr–PI, and B–PI the resins are subject to moisture absorption, which leads to strength losses at elevated temperatures. This is recognized to be the most serious environmental disadvantage of reinforced epoxies and polyimides.

Even though Gr–Ep and Gr–PI otherwise are highly corrosion-resistant, they can promote the corrosion of many metals. When they are in contact with metallic structures and moisture is present, these composites act like noble metals and promote the galvanic corrosion of any metal lower in the electrogalvanic scale.

The loss of strength at elevated temperatures of the epoxies and polyimides and the corrosion promoted by graphite-reinforced resin-matrix composites can both be alleviated by corrosion-prevention methods and sometimes good design practice. Little is presently known about the corrosion mechanisms of MMC. As with resin-matrix composites, it is the matrix and not the reinforcement that will be attacked.

Absorption of moisture by epoxy and polyimide resin composites can result in notable decreases in short-term elevated-temperature strength. The strength is restored upon release of the absorbed moisture. Moisture absorption can be attributed largely to the moisture affinity of specific highly polar functional groups in cured resins. The moisture acts as a plasticizer, which leads to swelling and lowering of the glass transition temperature. Decreasing the glass transition temperature leads directly to a lowering of the elevated-temperature capability of the resin matrix. Since moisture affects only the resin matrix, only matrix-dependent properties (flexural, compressive, and shear strengths) will be affected. The major fiber-dependent property, tensile strength, is very insensitive to adverse environments.

Moisture absorption can lead to degradation of elevated-temperature properties of the epoxy-resin composites. For example, if a Gr–Ep composite structure is exposed to severe moisture conditions and then subjected to approximately 350°F (177°C), there will be a notable loss in flexural strength. The effect of moisture and ambient aging is shown in Table 3.11.

More is being learned about the moisture absorption of polyimide-resin composites reinforced with graphite or boron. Some newer addition-type polyimides, e.g.. acetylene-terminated ones, appear to be less susceptible to moisture degradation. Tests conducted by governmental agencies and aircraft companies show that acetylene-substituted polyimide reinforced with a high-tensile-strength graphite fiber picks up 0.8% moisture after 1000 h at 120°F (49°C) and 95% relative humidity. Under the same conditions, moisture absorption[41] after 1400 h is 1.4%.

**Table 3.11** Effect of Moisture and Ambient Aging on Percentage Retention of Flexural Strength of Epoxy-Matrix Composites[†]

| Test temp | | Control | | 24-h H₂O boil | | | 6-wk humidity | | | 20-wk ambient | | | 52-wk ambient | | |
|---|---|---|---|---|---|---|---|---|---|---|---|---|---|---|---|
| °F | °C | ksi | MPa | ksi | MPa | % | ksi | MPa | % | ksi | MPa | % | ksi | MPa | % |
| | | | | | | | Boron epoxy | | | | | | | | |
| 75 | 23.8 | 300 | 2068 | 283 | 1951 | 94 | 308 | 2124 | 103 | 316 | 2179 | 105 | 330 | 2275 | 110 |
| 350 | 177 | 251 | 1731 | 57 | 393 | 23 | 99 | 683 | 39 | 149 | 1027 | 59 | 132 | 910 | 53 |
| 350‡ | 177 | 125 | 862 | 79 | 545 | 63 | 93 | 641 | 74 | 128 | 883 | 102 | 117 | 807 | 94 |
| | | | | | | | Graphite epoxy | | | | | | | | |
| 75 | 23.8 | 244 | 1682 | 244 | 1682 | 100 | 244 | 1682 | 100 | 268 | 1848 | 110 | 235 | 1620 | 95 |
| 350 | 177 | 188 | 1296 | 63 | 434 | 34 | 55 | 380 | 30 | 102 | 703 | 54 | 85 | 586 | 46 |
| 350‡ | 177 | 98 | 676 | 53 | 365 | 34 | 40 | 276 | 41 | 79 | 545 | 81 | 56 | 386 | 57 |

†Average of three tests. Except as noted, orientation 0°. Table based on *Mater. Eng.*, December 1976, p. 47.
‡For 0 ± 45° (0 ± 0.79 rad).

**3.77**

A potential method of preventing moisture pickup in epoxy-resin composites is ion-vapor deposition of aluminum, in which the part to be "plated" is placed in a vacuum chamber with molten aluminum. The part is made the negative electrode and the molten aluminum is made the positive electrode in a high-voltage dc circuit. By maintaining the proper inert-gas pressure in the vacuum system, a dc glow discharge is established about the workpiece wherever a portion of the evaporated plating material is ionized and accelerated toward the part. Although ion-vapor deposition of aluminum is relatively new, the process is already covered by military specifications.

Although Gr–Ep composites are not degraded by galvanic corrosion, they act like platinum or gold when coupled to ordinary metals in the presence of most electrolytes. There is considerable evidence that most aluminum alloys and many steels and cadmium-plated steels will corrode when coupled to Gr–Ep composites and exposed to an electrolyte such as sodium chloride. Even though actual service experience has not shown this corrosion problem, governmental and industrial firms recognize the potentially serious risk of corrosion under conditions of galvanic coupling to Gr–Ep. As a result of tests, the commonly specified structural alloys can be classified as acceptable, borderline, and unacceptable (Table 3.12).

Gr–Ep panels that must be drilled for fasteners, bushings, or other inserts are extremely corrosion-prone. Under these conditions the ends of hundreds of graphite fibers are exposed and can be brought into contact with the fasteners or bushings. Unprotected aluminum and cadmium-plated steel fasteners installed in Gr–Ep composites show evidence of corrosion after only 24 h exposure to 5% salt spray. In as little as 168 h, aluminum rivets will be half corroded and cadmium-plated steel fasteners will show great amounts of corrosion. The Air Force demonstrated that little more than 500 h is required for complete corrosion of aluminum rivets installed in Gr–Ep composites subjected to 5% salt spray.

**Table 3.12**  Galvanic Current Density of Alloys Coupled to Graphite-Epoxy†

| Material | Status | Current density | |
|---|---|---|---|
| | | $\mu A/in^2$ | $\mu A/cm^2$ |
| Ti–6Al–4V<br>Ti–6Al–2Sn–4Zr–2Mo<br>René 41 (Ni–19Cr–11Co–10Mo–3Ti–1.5Al)<br>Inconel X (Ni–15Cr–7Fe–2.5Ti–1.0Co–0.7Al)<br>Inconel (Ni–15Cr–7Fe)<br>AFC-77 (Fe–14.5Cr–13Co–5Mo–0.4V–0.15C)<br>PH17-7 (Fe–17Cr–7Ni–1.2Al–0.07C)<br>301 and 304 stainless steels<br>Cu–1.7Be | Acceptable | <32.25 | <5 |
| Al–graphite (6061Al–30% graphite)<br>MA-87 (Al–6.5Zn–2.5Mg–1.5Cu–0.4Co)‡<br>Aluminum alloys 2024-T3, 2024-T6, 7075-T6<br>440C stainless steel<br>1020 and 4130 steels | Borderline | 32.25–96.75 | 5–15 |
| 300M steel (Fe–5Mo–1.85Ni–1.6Si–0.9Cr–0.85Mn–0.42C)<br>10Ni steel (Fe–14Co–10Ni–2Cr–1Mo–0.14C)<br>4340 steel<br>Aluminum 2020-T651 | Unacceptable | >97 | >15 |

†Based on *Mater. Eng.*, December 1976, p. 48.
‡Powder metal alloy.

That the problem of corrosion induced by coupling of cadmium-plated steel and Gr–Ep is a real one is shown by the fact that the Air Force prohibits the use of cadmium-plated steel fasteners through Gr–Ep panels.[42]

Fasteners made from A-286 superalloy are considerably more corrosion-resistant in contact with Gr–Ep but still do not offer the ideal solution to the problem of fastener corrosion. The use of titanium fasteners is encouraged where Gr–Ep composites must be joined mechanically. When Gr–Ep must be in contact with metals and larger surface areas are involved, a good method of preventing corrosion of the metal is electrical isolation, e.g., with a layer of epoxy-saturated glass cloth.

Boron-reinforced composites in contact with titanium show practically no tendency to promote galvanic attack thanks to greater resistivity of the boron coating on tungsten filaments of boron-reinforced composites.

## Processing Effects

The effects of various processing methods on the transverse tensile strength of Gr–Al composites have been examined. The methods, which consisted of surface coatings applied to the graphite fibers in an attempt to limit fiber-matrix reactions and so prevent degradation of fiber properties, included the standard Ti–B vapor-deposition process (with two modifications), the sodium process, and nickel-plating process. Four types of graphite fibers were used from commercial precursors including rayon, pitch, and high- and low-modulus PAN materials. The 1100, 6061, and 201 aluminum alloys were used as matrix materials.

The transverse fracture strength was markedly influenced by the process method and type of graphite fiber but not by the matrix alloy. Fibers and processes that promote fiber-matrix reaction, e.g., low-modulus PAN material with the standard Ti–B coating, result in the highest transverse tensile strength. The fractured surfaces revealed fiber-matrix separation and fiber pullout as the dominant features of low-transverse-strength composites. Fiber splitting and absence of fiber pullout were characteristic of the high-transverse-strength composites.

Investigators have reported results on a study of the effects of fabrication and joining processes on the compressive strength of B–Al and BSc–Al structural panels. The nominal fiber volume[43] for each composite material was 48%. After hot-forming, cold-forming, and eutectic bonding the panels were subjected to axial compression tests at room temperature (20°C). The major findings of this study were as follows:

1. Fiber strengths of the boron and Borsic filaments were not significantly affected by the hot or cold stringer-forming processes but were degraded at least 25% for the boron fibers formed with eutectic bonding.
2. Compression tests indicated that the integrity of all five types of joints on the skin-stringer panels was maintained until panel failure.
3. Of the joining processes investigated, diffusion bonding and adhesive bonding produced panels which developed the highest average compressive stresses [61 to 94 ksi (421 to 648 MPa)].
4. Panels made with discontinuous joining methods (bolted and spot-welded) exhibited joint failure, whereas panels made with continuous joining (diffusion-bonded and adhesive-bonded) showed failures in the radius of the attachment flange and across the crown of the stringer.

## Protection against Lightning and Electromagnetic Interference

For optimum performance in a variety of hostile environments, aircraft designs require extensive use of lightweight, durable, and high-strength materials. Accordingly, advanced composites are being considered for extensive use on high-performance aircraft. Use of advanced composite structures at their design limits requires adequate protection of the structures against the degrading effects of moisture on strength. Properly chosen and applied metallic coatings can prevent moisture absorption and increase conductivity for better shielding against penetration of electromagnetic energy. In addition, suitable metallic coatings reduce the damaging effects of lightning strikes and also provide protection against paint strippers during aircraft refinishing operations.

In a program carried out to develop metallic coating systems for Gr–Ep laminated aircraft structures and to demonstrate the environmental resistance and serviceability of these coatings against moisture penetration, electromagnetic interference, lightning strikes, and paint strippers[44] three types of metallic protective coatings were applied to test panels fabricated from Gr–Ep preimpregnated tape:

1. Solid aluminum foil bonded in a secondary operation to cured Gr–Ep laminates
2. Perforated aluminum foil co-cured with pre-bled Gr–Ep laminates
3. Spray-and-bake metal-filled organic coatings applied to Gr–Ep laminates

The foil coatings provided a significant reduction in the moisture penetration and the associated strength loss of the laminate after exposure to humidity and humidity with thermal spiking. The foil coatings also exhibited good impact resistance and peel strength. The metal-filled resin coatings provided protection against strength loss of the Gr–Ep laminates from paint-stripper attack but were not resistant to moisture penetration.[45]

Lightning damage to resin matrix composites has been approached by different aircraft companies with different protection systems for large-area composites. Complete protection was provided from a direct 200-kA lightning strike by coating the surface with a 0.004-in (0.10-mm) thick aluminum foil or a 120-mesh aluminum screen wire. Most of an aircraft surface is susceptible to the swept-stroke phenomenon. On these surfaces, protection is provided by means other than screen wire mesh. For example, aluminum-foil strips placed at right angles to the airstream on a composite-skin external surface, accompanied by polyurethane paint over the fasteners, provides as much protection for much less weight than the 120-mesh wire, approximately 2.2 lb per 100 ft$^2$ (0.11 kg/m$^2$). Only about 12 lb (5.44 kg) of extra material is required to protect against lightning a 40,000-lb (18,140-kg) fighter aircraft, using composites extensively on external surfaces.

The proliferation of electronic equipment causes ever-increasing levels of electromagnetic radiation in the environment. Metallic shields can, of course, provide protection, but the high weight of solid-metal shields and housings, their relatively high costs, and the inconvenience of applying metal foils and coatings to plastic housings have goaded designers into searching for alternative methods for shielding against electromagnetic interference. The primary requirement for such a shielding material is that it conduct electricity. Much research in recent years has been directed toward inducing electrical conductivity into plastic materials. One of the most common ways to impart electrical conductivity to plastics involves incorporating conductive fillers in the plastic matrix. The most common fillers include conductive carbon blacks, carbon and graphite fibers, metal-coated glass fibers, metal particulates, and metal fibers.

In general, designers choose plastics for equipment housings because of their low density, their relatively low cost (as applied), their ease of forming, and their chemical inertness. In addition, the plastic chosen must exhibit physical properties adequate for the application. But the incorporation of large amounts of filler into a plastic usually will degrade its mechanical properties, such as tensile strength, and addition of heavy fillers will negate the desirable low density of the unfilled plastic.

In many applications, high thermal conductance is desirable to remove heat generated by the operating equipment. But plastics, by nature, are thermal insulators. The introduction of certain metallic fillers can improve the thermal conductance, as well as the electrical conductivity. Incorporating very-low-volume loadings of short aluminum fibers is an effective way to impart electrical conductivity to a plastic part. As the aspect ratio $L/D$ increases, a lower loading of conductive fibers is required for electromagnetic and radio-frequency shielding. Electromagnetic shielding of structural foam is effectively accomplished by incorporating C–Gr fibers in the plastic. For this type of conductive filler, long, high-modulus pitch fibers are best. A study showed that aluminum-coated glass fibers develop oxide surface layers which inhibit fiber-to-fiber conduction. Metal powders, also investigated, drastically reduced mechanical properties at effective conductive filler levels.

By using metallized glass fibers injection-molded plastic parts can be produced with electrostatic dissipation and a degree of electromagnetic and radio-frequency shielding. Other advantages of the conductive filler are improved heat transfer and improved mechanical properties. Although shielding is not as effective as metal, it is reasonable where lower weight, lower part cost, and easier fabrication are traded for some shielding effectiveness.

# NONDESTRUCTIVE INSPECTION AND EXAMINATION

Composite materials offer a number of potential advantages for aircraft applications, e.g., reductions in weight, fuel conservation, and life-cycle cost reduction, but they also introduce more requirements for inspection and quality assurance. For metallic structure, one is usually most interested in the detection of a crack or discontinuities that could coalesce to a crack, whereas for composites the list includes improper cure, resin-rich or resin-poor areas, fiber misalignment, unbonds, inclusions, machining damage, impact damage, fastener fretting and pullout, and environmental degradation. To detect and characterize this wide range of defects requires a number of specialized nondestructive inspection techniques.

Since the effect of a resin-poor defect is potentially different from that of an impact-damage defect, it is essential that nondestructive inspection techniques be able to differentiate between these and other types of defects. The quality assurance departments in most companies who use composites in quantity ensure that structural composite parts represent a level of quality consistent with design requirements. Their effort should include:

Qualification testing and source inspection of a subcontractor's facilities

Inspection or inspection verification of materials, parts, and assemblies received from sellers

Surveillance of manufacturing operations and processes, including inspection (examination and test) or inspection verification of the submitted part at predetermined operations and stations in the manufacturing cycle

Service-depot inspection

Field and in-service inspection

End-product acceptance

Some significant differences between metal- and resin-matrix composites affect not only their processing and applications but also their quality-control procedures:

*Fiber diameter:* The relatively large diameter of boron or SiC fibers [4 to 6 mils (102 to 153 $\mu$m)] facilitates inspection of fiber or tape by visual or x-ray techniques.

*Aging:* MMC materials are stable when stored by normal methods, which must preclude mechanical damage or corrosive atmospheres. No special temperature or moisture controls are required.

*Adhesion:* Metal-matrix tapes or sheets have surface characteristics generally similar to those of metal foils, and they can be shipped or stored without separation or backing materials (except where necessary to prevent mechanical damage). If they are shipped or stored in rolls, the roll diameter should be large enough to prevent overstressing the fibers.

*Cleanliness:* Metal-matrix materials can be subjected to moderate amounts of surface contamination (such as dirty fingerprints, contact with oily bench surfaces, or exposure to normally dirty factory atmospheres) and subsequently cleaned with methyl ethyl ketone or acetone. Once cleaned, however, their cleanliness must be maintained through the consolidation process.

## Requirements of Raw Materials and Processed Material

The manufacturer of resins, adhesives, prepreg tape, and other supplies must meet the rigid engineering and quality-assurance requirements prescribed in procurement specifications. When manufacturers meet them, they become approved sources. Usually one must be able to trace a manufacturer's identity, batch, lot, or serial number for composites and adhesives which are applicable to a composites program. Quality Assurance in most companies has the responsibility of preparing tests for each lot of composite and adhesive system; the test results then serve as controls relative to all material information. Typical tests conducted on prepreg tapes include resin content, volatile content, resin flow, flexural strength at 0 and 90° (0 and 1.6 rad), and horizontal shear. Tests on adhesives and primers include volatile content, metal-to-metal peel strength, and environmental [−67 to 160°F (−55 to 71°C)] lap-shear strength.

Each batch of preimpregnated material should be tested and certified by the manufacturer to the acceptance-test requirements of the applicable procurement specification. The aging characteristics of matrix resins require each shipment to be sent directly to refrigerated storage upon receipt. Samples of the preimpregnated material should be selected, and test panels should be molded for mechanical-property verification while the physical-property verifications are being obtained.

Physical and mechanical verification tests typically performed for boron, Kevlar, fiber glass, graphite and other preimpregnated materials include:

Resin content (percent) and fiber weight

Resin flow (percent)

Gel time

Tack [prepreg must adhere for 30 min to a steel plate held vertically at 70°F (21°C)]

Longitudinal flexure strength and modulus, kips per square inch (megapascals)

Longitudinal tensile strength and modulus, kips per square inch (megapascals)

Longitudinal short beam shear strength, kips per square inch (megapascals)

Physical properties: specific gravity, fiber content, resin content, and void content

Glass transition temperature

Chromatography, (chemical characterization tests include high-pressure-liquid and gel-permeation chromatography)

Figure 3.85 shows results of composite prepreg analysis. Figure 3.85*a* indicates good prepreg material. Each peak in that graph represents a different constituent. The two prom-

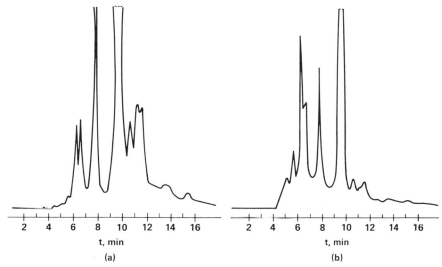

t, min

(a)

t, min

(b)

**FIG. 3.85** Results of composite prepreg analysis using high-pressure liquid chromatography: (*a*) acceptable; (*b*) unacceptable.[52]

inent peaks represent the epoxy resin monomer and the curing agent, the two major constituents of the prepreg. In Fig. 3.85*b* the third major peak represents a reaction product, indicating that the prepreg has aged beyond its useful life and is unacceptable. The test allows all material used in fiber-epoxy compounds to be characterized for acceptable parameters related to each application and to be analyzed before or during each batch use.

The previous discussion covered tests which Quality Assurance monitors for resin-matrix composites. The source approval and qualification testing apply equally to MMC. Metal-matrix tape should be properly identified by tape lot and by fiber lot. After the required testing is passed satisfactorily, the material is released to production with proper documentation. A representative specification used for metal-matrix tape covers the following items:

Visual appearance (no wrinkles, bubbles, or delamination; no missing, crossed, or broken fibers; no shipping damage or discoloration)

Uniformity of fiber spacing

Fiber diameter

Volume percent of fiber vs. matrix

Tape weight per unit area

Modulus

Tensile strength (longitudinal and transverse)

Longitudinal tensile strength can be characterized in three different ways (listed in order of increasing effectiveness):

*Fiber tests:*   Individual fibers are removed from the tape and tensile-tested. This technique defines fiber strength only.

*Tape tests:*   Small sections of tape are removed and tensile-tested. This is more difficult to control than the fiber test but provides a more representative value of the composite strength.

*Composite tests:*   Enough tape is removed to fabricate a multilayer panel; tensile specimens are machined from it and tested. Although this is the most expensive test of the three, it is the only one which will establish absolute confidence in the material strength.

## Tooling and In-Process-Critical Quality Checks

In the manufacture of structural composite assemblies, tooling fixtures and templates are the primary media of control used to assure dimensional and contour compliance. Quality-assurance engineers inspect and accept these tools as well as toolproofing articles. (Before a tool is used for the fabrication of parts, a complete thermal survey and contour check should be performed; this is toolproofing.)

The in-process steps which are normally inspected during fabrication according to most documentation procedures are:

Prefitting

Bagging

Curing and consolidation

Ply lay-up

Inspection ensures that during fabrication of composite structure the placement of reinforcing fibers is in the proper orientation. Close control is maintained over the number of plies in the buildup and the proper sequencing of the oriented fibers. Normally checked are filament and gap control, filament cross-over control, and foreign particles.

## Component Inspection

No one particular nondestructive technique can be used with certainty for all configurations. Test procedures and standards must be developed and tailored to each item. Geometry of parts and their inherent stress levels must be taken into account when determining the most appropriate test media. For metal-matrix parts, techniques include x-ray, ultrasonics, eddy current, and holography.

# Nondestructive Testing Methods

## Radiography

Defects normally associated with composites include delamination, undercure, fiber misalignment, damaged filaments, variation in resin fraction, variation in thickness, voids, porosity, fracture, contamination, and moisture pickup. As can be seen in Table 3.13, radiographic inspection detects almost every defect possible in a composite structure. The method can be used to inspect both internal and surface defects as well as parts covered or hidden by other parts or structure. Radiography includes a number of different techniques (x-rays, gamma rays, neutrons, radiation backscatter, fluoroscopy, and others), but they are all basically alike in that a penetrating beam of radiation passes through an object. As it does, different sections of the object, as well as discontinuities, absorb varying amounts of radiation so that the intensity of the beam varies as it emerges from the object.

In composites, radiography is used to determine fiber alignment in filament-wound and other structures, intimacy of contacts in bonded areas, and the fit of subassemblies. Radiography also is effective in reviewing core damage, moisture in core cells, and other defects in sandwich constructions.

Radiography requires special precautions to avoid hazards from radiation. Films are relatively expensive, and processing can require considerable time. Although radiography is perhaps the oldest of all nondestructive techniques, it has been used only sparingly for composite testing. Its main use has been in detection of boron cross-overs.

**Table 3.13**   Composite Defects Detected by Various Nondestructive Testing Methods†

| Defect discontinuity or variable | X-ray | Neutron | Gamma ray | Ultrasonic | Sonic | Microwave | Temperature differential | Heat, photosensitive agent | Penetrant |
|---|---|---|---|---|---|---|---|---|---|
| Unbound | x | | | x | x | | x | | |
| Delamination | x‡ | | | x | x | x | x | | x |
| Undercure | | | | x | x | x | | x | |
| Fiber misalignment | x | x | x | | | | | | x |
| Damaged filaments | x | | | | | | | | x |
| Variation, in resin | | | x | | | | | | |
|    In thickness | x | | x | x | x | x | | | |
|    In density | x | | x | x | x | | | | |
| Voids | x | | x | x | x | x | x | | x |
| Porosity | x | | x | x | x | x | | | x |
| Fracture | x | | x | x | x | | | | x |
| Contamination | x | | x | x | | | | | |
| Moisture | x | x | | | | | | | |

†From *Mater. Eng.*, March 1973, p. 69.
‡If oriented parallel to x-ray beam.

## Ultrasonics

Ultrasonics, like radiography, includes a number of different techniques. In ultrasonic inspection a beam of ultrasonic energy is directed into a specimen, and the energy transmitted through it (or the energy reflected from interfaces within it) is indicated. To direct the sound wave through the test material usually requires a liquid contact or sometimes liquid immersion of the part.

The four ultrasonic methods used in composite testing are (1) pulse echo, used to inspect fiber-glass–fiber-glass bonds, delamination in fiber-glass laminates up to 0.75 in (19 mm) thick, and facing-to-core unbonds; (2) pulse-echo reflector plate, used to inspect delamination in thin fiber-glass or boron laminates; (3) through transmission, used to inspect sandwich constructions and thick fiber-glass laminates; and (4) resonant frequency, used to detect unbonds in fiber-glass–metal bonds and fiber-glass–fiber-glass bonds where the exposed layer is not too thick. Typical part orientations and geometries and appropriate test techniques are shown in Fig. 3.86 and Table 3.14.

Immersion through transmission requires a pair of ultrasonic transducers on each side of the structure to be tested. The structure is immersed in a couplant (normally water) during the test. The transducer is a piezoelectric crystal which converts the electric energy into mechanical vibrations that are transmitted through the structure and received by a transducer on the other side. The display is shown on a cathode-ray tube. Through-transmission–reflection, like straight through-transmission, provides high sensitivity to small defects in parts with complex configurations. Under ideal conditions, defects as small as 0.062 in (1.6 mm) can be detected in complex B–Ep structures.

**Table 3.14**   Recommended Method in Fig. 3.86[1]

| Number | Recommended method |
|---|---|
| 1 | (*a*) Choice of pulse-echo angle-beam contact or squirter method on each skin, pulse-echo straight-beam contact method on each skin (not a ringing method), or through-transmission contact, squirter, or immersion method |
| | (*b*) Resonance frequency |
| 2 | (*a*) Choice of pulse-echo straight-beam contact method on each skin (not a ringing method) or through-transmission contact or squirter or immersion method |
| | (*b*) Resonance frequency |
| 3 | (*a*) Pulse-echo ringing-contact method |
| | (*b*) Resonance frequency |
| 4 | (*a*) Choice of pulse-echo straight-beam contact-method monitoring the thickness of the doubler, or pulse-echo ringing-contact, squirter, or immersion method on each skin of the doubler, or through-transmission contact or squirter or immersion method |
| | (*b*) Resonance frequency |
| 5 | Refer to 1; if these methods lack sufficient penetration power to detect reference defects in the reference standard, a pulse-echo ringing-contact, squirter, or immersion method can be used; otherwise the ringing method is not recommended |
| 6 | Pulse-echo straight-beam contact method monitoring the thickness of the doubler or tripler; if this method fails to resolve the back surface, a pulse-echo ringing-contact squirter or immersion method should be used; otherwise the ringing method is not recommended |
| 7 | (*a*) Through-transmission contact, squirter, or immersion method; dotted line represents beam direction |
| | (*b*) Resonance frequency |
| 8 | (*a*) Choice of pulse-echo reflector-plate immersion method from one surface, or through-transmission contact, squirter, or immersion method |
| | (*b*) Resonance frequency |

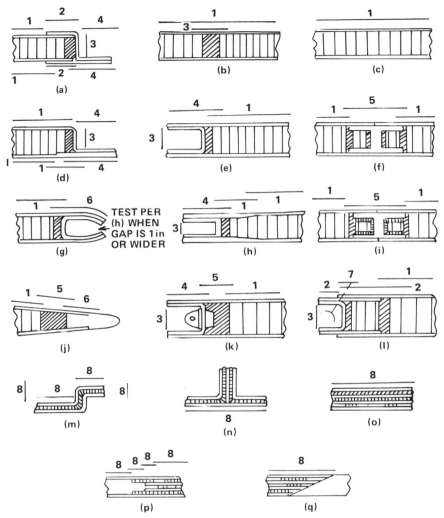

**FIG. 3.86** Typical test methods and scan planes; see Table 3.14 for applicable method according to number illustrated. When surfaces are symmetrical, illustrated method is typical for both sides. Shaded areas represent foam adhesive. Where the same acceptable method(s) appear in more than one scan plane, calibration is verified for each plane. (*a*) Core with Z closure member, (*b*) core and splice joint, (*c*) core, (*d*) core with Z closure member and land in skin, (*e*) core and closure member, (*f*) core and internal ribs with film adhesive bond to core, (*g*) core and closure member, (*h*) core closure member and land in skins, (*i*) core and internal ribs with foam adhesive bond to core, (*j*) core and closeout, (*k*) core and closure member with shear plates, (*l*) core, closure member, and doubler, (*m*) Z or hat laminates, (*n*) T web on channel laminated, (*o*) doubler, tripler, or composites, (*p*) composite splice joint, (*q*) composite splice joint.[1]

Contact through transmission is also used on stabilizer skins and on other complex B–Ep structures. Two transducers are manually held on each side of the part to be tested. A template is laid over both sides to ensure proper alignment of the transducers. Sound energy transmitted from one transducer passes through the specimen and is collected by the other transducer. The couplant used by these transducers is either oil or a solution of water and wetting agent. The display is shown on a cathode-ray tube. With this technique no permanent record is obtained since there is no resultant C scan, and 100% inspection is not possible. When the transducer is placed directly over the defect, the technique is sensitive to defects as small as 0.125 in (3.2 mm).

Resonance ultrasonic testing is performed with the Fokker bond tester, which depends upon transducer resonance theory, the transducer acting as a vibrator. The natural frequency is dampened by coupling a load to the resonating transducer, and the dampened effect is observed. If the coupled load is great, a shift in resonant frequency occurs. If the load is light, only a change in impedance occurs. When the transducer is applied to a bonded joint, a loading effect is imposed on the transducer. Extremes of quality ranging from 0 to 100% have been established.

Ultrasonics is a valuable inspection means for testing smooth-surfaced fine-grained materials. Flaw location and depth can be determined, and automation is common. Sonic testing uses frequencies in the audible region, from about 10 Hz (cycles per second) to 20 kHz. A clear, sharp ringing sound is indicative of a well-bonded solid structure. A dull sound, or thud, indicates a delamination or a relatively large void. Small voids generally cannot be detected by nondestructive sonic test methods. Structural defects which affect the response of composite materials to sonic vibration include large cracks or voids, evenly dispersed microcracks or voids, undercured resin areas, evenly dispersed foreign material, and dimensional variations.

## Microwaves

Microwave techniques can be used to locate defects and to measure thickness, moisture content, and dielectric properties of composites. Defects include voids, delaminations, porosity, foreign inclusions, resin-rich and resin-starved areas, and variations in degree of cure and moisture content. A microwave source directs the radiation at the test specimen. Energy reflected or transmitted by the specimen can then be used for the evaluation.

Experiments have been conducted with microwaves to determine the relative resin cure in boron laminates. There was no difficulty in using the through-transmission for laminates in which the boron filaments are essentially parallel. The $0°/+60°$ $(0/1.04\text{-rad})$ filament-orientation laminates proved to be opaque to the polarized through-transmission method; insufficient energy was transmitted to obtain any meaningful results. A scatter technique used with promising initial results is a single side test which involves bouncing microwaves off the surface of the laminate into a receiver horn. The method may be limited to the uppermost layers of boron tape. Since the microwaves are polarized, an optical angle must be obtained between the receiver and the specimen after alignment of the transmitter and specimen is complete. A different optimal receiver angle is necessary for each alignment configuration of the transmitter and specimen. The reflection angle and concentration of microwaves vary with resin cure.

## Heat and Liquid Crystals

The rate at which radiant energy is diffused or transmitted to the surface reveals defects in the part. Delaminations, unbonds, and voids are detected in this manner. Major applications are with filament-wound motor cases and aircraft sandwich structures. The most common name of the material, *liquid crystals,* seems self-contradictory, but since these materials are fluid and birefringent (an optical property of solid crystals), the name is descriptive.

The liquid crystals used in thermographic testing are cholesteric mixtures derived from organic components found in biological systems. Solvents such as chloroform and petroleum ether are used in liquid mixtures that are commercially available and prepared ready to apply. These mixtures are clear and have approximately the same mechanical properties as the solvents used. If this type of liquid-crystal mixture is applied to a surface, the solvents evaporate and the resultant coating is a thin, greasy layer of liquid crystals which remain clear above and below the transition temperature. Transition or melting-point temperatures can be obtained from 20 to 150°F ($-6$ to 66°C) by variations in chemical compounding.

Liquid penetrants are used on composites for finding discontinuities open to the surface such as cracks, porosity and edge unbonds, or delaminations. The penetrant is applied freely on a cleaned part and allowed to work into tight cracks. Penetrant is removed from all surface areas and the piece is sprayed with developer, which dries to a white coating while the penetrant bleeds up from any flaws through the developer, forming red or fluorescent indications on the white surface. The size of the defect is indicated by richness of color, speed of leak-out, and the dimensions of the indicator.

Commercially available penetrants are divided into two types, dye and fluorescent. The color of the dye is selected to give a high color contrast between developer and penetrant. Fluorescent penetrants must be viewed under ultraviolet or black light. The color changes between melting points; the color observed at a specific temperature is caused by scattering of incident light within the coating. The predominant wavelength depends upon the viewing angle, the angle of incident illumination, and the index of refraction of the liquid crystals. The appearance of a relatively cold discontinuity would be caused by an inclusion acting as a heat sink.

The most important property of liquid crystals is that with a change in temperature the index of refraction changes; thus it is reversible. With increasing temperatures the liquid crystals first appear colorless and with a response time on the order of 0.1 s change to red, then to yellow, green, blue, violet, and colorless again within a span of 1.5 Fahrenheit degrees. To see these colors, a black background is required since the liquid crystals remain transparent and do not absorb but reflect light. Color changes, that is red to blue, occur over a temperature range of 4 Fahrenheit degrees (2.2 Celsius degrees). A defect is observed by the color change at the surface nearest to the defect.

Cholesteric liquid crystals have been used to determine the extent of damage to B–Ep parts and to evaluate the integrity of subsequent repairs. Defective or disbond areas conduct heat at rates different from those of areas with good bonds. As heat is applied, the heat-conduction patterns surrounding the defective area will differ from the patterns in an area with a totally good bond. As a result, the defective area will be outlined by color difference.

Liquid crystals show promise as a field technique, since elaborate testing facilities are not needed. The technique is particularly effective on B–Ep parts, thanks to its low thermal conductivity, which allows sharp distinctions between bonded and disbonded areas. Video-tape recordings of the color changes are contemplated for permanent records, and work is being done on the encapsulation of narrow-range crystals in black Mylar tape, which would be reusable. This technique is severely limited by thick B–Ep face skins and complex structures. Liquid-crystal testing is most feasible in thin-skin structures. Only limited success has been achieved in the detection of voids or disbonds in a single or double titanium splice-plate configuration. On relatively thin-plied structures, however, defects as small as 0.125 in (3.2 mm) have readily been detected.

## Visual

Visual inspection is obviously the most convenient and widely used nondestructive test method for composites and other materials. Defects that can be seen include discoloration, foreign matter, crazing, scratches, dents, blisters, orange-peel surface, pitting, air bubbles, porosity, resin-rich and resin-starved areas, wrinkles and, to some extent, voids and delaminations. Aids to the eye can be helpful, like intense light and a magnifying glass.

## Future Methods

In addition to currently used techniques and refinements to them, other methods that show promise for boron, Kevlar, and other composites include infrared scanning, acoustic emission, holography, and microwave transmission. Recent advances in the state of the art of these techniques make them appealing for nondestructive testing of resin-matrix composites.

Infrared scanning is basically a thermal nondestructive testing technique that relies on the same basic defect-detection principle as liquid crystals, namely a disbond affecting the heat conduction of a structure. An infrared camera can detect thermal gradients as narrow as 1 Fahrenheit degree (1.8 Celsius degrees) or less with a display on either a cathode-ray tube or a C-scan recorder. The most likely use of the infrared camera is for thermal scanning of large thin structures. The part is thermally mapped by videotape recording of the display tube for a permanent record. This technique should significantly reduce testing time and should be as accurate as conventional ultrasonic techniques. The structure can be scanned in place without immersion or even direct contact. As with liquid crystals, testing with infrared cameras is limited to thin and rather simple structures.

Acoustic emission has been used in checking electron-beam welds and sandwich structures (brazed and diffusion-welded) and in static and fatigue tests. This technique involves placing a piezoelectric transducer or series of transducers about the specimen, applying a load, and "listening" for slippage and unbonding. Several studies have demonstrated the feasibility of acoustic emission for inspecting Gl–Ep and B–Ep parts.

Holography is a form of wavefront-reconstruction imaging in which two images of the same object are reconstructed and spatially superimposed. Any relative displacement between the object and the three-dimensional virtual image superimposed onto it will show the object covered by a pattern of interference fringes. Holographic interferograms measure differences either in shape or relative displacement. A feasibility study with B–Ep structures showed that a holographic inspection technique is quite promising. Disbonds between boron plies and between core and skin were clearly discernible under thermal stressing.

## Inspection and Maintenance

A total of 142 composite components have been under review and in flight service with 17 different operators, including foreign and domestic airlines and the U.S. military. The flight-service program initiated in 1972 by NASA was to determine the long-term durability of boron, Kevlar, and graphite composites in realistic flight environments. Over 2

**Table 3.15**   NASA Composite Component Inspection and Maintenance Results†

| Component | Inspection interval, months | Inspection methods | Status |
|---|---|---|---|
| CH-54B tail cone | 12 | Visual, ultrasonic | Minor disbonds, no repair required |
| L-1011 fairing panels | 12 | Visual | Minor impact damage, fiber fraying, and hole elongations |
| 737 spoiler | 12 | Visual, ultrasonic | Infrequent minor damage, repaired at Boeing |
| C-130 center wing box | 6 | Visual, ultrasonic | No defects after more than 5 years service |
| DC-10, aft pylon skin | . . . . . . | Visual | Minor surface corrosion on one skin |
| Upper aft rudder | 3, 12 | Visual, ultrasonic | Minor rib-to-skin disbond on two rudders |

†Data from Refs. 23, 24, 28, and 36.

million component flight hours have been accumulated, the highest aircraft time being 17,718 h. Table 3.15 lists the components being monitored. The Gr–Ep in the DC-10 upper aft rudder acquired 300 flight hours per month while the 737 Gr–Ep spoiler accumulated the greatest total component flight hours (1,438,000 in 6.5 years). The high-time spoiler had 17,500 flight hours, and approximately two-thirds of the spoilers accumulated over 10,000 flight hours. In addition to structural tests of the spoilers, tests were conducted to determine absorbed moisture content of the Gr–Ep spoiler skins from plugs cut near the trailing edge (Fig. 3.87). The plugs consist of aluminum honeycomb core, two Gr–Ep face sheets, two layers of epoxy film adhesive, and two exterior coats of polyurethane paint. About 90% of the plug mass is in the composite faces, including the paint and adhesive. The moisture content is determined by drying the plugs and recording the mass change. The data shown in Fig. 3.87 for plugs removed from three spoilers after 5 years' service indicate moisture levels in the Gr–Ep skins ranging from 0.66 to 0.75% for PAN graphite systems. Apparently, these moisture levels have not affected the room-temperature strength of the spoilers.

The composite components in the NASA flight-service evaluation program have been inspected at periodic intervals to check for damage, defects, or repairs that may occur during normal aircraft operation. The maintenance data in Table 3.15 were reported by the aircraft manufacturers who fabricated the various components. Minor disbonds found under small portions of the CH-54B B–Ep reinforcement were so small that they did not require repair. The Kv–Ep fairings on the L-1011 aircraft were visually inspected annually. Minor impact damage from equipment and foreign objects was noted on the wing-to-body Nomex honeycomb sandwich fairings. Fiber fraying, characteristic of Kevlar, and fastener-hole elongations were noted on all the Kv–Ep fairings, but no repair was required. The 737 Gr–Ep spoilers were inspected annually, and any defective spoilers were returned for repair. Infrequent minor damage included a mechanical interference problem and front-spar exfoliation-corrosion damage caused by an accidental breach of the corrosion-inhibiting system before final bonding of the Gr–Ep skins during fabrication. Visual, ultrasonic, and destructive testing found no evidence of moisture migration into the aluminum honeycomb core and no core corrosion.

The B–Ep-reinforced C-130 wing boxes were inspected every 6 months, and no defects were detected after more than 5 years of service. The B–Al aft pylon skins on the DC-10 aircraft were inspected annually; minor surface corrosion was reported on one panel. This

FIVE YEARS SERVICE

| GRAPHITE–EPOXY | MOISTURE CONTENT, PERCENT |
|---|---|
| LOW COST HIGH STRENGTH PAN, GRAPHITE (T300) | 0.67 |
| LOW COST HIGH STRENGTH PAN, GRAPHITE (T300) | 0.66 |
| LOW COST HIGH STRENGTH PAN, GRAPHITE (AS) | 0.75 |

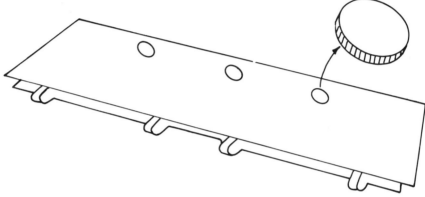

**FIG. 3.87**   Spoiler moisture levels determined from plugs.[24]

corrosion was believed to have been caused by improper surface preparation during fabrication of the panels. The Gr–Ep rudders on the DC-10 were visually inspected every 3 months and ultrasonically inspected every 12 months. Minor rib-to-skin disbonds detected on two rudders required no repair. These minor disbonds may have been caused by thermal stresses during cooldown after the manufacturing cure cycle. Overall, excellent performance has been achieved by the flight-service composite components.

Excellent experience has been achieved by utilizing state-of-the-art inspection techniques and developing maintenance procedures with approximately 200 composite components in flight service for 2.5 million total component hours. No significant degradation was observed in residual strength of composite components or environmental exposure specimens after 5 years of service or exposure. It is important to note that composites have matured to the point of acceptance by engineers and that quality procedures have been developed and accepted. Confidence in advanced composites technology has been developed to the extent that commercial and military manufacturers have made production commitments to composites for selected components.

## REFERENCES

1. "Advanced Composites Design Guide," 3d ed., Air Force Materials Laboratory, Wright-Patterson AFB, Ohio, 1973.

2. Kliger, H. S.: Design of Cost Effective Hybrid Composites for Automotive Structures, *SPI Reinf. Plast./Compos. 33d Annu. Tech. Conf., Washington, Feb. 7–10, 1978.*

3. Engineering Design Handbook, "Joining of Advanced Composites," DARCOM-P 706 316, U.S. Army Material Development and Readiness Command, March 1979.

4. Filament Wound Reinforced Plastics: State of the Art, *M/DE Spec. Rep.* 174, pp. 127–146, August 1960.

5. Palmer, R.: Composite Stiffened Panel Fabrication Experiences, Contr. NAS1-12675, Douglas Aircraft Co., March 1979.

6. Nadler, M. A., and L. J. Costanza: Tooling and Assembly Procedures, Serviceability Program Elements, *10th Natl. SAMPE Tech. Conf., Kiamesha Lake, N.Y., Oct. 17–19, 1978*, pp. 149–165.

7. Cohen, L. J., J. W. Young, and T. Muha: Design and Analysis of Advanced Composite Missile Structure, *10th Natl. SAMPE Tech. Conf., Kiamesha Lake, N.Y., Oct. 17–19, 1978*, pp. 898–923, Contr. F33615-77-C-5237.

8. Doble, G. S., and P. Melnyk: Air Bonded, FOD Resistant Metal Matrix Fan Blades, *TRW Rep.* AFML-TR-76-218, Contr. F33615-77-C-5222, December 1976.

9. Stabrylla, R. G., and R. G. Carlson: Boron/Aluminum Fan Blades for SCAR Engines, *General Electric Rep. NASA* CR-135184, Contr. NAS3-18910, June 1977.

10. Oller, T. L.: Fiber Composite Fan Blade Impact Improvement Program, *General Electric Rep.* CR-135078, Contr. NAS3-17836, November 1976.

11. Randall, D. G.: TF 30 Third-Stage Composite Fan Blade Service Program, *Pratt & Whitney Rep.* PWA-5141, Contr. F33657-70-C-0624, May 15, 1978.

12. F-100 3rd Stage Fan Blades, Pratt & Whitney, AFML Contr. F33615-78-C-5071, November 1980.

13. Titanium Composite Fan Blades, Allison Division of General Motors, Indianapolis, AFML Contr. F33615-69-C-1652, 1969.

14. Selective Stiffening and Low Cost Manufacturing of Turbine Blading Using Titanium Composites, Allison Division of General Motors, Indianapolis, AFML Contr. F33615-75-C-2044, 1975.

15. JT9D First Stage Composite Fan Blades, Pratt & Whitney, East Hartford, Conn., NASA Contr. NAS CR-134515, 1979.

16. Manufacturing Methods for F100 Composite Nozzle Flaps, Pratt & Whitney, West Palm Beach, Fla., AFML Contr. F33615-77-C-5099, 1977.

17. Stoltze, L., and E. M. Varholak; Manufacturing Technology for F100 Boron/Aluminum Vanes, United Technologies-Hamilton, IR-436-8(V), F33615-78-C-5071, November 1980.

18. Faddoul, J. R.: Preliminary Evaluation of Fiber Composite Reinforcement of Truck Frame Rails, *SAE Congr. Expos., Detroit, Feb. 28–Mar. 4, 1977, NASA Lewis Res. Cent. Rep.* TM X-73582.

19. Arnesen, D. A., and M. Marley: Detroit Brings Space Age Materials Down to Earth, *Iron Age,* Apr. 9, 1979, pp. 37–40.

20. F-15 Composite Wing Flight Test, AFML Contr. F33615-71-6-1576, McDonnell-Douglas Corporation, AFML-TR-75-78.

21. Advanced Development of Not-Critical-to-Flight-Safety Advanced Composites Aircraft Structures, AFML Contr. F33615-72-6-1781. Northrop, 1972, AFML-TR-74-38.

22. C-5A Wing Boron Leading Edge Slat Development Program, ASD-TR-72-62, AFML Contr. F33657-68-C-0900, Lockheed-Georgia Co., 1972.

23. Advanced Composite Selective Reinforcement of C-130 Wing Box, NASA CR-112126, -112272, -132495 and -145043, NASA/Langley, Contr. NAS1-11100, Lockheed-Georgia Co., 1972.

24. A Study of the Effects of Long Term Ground and Flight Exposure on the Behavior of Gr/Ep Spoilers, NASA/Langley, Contr. NAS1-11668, Boeing, October 1972.

25. Development of a Graphite Horizontal Stabilizer, NADC Contr. N00156-70-C-1321, McDonnell-Douglas, 1970–1972.

26. Boron Horizontal Tail Flight Test Qualification Program, AFML Contr. F33615-5257, General Dynamics/Convair/Ft. Worth, April 1971.

27. Birchfield, E. B., and R. Kollmansberger: Develop Fabrication/Processing Techniques for High Temperature Advanced Composites for Use in Aircraft Structures, AFML-TR-72-91, AFML Contr. F33615-70-C-1546, McDonnell-Douglas, July 1972.

28. Wooley, J. H., D. R. Pashal, and E. R. Crilly: Flight Service Evaluation of Kevlar Epoxy Composite Panels in Wide Bodied Commercial Transport Aircraft, *NASA Langley Tech. Rep.*, Contr. NAS1-11621, August 1972.

29. Foreman, C. R., and J. H. Pimm: Design of a Graphite Composite A-7 Speed Brake, *AFML/ AFFDL Conf. Fibrous Compos. Flight Veh. Des. Dayton, September 1972.*

30. Oken, S., et al.: Design of a Gr/Ep Metering Truss for the Large Space Telescope, *16th Struct., Struct. Dynam. Mater. Conf., Denver, May 27–29, 1975*, AIAA Pap. 75-784.

31. Brown, L. D., et al.: Composite-Reinforced Propellant Tanks, *NASA-Grumman Rep.* CR-134726, Contr. NAS3-14368, February, 1975.

32. Mitchell, S. C., and J. Goulding: Manufacturing Technology for Low Temperature Composite Engine Frame, *General Electric and Rohr Industries Interim Rep. 5, Sept. 30, 1979–Dec. 30, 1979*, AFML Contr. F33615-78-C-5086, December 1979.

33. McDanels, D. L., and R. A. Signorelli: Effect of Fiber Diameter and Matrix Alloys on Inpact-Resistant Boron/Aluminum Composites, *NASA Rep.* NASATN D-8204, November 1976.

34. Phillips, W. L.: Elevated Temperature Properties of SiC Whisker Reinforced Aluminum, *Proc. 1978 Int. Conf. Compos. Mater., Metall. Soc. AIME, Toronto, Apr. 16–20, 1978.*

35. Royster, D. M., R. R. McWithey, and T. T. Bales: Fabrication and Evaluation of Brazed Titanium-Clad Borsic/Aluminum Compression Panels, *NASA Tech. Pap.* 1573, March 1980.

36. "DOD/NASA Structural Composites Fabrication Guide," 2d ed., vol. II, Contr. F33615-77-C-5256, Lockheed Company Manufacturing Technology Division, Air Force Materials Laboratory, Wright-Patterson AFB, Ohio, May 1979.

37. Davis, L. W.: "Composites at Low Temperatures," Harvey Aluminum Inc., Torrance, Calif., 1979.

38. Toth, L. W., and R. A. Burkley: "Mechanical Response at Cryogenic Temperatures of Selected Reinforced Plastics Composite Systems," Goodyear Aerospace Corp., Akron, Ohio, 1979.

39. Burning Characteristics and Fiber Retention of Graphite/Resin Matrix Composites, *NASA Tech. Mem.* 79314, 1979.

40. Improved Fiber Retention by the Use of Fillers in Graphite Fiber/Resin Matrix Composites, NASA *Tech. Mem.* 79288, 1979.

41. Bilow, N., and A. L. Landis: "Recent Advances in Acetylene-Substituted Polyimides," Hughes Aircraft Co., Culver City, Calif., 1976.

42. Miller, B. A., and S. Lee: "The Effect of Gr/Ep Components on the Galvanic Corrosion of Aerospace Alloys," Air Force Materials Laboratory, Wright-Patterson AFB, Ohio, 1976.

43. Royster, D. M., et al.: Effects of Fabrication & Joining Process on Compressive Strength of B/ Al and BSc/Al Structural Panels, *LRC Rep.* NASA-TP-1121 (N78-20256), Hampton, Va., April 1978.

44. Staebler, C. J., Jr., and B. F. Simpers: Metallic Coatings for Gr/Ep Composites, *Grumman Aerospace Corp. Final Rep.* AD A069871, Contr. N00019-77-C-0250, Bethpage, N.Y., May 1979.

45. Patz, G. L., and D. E. Davenport: Conductive Prepregs, *10th Natl. SAMPE Tech. Conf., Kiamesha Lake, N.Y., Oct. 17–19, 1978*, pp. 498–503.

46. Kliger, H. S.: Customizing Carbon-Fiber Composites, *Mach. Des.*, Dec. 6, 1979, pp. 150–157.

47. Dreger, D. R.: Design Guidelines for Joining Advanced Composites, *Mach. Des.*, May 8, 1980, pp. 89–93.
48. House, E. E.: Integrally Stiffened Laminate Construction, *9th Natl. SAMPE Tech. Conf., Atlanta, Oct. 4–6, 1977.*
49. Dittmer, W. D., and P. R. Hoffman: Boron Composites: Status in the USA, *Interavia*, 28: 654 (June 1973).
50. Composite Rear Fuselage Manufacturing Methods & Technology Program, Sikorsky Aircraft, Division of United Technologies, Tech. Prop. SPB79-A2664-RD, Aug. 21, 1979.
51. Elkin, R. A.: Composite Nacelle Structures: DC-9 Nose Cowl, *3d Conf. Fibr. Compos. Flight Veh. Des., Williamsburg, Nov. 4–6, 1975*, NASA TM X-3377, pt. 1, April 1976.
52. Hagnauer, G., and D. Dunn: Quality Assurance of an Epoxy Resin Prepreg Using HPLC, *12th Natl. SAMPE Tech. Conf. Seattle, Oct. 7–9, 1980.*
53. Lubin G.: "Handbook of Fibreglass & Plastic Composites," Van Nostrand Reinhold, New York, 1969.
54. Jackson, A.: Development of Gr/Ep Covers for L-1011 Advanced Composite Vertical Fin, *12th Natl. SAMPE Tech. Conf. Seattle, Oct. 7–9, 1980.*

# BIBLIOGRAPHY

Blake, C., Electromagnetic Effects on Composites, *12th Natl. SAMPE Tech. Conf. Seattle, Oct. 7–9, 1980.*

Block-Onera, H.: Various Methods for Checking the Aging of Prepregs during Storage, *SNIAS Int. Conf. Adv. Technol. Mater. Eng., Cannes. Jan. 12–14, 1981.*

Botsco, R.: Advanced Ultrasonic Testing of Aerospace Structures, *26th Natl. SAMPE Tech. Conf. Los Angeles, Apr. 28–30, 1981.*

Cairo, R. P., and R. D. Torczyner: Graphite/Epoxy, Boron-Graphite/Epoxy Hybrid and Boron/ Aluminum Design Allowables, *Grumman Aerospace Corp. Tech. Rep. May 1971–June 1972*, AFML TR 72-232, Contr. F33615-71-C-1605, December 1972.

Carri, R.: The Effects of Lightning and Nuclear Electromagnetic Pulse on the Composite Aircraft, *10th Natl. SAMPE Tech. Conf. Kiamesha Lake, N.Y., Oct. 17–19, 1978*, pp. 1–13 (Grumman Aerospace Contr. F33615-77-C-5169).

Chamis, C. C., and J. H. Sinclair: 10° Off-Axis Tensile Test for Intralaminar Shear Characterization of Fiber Composites, *Lewis Res. Cent. NASA Tech. Note* D-8215 (N76-22314/NSP), 1976.

Christian, J. L., and M. D. Campbell: Mechanical and Physical Properties of Several Advanced Metal/ Matrix Composite Materials, *Adv. Cryogen. Eng.*, 18:175–183 (1973).

Cooper, C.: Characterization of Prepreg Resins for Incoming Inspection, *12th Natl. SAMPE Tech. Conf., Seattle, Oct. 7–9, 1980.*

Delgrosso, E. J., and C. E. Carlson: Composite Fan Blades Can Be Inspected Holographically, *Automot. Eng.* 85(12):62–65 (December 1977).

Dolowy, J. F., B. A. Webb, and W. C. Harrigan: Metal Matrix Composite Systems: Properties and Status, *SME 6th Maj. Conf. Struct. Compos. Manuf. Appl., Los Angeles, June 10–12, 1980*, EM80-429.

Fraser, W. F.: Radio Frequency Interference and Electromagnetic Interference Shielding, *11th Congr. Manuf. Qual., Brighton, November 1978*, pap. 13.

Haskins, J.: Advanced Composite Design Data for Spacecraft Structural Applications, *12th Natl. SAMPE Tech. Conf. Seattle, Oct. 7–9, 1980.*

Holehouse, I.: Sonic Fatigue Design Techniques for Advanced Composite Aircraft Structures, Rohr Industries Contr. F33615-77-C-3033, AFWALTR 80-301, April 1980.

June, R.: Commercial Airplane Composite Component Service Experience, *12th Natl. SAMPE Tech. Conf., Seattle, Oct. 7–9, 1980.*

Kasen, M. B.: Composite Materials for Cryogenic Structures, *Int. Cryogen. Mater. Conf. Boulder, Aug. 2–5, 1977.*

Knight, R., and E. Rosenzweig: S-3A Composite Spoilers Service Evaluation, *12th Natl. SAMPE Tech. Conf., Seattle, Oct. 7–9, 1980.*

Langston, P. R.: Overview of Kevlar Aramid Composites: Kevlar 49, Its Properties and Uses, *Kevlar Compos. Symp. Technol. Conf., El Segundo, Calif. Dec. 2, 1980.*

Leahy, J. D.: Carbon Fiber/LARC-160 Polyimide Composites for High Temperature Applications, *SME 6th Maj. Conf. Struct. Compos. Manuf. Appl., Los Angeles, June 10–12, 1980,* EM80-426.

Martin, J.: An Automated Ultrasonic Test Bed: Application to NDE in Graphite/Epoxy Materials, *26th Natl. SAMPE Tech. Conf., Los Angeles, Apr. 28–30, 1981.*

McGovern, S. A.: NDI Survey of Composite Structures, NADC-80032-60, Vought Corp., Naval Air Systems Command, Contr. N62269-78-M-6391, January 1980.

Reihner, J.: Acousto-Ultrasonic Inspection of Composites, *Int. Conf. Adv. Technol. Mater. Eng., Cannes, Jan. 12–14, 1981.*

Reynolds, W. N.: Non Destructive Examination of Composite Materials: Survey of European Literature, *Atom. Energy Res. Establ. Gt. Br.,* FR 10/79-2/80, DAJA37-79C0553, ADA086165, May 1980.

Rich, M. J., G. F., Ridgley, and D. W. Lowery: Application of Composites to Helicopter Airframe and Landing Gear Structures, Contr. NAS1-11688, Sikorsky Aircraft, NASA CR-112333, June 1973.

Schmid, T.: Turbine Engine Composite Components, *12th Natl. SAMPE Tech. Conf., Seattle, Oct. 7–9, 1980.*

Shelton, W. L.: NDI of Composite Materials, Non-Destructive Evaluation Inspection Practices, *AGARDGRAPH* 201, 2:581–592 (October 1975).

Shuford, R. J., and W. W. Houghton: Acoustic Emission as NDE Technique for Determining Composite Rotor Blade Reliability, Army Materials and Mechanics Research Center, ADA090440, June 1980.

Staebler, C. J., Jr., and B. F. Simpers: Metallic Coatings for Graphite-Epoxy Composites: Phase II, *Grumman Aerospace Corp. Final Rep.,* Contr. N00019-78-C-0602, Naval Air Systems Command, August 1980.

Stone, R.: Development of Repair Procedures for Gr/Ep Structures on Commercial Transports, *12th Natl. SAMPE Tech. Conf., Seattle, Oct. 7–9, 1980.*

Taig, I. C., A. August, R. Hadcock, and S. Dastin: Design of Structures in Composite Materials (Basic Data and Interdisciplinary Action), *AGARD Rep.* 639, January 1976.

———: Design Concepts for the Use of Composites in Airframes, *Joint Symp. AGARD Struct. Mater. Panel Propul. Energet. Panel, Toulouse, Sept. 20–22, 1972.*

Curing composites in an autoclave. *(Vought Corp.)*

# Processing

Low-cost manufacturing techniques are vital to greater use of structural composites in the future. Hand lay-up still seems to dominate the composites manufacturing industry as a means of ply orientation, stacking, draping, and consolidation. For this reason, the importance of ply placement, bleeding, curing, and consolidation procedures, and prevention of defects during assembly and bonding operations to ensure that design requirements will be met is always stressed by the industry. Mechanical properties of a laminate can be strongly influenced by processing.

Current attitudes dictate that composite structures be produced at a lower total cost than corresponding metal assemblies. Thus, a primary concern is to reduce material costs by using less expensive fibers. Low-cost high strength graphite and Kevlar aramid fibers have supplanted boron fibers used in early advanced-composite production hardware. In addition sheet-molding (SMC) and premix or bulk-molding compounds (BMC) have come into their own as promising materials.

Because aerospace-grade fibers are too costly for most commercial applications, less expensive grades have been developed. More efficient usable material forms include tape 48 in (1.22 m) wide, woven fabrics (unidirectional- and balanced-strength types), cross-plied sheets (permitting several plies to be laid up simultaneously) and narrow tapes (compatible with filament-winding and braiding processes). Resin-matrix materials that are easier to process include addition-curing polyimides, 250°F (121°C) curing epoxies, and various thermoplastics. MMCs such as Gr–Al, aluminum oxide, Mg–SiC, and Ti–SiC are being developed to complement B–Al.

Since lay-up and material handling are high-cost labor-intensive operations, automated processing is being stressed in all current and future operations and installations for composite production. Typical are numerical-control lay-up, laser or water-jet trimming, and mechanized (robotic) stacking. Rapidly installed prefabricated vacuum bags and shorter autoclave and press-cure cycles are also being implemented. These manufacturing processes, techniques, tooling and other operations, i.e., joining, machining, drilling, etc., will be discussed in detail in Chaps. 5 and 6.

Cost analyses of aircraft assemblies have shown that mechanical fastening and secondary bonding of precured details are costly operations. As a result, design, tooling, and processing concepts to produce integral structures have been developed to make better use of the unique properties of structural composites. Production acceptance has been obtained for monolithic internally stiffened rudders, hat-stiffened fuselage panels, and honeycomb sandwich structures with one or both skins cured over the core. Integral skin or spar structure with stitched joints, wing boxes with integral substructure, and integrally stiffened empennage structure (without honeycomb core) have been produced in development laboratories. A further reason to implement integral structure is to prevent aluminum corrosion, which normally occurs when aluminum comes into direct contact with Gr–Ep. This would eliminate the need for costly titanium fasteners and nonmetallic honeycomb core. Molding to net trim, automated drilling, and mechanized fastener installation, discussed later, are further evidence of operations which have been introduced into production.

## LAY-UP AND CURING

The two principal steps in the manufacture of laminated fiber-reinforced composite materials are lay-up, which consists of arranging fibers in laminae and laminae in layers or laminates, and curing, which is the drying, or polymerization, of the resinous matrix material to form a permanent bond between fibers and between laminae. Curing may occur unaided or consist of applying heat and/or pressure to speed polymerization. When pressure is used, processes include vacuum bag, autoclave, hydroclave, and tool press.

The three principal lay-up processes for laminated fiber-reinforced composite materials are winding and laying, molding, and continuous lamination. The choice of both the lay-up and the curing process depends on process effectiveness, part size, cost, schedule, familiarity with particular techniques, etc.

Winding and laying operations include filament winding, tape laying, and cloth winding or wrapping. Filament winding, discussed in detail in this chapter, consists of passing a fiber through a liquid resin and then winding it on a mandrel (Fig. 4.1). The fibers are wrapped at different orientations on the mandrel to yield many layers and hence strength and stiffness in many directions. Subsequently, the entire assembly is cured on the mandrel, after which the mandrel is removed. Tape laying starts with a tape consisting of fibers in a preimpregnated form held together by a removable backing material. The tape is unwound to form the desired shape in the desired orientations of tape layers. The tape is very similar to glass-reinforced, heavy-duty box sealing tape. Cloth winding or laying begins with preimpregnated cloth, which is unrolled and deposited in the desired form and orientation. The bidirectional character of the fibers in the cloth makes cloth winding or laying more inflexible and inefficient than filament winding or tape laying in achieving specified goals of strength and stiffness. Cloth layers are often used as filler layers where strength and stiffness are not critical.

Molding operations are less important than the various winding and laying operations but are used to make other common composite materials. Molding can begin with hand

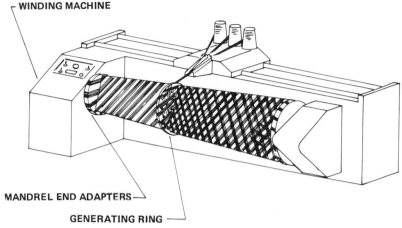

**FIG. 4.1** Filament-winding skins on a mandrel. *(Hercules, Inc.)*

or automated deposition of preimpregnated fibers in layers. The prepreg layers are often precut. Subsequently, the layers are compressed under elevated temperature to form the final laminate. Molding is used, for example, to fabricate radomes to close tolerances in thickness. A B–Ep fuselage frame assembly in a military bomber is another common molded composite part.

## Relationship between Material, Process, and Structure

Composite parts are sometimes classified according to the length of the fibers as filament-controlled laminates containing long continuous fibers in the form of tapes and fabrics or short random fiber-reinforced parts. Although mechanical properties of short-fiber compounds are significantly lower than those of filament-controlled laminates, these materials are widely used for moderately loaded parts which must be produced in large quantities at minimum cost. They are processed by injection molding with thermoplastic matrixes, transfer molding with thermosetting resins, and matched-die molding with thermosetting resins. A process similar to that used for glass-fiber compounds consists of the following steps:

1. Prepare, charge, weigh, or cut to size
2. Inject required volume
3. Introduce into mold
4. Apply heat and pressure
5. Perform secondary operations (machine, finish)

Random-fiber parts are usually molded to net size and shape. Thus, the material-process-structure relationship is bounded by the shape of the part, feasibility of producing the molds, molding characteristics of the material, and capacity of the molding equipment. For high-rate production using a maximum of automation, the part must be configured to be readily removable from the mold without dissembly. The configuration must also provide for a uniform distribution of the fibers to ensure adequate structural properties. For injection molding of filled thermoplastics, the plasticizing capacity and shot size of the molding machines limit part size. Parameters in compression molding include platen size, operation temperature, and pressure capability; cycle time is established by the resin gel time. Thus, polyester or vinylester resins are preferred to epoxy resins because they cure rapidly at moderate temperatures [250 to 350°F (121 to 177°C)]. Although the preforms used are typically cut from SMC, chopped-fiber preforms have been produced by the spray-up method. Mat-molding compounds with thermoplastic matrixes are also suitable for hot stamping to shape at high rates using high-temperature presses. The production of random-fiber composite parts by these processes is a specialized technology described later in this chapter.

For filament-controlled laminates, the relationship between material, process, and structure is far more complex as a result of the large number of options in materials, processes, and design concepts. Nevertheless several ground rules can be established:

Tooling to provide good fit of the most critical surfaces

Maximizing automated lay-up and processing

Reducing part and/or fastener count by molding integral assemblies

Considering curing as the most critical single step

Minimizing trimming cured parts by use of net molding

Automating end-item nondestructive inspection

Some of the interactions between material, process, and structure are shown in Table 4.1.

## FABRICATION PROCESSES

Many processes are available to produce the desired combination of design performance and economics of carbon, boron and glass fiber composites. Each process has its own usefulness for combining different kinds and amounts of boron, carbon, glass, aramid, and resin. The basic processes can be considered broadly in two classes: open- and closed-mold processes.

## OPEN-MOLD PROCESSES

Open molds are single-cavity molds, male or female, used in processes requiring little or no pressure. The principal characteristics of the molded object are that only one side is finished and complex shapes and large objects can be formed.

### Hand Lay-Up

Hand lay-up, or contact molding, is the oldest and simplest process for forming glass-fiber-reinforced plastic. The glass fibers and resin are placed in or on the mold, and entrapped

**Table 4.1** Relationships between Material, Process, and Structure[20]

| Material | Structure | Process |
|---|---|---|
| Matrix selection | Operating temperature-load intensities | Consolidation (temperature, pressure, cycle time) |
| Fiber selection | Stiffness bearing (if required) and strength of total load | Fiber controls machining parameters; load plus fiber controls part thickness, ply shades, and panel design concept |
| Adhesive or fastener selection | Joints and assembly | Integrally molded stiffened panels, mechanically fastened assembly, secondary bonded assemblies, co-cured bonded assemblies |
| Prepreg drapability | Contour requirements, mating surfaces, part size | Tooling concepts: Integrally stiffened panels (typically, air-passage tool) Fastened assemblies (air-passage or inner mold-line tools based on air-passage and fuel-sealing requirements) Secondary bonded structures (typically, tool from core surface) Co-cured bonded assemblies (typically, air-passage tool) Fiber placement (determined by size and shape): Large skins: machine lay-up or wide goods Substructure: process multiple parts concurrently by machine lay-up; plies cut from wide goods; braiding; filament winding |
| Protection Against external lightning and electromagnetic interference | Finishing and/or sealing requirements | Metallic coatings (flame spray); pretreat foils or mesh; secondary bonding may be required; coating application. |
| Paint, weather, and fuel sealing | | Sealant application: groove injection, thermoplastic sealing; faying surface sealant |
| Optimum material forms to reduce in-house handling | High production rate | Multiple sets of tools, multiple shifts or duplicate facilities, investment in more automated processing |

air is removed with squeegees or rollers. Layers of glass and resin are added to build up to design thickness. If a high-quality surface is desired, a gel coat (pigmented surfacing resin) is applied to the mold before lay-up. The lay-up normally cures at room temperature, but heat may be used to accelerate curing. The exposed side is generally rough but can be made smoother by wiping on cellophane or other suitable releasing films, e.g., Mylar or polyvinyl alcohol. Resins used in hand lay-up are usually polyesters or epoxies.

## Spray-Up

In the spray-up process glass fibers and resin are simultaneously deposited in or on a mold from special spraying equipment. Roving is fed through a chopper and into a resin-catalyst stream for deposit on the mold. Resin and catalyst may be combined in a single spray gun or from two guns with streams which intercept. After deposition, the glass-resin mix is rolled with a hand roller to remove entrapped air, lay fibers down, and smooth the surface. The lay-up is then cured at room temperature or with an accelerated heat cure. Resins used in spray-up are usually polyesters or epoxies (Fig. 4.2).

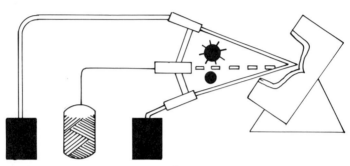

**FIG. 4.2**   Sketch of spray-up technique.[1]

## *Vacuum-Bag Molding*

The vacuum-bag molding process, a refinement of hand lay-up, was developed for fabricating a variety of components and structures. Complex shapes, including double contours, and relatively large parts are readily handled. The process is principally suited for cases where higher-pressure molding is not feasible. It uses a vacuum to eliminate voids and force out entrapped air and excess resin. A suitable film such as cellophane, polyvinyl alcohol, or nylon is placed over the lay-up and sealed at the edges with a special sealing compound. A vacuum is then drawn on the bag formed by the film, and the laminate is cured.

The essential steps in the process are the lay-up, preparation of a bleeder system, and bagging. The necessary number of plies for a lay-up are precut to size and positioned in the mold one ply at a time. Each ply is separately worked to remove trapped air and wrinkles to ensure intimate contact with the previously laid-up ply. The completed lay-up is covered with the porous nonadhering film mentioned above, which provides easy release and vacuum access to the lay-up.

A controlled-capacity bleeder system is normally required to absorb excess resin and permit the escape of volatiles while maintaining the specified fiber-volume ratio of a particular laminate. Bleedout may be through the edges (edge bleedout), through the top surface (vertical bleedout), or both. An edge bleedout is constructed by placing a narrow width of bleeder cloth [about 1 in (25.4 mm) of uncoated fiber glass, burlap, or similar material] around the periphery of the lay-up. Vertical bleeder plies of the same materials are stacked directly over the perforated film. Edge-bled parts usually require trimming to remove resin-rich edges. When parts cannot be trimmed, flexible dams are used in conjunction with vertical bleeders to minimize edge bleedout. Bleedout of unidirectional tapes or wide goods normally requires the addition of a thin, flexible caul plate over the bleeder plies to prevent excessive resin washout and provide a smooth bag-side surface. The lay-up may contain peel plies when subsequent bonding operations are to be performed. The peel ply is placed in direct contact with the lay-up, below the release film.

Vacuum connections are placed over the vertical bleeders or outside the edge bleeders. Enough ports are provided to ensure a uniform flow of resin and volatiles. The bagging film, tailored to fit the part, is placed over the lay-up, bleeder system, and vacuum connectors and is sealed to the mold plate. A partial vacuum is usually drawn to smooth the bag surface before applying full vacuum and heat. The bagged mold is transferred to an oven for curing with full vacuum applied. Vacuum is usually maintained during the entire heating and cooling cycle.

The process uses the simplest and most economical molds and equipment. In addition to the shapes and parts the method is primarily applicable where:

Only a few parts are required.

Design changes are frequent.

High structural properties are not required.

Configuration of the part does not lend itself to press molding.

## Pressure-Bag Molding

Pressure-bag molding involves placing a tailored bag, usually a rubber bag, over the lay-up and then using air or steam pressure to eliminate voids, force out entrapped air, and drive out excess resin (Fig. 4.3). This method, which is applicable to molding integrally

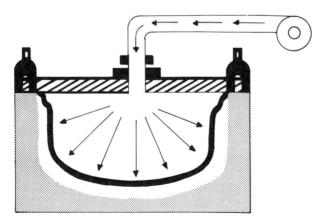

**FIG. 4.3**   Pressure-bag molding.[1]

stiffened structures having complex shapes, consists of wrapping prepreg layers over blocks of rubber and restraining the lay-up and rubber blocks in a metal cavity. As temperature is increased, the rubber blocks or mandrels expand more than the metal restraining tool; this generates curing pressure, eliminating the need for external pressure application, e.g., an autoclave. Pressures up to 30 lb/in$^2$ (207 Pa) can be used in pressure-bag molding to achieve higher strengths than obtainable with vacuum-bag molding. Higher pressures can be applied through the use of a hydroclave or a heated air-circulating autoclave.

## Autoclave

Autoclave molding is similar to the vacuum-bag method and is a modification of pressure-bag molding. The lay-up is subjected to greater pressures, and denser parts are produced. The bagged lay-up is cured in an autoclave by the simultaneous application of heat and pressure. Most autoclave processes also use vacuum to assist in the removal of trapped air or other volatiles. The vacuum and autoclave pressure cycles are adjusted to permit maximum removal of air without incurring an excessive resin flow. Vacuum is usually applied only in the initial stages of the curing cycle; autoclave pressure is maintained during the entire heating and cooling cycles. Curing pressures are normally in a range of 50 to 100 kips/in$^2$ (345 to 690 MPa), and pressure can be controlled to give the desired degree of quality. Compared with vacuum-bag molding, this process yields laminates with closer control of thickness and lower void content.

## Filament Winding

In this process fibers or tape are usually wound onto a rotating mandrel from a relatively stationary position. Roving or single strands are fed from a creel through a bath of resin; preimpregnated roving may also be used. Fibers are dispensed from the translating head at controlled angles normal to the rotating mandrel axis. Lay-up compaction can be applied by the degree of tension maintained on the fibers during the application. When the desired number of layers has been applied, the wound mandrel is cured at room temperature or in an oven.

## Centrifugal Casting

Round objects such as pipe can be formed by this process, in which chopped strand mat is positioned inside a hollow mandrel. The assembly is then rotated, and centrifugal action forces glass and resin against the walls. Cure is at room temperature, or it may be accelerated by placing the assembly in an oven.

## Continuous Pultrusion

With pultrusion, the composite equivalent of metallic extrusion, low-cost constant-cross-section structural shapes have been produced. Channel sections, Z sections, and flat bars up to 40 ft (12.2 m) long have been pultruded using several resin systems. Continuous strands in the form of roving or other reinforcement are impregnated in a resin bath; alternatively preimpregnated uniaxial tapes or sewn cross-plied strips are drawn through a ceramic die which sets the shape of the stock and controls the resin content. Final cure is effected in an oven or by infrared, radio-frequency, or microwave energy, through which the stock is drawn by a suitable pulling device (Fig. 4.4).

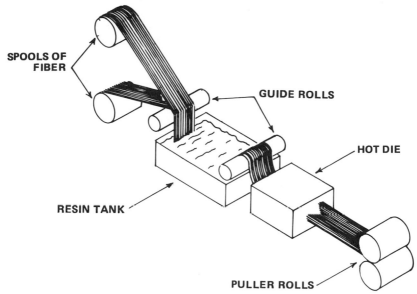

SPOOLS OF FIBER

GUIDE ROLLS

HOT DIE

RESIN TANK

PULLER ROLLS

FIG. 4.4   Continuous pultrusion.[22]

## CLOSED-MOLD PROCESSES

Closed molds are two-piece male and female molds, usually of metal. The principal characteristics of the molded object are a controlled surface finish, finishing on two sides, and excellent reproduction of detail from part to part.

### Matched-Die Molding

Matched-die molding combines mat, fabric, or preform with resin at the press just before or just after placement in the mold. In the mold, matched metal dies under pressure form and cure the part at a preselected temperature. Depending on part and shape and the resin performance, cure cycles range from less than 1 min to as much as 5 min.

### Injection Molding

Injection molding is a high-volume process that involves (1) softening a thermoplastic (in most cases) or thermosetting plastic material, which may contain fillers, by heating and/ or mechanical working with an extrusion-type screw; (2) forcing the softened plastic into a cooled mold under high injection pressure from a ram or screw; (3) solidifying the plastic in the pressure-clamped mold; and (4) ejecting the finished part. The runner and sprues that channel the hot plastic from the injection cylinder to the mold cavity can be reclaimed and remolded if the material is thermoplastic. Parts of highly intricate configurations, ranging in size from small washers to fairly large parts, can be injection-molded at high production rates. This process is widely used by the toy, appliance, and automotive industries which deal in volumes of millions of parts. Injection molding can also be used successfully by the aerospace industry through innovative mold design and careful planning of material use. Parts that are normally made by machining from prefabricated sheet, rod, or bar stock often lend themselves to fabrication by injection molding. Typical materials are polyethylene, polystyrene, polypropylene, polycarbonate, polyvinyl chloride, and polymethyl methacrylate.

## Continuous Laminating

Continuous laminating consists of layers of fabric or mat which are passed through a resin dip and brought together between cellophane covering sheets. The lay-up is passed through a heating zone, and the resin is cured. Laminate thicknesses and resin contents are controlled by squeegee rolls as the various plies are brought together.

# OTHER PROCESSES

## Braiding

Braiding is a mechanized textile process in which a mandrel is fed through the center of a machine at a uniform rate and fibers from numerous rolls mounted on a frame are deposited on a tool passing through the frame. Typically, 45 and 90° (0.79- and 1.6-rad) plies are braided and 0° (0-rad) plies are laid up from tape. The process is applicable to circular, box, and U-channel sections and webs.

## Compression Molding

Compression molding is used to fabricate a wide range of strong and high-temperature-resistant components from filled and unfilled thermosetting plastics. Preconsolidated plugs of the molding material or the loose granules themselves are placed directly in the machined mold cavity or in a separate melting area adjacent to the cavity, heated after the mold halves are closed, and forced to take the shape of the mold cavity. In contrast to injection molding, the mold must be heated to the curing temperature of the material to bring about cross-linking and a permanent set. Typical materials molded are phenolics, melamines, polyesters, and epoxies.

## Casting, Vacuum Infiltration, and Diffusion Bonding

Casting, vacuum infiltration, and diffusion bonding are processes used to produce metal-matrix parts. The use of casting has been limited to prototype production and laboratory development. Vacuum infiltration involves placing the reinforcing fibers in a die, applying vacuum at one end of the die and a source of molten matrix at the other, and then using the vacuum to draw the metal through the die to impregnate the fibers. This process has been used to produce structural shapes and selectively reinforce metallic parts.

Most metal-matrix parts are consolidated using diffusion bonding to fuse the foil or foil–flame-spray matrix into a unified mass. The critical process requirements are the application of adequate bonding pressure to the entire surface of the part and prevention of internal oxidation, which reduces bond strength. Matrix melting must also be avoided unless the fiber being used is resistant to degradation in the presence of a molten matrix. The consolidation is typically performed in presses fitted with ceramic dies.

Some fabricators use a *green tape,* consisting of aluminum foil, boron filaments, and flame-sprayed aluminum to hold the fibers in place. Plies cut from this material can be spot-welded to hold the lay-up together. The lay-ups can then be consolidated by pressing without applying vacuum (air-bonding process). This process can also be performed using prefabricated monolayers for the plies. The monolayer is produced by pressing foil or fiber tapes or using heated high-pressure rollers. The foils and filaments are usually bonded together temporarily with an organic binder. The lay-up and associated tooling are placed in a retort (Fig. 4.5). The assembly is heated to various intermediate temperatures under high vacuum to remove the organic binder by volatilization. The part is then consolidated at pressures of 4 to 5 kips/in$^2$ (28 to 35 MPa) at temperatures of 950 to 980°F (510 to 527°C) for B–Al. Gr–Al has been processed in similar fashion. Titanium-matrix composites are diffusion-bonded at higher temperatures, typically 1100 to 1200°F (593 to 649°C).

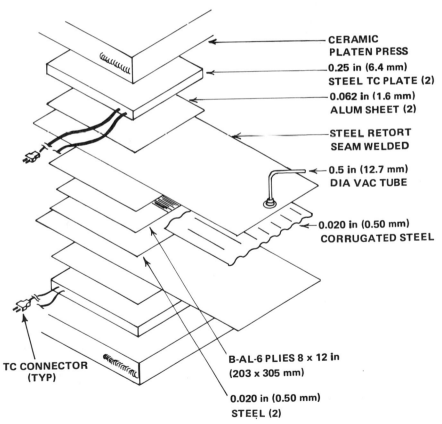

CERAMIC
PLATEN PRESS

0.25 in (6.4 mm)
STEEL TC PLATE (2)

0.062 in (1.6 mm)
ALUM SHEET (2)

STEEL RETORT
SEAM WELDED

0.5 in (12.7 mm)
DIA VAC TUBE

0.020 in (0.50 mm)
CORRUGATED STEEL

TC CONNECTOR
(TYP)

B-AL-6 PLIES 8 x 12 in
(203 x 305 mm)

0.020 in (0.50 mm)
STEEL (2)

FIG. 4.5    Preparation of metal-matrix monolayers.[20]

In the rest of this chapter both common methods and specialized techniques for composite fabrication will be described in detail.

## HAND LAY-UP TECHNIQUES

Hand (or contact) lay-up is a method of molding room-temperature-curing thermosetting polyesters and epoxies in association with glass fibers (usually mat or woven roving) or mineral or fiber reinforcements. A chemical reaction initiated in the resin by a catalytic agent causes hardening to a strong, light final part in which the resin serves as the substrate and the fibers as the reinforcement. In practice, it is much like pouring concrete over steel reinforcing rods.

In fabrication, an open mold is made. For a high-quality surface, a pigmented gel coat is first applied to the mold by spray gun (Fig. 4.6). The fiber reinforcement and resin are

**FIG. 4.6**   Gel coat covers mold surface. *(Owens-Corning Fiberglas Corp.)*

placed in or on the mold. Entrapped air is removed with squeegees, rollers, and/or brush-dabbing (Fig. 4.7). Layers of fiber reinforcement and resin are added to build up to the design thickness. Catalysts or accelerators may be added to the resin system, depending on the resin specifications, to allow the resin to cure without application of external heat. Thus, the typical hand-laid-up part cures at room temperature.

The two most common resins used in contact lay-up are polyesters and epoxies. Polyesters are favored because of low cost, wide availability, and ease of handling. Epoxies are more expensive and more exacting in their formulation.

Typical applications include boats and boat hulls, radomes, ducts, pools, tanks, furniture, corrosive-environment equipment, automotive components, housings and guards, statuary, corrugated and flat sheets, prototype and production molds, jigs and fixtures, and stage properties. The major advantages of polyester and epoxy systems using lay-up techniques are low-cost tooling, easy methods of fabrication, wide range of available colors, light weight, high strength, and large size. Hand lay-up techniques are best used in fields where production is low and other forms of production would be prohibitive because of costs or size requirements.

**FIG. 4.7** Rolling out entrapped air. *(Owens-Corning Fiberglas Corp.)*

When unfilled polyesters cure, a large warpage distortion and/or a rippled surface results from the large volumetric shrinkage. With epoxies, however, because of the low shrinkage factor, the surface will cure almost ripple-free, with no visible warpage.

Polyesters can be easily removed from a multiplicity of molds using a variety of mold releases. Epoxies, because of their tenacity, require more specific types of mold releases. Both materials can be made flexible or resilient. They can be made flame-retardant, non-burning, or self-extinguishing.

The most common reinforcing material for epoxy and polyester with hand lay-up is glass in fiber form. The fibers are made of bundles of fine filaments which have been pretreated with a chrome or silane finish to provide good capillary wetting. In hand lay-up the most common glass-fiber forms are roving, woven roving, cloth and mat, chopped and milled fibers, woven and knitted tapes, and veils. The fiber form is not so critical as it is in compression molded or filament-wound parts. Epoxy parts are used mainly for dimensionally stable, high-strength applications, mentioned above. The polyesters are used more widely for less critical applications.

Hand lay-up techniques started in the forties, and raw materials and techniques have not basically changed since then. Hand lay-up still depends upon the individual skill of the operator. The techniques have become so refined that entire industries such as boat building rely on these methods for production. For some time, the auto-body business was also a hand operation. Today, many companies in many fields use hand lay-up to build models, prototypes, and production parts where limited numbers are required or where parts are too large for press molding.

A simple hand lay-up can be made over a male plug or in an open female cavity (Fig. 4.8). The basic lay-up techniques are the same. The choice of the mold, however, depends upon which surface is more critical in appearance. If the outer surface is critical, a female mold is used; conversely, if the inner surface is critical, a male mold is used.

Some variations of hand lay-up techniques are bag molding, drape molding, vacuum-table molding, vacuum draw, spray-up molding, and combinations of wet lay-up and low-compression molding.

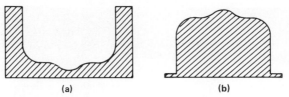

**FIG. 4.8**   Sketch of (*a*) female mold and (*b*) male plug.[1]

## Drape Molding

Drape molding is a vacuum technique associated with hand lay-ups. It is assumed that a part is a thin lay-up with a low relief surface such as a contoured map. A pattern for this relief is prepared in hard plaster or other pattern material. The pattern will act as the male mold plug and should be treated accordingly. A vacuum apparatus pulls a flexible cover down over the molding.

With any lay-up techniques, either gel coating or pigmenting resins can be used. After the mold has been prepared, a simple lay-up, as described earlier, is applied. Up to 60 wt % fillers can be added to extend the resins if desired. The more filler added the more viscous the polyester and the more difficult to wet out the glass reinforcement. A slow catalyst system must be used with the resin because vacuum systems require more time to cure.

Drape molding is used in making domes, tanks, drums, boat hulls, and many other forms. In some of these cases, however, it may be necessary to shape or gusset the drape for better conformity to the mold to eliminate creasing or bunching.

## Wet-Lay-Up Low-Compression Molding

In low-compression matched-metal-die molding or in low-pressure laminating occasional parts are too complicated to be made by normal molding techniques. The only way to get results is to combine hand lay-up methods with machine techniques. Examples are laminating complicated electrical systems, molding special radome sections, or laminating oversized transparencies as traveling display units. Specialized parts larger than 60 by 72 in (1524 by 1829 mm) have been successfully made by this method. Two other methods should be mentioned, moldless lay-ups and direct lay-ups or one-shot techniques.

## SPRAY-UP

In an effort to mechanize the hand lay-up method, various types of spray-up techniques and equipment have been developed. They reduce the overall cost of the laminate by using roving instead of more expensive material or cloth, by decreasing application and impregnation time of the reinforcement to save labor, and by eliminating scrap material and catalyzed unused pots of resin.

Resin-handling equipment can be divided into (1) systems mixing equal parts of resin from a catalyzed-resin pot and an accelerated-resin pot and (2) systems using accelerated resin from one or two pots and the pure catalyst solution from another. These systems in turn may be of the air-atomizing or the airless-spray-type, the latter operating with high or low pumping pressure. Small pots supply the resin to the spray gun by air pressure on the pot; larger single-batch systems use metering displacement pumps. Large production facilities have circulating resin lines to several spray guns. Over a dozen spray guns on the market are designed especially for use with spray-up systems.

For ease of handling, the spray gun is usually suspended from an all-position counter-balanced boom. Also necessary are hand rollers of various lengths and diameters to compact the sprayed fiber filaments into the resin and squeeze out the entrapped air.

The spray-up process is ideal for low- to medium-volume applications. The spray-up technique is an open-mold method in which fiber reinforcements and catalyzed resin (and in some cases filler material) are simultaneously deposited in the mold from a combination chopper and spray gun. Additional layers of chopped roving and resin may be added to build thickness. Usually room-temperature curing is required, but in some applications moderate external heat is used to accelerate the production process. As with hand lay-up, an excellent surface finish can be obtained by first spraying gel coat onto the mold before spray-up of the substrate. Woven roving can be added to the laminate for specific strength orientation. Molds for spray-up may be of wood, plaster, vinyl, silastic, polyester, epoxy, rubber, or steel.

In determining whether to choose hand lay-up or spray-up one's needs should be compared with the advantages offered by each method (Table 4.2). Spray-up, like hand lay-up, can be handled with simple, low-cost tooling and offers simple processing. In addition, spray-up is particularly well suited to efficient fabrication of complex shapes (Fig. 4.9). Spray-up systems are particularly effective in a wide variety of fabrication applications: bathroom tub and/or shower units, recreational-vehicle components, truck roofs and housings, boats and boat hulls, automobile and truck bodies, tank linings, large integral moldings, large panels, containers, housing components, furniture, motion-picture sets, and simulated masonry. Spray-up is effective where large matched metal molds would be too expensive to machine.[1]

## STRUCTURAL-LAMINATE BAG MOLDING

Bag molding is one of the oldest and most versatile processes used in manufacturing parts. It is a process for molding reinforced preimpregnated material containing a thermosetting resin. The lamina are laid up in a mold, covered with a flexible diaphragm, and cured with heat and pressure. After the required curing cycle, the materials become an integrated molded part shaped to the desired configuration.

Bag molding is considered to be a state-of-the-art process because the skill and know-how of the workers laying up and bagging the part largely determine the quality of the finished part. For a company to remain a leader in bag-molding composites, it must have well-trained production personnel and an aggressive composites-development program. The size of a part that can be made by bag molding is limited only by the curing equipment, i.e., the size of the curing oven or autoclave.

### Description of Processes

The general process of bag molding can be divided into three basic molding methods: pressure bag, vacuum bag, and autoclave (Fig. 4.10). Vacuum-bag and autoclave methods

**Table 4.2**  Guide to Selection of Hand Lay-up or Spray-up†

| Characteristic | | Hand lay-up | Spray-up |
|---|---|---|---|
| Minimum inside radius | | 0.250 in (6.4 mm) | |
| Molded-in holes | | Large | |
| Trimmed in mold | | Yes | |
| Undercuts (split mold) | | Yes | |
| Minimum draft recommended | | 2° (0.035 rad) | |
| Minimum practical thickness | | 0.030 in (0.76 mm) | 0.060 in (1.5 mm) |
| Maximum practical thickness | | Virtually unlimited | |
| Normal thickness variation | | ±0.020 in (±0.50 mm) | |
| Maximum thickness buildup | | As desired | |
| Corrugated sections | | Yes | |
| Metal inserts | | Yes | |
| Surfacing mat | | Yes | |
| Limiting size factor | | Mold size | |
| Metal edge stiffeners | | Yes | |
| Bosses | | Yes | |
| Fins | | Difficult | |
| Hat sections | | Yes | |
| Molded-in labels | | Yes | |
| Raised numbers | | Yes | |
| Translucency | | Yes | |
| Strength orientation | | Random or directional | Random‡ |
| Typical fiber-glass reinforcement loading | | 25–65 wt % | 20–50 wt % |

†Data from *Owens-Corning Fiberglas Corp.*
‡Many molders combine spray-up with woven roving and unidirectional reinforcements.

FIG. 4.9    Sprayed-up large parts efficiently produced. *(Owens-Corning Fiberglas Corp.)*

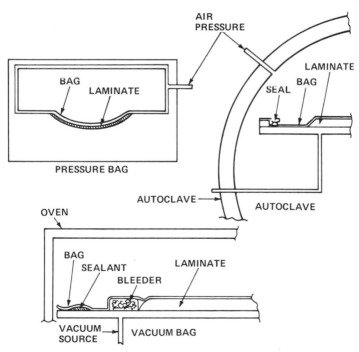

FIG. 4.10    General processes of bag molding.[1]

are used to produce most bag-molded parts. Their main advantages are that the tooling is relatively inexpensive and the basic curing equipment (oven and autoclave) can be used for an unlimited variety of shaped parts. The disadvantage of the pressure-bag system is that the tooling is relatively expensive because it is combined with the curing pressure system and can be used only for the specific part for which it was designed. Since the principles of vacuum-bag and autoclave molding are similar, they will be discussed together. It is important to select the correct material and mold design for tools used in the fabrication of bag-molded reinforced plastic parts. Some general comments on bag-molding tools follow.

Wood, plaster, plastic-faced plaster, and reinforced-plastic tools should be limited to oven-cure and wet-lay-up resin systems.

Steel has about the same coefficient of expansion as reinforced plastic; aluminum has about 3 times the thermal expansion of reinforced plastic.

The surface of aluminum tools can be hardened by anodizing it.

Low-melting metals and salt compounds can be used for the mandrel of ducts.

Electroformed nickel, copper, and other metals produce smooth tool surfaces.

Cast-aluminum tools may have surface defects or internal defects which can cause vacuum leaks or pitted surfaces.

## *Tooling for Bag Molding*

Many types of tools are used for bag molding:

Aluminum breakaway mold

Tools for cylinders, shell-type wash and plaster

Typical plaster with plastic face

Typical glass or carbon-reinforced-plastic mold with steel reinforcement

or plastic eggcrate (Fig. 4.11)

Large steel-plate mold (Fig. 4.12)

Cored aluminum mold for steam

Solid-steel kellered mold

Aluminum mold

Castable-ceramic mold

For many complicated parts, a family of accessory tools is also necessary. The following list is typical:

Location tools for buildups, honeycomb core, and steel inserts

Trim and drill shells

Secondary bonding tools

Inspection tool (for contour)

Postcure tool

**FIG. 4.11** Graphite prepreg fabric used in laminated fiber-glass mold. *(Vought Corp.)*

## Tooling Materials

Through the proper use of plastics in tooling, several definite advantages can be realized over tools of metal and/or wood:

1. Tools can be cast or laid up to the desired final shape and dimensions in one operation.
2. Relatively inexpensive equipment and labor and fewer worker-hours are required.
3. Delivery of tools is faster, reducing the required lead time between design and production of the end-use article.
4. More frequent changes in design are feasible.
5. Duplication is easier where several tools are required.
6. Revisions and repairs are simpler, promoting efficient development of design and reducing downtime.
7. The tools are relatively light in weight and easy to handle.
8. Since they are resistant to corrosive atmospheres, lubricants, and weather, they can be stored outdoors.

However, careful evaluation of the end use must dictate the selection of material for each specific application.

**FIG. 4.12** Largest compound-contoured steel mold being fabricated.[21] (*Sikorsky Aircraft, Division of United Technologies.*)

Plastic tools are usually made from viscous liquids (resins), which are converted into solids by the action of a chemical (catalyst, hardener, accelerator, etc.), sometimes assisted by heat. Pressure may or may not be required. Only a few basic raw materials are suited to plastic tooling. Strict requirements for stability and dimensional tolerance limit the choice of materials. Epoxy, phenolic, polyester resins, and urethane are the four major types. Polyamides and polysulfides are seldom used alone but are used in combination with epoxy resins to provide several excellent flexible tooling materials. Vinyls and silicones find application as parting or release agents. When properly formulated, the epoxy and polyester resins have good dimensional stability.

The need for careful and expert chemical control of resins used for tooling by the fabricator of the finished tool has led to the establishment of special industrial firms, known as *formulators,* who are an essential link between the manufacturer of the uncompounded base resin and the toolmakers, who design and produce tools from the resins. In most cases, *fillers* constitute an important part of the finished tool. Glass fibers, metal, sand or gravel, etc., may be used. Sapphire and carbon fibers offer extremely high physical properties. Core structures of foamed plastics may be used to reduce weight where high strength is not required.

**Phenolic Resins:**   Phenolic specialty resins were among the earliest raw materials used to make plastic tools. Chemically engineered to overcome most of their natural shrinkage on curing, the materials are cast to shape. Fairly strong acids have been used as catalysts to cure the resin.

**Polyester Resins:**  Although the term "polyester" is applied to various plastics, certain specific compounds under this designation have been manufactured and formulated for tooling with plastics. As a rule, monomeric styrene is one of the reactants. Proper reinforcements, such as glass fibers, are used to impart strength and toughness, especially in thinner sections. These structures have good resistance to wear.

**Epoxy Resins:**  The favorable shrinkage and stability characteristics of properly formulated and cured epoxy resins are mainly responsible for the increasing use of these materials. Fillers are used to improve impact strength and resistance to heat or abrasion. Since the development of heat by the exothermic reaction of cure must be restricted to minimize shrinkage, recommended and established procedures must be strictly followed. Normal shrinkage can be further minimized by applying face casts to rigid cores and by using special formulations containing large proportions of fillers.

### Tooling Methods

Although individual skill has created auxiliary techniques in making and repairing plastic tools, there are only three basic methods in use today, singly or in combination: casting, laminating, and troweling and splining.

**Casting:**  The most common process is casting. The molds used for casting plastic tools can be made from almost any rigid or flexible material strong enough to hold the resin. The type of resin and the size and contours of the casting determine the choice of the mold. It is common to use a wood master or a plaster splash taken from the master. Sometimes, an actual metal part may serve as a mold. The strong adhesion of epoxies to many types of materials makes it necessary to apply parting agents to the mold surface. Such agents include waxes, vinyl alcohols, silicones, polyethylene, and fluorocarbons. They are applied by brushing, wiping, slushing, or spraying.

Before casting, molds made of plaster or other porous materials must be thoroughly dried or sealed with moisture-barrier sealants. Since the cast resin is capable of reproducing minute imperfections on the mold surface, great care must be taken in preparing the mold. Flaws must be corrected, and mold preparation should include sanding off blemishes, filling in hollows, and smoothing down any weaves or wavelike patterns on the surface of the mold. The resin-hardener mix is poured into the prepared mold and cured at room or elevated temperature, depending on the nature of the resin and the size of the casting.

Advantages of the casting process include low cost, speed, handling ease, and excellent pick-up of detail. It is common to reinforce plastic tools to impart strength and to sustain the wear of longer runs. Glass and synthetic fibers are the most popular reinforcing materials. To prevent serious erosion of contour, metal inserts are used at critical locations to strengthen plastic tools. The inserts can be set in place before casting or fitted into cavities milled into the cured and hardened tool.

**Laminating:**  Reinforcing the resin with glass fibers or other synthetic fiber cloth, called *laminating,* is done by first applying a plastic gel coat to a prepared form and allowing the gel to become tack-free. The gel coat provides the tool with its surface properties like abrasion resistance and chip resistance. The coat is backed up to a predetermined thickness by successive layers of fibrous glass cloth or other fabric impregnated with a laminating resin that can be brushed directly onto the consecutive layers of the reinforcing material.

**Splining and Troweling:** These methods use pastelike resins, i.e., thixotropic compounds that allow work on vertical surfaces. They are applied over the outside of male forms or on the inside of female forms.

### Factors in Choice of Tooling

The following are the major factors considered in the design and fabrication of tooling for structural composite parts:

Control of fiber orientation
Contour and size of detail part
Location of detail parts in a structurally reliable assembly to give lowest possible cost
Control of dimensional tolerance and configuration stability

The tooling required to fabricate most structural composite parts can be broken down into several major categories, including ply lay-up tools, skin or cover mold forms, curing aids, handling tools, trimming tools, and assembly tools. The last three types will be discussed in Chaps. 5 and 6.

**Ply Lay-up Tools:** Ply lay-up tools, which are used by many firms to control fiber orientation and the relative position of individual plies, consist of Mylar film templates made by photoreduction of master drawings.[2] The tools should be coated with adhesive-backed polyethylene film or transparent linear polyurethane to permit easy release of the laid-up plies (Fig. 4.13). The typical tolerance of fiber orientation which is controlled is $\pm 1°$ (0.017 rad); several approaches to fiber orientation and the relative positions of partial plies are shown in Fig. 4.14. Coordination of Mylar templates to mold forms is achieved by

FIG. 4.13 Lay-up tool for upper shell ply of F-5 aircraft.[2] (*General Dynamics, Ft. Worth Division.*)

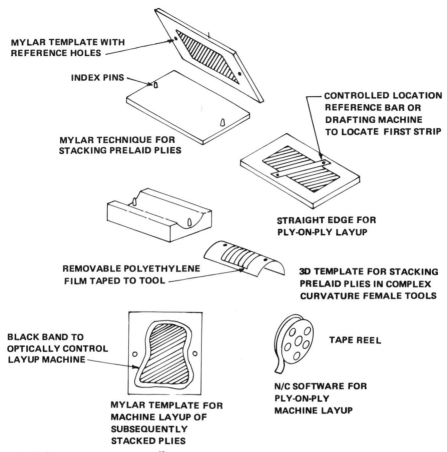

MYLAR TEMPLATE WITH
REFERENCE HOLES

INDEX PINS

MYLAR TECHNIQUE FOR
STACKING PRELAID PLIES

CONTROLLED LOCATION
REFERENCE BAR OR
DRAFTING MACHINE
TO LOCATE FIRST STRIP

STRAIGHT EDGE FOR
PLY-ON-PLY LAYUP

REMOVABLE POLYETHYLENE
FILM TAPED TO TOOL

3D TEMPLATE FOR STACKING
PRELAID PLIES IN COMPLEX
CURVATURE FEMALE TOOLS

BLACK BAND TO
OPTICALLY CONTROL
LAYUP MACHINE

TAPE REEL

N/C SOFTWARE FOR
PLY-ON-PLY
MACHINE LAYUP

MYLAR TEMPLATE FOR
MACHINE LAYUP OF
SUBSEQUENTLY
STACKED PLIES

FIG. 4.14   Lay-up tooling concepts.[27]

positioning the templates so that index pins in the mold forms protrude through matched holes in the templates.

A three-dimensional template reduces lay-up flow time and improves the accuracy of complex parts laid up in compound-curvature female molds. In production, three-dimensional templates can be made as follows. A three- to five-ply fiber-glass lay-up is cured in the female mold to provide the required contour. The ply outlines are applied to the fiber-glass template by transferring them from a laid-out male model. The fiber-glass template is then coordinated to the mold form by pickup points and covered with a spliced polyfilm slip sheet to provide a releasing surface. The composite prepreg is laid up directly on the polyfilm surface. The prepreg-template-polyfilm assembly is then inverted and placed in the female mold. The template and polyfilm are removed sequentially to complete the ply-stacking process. Multiple three-dimensional lay-up templates can be built to reduce the lay-up flow time to levels acceptable for high-rate production.

**Mold-Form Tools:**   Mold-form tools on which individual plies are located and stacked can be made from a wide variety of materials. Characteristics of several candidate materials

**Table 4.3** Tooling-Material Alternatives[27]

| Material and coefficient of linear expansion† | Molding skins | | Molding wing beams and webs | |
|---|---|---|---|---|
| | Advantages | Disadvantages | Advantages | Disadvantages |
| Steel 6.3 (11.3) | Dimensionally stable with temperature; compatible coefficient of expansion; molding surface obtained directly from numerically controlled program; hard surface provided for jig pickup points; readily adaptable to combination-type fixtures; can be brake-formed to reshape | Machining slow; tool weight and mass high; slow heat-up rate | Coefficient of linear expansion compatible with boron laminate; easily adapted as combination fixture | Tool weight and mass high |
| Aluminum with steel slip sheets 12.9 (23.2) | Easily machined; molding surface obtained from numerically controlled program; tool weight and mass low; readily adaptable to combination-type fixtures; stable with temperature; good heat-up rate | Incompatible coefficient of expansion; jig pickup points must be slotted; hard jig pickup points impossible | Tool weight and mass low | Thermal expansion incompatible |
| Iron or nickel electroform Fe 6.6 (11.9) Ni 7.0 (12.6) | Compatible thermal expansion; lower tool cost if model available; hard jig pickup points can be provided; good heat-up rate | Few manufacturers have plating tanks for full-length skin tools; plaster model needed for contour | May be cheaper for duplicating tool; compatible thermal expansion | Model needed |
| High-temperature plastic 11 (20) | Less time to produce tool if model available | Incompatible thermal expansion; not adaptable to combination-type fixtures; high material cost; plaster model and splash needed first; slotted rather than hard jig pickup points; slow heat-up rate | None | Model needed; incompatible thermal expansion |

| Material | | | | |
|---|---|---|---|---|
| Cast ceramic (Glasrock) 0.45 (0.81) | Can make self-heated tools; very stable | Incompatible thermal expansion; not adaptable to combination-type fixtures; model and splash needed first; cannot be drilled after casting; chips easily; slow heat-up rate | None | Not feasible |
| Cast Nicrosil or Meehanite 7.5 (15.5) | Costs one-third less than machined-steel hot-sizing dies; has $\pm 0.020$-in (0.51-mm) tolerances on casting of small parts up to 12 in (305 mm); good surface finish | Requires model using contour templates or machining program for numerically controlled casting; has $\pm 0.030$-in (0.76-mm) tolerance on large dies 24 in (610 mm); very difficult to machine; slow heat-up rate | None | Not feasible |
| Plastic-faced plaster | None | Not feasible | None | Not feasible |
| Silicone rubber 45 to 200 (81 to 360) | Not feasible | Pressure hard to control, predict, or measure; loses stability with repeated use | No shape restrictions; transmits pressure readily; may be cheaper for duplicate tooling | Model needed; design critical |
| Graphite epoxy $-0.5$ to 1 ($-0.9$ to 2) | Excellent dimensional stability; hard points can be provided; good heat-up rate; potential low cost | Durability not known; plaster model needed | Not feasible | |

†Per $10^{-6}$ Fahrenheit degree ($\mu K^{-1} = 10^{-6}$ Celsius degree).

are listed in Table 4.3. The selection of a material for a given mold form is usually based on a trade-off between total tool cost, required tolerances, and part design criteria.

*Thermal Limitations:* Mylar-template ply lay-up tools are used at room temperature only. They are not involved with the subsequent high-temperature curing operations. Most of the mold-form tooling materials listed in Table 4.3 can be reliably used at curing temperatures up to 350°F (177°C). Silicone-rubber vacuum bags have also been designed for use at 350°F temperatures.

## Ply Stacking

In addition to maintaining contour, mold-form tooling must also provide for precise location of partial plies and metallic inserts and uniform pressure application to the part surface.

Large skin molds have usually been made from tool steel in either the machined-cavity or eggcrate design (Fig. 4.15). Mold design selection should be based on autoclave capability, method of heating, handling problems, and cost. The machined-cavity concept is often limited to heated platen-type autoclaves due to the slow heat-up rate of these tools in circulating-air autoclaves. The high cost of machining and handling problems also favors the eggcrate design. The tool face can be fabricated from mill-polished plate stock rolled to contour to provide a smooth surface at low cost. The tool face is typically 0.25 to 0.5

**FIG. 4.15** Compound-contoured mold form in eggcrate design.[21] (*Sikorsky Aircraft, Division of United Technologies.*)

in (6.4 to 12.7 mm) thick. The design of the eggcrate tool-and-bag surface installation system developed during the heating survey must be coordinated so that the tool side of the part is warmer than the bag side to ensure that the entrapped volatiles and excess resin will be removed from the skin before resin gelation.

The mold form must also provide a high vacuum flow to the entire surface of the part. A vacuum groove, spring, or perforated pipe should be provided around the entire perimeter of the part. Several vacuum ports, each connected to a separate vacuum system, must be provided to ensure an adequate vacuum flow. To minimize the number of vacuum hoses the tool should be supplied with permanent piping wherever possible.

**Substructure Fabrication:** The major concerns in the design of substructure tooling are to maintain chord height and to apply pressure uniformly over the entire surface of the part. In matched-die tooling the proper application of molding pressure is mandatory. Single-channel-section parts and channel sections that are subsequently bonded to cap plates to form I-beam sections are traditionally made in female molds to control the chord height and faying surfaces. Silicone-rubber pressure intensifiers are used in the radii to provide adequate molding pressure. Bonding fixtures for substructure components may either consist of frames to support pinned details or be of conventional design, with one or more silicone-rubber plugs.

Most I-beam-section substructure parts, however, are made with a single curing process to reduce costs. Thick silicone-rubber tools (Fig. 4.16) have been used extensively both

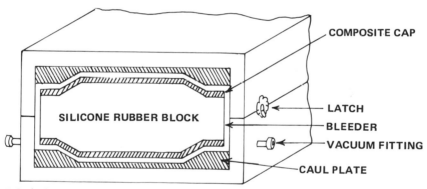

FIG. 4.16  Frame to control height of parts made on silicone-rubber tooling; if the joint in the frame is sealed, the assembly can be bagged to the frame.[27]

with and without restraining frames to control chord height. The rubber blocks must be sized to include the significant thermal growth of the rubber during cure. Although silicone-rubber tools usually produce structurally sound parts, they eventually lose dimensional stability. Thus, their use is limited to production runs having fewer parts than metal tools.

For larger-scale production of corrugated web spars one method uses hollow silicone-rubber molds, ceramic caul plates, and a rigid restraining frame; the second method uses matching aluminum blocks faced with silicone-rubber sheet applied to the tool surface as part of the bleeder system. The bleeder system is a Dacron mat capable of decreasing significantly in thickness. Thus the rubber facing and mat deform sufficiently to allow the rigid aluminum tools to compensate for the thickness variations in successive lay-ups of the same part. The gap between the tools is designed on the basis of the part plus the thickness of the rubber face and the resin-filled bleeder systems.

## Tooling by Electroforming

Electroforming consists of electroplating against a conductive surface (called a *pattern* or *mandrel*) for a long time to reproduce a reverse of that surface. The electroform cavities or cores thus formed can be backed up as required. Patterns, or mandrels, may be made of almost any material that does not absorb moisture and does not expand excessively through a 100°F (37.8°C) heat differential. The adhesion of the plating to the mandrel is part of the state of the art since it must be strong enough to retain the surface detail and contour but weak enough to be removed after plating. The adhesion is determined by the film deposited on the pattern or mandrel.

The type of plating used in electroformed molds is normally nickel, copper, or a laminate of both. For molds requiring machining, such as most compression molds or die inserts, a nickel face with copper backup is used because of the poor machinability of nickel. The copper backup can then be easily machined using the nickel surface as a reference.

The normal plating rate of 0.012 to 0.015 in (0.30 to 0.38 mm) per 24-h day can be increased when plating copper, but stress and brittleness also increase. Plating goes on 24 h, 7 days a week, requiring 9 to 10 days to produce a mold 0.125 in (3.2 mm) thick.

Molds are separated from mandrels in three basic ways. One-piece molds with undercuts require a disposable mandrel, which is usually melted or etched out of the mold. A mold with sufficient draft is usually pried loose from the mandrel mechanically at one or two points and then removed completely by using air pressure. An electroformed mold with little or no draft can normally be released by cooling the pattern while heating the electroform. The expansion of the electroform will usually be sufficient to remove it from the mandrel.[3] Advantages and disadvantages of electroformed molds are as follows:

1. Advantages
   a. Extremely accurate reproduction of detail in each cavity compared with steel for a given price.[4]
   b. Zero porosity compared with aluminum castings.
   c. Zero shrink compared with beryllium copper castings.
   d. The only way certain molds can be made.
   e. Less costly than steel, comparable to beryllium copper (more costly than aluminum).
2. Disadvantages
   a. Relatively soft compared with steel.
   b. Limited configuration. Configuration plays an important part in whether or not an electroformed mold can be used, because of the difficulty in electroforming to a uniform thickness over the complete surface of the mold.
   c. Recesses deeper than they are wide must be avoided.
   d. Sharp corners and most internal bosses or projections should be avoided, although some bosses can be inserted into the mandrel as "grow-ins," and the plating will attach itself securely to the bottom of the boss and the grow-in becomes an integral part of the mold (Fig. 4.17).

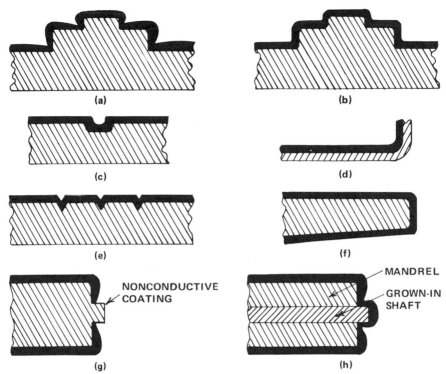

**FIG. 4.17** Matrix manufacture of electroformed molds: (*a*) electrodeposited metal builds up on outside corners and thins out on inside corners; (*b*) breaking sharp corners and providing fillets minimizes variation in metal deposit thickness; (*c*) recesses should be wider than deep; it is difficult and sometimes impossible to electrodeposit into deep, narrow recesses; (*d*) fillets at least equal to metal deposit thickness should be used for strong inside corners; (*e*) holes can be spotted for later drilling by providing depressions in patterns; (*f*) when feasible, provide slight taper to make mandrel removal easier; (*g*) eliminate drilling and reaming operations by providing masked or nonconductive studs on pattern; hole diameters can be held to ±0.0002 in (0.005 mm) and have excellent surface finish; (*h*) extend internal piece beyond end of surrounding part to assure deposition on sides as well as end on internal grow-in piece and bonding of internal shaft to outside cylinder. (*Electro Mold Corp.*)

### Secondary Materials for Bag Molding

In addition to the basic bag-molding structural materials, several secondary materials are required for production applications (Table 4.4). Even though all these materials are used in production applications, bagging, sealing, and bleeder materials require further development; separator films also need greater range in numbers and sizes of perforations. With

**Table 4.4** Common Secondary Materials for Bag Molding[27]

| Material | General types |
|---|---|
| Mold-parting agents | Silicone liquid high-temperature waxes |
| Films | Cellophane |
| Separator sheets | Polyvinyl alcohol |
| Bag films | Polyvinyl chloride, polyester (Mylar), silicone rubber |
| Bag sealants | Zinc chromate, butyl rubber |
| Bleeder material | Jute |

the increased use and development of polyimides high-temperature films are required for polyimide processing. Other requirements for further development according to the 47 companies surveyed include:

High-temperature bagging and sealant material

High-pressure leak-resistant reusable bags

Higher tear strength for reusable silicone-rubber bags and separator films

## Bagging and Sealing for Autoclave Curing

Vacuum bagging of parts for autoclave cure is the most critical operation for quality of parts produced and probably the most important cause of scrap. Three requirements must be met to ensure success:

1. The bag must apply curing pressure uniformly.
2. The bag must not leak under autoclave conditions.
3. A good high-capacity vacuum path must be provided.

Because of the high overall cost of composite parts and the requirement for pressure intensifiers silicone-rubber vacuum bags which could be reused have been developed by government and industry. It was imperative to reduce the risk of losing production parts due to failure of nylon-film bags during curing. For temperatures up to 380°F (193°C) nylon film has been widely used once and discarded. Kapton polyimide film, sealed with silicone rubber, is applicable to temperatures in the 550°F (288°C) range; metal foils and mechanical clamping are used at higher temperatures.[5]

Typical bag concepts are shown in Fig. 4.18. An annulus must be provided in either the bag or the tool so that vacuum can be applied to seal the bag to the tool. Another vacuum system is used to evacuate the air between the bag and the part to be cured or bonded. Although silicone-rubber vacuum bags (Fig. 4.19) involve an initial fabrication cost, it is more than offset by the sharp reduction in bagging labor and the elimination of recurring film and sealant material costs. Silicone-rubber vacuum bags are repairable, essentially self-sealing with respect to pinholes, and have a service life in excess of 400 autoclave cycles. Silicone-rubber vacuum bags can be fabricated by two basic techniques.

## *Method 1*

A part is cured by conventional means and then the part (or a dummy part made of wood) and its tool are used as a mold for curing a vacuum bag made from calendered-knit-fabric-reinforced silicone rubber. In this process the cured part and the tool are covered first with Tedlar film and then with precut uncured silicone-rubber sheet stock. Joints are made by solvent-tacking the faying surfaces of adjoining sheets and stitching them together with a serrated roller. If necessary, seams and highly stressed areas can be locally reinforced. The assembly is then vacuum-bagged and cured for 1 h at 50 kips/in$^2$ (345 MPa) at a temperature of 300 to 350°F (149 to 177°C).

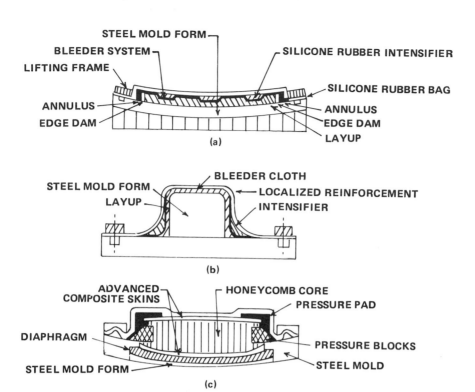

FIG. 4.18 Silicone-rubber vacuum-bag techniques: (*a*) complex skin in air-passage tool, (*b*) hat section on male tool, (*c*) bonding core assembly on diaphragm tool.[27]

FIG. 4.19 Reusable silicone-rubber vacuum bag. (*Sikorsky Aircraft, Division of United Technologies.*)

## Method 2

An uncured part is bagged with conventional nylon film, which is then coated with a releasing agent. Any required pressure intensifiers are cast in place from room-temperature-vulcanizing silicone rubber. Fiber-glass cloth which has been brush-impregnated with a solvated silicone resin is laid up as required. The bag assembly is cured at room temperature for 3 days and then postcured in an oven to remove residual volatiles.

# PRODUCTION PROCEDURES FOR BAG MOLDING

The typical shop for producing composites contains several areas for bag-molding fabrication including preparation, lay-up, bagging, curing, part removal, finishing, and final inspection.

## Preparation

In this area tools are cleaned and inspected for damage, and necessary repairs are made. Parting agent is applied to the surface of the tool. An instructional paper, consisting of detailed lay-up procedures and drawings depicting the number and location of each ply of reinforcement, is attached to its respective tool. Flat rectangular patterns are normally developed from the shop drawings, and the reinforcement (dry glass cloth or prepreg material) required for each particular part is cut to pattern and placed with the tool.

The prepreg material, whether tape or woven cloth, is normally stored with no detrimental effects in a freezer in 0°F ($-17.8°C$) environment for up to 12 months, depending on the material. Once thawed to room temperature, it must be used and cured in approximately 20 days. The next step from prepreg or dry glass cloth to component is cutting.

## Cutting

### Manual

Manual methods for cutting and trimming composite prepreg materials are slow, laborious, and subject to human error. Manual cutting of composite material was the primary method of prepreg cutting in the early stages of composite fabrication. This method uses a knife or scissors and an appropriate guide (template or straightedge), and the actual cutting is accomplished at an estimated 111 in/min (282 cm/min) by a composite technician. This technique is still employed for the pattern cutting of medium and small components that are complex and have sharp bends and contours. It is also practical to cut material manually when the width of material is narrow, e.g., 3- or 12-in (76- or 305-mm) unidirectional tape. Drawbacks of manual cutting are listed in Table 4.5.

With increasing commitments by more companies to composites, the manual method was no longer deemed adequate to meet production requirements in some cases and was replaced by automated cutting where possible. These more effective methods provide speed, accuracy, reproducibility, high quality, and reduced scrap and waste unmatched by conventional methods.

**Table 4.5** Material Cutting Techniques[1]

| Technique | Advantages | Disadvantages | Application |
|---|---|---|---|
| Manual | Flexible, limited set-up time, economical for narrow tape | Slow, tedious, labor-intensive; requires Mylar and templates; costly inspection; difficult to cut multiple plies | 3-, 12-, 48-in wide (76-, 305-, 1220-mm) tape and broad goods, fiber glass, B–Ep, Gr–Ep, and Kv–Ep |
| Gerber knife | Fast, 720–1200 in/min (305–508 mm/s; computer-controlled; reliable for textiles; clean cuts; reliable up to 20 plies; accuracy ±0.030 in (0.76 mm) | Accuracy less than that of laser and water jet; knife gumming with some resins | Cutting uncured prepreg composite materials |
| Laser | Cuts B–Ep at 540 in/min (228 mm/s) and Gr–Ep at 780 in/min (330 mm/s); computer-controlled; takes wide material | Basic cost and energy costs high; limited number of plies; eye protection required | Cutting cured and uncured composite materials |
| Water jet | Generates no dust; multiple-ply cutting (to 40 plies); no heat-affected zone at edges; computer-controlled; takes wide materials | Slight moisture absorption in uncured prepreg materials; limited number of plies | Cutting cured and uncured composite materials |
| Steel-rule die | Generates no dust; multiple-ply cutting (to 15 plies); no heat-affected zone at edges; takes wide materials; clean cuts | Design change requires die change | Cutting uncured prepreg composite material; 3-, 12-, 48-in (76-, 305-, 1220-mm) widrhs |

### Reciprocating Knife

The reciprocating-knife system is particularly effective in cutting prepreg details in production or prototype quantities (Fig. 4.20). The Gerber fabric-cutting machine, a spinoff of those used in the textile industry, is designed to handle material up to 60 in (1524 mm) wide. The cutting-table length is 20 ft (6 m). The cutter is computer- or numerically controlled; i.e., flat patterns are developed and nested by computer for efficient use of materials and are cut by the reciprocating knife cycling up to 5000 strokes per minute. The machine is capable of cutting at speeds up to 1200 in/min (3048 cm/min) with a high degree of accuracy. Single-ply cutting rates up to 900 in/min (2286 cm/min) and 300 to 600 in/min (762 to 1524 cm/min) for multiple plies are common with no deterioration in quality. Different cutting-blade geometries provide a choice of cutting action (chopping or slicing) best suited for a particular prepreg.

The basic procedure in reciprocating-knife cutting is as follows. The program to cut the material to the desired configuration is called up on the computer console by the operator, who then unrolls, say, a prepreg material 48 in (1.22 m) wide from a spool at the end of the table over the length of the table and cuts it off at the spool. A thin piece of plastic

**FIG. 4.20** Gerber reciprocating-knife system with Gr–Ep prepreg. *(Sikorsky Aircraft, Division of United Technologies.)*

film is placed over the prepreg, allowing a vacuum to be pulled on the material to maintain its position during cutting. The cutting head is placed in the start position at the end of the table, and plies are cut. Upon completion of the programmed cut, the head returns to the end of the table, the vacuum is released, and the cut plies are identified and moved into storage. The larger pieces of scrap are kept for use in making smaller detail parts.

### Die Cutting

Die cutting is particularly effective where close tolerances are required, and is cost-effective for high-rate production of repetitive parts and complex parts with many holes and cutouts. Kv–Ep prepreg, once the most difficult to cut, is cut cleanly, efficiently, and without difficulty by this method, as are all the prepregs. Figures 4.21 and 4.22 show the steel-rule-die tooling and the completed instrument panel. The 13-ply one-piece Kv–Ep preform with 22 cutouts for instruments and 110 holes for fasteners is produced in a single pass.

This process, applicable to both resin and metal matrixes, although somewhat new to composites, is a standard and accepted method for production blanking of metallic materials and is used for production of helicopter composite rotor-blade fabrication (Fig. 4.23). This method is especially attractive when a large number of identical parts and/or plies are to be made. The plies can either be cut individually or stacked. Initially, the blanked material is removed from the die manually. (The part or material can also be forced from the die by a mechanically actuated knockout plate.)

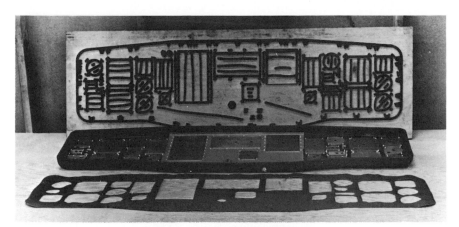

**FIG. 4.21** Instrument-panel tooling. *(Sikorsky Aircraft, Division of United Technologies.)*

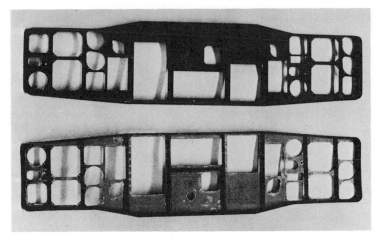

**FIG. 4.22** Instrument panel for a helicopter. *(Sikorsky Aircraft, Division of United Technologies.)*

The steel-rule dies can also be inserted into (cutting) boards on the wooden base (table) of the press after the configuration has been routed. Then the plies of material are stacked on top of the rule-die pattern. The boards and the prepreg are positioned under a heavy urethane block (press head), which is actuated by a cam on a (clicker) press to force the steel-rule die through the prepreg. The pattern layout is designed to produce the required size, shape, and fiber orientation with the minimum of scrap.

The uncured material is placed on the press table; the press is activated with the part being blanked. A limiting factor with steel-rule dies is the squareness of the cut required; i.e., as the number of plies is increased, the squareness of the edge diminishes. Also, some materials such as Kv–Ep may require higher press pressure to get a good impression and sever the fibers, placing a restriction on the maximum number of plies to be cut. Generally, die life is not a critical problem in the aircraft industry because of the limited quantities that are produced compared with mass-produced consumer durables.

**FIG. 4.23**  Identical parts produced by steel-rule die. *(Sikorsky Aircraft, Division of United Technologies.)*

## Water Jet

Water-jet cutting uses a focused stream of high-pressure [50 kips/in² (345 MPa)] water through a very small orifice [0.01 in (0.25 mm)] to cut prepregs and cured composites. Advantages include high cutting rates (single-ply cutting speeds up to 3000 in/min (7620 cm/min), high quality of the cut edge, no heat-affected zone when cutting multiple plies (up to 40), and no slowdown of cutting due to resin buildup on the cutting edge. The system produces a finish cut similar to that of a shear or scissors. Also, the operation produces no dust. Some users express concern over the actual or probable problem of moisture absorption at the cut edges in prepreg material. Water-jet cutting of cured composite materials and systems is discussed in more detail in Chap. 6.

## Laser

Laser cutting of Kv–Ep, Gr–Ep, and B–Ep prepregs has proved successful. A 500-W $CO_2$ laser with direct numerical control has cut B–Ep at approximately 540 in/min (1372 cm/min) and Gr–Ep at approximately 780 in/min (1981 cm/min). Layers of the composite are cut to their proper plan shape from either tape or cloth as wide as unidirectional broad goods [48 in (1.22 m)].

Prepreg material drawn off bulk rolls passes through a mechanism that lays a continuous paper sheet over the upper surface (Fig. 4.24). The material, with the paper on top, then feeds into the enclosure surrounding the laser-cutting assembly, a device borrowed from the textile industry. The laser beam, generated in an adjacent cabinet, travels through an optical train into the cutting enclosure, where it is focused on the material and moved through its programmed cutting pattern by an optical cutting head. The laser beam, focused to a point, vaporizes the composite at the point of contact, effectively cutting the material with little or no kerf. The cut edges are clean and sharp but do exhibit a heat-

**FIG. 4.24** Prepreg material cut into planform patterns by laser.[23]

affected zone. Once each length of paper-covered prepreg stops beneath the cutting head, the system runs through a cutting routine, or pattern, designed to minimize material waste by clever nesting of components, rather like a jigsaw puzzle.

Each length of material, with the appropriate lay-up pattern burned through it and its matching layer of paper, emerges from the cutting enclosure and comes to rest beneath a ceiling-mounted slide projector. An image of the cut pattern with an alphanumeric code superimposed on each piece projects onto the cut material, identifying each element according to its place in the composite component to be laid up. Once marked, the composite pieces are stripped off the cutting-system conveyor along with their paper doubles and are filed in coded drawers in special cabinets for transport to the lay-up area.

The paper is fed onto each side of the prepreg to prevent it from sticking to the conveyor and to keep it clean. The paper simplifies the transfer operation. Table 4.5 compares the five cutting techniques and lists applications.

## Composite Orientation

After the composite prepreg material has been cut to the desired configuration using one of the cutting techniques, the material must be transferred and stacked in the required orientation. Some of the various methods depend on which cutting technique is used.

### *Manual Transfer*

This process of applying the unidirectional tape or woven material by hand in a template or tool is the most commonly used in the industry today, primarily because a robot (Chap. 5) or flip roller cannot yet transfer material from complex shapes, such as corrugations, stiffening elements, speed brakes, etc. These shapes lend themselves to manual positioning, where digital dexterity is required. Automated transfer techniques are most applicable and cost-effective for relatively large, simple flat or gently curved components, e.g., stabilizers, wing skins, doors and panels. The built-in flexibility of manual lay-up allows composite technicians to form the prepreg in the lay-up tool.

Manual lay-up is commonly used for plies that were cut manually and via automatic techniques or from steel-rule dies. The manually cut plies are deposited ply on ply directly on the lay-up or curing tool (when composite strips are laid up directly on tape of a previously laid-up ply, the procedure is called *ply-on-ply lay-up;* Fig. 4.25). When the first

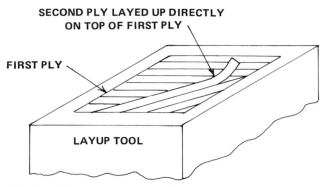

**FIG. 4.25** Manual ply-on-ply lay-up.[27]

ply is laid up directly on the lay-up tool and subsequent plies are laid up on the preceding ply, the procedure is called *direct on-tool lay-up.* This is accomplished by using ply-on-ply (Fig. 4.25) or ply-on-Mylar (Fig. 4.26) deposition. The method is recommended for manufacturing complex parts with sharp bends, especially when accuracy of ply orientation is critical. This technique is called *preplying* (Fig. 4.27) when the ply-on-ply or ply-on-Mylar technique is used to form a stack of plies transferred to the lay-up tool as a group.

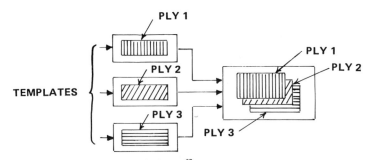

**FIG. 4.26** Manual ply-on-Mylar lay-up.[27]

## Lay-Up

If a part is to be made using prepreg, it is prepared in the following basic lay-up steps:

1. Place the first ply of fabric on the mold (parting film side up) by laying one end of the fabric down first and working it smoothly toward the other end.
2. Remove the parting film from the preimpregnated fabric.

PREPLIED STACK LAYED UP FLAT
(PLY-ON-PLY OR PLY ON MYLAR)

PREPLIED STACK FORMED
TO CONTOUR ON LAYUP TOOL

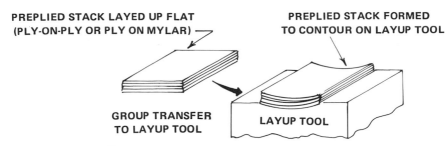

GROUP TRANSFER
TO LAYUP TOOL

LAYUP TOOL

FIG. 4.27   Preplying.[27]

3. Cut and dart the fabric as required so that it can be worked into intimate contact with the mold surface.
4. Keep the size of the overlaps 0.75 in plus 0.25 in (19 mm plus 6.4 mm).
5. Keep the number of overlaps to a minimum and never superimpose laps.
6. If necessary, use a heat gun to make the prepreg more pliable and give it more tack so that it can be worked into small radii.
7. Work the preimpregnated fabric with a Teflon or steel rubbing tool until all wrinkles and air pockets are eliminated.
8. Repeat the same procedures with each ply until all the material is laid up. In areas of thick buildup dry plies of glass fabric can be used between plies of prepreg for internal bleeding (approximate ratio of 1:4 dry to wet plies).

The part is now ready to be bagged.

If a part is to be made using a wet lay-up system, it is prepared in the following basic lay-up steps:

1. Secure the correct amount of catalyzed resin from the resin-mixing crib with the starting resin content by weight at 45%.
2. Coat the surface of the mold with a thin uniform layer of catalyzed resin.
3. Uniformly impregnate all plies of glass fabric one at a time on a clean flat table. Use a Teflon rubbing tool to force the resin into the fabric.
4. After the plies have been impregnated, allow about 15 min for the resin to continue to wet the surface of the fabric.
5. Place the first ply of impregnated fabric on the mold by laying one end of the fabric down first and working it smoothly toward the other end. Work the fabric with a Teflon rubbing tool until it adheres to the mold without air pockets or wrinkles.
6. Lay up each successive ply in the same manner, making sure not to disturb the plies already laid up.
7. After all the plies are laid up, cover with polyvinyl alcohol bag.

## Bagging

Bagging is an important step in bag molding. Many parts which have been laid up very carefully are rejected because of improper bagging. The method and materials used in bagging a part are based on the method of lay-up (preimpregnated or wet), the type of resin used, the thickness of the part, and curing pressure and temperature (autoclave or oven). A typical bagging system used for a preimpregnated lay-up system consists of the following steps:

1. Cover the lay-up with a perforated parting film or separator cloth. Then lay up a layer or layers of bleeder material. The combination should be such as to ensure adequate bleeding of air and excess resin out of the part so that the cured part will have the desired resin content.
2. Place a strip of jute just beyond the edge of the lay-up and put bag-sealing compound 4 in (102 mm) beyond the edge of the lay-up.
3. Cover the lay-up, jute, and sealing compound with a flexible-film diaphragm and seal the diaphragm to the mold with the sealing compound.
4. Connect the vacuum lines and slowly apply the vacuum pressure while working the wrinkles and excess air out of the lay-up, bleeder material, and vacuum bag (Fig. 4.28).

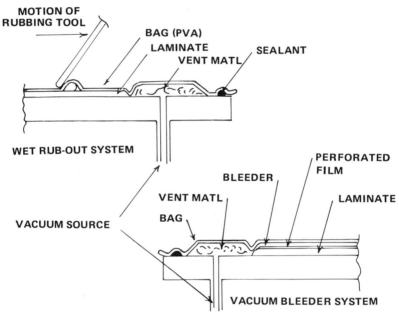

FIG. 4.28   Schematic of wet and dry systems.[1]

5. Check system for vacuum leaks.
6. Keep the part under vacuum while it is waiting to be cured in the oven or autoclave.

A bagging system for a wet-laid-up part must be as free as possible of entrapped air. The working life of this type of resin system is important because the bagging and rub-out procedure must be accomplished before the resin advances too far and becomes semigelled. As soon as the part is laid up, it should be bagged. The following procedure is typical:

1. Place a 2 in-wide (50.8-mm) band of jute bleeder material 1 in (25.4 mm) beyond the edge of the wet lay-up. This allows for 1-in-wide (25.4-mm) resin seal, which prevents air in the jute from flowing back into the part.
2. Place one or two rows of bag-sealing compound 1 in (25.4 mm) beyond the jute bleeder material.
3. Tape vacuum funnels on top of the jute.
4. Cover the entire lay-up, jute, and sealing compound with a flexible clear polyvinyl alcohol bag and seal it to the mold surface with the bag-sealing compound.
5. Connect the vacuum hoses to the mold and slowly apply a vacuum. Work the wrinkles out of the wet lay-up and polyvinyl alcohol bag.
6. Apply full vacuum; starting at the center of the part, work the large air bubbles and excess resin out of the lay-up and from underneath the polyvinyl alcohol bag with a Teflon rubbing tool. Continue the same rubbing process, starting from the center of the part and working the air and resin out into the jute bleeder.

In most epoxy-resin systems heat should be applied to the wet lay-up to help reduce the viscosity of the resin. A heat gun may be used or the whole part may be placed into an oven heated to approximately 150°F (66°C). A thumb-pressure test is sometimes used to determine the correct amount of resin remaining in the part. If medium thumb pressure applied to the part leaves a small depression, the part has been rubbed out adequately. If the thumb print remains white, too much resin has been rubbed out. Maintain full vacuum pressure until the resin has gelled hard. It is ready to be cured in the oven. Typical setups of bagging systems are shown in Fig. 4.29.

## Compaction (Debulking) and Prebleeding

Debulking, the process of densifying and forming a multiple-ply lay-up, is usually accomplished through the application of pressure and (if required) heat to a stack of plies enclosed in a vacuum bag on tooling. The bag used may be reusable or disposable. Many company specifications today require composite materials to be debulked before being cured. A lay-up with as few as three plies may be required to be debulked, depending upon the company, the type of material, and the complexity of the part being prepared. In fabricating a relatively thick laminate, as many as 10 to 15 debulk cycles could be required, resulting in a substantial cost increase in fabricating the composite structure. While most debulk cycles do not require a temperature excursion, requiring only some designated pressure to be applied to the composite plies, this means that the component must be vacuum-bagged and unbagged for each debulk cycle. Some companies require the debulk cycle to be performed at an elevated temperature; i.e., the component must be placed in an autoclave to accomplish the debulk cycle.

During cure, a typical Gr–Ep prepreg tape is reduced in thickness by approximately 25%. Wrinkling of thick parts with many partial plies can occur unless the lay-up is debulked periodically to ensure proper nesting of the plies. Debulking is normally done by heating the lay-up with one bleeder ply in place while under vacuum. Hand rolling is frequently used to supplement the vacuum. A minimum of resin is removed during compacting. Gr–Ep tape lay-ups are typically compacted once every 10 to 20 plies. Kv–Ep and Gr–Ep fabric parts typically are compacted to ensure that the lay-up will be snugly

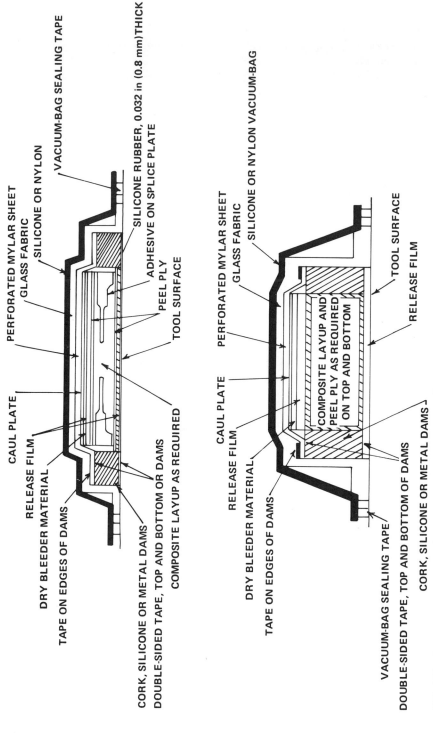

**FIG. 4.29** Component cross sections of typical vacuum-bagged composites.[23]

seated on the tool. Compaction cycles are not required for B–Ep parts. By contrast, pre-bleeding is used to remove most, if not all, of the excess resin from the lay-up so that either no bleeder or a minimum bleeder system is used during cure. Prebleeding is used in special situations, e.g., parts with stitched joints or assemblies cured in trapped tools. A breather system should still be used to facilitate removal of volatiles. Prebleeding is normally accomplished under vacuum at temperatures of 180 to 225°F (82 to 107°C) for 1 h with a full bleeder system in place. Prebleeding is an extra operation and should be used sparingly to reduce costs. Also, the probability of voids is increased due to the absence of additional resin to flush out volatiles during cure.

Over the past several years, the aerospace industry has extensively evaluated the requirement of debulk before the final cure cycle on various thicknesses of laminates, shapes of components, types of composite materials, and combinations of materials. These evaluations have demonstrated that in many cases debulking requirements can be accomplished at room temperature or eliminated through the mechanical lay-up and compaction of composite components.

The pressure application for the debulking and compaction of the resin and fiber has been applied quite differently for rotor blades than for other composite component structures. For example, a small rotor blade 17 ft (5.18 m) long and weighing less than 100 lb (45.4 kg) has been fabricated in a single major cure cycle. The blade's uncured structural fiber-glass shell is filled with a polyvinyl chloride foam which is machined over the required cured thickness of approximately 5%. Compression of the foam core during cure provides the necessary pressure against the rigid aluminum full-cavity tool (Fig. 4.30).

Other blades with tubular fiber-glass spars have been wound on a solid-aluminum mandrel, which is then enclosed in a steel spar tool. The resulting relative expansion coefficients compact the fibers and resin during cure (Fig. 4.31). The tubular composite spars have

FIG. 4.30 Foam-filled blade.[27] *(Boeing-Vertol Co.)*

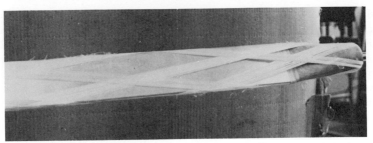

**FIG. 4.31** Tubular fiber-glass spar wound on solid mandrel.[27] *(Boeing-Vertol Corp.)*

been built by placing the uncured composite material around a thin silicone-elastomer tube supported on an expendable styrofoam mandrel. The assembly is then enclosed in a steel cavity tool and the silicone tube pneumatically inflated during blade assembly and cure.

Most main-rotor composite blades produced in the United States currently use an aramid composite material, Nomex, as the honeycomb core to fill and stabilize the rear portion or trailing-edge fairing, aft of the blade spar (Fig. 4.32).

**FIG. 4.32** Full-depth Nomex core blade.[27] *(Boeing-Vertol Corp.)*

## Bleeder and Breather Systems for Autoclave Cure

The bleeder system, which is material used to absorb excess resin during cure, and the breather system, which is used to provide a vacuum path over the surface of the part, represent a recurring expense which should be minimized by the reduction of the number of plies and cost of the materials used. Typical bleeder materials are glass and mat, which can stretch over contours and provide a cushioning effect to matched metal tools. The choice of bleeder material appears to be a matter of individual preference.

The bleeder-prepreg ratio is controlled by the selection of the prepreg. Again, the resin content of the material should be as low as possible compatible with prepreg handling characteristics, the lay-up process, and the ability to mold void-free parts. Based on the specific prepreg characteristics, the bag seal next to the dams and the Tedlar sheet might be omitted. For thick parts (above 50 plies) the Tedlar should be slitted on 2- to 3-in (51- to 76-mm) centers to facilitate volatile removal.

For fabric prepregs, the bleeder system is typically glass, which actually serves as a breather. B–Ep parts typically use one ply of glass for five prepreg plies. For Gr–PI bleeder-system requirements have not been rigidly established, and reference to applicable reports and material suppliers is recommended.

## No-Bleed System

Over the past few years several aerospace companies have evaluated and developed a manufacturing method in which graphite structures are fabricated without resin bleeding before or during the cure cycle. Normally, in fabricating composite structures, the resin must be bled during curing in order to obtain relatively void-free assemblies which meet precalculated thickness tolerances (inches per ply) and to contain the required fiber volume for structural integrity. The industry has demonstrated that with thin-skin honeycomb sandwich structures and relatively thick monolithic structures, as in fuselage and wing components, the requirement for resin bleeding can be eliminated. The elimination of the bleeder cloth not only affects the total acquisition cost of the structure by eliminating the labor hours associated with the location and positioning of the bleeder cloth on the composite surface and associated material costs but also provides improved producibility for obtaining smooth aerodynamic surfaces. Normally in a honeycomb sandwich structure, both inner and outer skin assemblies must be resin-bled during the cure cycle, subsequently requiring bleeder cloth both on the tool surface and inner surface of the component. Any movement occurring during the cure cycle could result in a markoff to the composite structure, with potential rejection of the finished component.

## Zero vs. Controlled Resin Bleed

A major process factor which imposes unique tooling requirements is blade weight control. The tooling must control the final blade material volume, distribution, and resin content if repeatable consistent dynamic performance and blade weight are to be obtained. Two approaches in current use are zero resin bleed and controlled resin bleed. The latter has produced the most blades, over 10,000 with consistent results.

In the controlled-bleed process, the resin content of the material placed in the tooling is closely controlled by weighing samples of each tape placed in the mold and adjusting the resin content as lay-up progresses to achieve a final uncured assembly weight. The tool must be capable of being essentially sealed so that minimal resin is lost during cure. Tooling factors which control resin flow (pressure, temperature, and time-temperature rate change) are controllable and repeatable. This process is in production use on blades and components fabricated by lay-up of on-site preimpregnated unidirectional and fabric materials.

In a different controlled-bleed approach developed for blade components fabricated by wet filament winding a fluid resin is used in excess of the amount desired in the cured component. The tooling is designed to accept the fluid system and precisely control the volume of the cured components while allowing relatively free escape of the resin system. Component volume (and hence weight) is the result of the tool internal volume at the gelation temperature of the resin. The most successful method uses an aluminum winding mandrel enclosed in a steel female die. The volumetric expansions of the aluminum mandrel and steel die react to provide the pressure necessary to force out the excess resin. This process continues during heat up until the resin viscosity has increased to a virtual no-flow condition. The resin content and hence component weight are essentially set at this point in the cure cycle. The final cure cycle completes resin polymerization and sets the component dimensions. Over 200 main-blade spars have been fabricated by this process.

## Curing

Generally speaking, most prepreg reinforced plastic parts made on good permanent tools can be cured either in an oven or autoclave. Processing today is about equally divided. Manufacturers who cure exclusively in the autoclave claim that autoclave parts are more uniform in thickness, reproduce better-detailed configurations, and have a general overall better appearance than parts cured in the oven. The big advantage of oven curing is the lower cost of equipment. Parts which are laid up using the wet system should be cured in the oven because after a part has gelled at room temperature, there is no advantage to curing in an autoclave.

A *cure cycle* is a series of time-temperature steps at a constant pressure to change a liquid resin and reinforcement into a cured part of desired configuration and structural strength. The general cure cycle of a resin system is based upon the actual temperature of the resin during the cure cycle. The specific cure cycle for a part is based upon the resin system, thickness of the part, the type of mold material, and the thickness of the actual mold. Standard cure cycles are established for groups of similar parts in an assembly-line curing system. Some typical cure cycles for structural parts are listed in Table 4.6.

A precure cycle changes the resin from a liquid or B stage to a completely hard condition so that the part can be removed from the mold and handled without damage. A postcure cycle is additional curing of the resin so that the part will meet its structural requirements. In some cases the cure cycles are combined and the complete cure is accomplished while the part is on the lay-up mold. In other cases, when two cure cycles are used, the parts are postcured after removal from the lay-up mold. If warpage is a problem, the part must be postcured on a postcure fixture, which holds the part to contour. Parts that require postcuring may be postcured in an oven or autoclave under atmospheric pressure.

The three parameters traditionally controlled in the consolidation or cure of composite structures are pressure, temperature, and time. In a conventional curing system, in which an autoclave is used to supply the pressure and temperature required, little or no instrumentation is built into the tooling. The autoclave control system usually contains vacuum gages to monitor the under-bag vacuum source and often separate gages to monitor the vacuum pressure under each part bag or in a manifold connecting several tools. Autoclave pressure is usually maintained automatically, as is autoclave fan or heat-exchanger outlet air temperature. Set-point timers or clocks are often incorporated in the system so that a temperature-time program can be established and automatically conducted by the autoclave system. Normally an autoclave system contains one or more multipoint temperature recorders permanently connected to a thermocouple junction box inside the pressurized area.

Inherent in the use of time and temperature as a means of establishing and controlling a proper cure cycle is the tacit assumption that a part is cured if it has been exposed to the established time-temperature parameters. At best, because no one wants to err on the short-cycle side, all parts are usually overcured. The main fault in this approach is excessive use of expensive curing facilities and available time.

Two different cure-monitor techniques being used in conjunction with thermocouple

**Table 4.6** Typical Cure Cycles for Structural Parts[1]

| Resin type | Precure cycle | | | Postcure cycle | | |
|---|---|---|---|---|---|---|
| | Temp | | | Temp | | |
| | °F | °C | Time, h | °F | °C | Time, h |
| Approximate pressure 10–25 lb/in² (69–172 kPa) | | | | | | |
| General-purpose polyester | 150 | 66 | 0.5 | † | † | † |
| | 200 | 93 | 1 | † | † | † |
| | 250 | 121 | 0.5 | † | † | † |
| Modified polyester | 200 | 93 | 2 | 200 | 93 | 0.5 |
| | 250 | 121 | 0.5 | 300 | 149 | 0.5 |
| | 300 | 149 | 0.5 | 350 | 177 | 1 |
| | | | | 400 | 204 | 1 |
| | | | | 450 | 232 | 1 |
| | | | | 500 | 260 | 1 |
| Epoxy, medium-temperature | 200 | 93 | 1 | 250 | 121 | 0.5 |
| | 250 | 121 | 0.5 | 300 | 149 | 0.5 |
| | | | | 350 | 177 | 2 |
| High-temperature | 200 | 93 | 1 | 200 | 93 | 0.5 |
| | 300 | 149 | 1 | 300 | 149 | 1 |
| Approximate pressure 10–45 lb/in² (69–310 kPa) | | | | | | |
| Phenolic | 200 | 93 | 0.75 | 300 | 149 | 0.5 |
| | 250 | 121 | 0.5 | 350 | 177 | 2 |
| | 275 | 135 | 0.5 | | | |
| Polyimide | | ‡ | | 350 | 177 | 4 |
| | | | | 425 | 218 | 4 |
| | | | | 460 | 238 | 4 |
| | | | | 520 | 271 | 4 |
| | | | | 620 | 327 | 4 |
| Silicon | 350 | 177 | 1 | 200 | 93 | 16 |
| | | | | 250 | 121 | 2 |
| | | | | 300 | 149 | 2 |
| | | | | 350 | 177 | 2 |
| | | | | 400 | 204 | 2 |
| | | | | 480 | 249 | 16 |

†No postcure.

‡The precure cycle for polyimide consists of raising from room temperature to 350°F (177°C) in 140 ± 20 min, holding at 350°F (177°C) for 120 ± 15 min, and cooling under vacuum to 150°F (66°C).

control use small foil gages to sense changes in resin electrical properties as a function of the cure process. One such system uses equipment to sense changes in the resin dielectric properties as a function of cure; when no further change is noted, the part is said to be cured. The system senses changes in resin electrical resistance as a function of cure. The result, as far as tool design and construction is concerned, is the same. The sensors must be easily and reliably installed and must not interfere with other tool or component functions.

## Oven Curing

A typical vacuum curing oven (a vessel that provides heat by convection) is a large metal thermally insulated gas-heated forced-air circulating oven with large doors at one or both ends. An example of standard size is 8 ft (2.4 m) high, 12 ft (3.7 m) wide, and 30 ft (9

m) long. The following minimum requirements should be met by ovens and associated equipment for efficient curing of reinforced composite parts and assemblies:

1. A large thermal input source is necessary for the quick temperature adjustment of the oven. The oven should be able to operate through a temperature range from room temperature to 600°F (316°C), which includes the precure and postcure for polyester, epoxy, phenolic, silicon, and polyimide resin systems.
2. A high-volume uniformly controlled forced-air circulating-air system is required. The air must be well baffled so that temperature in the oven will be maintained at a specific curing temperature with a tolerance of ±15 Fahrenheit degrees (±8.3 Celsius degrees).
3. A high-capacity vacuum system with a number of independent vacuum headers, gages, and valves to maintain a minimum of 21 inHg (71 kPa) on all parts during cure is necessary.
4. Each vacuum gage should be connected to indicate the actual pressure on the part being cured.
5. There should be a stand-by vacuum system that can be immediately connected into the system if the primary system fails.
6. The oven should include an automatic recording control to produce a permanent record of each thermal-time cure cycle.

### Autoclave Curing

A typical autoclave consists of a large cylindrical metal pressure vessel pressurized with air and/or $CO_2$, thermally insulated, steam-heated with forced circulating hot air and a large circular door at one or both ends. A typical size is 12 ft (3.7 m) in diameter and 55 ft (16.8 m) long (Fig. 4.33). An autoclave system used for curing reinforced plastic should meet the following minimum requirements:

1. Sufficient thermal input source (oil or steam) to provide quick temperature changes from one temperature to another during the cure cycle. In general the autoclave should

**FIG. 4.33** Autoclaves used in curing aircraft and rotor-blade components. (*Sikorsky Aircraft, Division of United Technologies.*)

be capable of operating through a temperature range of room temperature to 350°F (177°C). If additional temperature is required for postcuring, it is more economical to postcure the parts in an oven.

2. The autoclave must include a means for circulating the air inside since temperature in the curing area should be maintained at a specific curing temperature with a tolerance of $\pm 15$ Fahrenheit degrees ($\pm 8.3$ Celsius degrees).

3. A high-capacity pressurization system is required to pressurize the large volume of the autoclave quickly. In many installations $CO_2$ is used as the pressurizing medium because of its fire-retardant characteristics. The $CO_2$ requires a large storage pressure vessel because the cost of $CO_2$ requires it to be recycled. The pressure used during cure usually does not exceed 100 lb/in$^2$ (690 Pa).

4. An adequate vacuum system is required to maintain vacuum pressure on the parts before cure and during cooldown after cure. During cure, the pressure is maintained on the parts by air or $CO_2$.

5. The autoclave should provide an automatic recording system for a permanent record of pressure and temperature for each cure cycle.

**Steps in Oven and Autoclave Curing:** The following general steps are required for curing operations in oven or autoclave:

1. All parts that have the same standard cure cycle are placed together on the same cure dolly (Fig. 4.30).

2. The dolly is placed in an oven or autoclave, and vacuum hoses are connected so that vacuum pressure is registered on gages outside the curing chamber.

3. Parts are checked for vacuum leaks.

4. Doors are closed, and the cure cycle begins.

5. For oven curing vacuum must be maintained between 21 and 28 inHg (71 and 95 kPa) during the cure.

6. For autoclave curing, the air pressure and heat are increased as required by the first step of the cure cycle.

7. After the particular standard cure cycle has been completed, the air pressure in the autoclave is released. The doors of the oven or autoclave are opened slightly to help cool the part.

8. All parts should be cooled below 150°F (66°C) while still under vacuum pressure to minimize thermal shock and warpage.

## Part Removal, Finishing, and Subsequent Operation

Many parts that have been laid up, bagged, and cured correctly are rejected because they are damaged while being removed from the tool or are mistrimmed or misdrilled. The bagging materials and parts must be removed from tools carefully. The tools are inspected for damage and returned to the tool area for repair or to be cleaned and prepared for the next molding cycle. A polished tool surface is generally not required for a composite structure and is usually a waste of time and effort. Almost invariably, the tool surface is covered with a parting film such as Teflon, fluorinated elastomer, or Teflon-impregnated fabric, which acts as a tool surface seal and parting agent and is applied before every new part is made. These materials impart their own surface finish to the component. The use of peel plies, a sacrificial fabric such as nylon cloth, is becoming more prevalent. This material acts as a surface bleeder in net-cured resin systems and is removed just before bonding or painting. Peel ply protects the laminate surface, gives a clean roughened resin surface suitable for subsequent bonding or painting, and reduces the amount of surface chipping during drilling and countersinking. Nylon peel plies are usually 0.12 to 0.24 in (3 to 6 mm) thick and tend to blend out any cutter marks remaining on the machined tool surface. Trimming, drilling, and other postcuring operations will be discussed in Chap. 6.

## Other Curing Methods

### Hydroclave

A hydroclave can exert a pressure of greater than 1 kip/in$^2$ (6.9 MPa) in a bagged composite component during the cure cycle. This method allows higher pressures to be applied to a component than the traditional autoclave, a fact that becomes important for components that are to be cured outside the cure envelope of the autoclave. As the name suggests, a fluid provides the pressure.

### Self-Contained Tools

Self-contained consolidation tools (press curing, matched metal dies, etc.) can also be used to cure composite components in a portable environment. Each unit acts as a miniautoclave in that it provides the heat and pressure for curing. The use of self-contained tooling is increasing since many components are too large for existing autoclaves or manual operations must be performed on the parts and tooling during the cure cycle (Fig. 4.34). This method is most applicable if the composite material to be cured falls outside the cure profile of the autoclave, e.g., MMCs. This procedure reduces component handling, especially where lay-up and fabrication operations are not near the curing facilities.

## Co-curing

Co-curing, the ability to cure and bond structural composite systems in one operation, plays a major role in the fabrication and assembly of composite structures. Its use is industry-wide. Extensive evaluations and developments of co-curing techniques on many government-sponsored and industrial programs have led to the application of this cost-competitive manufacturing technique in the fabrication of approximately 85% of the composite structures for two new lightweight fighter aircraft (Fig. 4.35), equal to 850 lb (386 kg) and approximately 2300 lb (1043 kg). The use of co-curing techniques has demonstrated the potential for composite manufacturing cost to be reduced 30 to 45% through elimination of costly repetitive fit-up and assembly operations, reduction in number of detail parts, and the simplification of tool requirements. Co-curing techniques are currently employed in primary and secondary structures, radomes, fairings, rotor blades, empennage structures, and fuselage assemblies.

**FIG. 4.34** Computer-controlled self-contained tooling system.[27] *(Boeing-Vertol Corp.)*

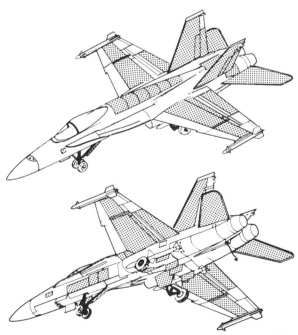

**FIG. 4.35** Shaded areas indicate Gr–Ep composite components.[23]

# FILAMENT WINDING

Filament-wound items are produced in quantity for aerospace applications and for a commercial market. Winding technology and design are different for these two fields. Formerly, when only a few fabricators were engaged in the winding, most commercial and federal aerospace needs were supplied by the same manufacturer. The differences, if any, were slight, and the same equipment was used for both purposes. Today the difference has been magnified. Aerospace companies have taken over their own winding operations, and the industrial winders have extended into the growing commercial market. Aerospace winding has become highly specialized and is geared to superior performance at a premium price. The major efforts are directed to such military applications as rocket-motor cases, radomes, pressure bottles, small specialty missiles, and helicopter blades, pylon, tail, and stabilizer. Here the strength-to-density ratio is of prime importance, and the aim is to orient the fiber along the direction of the applied stress.

The commercial sector, on the other hand, stresses lower production costs while taking advantage of other composite properties besides high strength. The trend is toward cheaper grades of glass rovings, substitution of polyester for the more expensive epoxys, and simplification of winding equipment to favor higher production speeds. Corrosion-resistant and electrical-grade materials are exploited in developing new markets for storage tanks, reinforced pipe, unfired pressure vessels, chimney liners and stacks, cherry-picker booms, lightning arresters, chemical-storage and processing tanks, springs, drive shafts, and wind-turbine blades.

The differences between the two divisions of the industry are manifest in the choice of raw materials, the precision of the winding equipment, and the details of design and analytical procedures. The filament-winding process itself is comparatively simple. It consists of wrapping bands of continuous fiber and/or rovings or strands over a mandrel in a single machine-controlled operation. A number of layers of the same or different patterns are placed on the mandrel, and the repetitive patterns and reinforcement spacings are subject to close control. The fibers may be impregnated with resin before winding (wet winding), preimpregnated (dry winding), or postimpregnated. The first two winding sequences are analogous to wet or dry lay-up in other reinforced-plastic fabrication methods. The process is completed by curing the resin binder and removing the mandrel. Curing is normally conducted at elevated temperatures without pressure. Finishing operations such as machining or grinding are usually not necessary.

## Winding Methods

From the raw-materials standpoint, three methods of winding can be used: (1) wet winding, in which the roving is fed from the spool, through an impregnating resin bath, and onto the mandrel (Fig. 4.36), (2) dry winding, in which preimpregnated B-staged roving is fed either through a softening oven and onto the mandrel or directly onto a heated mandrel, or (3) postimpregnation, in which the dry roving is wound on the mandrel and the resin is applied to the wound structure by brushing or by impregnating under vacuum or pressure. This last technique is usually limited to relatively small parts, such as shotgun barrels, since thorough impregnation without using pressure or vacuum is difficult.

At present wet winding is the most common method. It is lowest in terms of materials cost, and for users and producers equipped with plastics-formulating facilities it offers the benefits of flexibility of resin formulation to meet specific requirements for different parts. On the other hand, wet winding imposes definite limitations on the resin systems that can be used. Systems must (1) be low in volatile content, to prevent gassing and bubbling within the wound structure, (2) be available within specific viscosity limits at room temperature to provide thorough saturation (heating must be used with some systems), and (3) have pot life long enough to permit preparation of large impregnation baths for optimum economy of production. In wet winding, tension of the roving must be altered as the diameter of the part increases if accurate control of resin-glass ratio is mandatory. If winding tension is not altered, resin content varies directly with diameter.

From the standpoint of the user, the most important benefit of dry or prepreg winding is the variety of resin types that can be wound. Use of prepreg phenolic, phenylsilane, and silicone resins may offer a number of benefits in high-temperature structures or where high-temperature elec-

**FIG. 4.36** Fibers leaving matrix-resin bath being laid on rotating mandrel. *(Celanese Corp.)*

trical properties are required. Although materials costs for dry winding are naturally higher than those for wet winding, dry winding offers additional benefits in that (1) smaller shops do not require plastics-formulation facilities, (2) close control over consistency of resin system and resin-glass ratio can be maintained, and (3) preimpregnated roving can be quality-controlled before winding, providing a high degree of reproducibility of quality from part to part. An additional benefit in comparison with wet winding is that material can be supplied with a relatively high tack, to permit winding on steeper slopes without the roving's sliding off.

## Basic Winding Materials

### Reinforcements

At first fiber glass was the only reinforcement used in filament winding. The designer had two choices of fiber glass: the relatively cheap E glass or the stronger and more expensive S glass. Both materials have a high specific tensile strength, but their low modulus of elasticity constitutes an inherent weakness. The maximum modulus transmitted to the composite is on the order of 7000 to 8000 kips/in$^2$ (48 to 55 GPa) for unidirectional windings. In many motor-case designs, buckling, bending, and torsional loads are not excessive and the low-modulus glass can be used without appreciable weight penalties. A higher-modulus reinforcement, however, is of real interest and can be used in wider areas of application. The exceptionally high specific moduli of boron, graphite, beryllium, and other new materials make them suitable. Representative filamentary materials are compared in Table 4.7 on the basis of specific tensile strength and specific modulus.

**Table 4.7** Comparison of Filamentary Reinforcements

| Reinforcement | Density lb/in$^3$ | Density kg/m$^3$ | Specific tensile strength ksi | Specific tensile strength MPa | Specific modulus $10^6$ psi | Specific modulus GPa |
|---|---|---|---|---|---|---|
| E glass | | | 500[a] | 3447[a] | | |
| | 0.092 | 2547 | 400[b] | 2758[b] | 10.5 | 72.4 |
| | | | 500[c] | 3448[c] | | |
| S glass | 0.090 | 2491 | 665[a] | 4585[a] | 12.4 | 85.5 |
| | | | 550[b] | 3792[b] | | |
| $SiO_2 \cdot Al_2O_3 \cdot MgO$ | 0.091 | 2519 | 700 | 4827 | 13.5 | 93 |
| $SiO_2 \cdot Al_2O_3 \cdot MgO \cdot BeO$ | 0.090 | 2491 | 715 | 4930 | 14.9 | 102.7 |
| Boron | 0.095 | 2630 | 450 | 3103 | 60 | 414 |
| Graphite, T-50 | 0.058 | 1605 | 300 | 2069 | 50 | 345 |
| Gy-70 | 0.072 | 1993 | 250 | 1724 | 40 | 276 |
| T-300 | 0.063 | 1744 | 400 | 2758 | 60 | 414 |
| Beryllium | 0.066 | 1827 | 160 | 1103 | 35 | 241 |
| Steel music wire | 0.284 | 7861 | 606[d] | 4178[d] | 29 | 200 |
| | | | 575[e] | 3965[e] | | |

[a]Monofilament.  [b]Strand.  [c]Hollow fiber.  [d]Single.  [e]Tape.

## Resin Systems

Most strength-critical aerospace structures are wound with epoxy-based resin systems. The polyesters, silicones, and phenolics are limited to special applications—the polyesters to airborne radomes and phenolics and silicones to high-temperature use. The service requirements of supersonic transport and high-performance tactical aircraft have been met using the heat-resistant polymers, particularly the polyimides. The problems which delayed adoption of polyimides, namely, reduction of the volatiles evolved during cure and elimination of porosity in finished composites, have been overcome. Void content has been lowered below 5% for unidirectional windings of 1% for laminates by preimpregnating of B staging before the final cure.

In commercial winding the resin systems have been either epoxy- or polyester-based. The trend to the polyester is motivated by their easier handling, corrosion resistance, and lower cost. For aerospace applications the selection of epoxys has been attributed to their superior mechanical properties, fatigue performance, heat resistance, and adhesion to glass, coupled with a lower curing shrinkage. Whether all these claims can be verified is open to question. In general, aerospace fabricators place greater reliance on the epoxys, which appear to exhibit somewhat higher compressive strengths and a higher threshold before the initiation of crazing.

The choice of a resin system depends on its processing characteristics, curing requirements, and physical properties of the resin which affect composite properties. Viscosity and pot life of the catalyzed system are the important processing considerations. Gel time and flow are secondary factors. A low viscosity is required for complete wet-out of the strands and for removal of entrapped air. In wet winding this condition is best reached without adding nonreactive diluents or with only small percentages of reactive diluents. In optimum systems it is often necessary to heat the resin to lower viscosities. The effect of time-temperature histories on pot life and viscosity are evaluated and temperature ranges established for optimum winding conditions. A viscosity between 360 and 840 lb/ft·h [roughly 150 and 350 centipoises (cP) or 0.15 to 0.35 N·s/m$^2$] is considered to be a low range, while a viscosity greater than 2400 lb/ft·h (1000 cP, 1 N·s/m$^2$) begins to be excessive. A pot life of at least several hours (sometimes days) is required, since it is not generally advisable to wind over gelled or partially gelled resin.

The curing conditions are sometimes limited by the particular application. For example, in most instances motor cases cannot be cured at temperatures above 300°F (149°C). Such limitations are imposed by the presence of lining materials which deteriorate above this temperature. Higher-temperature cures also place greater structural demands on the mandrel materials and tend to increase mandrel costs. The effect of the resin on composite performance has been the subject of numerous studies.

In summary, micromechanics offers the best approach to determining the gross effects of resins on the composite. It establishes limiting values to be anticipated for any set of resin properties. Resin systems can be optimized for external pressure, cryogenic, elevated-temperature, electrical, or other applications. Generally optimization can be made for only one property at a time. When more than one property is of concern, trade-offs become necessary.

## The Winding Process

Each of the two basic filament-winding methods produces a distinctive winding pattern.

### *Hoop (Circumferential) Winding*

Bands, or rovings, are wound almost perpendicular to the mandrel axis. Each revolution of the mandrel advances the carriage of material-delivery eye one bandwidth. Only a lathe-like capability with slow carriage motion to fast mandrel rotation is needed (Fig. 4.37).

### *Helical (Longitudinal) Winding*

This type (Fig. 4.38) requires a machine capable of laying down a band in a helical path over the surface of the mandrel, turning it around on the end, and returning to the starting position to repeat the cycle. In a typical 10-circuit pattern (Fig. 4.39) the fiber path advances one-tenth of the circumference plus one-tenth of the bandwidth per circuit. The eleventh fiber path then falls

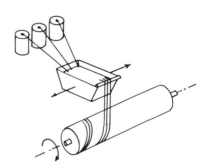

FIG. 4.37 Hoop or circumferential winding. *(From Am. Mach., May 1981, p. 127.)*

parallel to the first. The usual problem is to lay down the helical path at a fixed winding angle $\alpha$, equal to the arcsine of the pole diameter divided by the part diameter. In other words, it is the angle between the filament path and the centerline of the part through the cylindrical section. If a flat, narrow band were to lie at this angle across the cylindrical section, its natural geodesic path would pass over the dome, tangent to the pole piece, and reverse itself along the opposite side at the same helical angle. One complete path of filament beginning on the tangent line at point 1 in Fig. 4.38 and following around through a natural geodesic path to point 2 on the tangent line is called a *circuit*. Once geometric factors, such as part diameter, pole diameter, and the length of the cylindrical section, are fixed, there is only one true filament path between points 1 and 2 on the tangent line. It

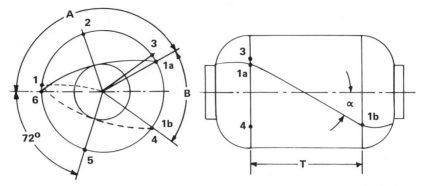

**FIG. 4.38** Helical winding machines lay down bands in helical pattern; $72° = 1.26$ rad. *(From Am. Mach., May 1981, p. 127.)*

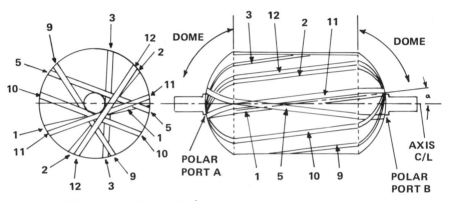

**FIG. 4.39** Multiple-circuit helical winding.[1]

is then up to the machine to position the delivery eye of the band-delivery system with respect to the mandrel so that the filaments will follow this path through a circuit, repeating circuit after circuit.

## Polar Winding

Polar winding is a special case of helical winding in which the filament path can be described by the intersection of a plane passed through the part. A single band, as it is laid on the mandrel, is contained in this plane. The plane would be tangent to the pole piece on one end of the part and one bandwidth away on the opposite pole piece. During winding, the mandrel is usually stationary while the delivery system revolves around it. Or both delivery system and mandrel may be fixed while the machine holding the mandrel revolves. After each rotation, the mandrel advances one bandwidth. Polar winding is used mainly for bottlelike shapes rather than for open-end structures (Fig. 4.40).

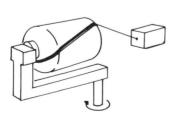

**FIG. 4.40** Typical polar winding. *(From Am. Mach., May 1981, p. 127.)*

## Filament-Winding Machines

Winding machines are designed to perform either polar or helical winding. Variations in each type add versatility and compensate for some of the weaknesses. Polar machines normally have the mandrel mounted in a vertical position, which eliminates mandrel deflections and allows for easier construction of a rotating arm. Such an arrangement is satisfactory for windings of moderate size but becomes impractical for large windings. In this case the mandrel is supported in a horizontal position, and the rotating arm is replaced by a "racetrack" feed drive. A machine of this type has been built to wind 260-in-diameter (6.6-m) motor cases; similar versions are commercially available for smaller windings. Another variation of polar winding is the *tumbling process,* in which the feed remains stationary while the mandrel is revolved or tumbled in a nearly horizontal plane. The inclination of the plane to the horizontal determines the winding angle, and indexing is accomplished by mandrel rotation as before. A combination of tumbling and continuous rotation adds to the flexibility of this machine.

To the two fundamental motions described for the helical winding machine (mandrel rotation and reciprocating carriage motion) a third cross-feed motion perpendicular to the mandrel axis can be imparted for more accurate fiber placement over the end dome. In some instances a rotational motion in the cross-feed is also added to keep the feed tangential to the point of fiber placement. Without cross-feed controls on the filament guide, it is doubtful whether strict geodesic paths could be maintained over the end dome.

Mechanical or numerical means are used to control the motions of helical machines. Mechanical machines usually have one drive; the relative speeds of mandrel rotation and carriage movement are held constant by gear trains, chains, and sprockets or feed screws. Cross-feed can be controlled by cams or by arrangement of chain drives to fix the dwell time at each end of the mandrel. Some mechanical machines are equipped with a separate servo drive on the cross-feed. Other features are added to the mechanical winders for easy change or selection of winding angles and for establishing the winding pattern empirically.

With numerical machines, the three or four axes of motion are powered by separate hydraulic servo systems, and operation is by digital punch-tape control. Numerical tape control increases the versatility of the winding machine, and nearly all pattern types can be programmed. Computerized filament-winding machines are discussed in more detail in Chap. 5.

### Comparison of Patterns and Machines

In comparing helical and polar winding one should first weigh the advantages of the patterns associated with each method and then determine how pattern selection is influenced by factors related to actual machine operation. The influence of winding patterns on the total performance, however, can be reduced to two possibilities: differences in the internal volume of comparable planar or geodesic heads and the effect of helical-fiber crossovers on burst strength. Since volume differences between planar and geodesic end closures appear to be minimal, particularly with small center-port openings, the filament crossover effect is the most important factor in comparing helical and polar winding patterns. It has been assumed that the crossover points act as stress risers and should result in lower burst pressures. Three conclusions can be drawn:

1. While differences may exist because of filament crossover, their exact interpretation depends on a detailed analysis of the local stress condition.

2. Deviation of fiber paths from theoretical positions may be more decisive than filament crossover.

3. For practical purposes there is no significant difference between polar and helical patterns in ultimate burst strength.

A major advantage of the polar winder is that the fundamental machine motions require relatively simple controls. There is no need for carriage reversal, and uniform speeds can be maintained. Its disadvantages are that it is limited to the use of prepregs to prevent fiber slippage and that port openings usually cannot be in excess of 30% of the diameter of the item being wound.

Helical winders have the advantage of greater flexibility. An inherent weakness is the inertial effects on carriage reversal, which hinder control over the end region. In general, helical winding is more widely accepted than polar winding for commercial applications. Much of the standard-diameter pipe [2, 4, and 6 in (50.8, 102, and 152 mm)] and large-diameter pipe is wound this way. Continuous lengths of standard pipe are made by a combination of 0° (0-rad) or low-angle longitudinals, plus near 90° (1.6-rad) hoop winds. This process is similar to pultrusion (discussed later in the chapter), with the addition of a rotating head for wrapping hoop winds. Polar winding is impractical for small pipe and most large-diameter pipe. Storage tanks and pressure vessels are wound as single-angle helicals or as combination winds of longitudinals and hoops, where the longitudinals may be polar windings.

## Mandrels

Although most phases of winding technology have been subjected to intensive development, until recently mandrel materials and design have been more or less neglected. Poor mandrel design can result in damage to fibers, deviations in dimensions, and excessive residual stresses. Mandrel construction can be simple or sophisticated, depending on the product—shape, weight, finish, cure cycle, and production volume. Cardboard and wood tubes may suffice for winding routine pipe. Blow-molded thermoplastics can serve to form pressure vessels. Steel and aluminum mandrels provide long life, and some are designed with internal steam ports to speed curing. Even air-filled elastomeric bladders, plaster, and water-soluble salts have been used.

The following conclusions have been reached regarding materials and concepts after accelerated study, development, and experimentation:

**Low-Melting Alloys:** These have an undesirable combination of high density and a tendency to creep under moderate loads; use for small vessels, 12 in (305 mm) diameter by 12 in long.

**Soluble or Meltable Salts:** Mandrels up to 60 in (1525 mm) in diameter have been made with sand and polyvinyl alcohol binder. The polyvinyl alcohol binder is readily soluble, and removal characteristics are excellent.

**Eutectic Salts:** These are better suited than the alloys and are applicable to diameters up to 24 in (610 mm).

**Soluble Plasters:** These are not recommended since they exhibit low resistance to winding loads and curing and washout characteristics are poor.

**Collapsible Metal:** Segmented aluminum mandrels have been successfully fabricated and used when the quantity of parts is over 25. Diameters of 36 to 60 in (915 to 1525 mm) are well suited to these mandrels. Care must be taken when small polar openings are present since they may complicate segment removal.

**Segmented Aluminum Sandwich:** This concept was analyzed for diameters of 144 to 240 in (3.66 to 6.1 m). Costs using this method are high because curing complicates structural problems. To be economical fairly large quantities of parts must be produced. Conventional curing methods cannot be used; heating shrouds or lamps are required.

**Plastics and Foams:** Most foams are not satisfactory because of poor mechanical properties, particularly at elevated temperatures. The use of laminated-plastic tooling is normally restricted to small loads and below 200°F (93°C).

**Plaster-Chain:** This combination of breakout chains embedded in plaster has been widely used but is not recommended except for large diameters with easy access.

**Metallic Substructure with Plaster Facing:** This combination is applicable to single parts or even small quantities for diameters of 144 to 240 in (3.66 to 6.1 m). An added advantage is the ability to handle changes in diameter or other dimensions if anticipated. A disadvantage is that mandrels over 240 in (6.1 m) in diameter or length cannot be used since high shear loads are imposed on the mandrel shaft and bearing loads on the equipment.

**Pressurized Segments of Aluminum:** This design has been used successfully for large mandrels of 144 to 240 in (3.66 to 6.1 m) in diameter. It offers simpler designs, ease of fabrication, and reduced weight; care must be taken in pressurization, which may pose a safety problem.

**Inflatable Mandrels:** If no supporting structure exists, these mandrels are impractical, but past problems of torque transmission and dimensional control have been overcome with improved designs to make this technique currently attractive.

## Test Methods

The testing and evaluation of composites and basic raw materials constitutes a major phase of filament-winding technology. Tests have been conducted for a number of purposes, e.g., determining reinforcement and resin properties and design allowables, quality control, verifying micromechanics theory, proof testing, and hydroburst evaluations of finished pressure vessels.

Tests based on wound rings have been used to obtain data on composites and constituent materials. Originally developed to measure the effects of coupling agents and finishes on the strength of fiber glass, graphite, etc., the rings have been adapted to tests for other mechanical properties such as tension, compression, and interlaminar shear. Evaluations based on ring testing have been found useful for quality control, qualitative screening of raw materials, and comparisons of new reinforcements and resins.

A test to determine the compressive strength of a cylinder subjected to external pressure has also been standardized. Other standard tests include one for strand tensile strength. Properties other than burst strength have also been obtained from testing small-diameter cylinders. They include uniaxial tension, uniaxial compression, bearing strength, torsional strength, and shear stiffness.

Design burst levels established by testing subscale models appear to be fairly reliable, particularly when large-diameter vessels are employed in the tests. Results may not be in complete agreement with full-scale performance, and scaled correction factors are applied to account for the differences. Special attention is paid to filament-wound motor cases since it is customary to proof-test all of them. Proof pressure is usually 75 to 80% of the design ultimate burst strength. Although proof testing can be detrimental and result in resin crazing, it is justifiable on the basis of increased reliability. Proof testing and time held at proof pressure can influence subsequent burst tests, but only in rare instances have final bursts occurred below proof pressure.

## Variation in Properties

The following major processing factors influence composite properties:

Resin content and distribution

Winding tension

Condition of impregnated strand

Variation in strand bandwidth

Position of the strand

Curing cycle

Individual effects are difficult to isolate and for the most part closely interrelated. The observed weaknesses or defects are manifested as voids, dry spots, resin-rich areas, premature failures, localized failure regions, and nonuniform or incomplete cures.

Resin content and distribution are of importance for weight and thickness control. Most aerospace components are usually held to within 2% or better weight tolerance. From a strength standpoint, excessive variation results in uneven stress distribution and areas where failures can be initiated. Some filament-winding fabricators have found it easier to control resin content by using prepregs. Here a variation in prepreg resin content within $\pm 2\%$ is about the best being currently achieved. Note that the roving itself varies in weight by $\pm 1.5\%$. To attain more uniform resin contents prepreg tapes in lieu of strands have been used by fabricators.

Figure 4.41 illustrates a numerically controlled tape-wrapping machine for prepreg tapes with both single and dual wrap-head capabilities. Figure 4.42 shows the closeup of the wrapping head.

Resin content, especially in wet winding, is closely related to the winding tension. A certain amount of flow is beneficial in removing trapped air or other volatiles, but excessive tension can produce marked differences in the amount of resin in the inner and outer layers. This is in addition to the effect of winding tension on fiber prestress and mandrel deflection. The optimum winding tension tends to vary from fabricator to fabricator. The limits generally accepted are from 0.25 to 1.0 lb (0.11 to 0.45 kg) per end. It should be noted that heated mandrels lower the optimum tension. Winding tension is the critical parameter in controlling and limiting the void content. Excessive voids are responsible for decreased interlaminar shear strength. Indirectly they cause lower compressive strength and resistance to buckling. It was once thought that interlaminar shear varied with resin content; but if

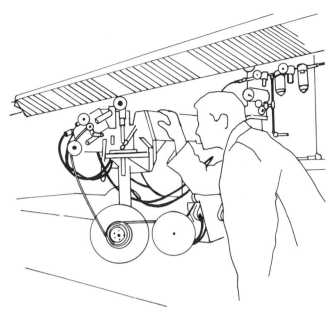

**FIG. 4.41** Automated numerically controlled tape wrapping of full-scale composite-sandwich box-beam structure.

**FIG. 4.42** Numerically controlled head.

the voids are kept at a low level, high interlaminar shear can be obtained over the normal range of resin content.

The optimum strand condition requires the fibers to be *collimated;* i.e., the monofilaments are uniformly coated with resin and in uniform tension. Such a condition is difficult to obtain without specialized impregnating equipment. Fabricators tend to rely on the prepreg or strand suppliers to furnish low-catenary low-twist fibers for even tension. Well-collimated fibers, however, combined with low voids, are known to improve tensile performance. The strand bandwidth is also difficult to control and is mainly a function of the winding tension. A typical prepreg strand will vary in bandwidth from 0.050 to 0.125 in (1.3 to 3.2 mm). Uneven bandwidth results in resin-rich areas between strands, which are structural weak points.

The strand position is a function of winding-machine precision and accuracy in machining the mandrel. Particularly sensitive areas of pressure vessels are at the dome-cylinder juncture and in the end domes, where the tendency for fiber slippage is greatest. Prepregs with sufficient tack, as already mentioned, are less likely to slip than wet-wound fibers. Misplacement of fibers constitutes a weakness and is known to result in premature failure.

Controlled curing cycles are required to ensure sufficient (but not excessive) resin flow and to minimize fiber movement. In critical applications, vacuum-bag and autoclave techniques provide improved control of resin flow and fiber movement, as well as reducing void content.

### Environmental Factors

Filament-wound composite properties are affected by temperature, humidity, weathering, aging, and other environmental conditions. In general, the response to these factors is similar to that of other reinforced plastics. One complication is that the lower resin content of filamentary composites may render them more susceptible to moisture penetration and damage by exposure to high relative humidity. Investigation of moisture and humidity exposures has indicated that filament-wound composites lose little strength in water with no load or under static load. An appreciable loss of strength, sometimes as high as 30% for some composite materials (S glass), occurs after a cyclic exposure to 160°F (71°C) at 95% relative humidity. Both tensile strength and interlaminar shear strength decrease under these conditions. Similar results have been noted for fabric-base laminates, but the strength degradations are not as great as 30%. The deleterious effects of temperature-humidity cycling can be minimized by using superior finishes, less moisture-sensitive resins, and improved cure conditions.

## PULTRUSION

The generic term *pultrusion* was coined for a continuous processing technique which made its debut in the early fifties. At first the process was built almost entirely around the fishing-rod-blank industry, and all production was aimed at making various diameters of round bar from a combination of glass fibers and polyester resin. Three quite different methods for meeting this rod-stock demand evolved almost simultaneously, parallel development of each of these techniques began in that era, and there are current practitioners of each of these methods today.

Probably the most widely known is the *tunnel-oven process* (Fig. 4.43). It uses fiberglass rovings or similar reinforcements, which are drawn through a resin bath and then through a sizing bushing to remove entrapped air and excess resin and to impart the desired diameter. From here the composite is drawn through a tunnel oven and cured continuously in a free state to produce the desired end product.

A decade ago the second approach was not a true continuous process since it employed intermittent or step molding to produce a substantially continuous final product (Fig.

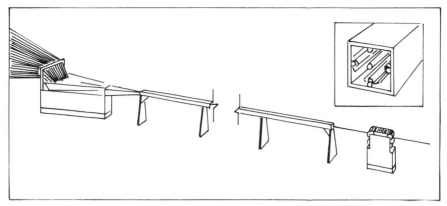

**FIG. 4.43**    Schematic of tunnel-oven pultrusion.[1] *(Goldsworthy Engineering, Inc.)*

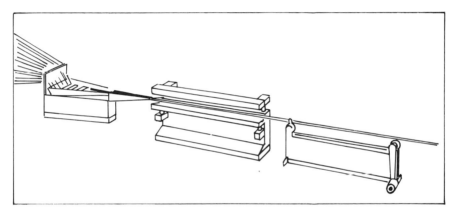

**FIG. 4.44**    Schematic of step-molding pultrusion.[1] *(Goldsworthy Engineering, Inc.)*

4.44). In this instance, the reinforcement was drawn through the impregnating bath and then through a split female die, where it was statically cured by external heating of the die. Usually the input end of the die was cooled at its extremity in order to prevent curing at this point, so that the next discrete length could be drawn into the die as it opened after completing the first cure.

The third pultrusion technique involves the same process of reinforcement impregnation, after which the impregnated material is drawn through a die tube whose inside configuration forms the desired profile. The material is simultaneously subjected to dielectric heating by an energy source of radio or microwave frequency. Cure is so rapid with this type of energy that the complete cure is achieved within the confines of the die tube, even though it is normally no more than about 18 in (457 mm) long.

## Variations of Pultrusion Process

A number of ingenious variations and modifications of the three basic types have evolved over the last several years. The advent of translucent building panels and corrugated and flat sheet and their acceptance by industry have produced the largest single market for polyester resin and glass reinforcement.

Another segment of the plastics industry to convert from batch to continuous processing was the filament-wound pipe industry. Designs for continuous-pipe machines, some involving highly ingenious mechanical devices, reached the hardware stage and a limited number went into actual pipe production. They are limited, however, by the difficulty of providing a mandrel upon which to wind the composite structure and which can either be withdrawn or retained as an integral part of the structure. Four techniques have been refined for experimental production hardware:

1. Feeding metal mandrels end to end through the winding heads, subsequently cutting the cured pipe at the mandrel joints, and withdrawing the mandrel to be fed back through the machine again.

2. Creating a thin thermosetting liner in a primary operation and overwrapping it in a continuous winding head to obviate the need for a mandrel.

3. Utilizing the very short cure zone provided by microwave heating to allow the winding to take place on a fixed mandrel and be continuously withdrawn as it is wound and cured.

4. Putting a thermoplastic extrusion machine in line with the filament winder and overwrapping the extruded thermoplastic tube as an in situ mandrel.

### Advantages and Disadvantages of Pultrusions

Pultrusion is analogous to extruding metal. Solid shapes with constant cross section and unidirectional strengths are formed by running continuous fibers (tow) through an epoxy-resin tank and then through a heated die. The feed material can also be preimpregnated. The technique's main advantage is speed of fabrication. Pultrusion advocates also point out that material utilization is 95%, compared with approximately 75% for manual lay-up. Problems which still crop up in production and have delayed universal acceptance are fiber orientation and resin control. Epoxies tend to stick to the die, resulting in parts of inconsistent quality. Finished resin content is not easy to control.

### Equipment

There are two basic types of pultrusion equipment. In the *wet method* (Fig. 4.4), the fiber is pulled from spools through a resin tank, a wiper-shaper, and a heated die and curing chamber. In some cases, a postcure is required. This conventional method used with fiber glass has been adapted to other low-cost materials. The wet method is normally used where cost is the major consideration, since longer set-up time is required, there is potential contamination, and risk to the structure being formed is greater. Finally, temperature, viscosity, composition, and the quantity impregnated must be carefully monitored.

The second method uses preimpregnated material as the feed-in. This may be in the form of unidirectional tape, but generally combinations of preplied tape with various fiber orientations, 0, 90, $\pm 45°$ (0, 1.6, and $\pm 0.79$ rad) are used to secure the desired strength orientation in the finished product (Fig. 4.45). Woven fabric may be used in sandwich panels. Material costs in this system are higher and the feed system more complex (Fig. 4.46). Process developers and users, however, claim better material uniformity and quality with the second process. Shaping and curing by microwave occurs in a production-line

**FIG. 4.45** Forty-two spools of Gr–Ep prepreg tape ready to be fed into shaper. *(Boeing Commercial Airplane Co.)*

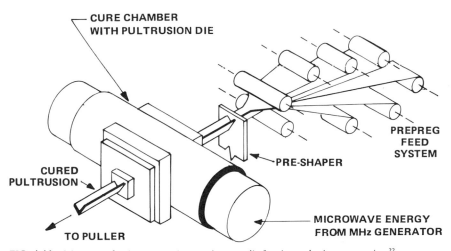

**FIG. 4.46** Schematic of pultrusion equipment that uses die forming and microwave curing.[22]

chamber immediately following a single or multiple preshaper. Each shape requires a single or multiple-part ceramic die costing (1970) from $3000 to $6000. Die material is aluminum oxide. Such accessory equipment as preformers, clamps, and die and feed accessories cost $500 to $2000, depending on size and complexity. Thus, large production runs are essential to keep unit tooling costs low. Die design is critical to successful pultrusion.

Curing manually laid-up parts has conventionally been done in an autoclave. Microwave curing is not as sensitive to part shape as the autoclave. This places no restrictions on part size whereas the microwave cure chamber is the principal factor limiting part size. Every significant increase in part dimensions is usually accompanied by a new cure-chamber design. The problem of die sticking is handled by using a Teflon-impregnated woven-glass slip-tape as the top and bottom ply of the tape feed.

**Material Properties:** Gr–Ep pultrusions cured by microwave show 14 kips/in² (97 MPa) in interlaminar shear, which is equal to that of autoclave-cured laminates. Ultimate flexure tests show 220 kips/in² (10.5 GPa), or slightly more than autoclave parts. Test results with a polyimide matrix have not been as good as for epoxy. Ultimate flexure in particular was far lower for pultruded than for autoclave-cured material. As a result, a two-stage cure is needed to improve polyimide properties. Another matrix material, polysulphone, has also been evaluated; a complex curing system is required, incorporating a set of preheat, shaping, and chill dies. The preheat die operates at 600°F (316°C) with 10 lb/in² (70 MPa) pressure and the shaping die at 550°F (288°C) with 150 to 200 lb/in² (1034 to 1379 Pa) pressure. The part leaves the chill die at 350°F (177°C). These cure temperatures compare with 250 or 350°F (121 or 177°C) for epoxy. Typical types of parts produced include Gr–Ep bar stock up to 0.3 in (7.6 mm) thick and 3 in (76 mm) wide. Small z and hat shapes 0.06 in (1.5 mm) thick and sandwich panels 0.075 in (1.9 mm) thick and 12 in (305 mm) wide have been produced. The honeycomb core was Nomex or fiber glass, and the facings were Gr–Ep or Gl–Ep. Other companies have investigated continuous roll forming, utilizing preplied tape, to produce hat stiffeners by pultrusion (Fig. 4.47).

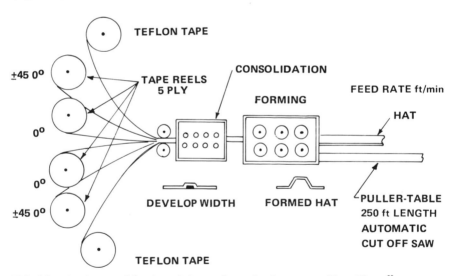

**FIG. 4.47** Continuous-roll forming technique to form pultrusion automated hat stiffeners.[22]

## Pultruded Parts

Whereas a steel beam in an automobile door can weigh up to 17 lb (7.7 kg), tests indicate that a pultruded composite structure can perform similar duty and weigh only 3 lb (1.4 kg). Such substitutes for door beams and bumper reinforcements are being tested, with mixed results. Because current designs in steel depend on the deformation of the beam to absorb a crash and because deformability is not typical of simple fiber-resin composites, door beams will probably have to be redesigned for the material. For example, a new design might combine a Gr–Ep composite with structural foam and steel in the end mountings. Generally, the same design constraints hold true for bumper reinforcements and even bumper faceplates.

Pultrusion process may offer more immediate weight savings in the form of mounting brackets. Pultrusions are currently available in a variety of cross-sectional shapes that can be directly substituted for shapes holding auxiliary components to an engine, for example. The following applications have also been considered by aerospace and automotive engineers:

Tees, 2.5 by 3 in (64 by 76 mm) would serve as 22-ft-long (6.7-m) floor-beam chords in freighter and passenger aircraft. Since decompression buckle is critical in the design of the part, Gr–Ep is an excellent choice, being stronger and stiffer than aluminum, currently in use. The weight savings would be 809 lb (367 kg) per aircraft.

Compartment flooring for future cargo aircraft.

Curved bumper beams for automobiles fabricated from glass-reinforced polyester with graphite reinforcement on the front face. The unidirectional nature of the material presents no problem. There is a weight saving of 40% over the metal beam.

Citizen-band radio antennas of approximately 0.125 in (3.2 mm) diameter fabricated from Gr–Ep.

## PULFORMING

Whereas pultrusion generally can produce only straight, constant-volume products or profiles, using continuous reinforcements, pulforming does not have this limitation. Continuous reinforcements are used in pulforming as in pultrusion, but since other reinforcements can be added, pulformed products can be curved or straight, having either constant volume and changing shape or changing volume and changing shape. This definition does not include pultrusions (commonly defined as profiles) or constant-volume constant-shape products.

### Early Developments

Several basic techniques are common to pulforming and pultrusion: continuous fiber reinforcements are necessary; resins and resin-impregnation methods are identical; heat curing of the resin matrix is virtually the same; and both processes are continuous; i.e., the products emerging from both types of equipment are continuously cured and of unlimited length. The significant difference is that pultrusion produces only straight profiles. Equipment produced during the years of early development (1972) set the stage for curved pulforming, a mechanized process that continuously forms and cures curved profiles. The original machine demonstrated the feasibility of continuously producing by pultrusion methods a flanged hat section 1.5 by 1.5 by 0.75 in (38 by 38 by 19 mm) curved to a 10-ft (3-m) radius. The hat section was continuously "built" and cured in a giant unlimited-length

FIG. 4.48   Curved pultrusion equipment. *(Goldsworthy Engineering, Inc.)*

spiral. Its end use as a cylindrical space-structure rib required the continuous length to be cut at every complete circumference and butt-joined to form the finished rib. Figure 4.48 shows the equipment.

## Pulforming Equipment for Springs

Gl–Ep automobile springs can achieve mileage improvements, saving 81% weight and 60% raw material over a comparable steel spring. A low-labor manufacturing process for composite monoleaf or multileaf vehicle springs is necessary to capitalize on the other advantages of composite springs for weight and cost savings. While springs can be made by filament winding, the process is not continuous, requires separate curing facilities, and is relatively slow. Curved pulforming provides economic production of the large quantities common in the automotive industry. Continuous processing, in-line curing, and production speeds common to pultrusion are important features of curved pulforming (Fig. 4.49).

The constant-volume–constant-shape curved pultrusion monoleaf spring is not particularly useful as an automobile spring. In a monoleaf spring, a better design is a constant-volume–changing-shape geometry in which the center attachment point is essentially square in cross section with flattened rectangular ends, the longest dimension of the cross section being normal to the vertical load on the spring. This type of bow-tie curved spring (Fig. 4.50) can be made on curved pulforming equipment by die-sinking the shape in the circular rotating machine component which becomes the female mold. The male part of the tool is a stationary plate or shoe, curved to the convex radius of the spring.

The components of the equipment include racks for dispensing filament reinforcements, a resin tank for impregnation, roving-guidance tooling, radio-frequency preheater, rotating die set, stationary die shoe, die-heating capacity, and a saw to cut each spring to length.

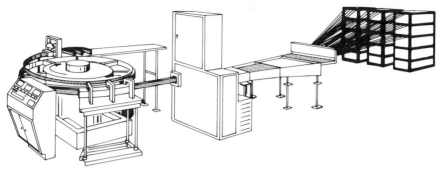

FIG. 4.49   Curved-spring pulformer. *(Goldsworthy Engineering, Inc.)*

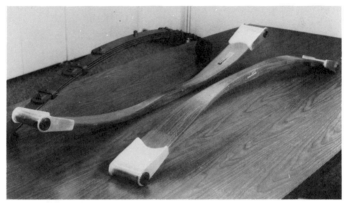

FIG. 4.50   Typical composite monoleaf bow-tie automotive spring with metal counterpart in background.[6] *(Goldsworthy Engineering, Inc.)*

Typically, glass-fiber roving and/or graphite tow are used for reinforcement. Polyester, vinyl ester, or epoxy resin can be used, although the last cures more slowly in production.

To understand the process it may be helpful to trace the machine's operation for a typical constant-radius spring. First, a number of spring cavities are mounted on the rotating table to form a complete die circle. The number of spring cavities depends on the arc length of the spring. The stationary die shoe is brought into close contact with the convex surface of the rotating die-circle face. After the necessary rovings have been threaded through the resin tank, the radio-frequency preheater, the roving-guidance tooling, and the orifice formed between the rotating die and the stationary die shoe, the rotating table on the machine is started. Like a capstan, the rotating table pulls the rovings into the female molds in the rotating die circle. As the resin-wet reinforcement is drawn past the stationary die shoe (which forms the closed cavity with the rotating female mold), the material cures as it does in a pultrusion die. In this continuous process, the cured springs remain in the female rotating dies until they reach the flying cutoff saw, where each is automatically guided out of the die, cut to length, and removed. The die circle rotates the emptied die cavity into position to receive a new charge of material, and the process is repeated. Important considerations in machines of this type are precise guidance of each roving into its proper position and the use of radio frequency for cure initiation because of the relatively large cross sections of material to be cured.[6]

## Future of Pulforming

Since it is unlikely that automotive springs will exceed a 60-in (1.5-m) radius, there probably will be no need for larger machines, but operating speeds are likely to increase. For multiple-stream operation, dies can be stacked vertically on single machines. The next-generation of curved pulformers will include those for the production of other than constant-radius springs. Within reasonable variations from a constant radius, constant-volume curved springs can be produced on similar equipment simply by substituting a "flexible" stationary die shoe to compensate for radius changes in the spring length. While the rotating die component will not be a perfect circle, the compensating die shoe will make it possible to exploit the principle of current machines.

## MATCHED-DIE MOLDING

Matched-die molding is a molding process in which the cure is obtained while the material is restrained between two mold surfaces. It is an automated, high-production process in which the resin can be added to a precut and shaped sheet of mat, fabric, or preform positioned in a heated metal mold. The pressure on such molds is normally applied by a hydraulic press. A female mold or cavity is the concave part of the set of matched dies. The male mold is the convex component. The male mold is matched to the female so that when the dies are mated, a controlled space results.

Matched-die molding processes are the extensions of the bag-molding methods developed as mass-production techniques for an operation which was originally batch molding of a limited number of parts. The design of composite parts, however, is considerably different for matched-mold parts than for bagged parts. The greatest difference is economic: the molds are more expensive but can be justified by the larger number of parts to be produced in a shorter length of time, thus reducing the unit cost.

One common application for matched molds is in manufacturing complex specialized parts where tight tolerances and contour problems necessitate the use of matched molds almost without regard to cost. These parts are used primarily in aircraft and space applications, where the number of units is small, the contours and tolerances are critical, and the weight savings with composites are such that the added cost of matched molds (usually metal) is economically justifiable.

## Matched-Molding Technology

Matched-die processes include:

| | |
|---|---|
| Premix molding | Vacuum-injection molding |
| Prepreg molding | Cold molding |
| Preform molding | Displacement molding |
| Wet-fabric molding | Flexible-plunger molding |
| Sheet-molding compounds | |

Premix molding is used where a large number of relatively small parts [12 in (305 mm) or less] is required. The process suits parts with variable wall thicknesses, inserts, and sharp contour changes. The premixes consist of chopped fibers and resin usually extended with fillers and pigments and are the cheapest of all molding materials. They may have good appearance and better than average chemical resistance, but the mechanical properties are lower than those of continuous-fiber-reinforced parts. The prepregs are among the most versatile of molding materials and result in parts having the highest mechanical strength of such composites. They are limited only by their higher initial cost and by the cost of the molds. The maximum size of the molding is limited only by the size of available presses and molds.

The complexity of current composite technology makes rigid distinctions in molding techniques impractical. Many processes combine several unit operations before the final part is produced. As an example, nozzles for spacecraft engines may be fabricated from asbestos or glass tapes which are first wound on a mandrel, like other filament-wound parts. Before curing, however, they are compacted by autoclave or hydroclave pressure or placed into a matched metal mold and cured in a press.

## Mold Design

The design of matched-molded parts is closely related to the types of molds to be used and the methods of applying pressure. For most laminated or premix molded parts, the pressures are high, ranging from hundreds to thousands of pounds per square inch. The molds, usually plastic or metal, are designed to withstand these pressures and the frictional abrasion of repeated molding cycles. The size of the part is one of the first considerations to be investigated by the designer. If a part is small [24 in (610 mm) or less], a matched metal mold is feasible and the focus shifts to material selection, where the choice is between mat or fabric laminates and chopped premix sheet or premix molding compounds. For parts over 24 in (610 mm) in any dimension matched molds, metal or ceramic, are still feasible, but the use of molding compounds may be impractical and the designer is limited to mat or fabric for reinforcements.

### Molding Materials

Matched-tool molding uses glass and asbestos materials, chopped-glass preforms, and woven, nonwoven, and roving cloth fabrics. These mats must have a binder which is reasonably insoluble in the liquid resin and which will allow the mat to assume the contour of the mold without disintegration or separation of the individual fibers when the resin is added during the molding cycle. Depending on the depth and contour of the mold, the mat or layers of mat may be used flat and assume the final shape during molding or the mat may be tailored into patterns and contoured dry before the resin is added. Figure 4.51 shows the position of the mold in the press, which is important.

The molding technology for continuous-fiber mats is similar to that for wet mats. With impregnated asbestos mat phenolic resin is normally used as a binder, and the parts are designed for high-temperature and ablative applications. Woven or nonwoven fabrics may be used instead of mat when higher strengths are required for finished parts. Fabric laminates usually have higher glass content and are used for nonconsumer items such as equipment housings, aerospace parts, and industrial battery cases. The fabrics can be put into a mold dry and the resin added, as in mat molding. The dry fabric can also be tailored or

FEMALE CAVITY →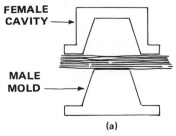

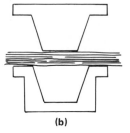

MALE MOLD →

(a)

(b)

FIG. 4.51   Matched molds with flat mat.[1]

sewn into a preform and placed in the mold. This type of the preform is used mainly in vacuum-injection molding. By far the greatest majority of matched-molded fabric parts use prepreg fabrics, i.e., rolls of fabric impregnated with partially polymerized (B-stage) resin and made up into preforms by tailoring a pattern and putting it on a male mold. The use of prepregs ensures product uniformity, adherence to tight tolerances, uniform resin content, and high strength. This molding method is used primarily for smaller parts since the cost of the molds is high and large parts can be molded more economically in an autoclave.

## Mold Processes

### Vacuum-Injection Molding

This process is designed primarily for large complex parts requiring tight tolerances when both inside and outside surfaces must have a smooth finish. Parts with integral partitions, complex inserts, and uneven wall thickness can readily be accommodated by this method. The molds are usually inexpensive plastic molds, and the curing is generally accomplished at room temperature or slightly above. If high curing temperatures are needed for higher strength, thin sheetmetal molds can be used. The main disadvantage of the process is the long time (up to 2 to 3 h) required for each cycle. Economic considerations limit it to applications requiring not more than a few hundred parts.

The dry preform (made of glass fabric, mat, or a combination of both) is placed over the male mold and the female mold placed over it and allowed to close. The molds can be made of any inert material; the only requirements are that they be reasonably rigid and completely nonporous. The male mold usually has a trough or a channel around its base for the resin, and the air-removal outlet at its highest point for the vacuum connection (Fig. 4.52). After the molds are closed, resin containing catalyst and promoter is poured into the trough and a vacuum is drawn, which evacuates the air in the reinforcement. When the resin is running clear, the vacuum is turned off and the resin allowed to set at room temperature or by heat if the mold has heating elements. After the resin has hardened, the back-off bolts are screwed down against the part on the female mold (Fig. 4.52) and the mold pops off and is removed. The part can be removed from the male mold either by using additional back-off bolts or by air pressure through a valve. Parts made by this process are usually homogeneous and nonporous.

The vacuum-injection process has found limited use. The few firms specializing in it in the United States can produce complex, high-precision parts, which include vanes, engine-cowling defrosters, and aircraft radomes.

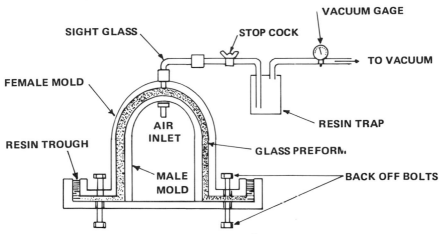

**FIG. 4.52** Schematic of vacuum injection-molding process.[1]

### Flexible-Plunger Molding

In this process, the female cavity is a precision-made mold to the outside contour of the part, while the male is a flexible plunger, usually fabricated from solid rubber. The matched mold is mounted in a hydraulic, cam, or air-operated press. The preform is placed in the female mold, resin is added, and the plunger is slowly lowered until it touches the bottom of the mold. At this point, further motion of the press compresses the plunger, forcing the resin to flow and impregnate the preform. When the resin appears at the top of the female mold, the press is stopped and the resin allowed to cure. The motion of the resin drives the air ahead of it, and the part usually comes out reasonably free of entrapped air. The cycle of this process is necessarily slow since only the female mold is heated. (The plunger is rubber and is difficult to heat.) The advantages are the lower cost of the molds and the ease of molding irregular wall thicknesses. The disadvantages, besides the slow cycle, are a limited number of shapes and the short potential life of the rubber plunger.

### Matched-Metal-Die Molding

This process is chosen for large-scale production and fabrication of parts requiring the highest dimensional accuracy. A wet or a prepreg lay-up or preform is placed on one of the matched molds and the other mold is pressed against it, applying direct pressure on the lay-up. The molds are usually heated, liquefying or reducing the viscosity of the resin and enabling it to wet the graphite, fiber glass, and/or other materials and to squeeze out excess air. Since air is much more soluble in liquids under pressure, and since this condition exists in the mold, much of the air is dissolved permanently in the resin, resulting in a nonporous product; however, since the amount of air that can be dissolved is limited, the mold should be designed to avoid trapping large quantities of air. The pressures in this type of molding vary from 15 to 1000 lb/in² (103 to 6895 Pa) or more, the usual range being 50 to 200 lb/in² (345 to 1379 Pa). Curing temperatures range from 240 to 280°F (116 to 138°C) for polyester, 260 to 350°F (127 to 177°C) for epoxies, and in excess of 300°F (149°C) for phenolics and silicones.

The molds are heated by steam or electricity, whichever is more practical. For reinforced plastics, usually no cooling is required unless the part is so intricate that it distorts if removed from a hot mold; in such cases, the molds may be water-cooled to room temperature before the part is removed. If one of the matched molds is 10 to 30 Fahrenheit degrees (5.6 to 16.7 Celsius degrees) hotter than the other, the part will stick to the hotter half of the mold and removal may be easier.

Articles made by matched molding are denser, stronger, and less porous and have a better finish than parts molded by vacuum-bag methods. The size of the molds is limited only by practical limitations in mold making or press size. Matched molds made from plaster, concrete, and plastics have been used for short runs, but most are made from aluminum, bronze, kirksite, beryllium, various steels, and other alloys. These metals can be machined, cast, hobbed, or electroformed (iron, nickel, or copper) to shape and are normally heated by steam, hot water, or oil and electricity.

## Positive Molds

The basic type of positive mold is shown in Fig. 4.53*a*. A positive mold is one into which the charge of material is put and the full pressure of the press applied. Since thickness of

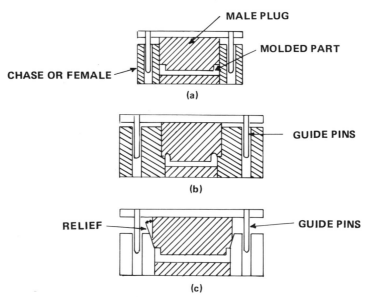

**FIG. 4.53** (*a*) Matched-mold materials and designs, (*b*) positive mold with land, (*c*) positive mold with alignment.[1]

the part is determined by the amount of material in the mold, this type of mold requires precise measurement of the charge. It does not readily permit air or gas to escape from the cavity. Positive molds are mostly suitable for molding epoxies and polyesters, where no gaseous products of polymerization are generated. They are not practical for phenolics, silicones, or polyimides, where water or steam is generated. When a finished rounded edge is required, a landed positive mold (Fig. 4.53*b*) is used. In an improved type (Fig. 4.53*c*) the walls are tapered, reducing friction and wear on the mold without sacrificing full pressure on the part when the mold is closed. Dowels and bushings are of prime importance to keep molds in alignment and prevent excessive wear. To improve wear, hardened steel has been used in molds with chrome plating on the faying surfaces. Continuous wear on positive molds wears the metal away and opens gaps between the male and the female halves. Such gaps reduce the effective pressure on the mold, allow for some loss of material, and increase further wear. Positive molds normally produce individual or single parts where the charge is placed into the mold cold and the entire heating and curing cycle takes place in the mold cavity.

## Transfer Molds

The transfer-molding process is used for rapid production in single and multiunit molds. Preheating and liquefaction or plastication of the charge are carried out in a separate chamber, and then the fluid material is forced by ram pressure into the actual mold through a narrow orifice. This type of molding is used for intermediate and small parts, for large production runs, and for parts with inserts and thin walls where a very fluid resin stage is required. If the parts are small, one preheating cavity can service several molds. This operation is similar to injection molding of thermoplastics except that a separate charge must be placed in the preheating cavity for each molding cycle and the whole system must be purged of cured resin flash before recharging.

The special transfer press used in this method has the usual double-acting ram to open and close the mold and a secondary ram to push the heated charge from the preheater into the final mold. An elementary type of transfer mold is shown in Fig. 4.54. The cycle starts

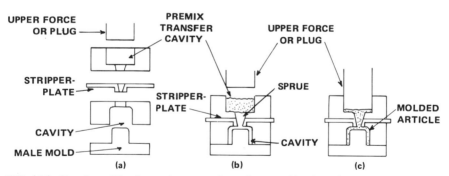

FIG. 4.54 Transfer molding for powders or premixes only: (*a*) mold and transfer cavity open, (*b*) mold closed, transfer cavity open, (*c*) plunger forcing preheated composite into mold.[1]

with the mold and preheating cavity open (Fig. 4.54*a*). The molded part from the previous cycle is removed at this point, the mold is closed (Fig. 4.54*b*), and the charge is placed in the preheating chamber. The upper force or plug is lowered to touch the charge and the preheating cycle is started (Fig. 4.54*c*). When the material has been sufficiently softened, additional pressure of the plug forces it down into the cavity around any required inserts. When the resin has polymerized, the mold is opened, the stripper plate breaks off the sprue from the preheating chamber, and the part and the waste material from the transfer cavity are removed.

### Open Flash Molds

Positive molds are generally used to premix molding of chopped glass–resin formulations where a preweighed charge is placed in the mold and allowed to flow out to the final contour. For molding sheet materials or blown preforms the open flash mold (Fig. 4.55)

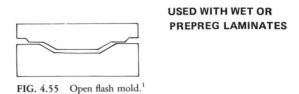

**USED WITH WET OR PREPREG LAMINATES**

FIG. 4.55    Open flash mold.[1]

is required. It is used for sheet materials, fabric or mat, dry or prepreg. Oversize patterns are used, and excess resin and air are squeezed out during molding. When the cured part is removed from the mold, it requires machining to finish the edge and remove the extra material. This type of mold is the most inexpensive, has minimum wear, and can be fabricated from many materials besides metals. Ceramics and plastics have been used successfully, the only problem being the poor heat transfer. Labor costs for this type of molding are high and for any sizable production a modified mold, the cutoff mold, must be used.

### Cutoff Molds

This modified version of the open flash mold has a shearing edge to cut excess off the preform during the last stages of mold closing. Figure 4.56 shows a simplified cutoff mold used for shallow parts. For deeper parts requiring mechanical removal from the mold, the type with a stripper plate (Fig. 4.56) is used. The mold temperature and mold contour are

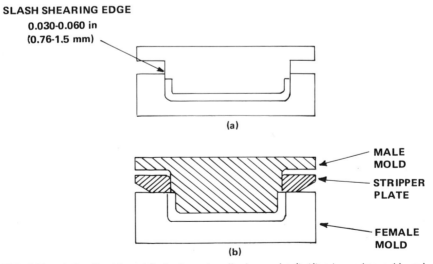

**SLASH SHEARING EDGE**
**0.030-0.060 in**
**(0.76-1.5 mm)**

(a)

MALE MOLD

STRIPPER PLATE

FEMALE MOLD

(b)

FIG. 4.56    (a) Cutoff mold, used for laminates (usually chrome-plated); (b) stripper-plate mold, used for premix or laminates.[1]

adjusted so that the part comes off onto the male and is removed by action of the stripper plate. The shearing edge is usually 0.030 to 0.060 in (0.76 to 1.5 mm) deep, and the lateral spacing of the gap between the shear edges is 0 to 0.003 in (0 to 0.08 mm). The cutting edge is normally hardened or chrome-plated for longer wear. The slight positive action at the end of the stroke provides the necessary backpressure and ensures a sound molding. All blown preform molds are a cutoff mold in one form or another.

### Complex and/or Combination Molds

Numerous modifications of these simple basic molds are possible. Parts with undercuts can be molded by using a separate hydraulic cylinder or pressure clamp. Four-sided moldings can be made by using two side plungers in addition to the vertical ram. There are no rules for these types of molds, and their design is limited only by the ingenuity of the mold engineer. In general, these molds are used only for molding prepregs which maintain their shape during any movement of the mold. Using these molds naturally is more expensive and takes longer than simple molds. Many companies find it faster to mold smaller components separately and assemble them using adhesives or fasteners.

## PRESS MOLDING

Press molding, one of the earliest techniques for molding and curing plastics, was long overlooked as a practical process for fabricating lightweight composite structures with highly oriented continuous-fiber reinforcements. Yet despite the higher initial tooling costs, this is one of the most practical and efficient methods. Lower unit production costs allow tooling costs to be amortized and recovered in even moderate production runs.[7]

Press-molding techniques include matched-die molding (already discussed), elastomeric-plug molding, and matched-cavity molding, which offer the following significant advantages:

Finished or near finished parts with close tolerances

Faster rates than most other processes

Part-to-part uniformity and reproducibility

Better material properties due to higher curing pressures

Excellent control of fiber-to-resin ratios and void content

Shorter process cycle times resulting from use of preforms

Avoidance of heat lag during cure since part is heated by mold directly

Loading and unloading usually with mold at operation temperature

The different press-molding processes do not depend on operator skill; labor is used chiefly in preparing and loading preforms and unloading cured parts, which require only deflashing or light trimming for finishing. The processes lend themselves readily to automation, since most presses can be equipped with controllers to regulate press closure, dwell, pressure, temperature, and press opening. Molds are fabricated with part ejectors or parting lines to facilitate removal of the cured parts.

The three types of press molds are illustrated in Fig. 4.57. Elastomeric-plug molding, as the name implies, uses an elastomeric plug slightly oversized in one dimension. As the

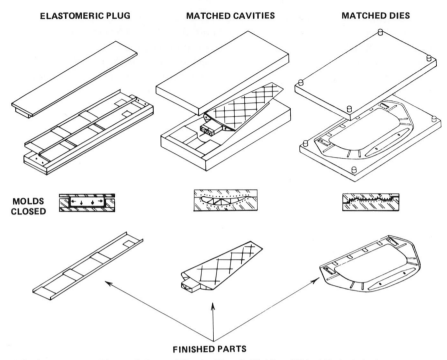

ELASTOMERIC PLUG     MATCHED CAVITIES     MATCHED DIES

MOLDS CLOSED

FINISHED PARTS

**FIG. 4.57**  Press-molding techniques. *(Sikorsky Aircraft, Division of United Technologies.)*

mold is closed, the plug is deformed and fills the cavity, pressing the composite material against the face of the mold. A variation of this technique, thermal expansion or captive elastomer molding, depends upon thermal expansion of the rubber for pressurization.

## Press-Molding Resins and Reinforcements

The specific requirements of resins and reinforcements for press molds are as follows:

1. Low viscosity during molding cycle, in order to wet reinforcements and fill in the cavity of mold (200 cP or lower for polyesters, under 500 cP for more viscous epoxies and phenolics)
2. Fast curing cycle at elevated temperatures to reduce the time in the mold (for polyesters, 1 to 3 min: longer for epoxies and phenolics)
3. Optimum cure in a short time, eliminating postcure
4. Rigidity at high temperature, enabling the molding to be removed from the mold without fracture
5. Dimensional stability, minimum warpage on cooling
6. Good surface appearance of the final part (no fiber pattern)
7. No adhesion to the mold (easy release)
8. Toughness and resistance to crazing

9. Compatible surface treatments to ensure proper wetting by resin
10. Binders for glass mat or preforms insoluble in resin and capable of restraining fibers from moving

Polyesters are used for general applications for lower cost, quick molding, chemical resistance, and attractive appearance. Epoxies are used for higher strength and dimensional stability. Phenolics are used mainly for higher-temperature applications.

## Elastomeric Press Molding

Press molds have been constructed from aluminum and steel; the designs compensate for the thermal coefficient of expansion for a curing temperature of 250 or 350°F (121 or 177°C). Shown in Fig. 4.58 is a press-laminating elastomeric mold. The male portion of the die is made of cast silicone elastomeric compound, and the base is made of aluminum, 0.75 to 1 in (19 to 25.4 mm) thick. The cavity is composed of machined-aluminum fences, which are secured with a minimum number of bolts and use a steel key to matched keyways in the base, making it possible for the fences to absorb side-load molding pressure.

Resin traps under the fences are machined into the base plate to provide a path for resin flow, thus protecting the bolts and keyways from excess resin. Cast elastomeric inserts provide the hydraulic molding pressure required to cure the laminate. The curing heat is

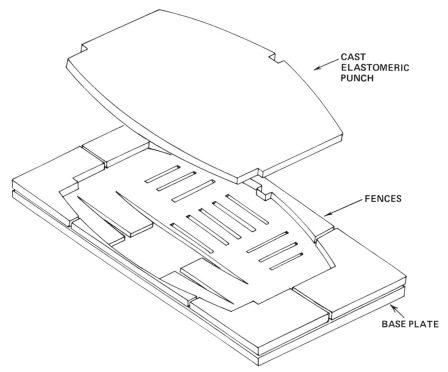

CAST
ELASTOMERIC
PUNCH

FENCES

BASE PLATE

**FIG. 4.58**  Press-molding elastomeric tool. *(Sikorsky Aircraft, Division of United Technologies.)*

derived from the heated press platens. After curing, the mold is removed from the press, the fence bolts are removed and lifted from the mold base, the part and elastomeric punches are removed, excess resin is removed from the tool, and the mold is reassembled for the next run.

## *Applications*

Composites have been proposed as reinforcement for truck-frame rails in an attempt to replace traditional steel reinforcing sections on the flanges of C-sectioned steel rails with Gr–Ep composite. Composites increase stiffness, provide good resistance to fatigue failure, and increase the buckling-load capability; tooling costs are reasonable. Reinforcement sections are formed from graphite prepreg, a single layer of collimated high-modulus [50,000 kips/in$^2$ (345 GPa)] fibers in a prepregged epoxy resin that cured in 1.5 h at 250°F (121°C) (Fig. 4.59). The molding process involved encasing the prepreg within silicone-rubber-faced steel plates so that differential thermal expansion during cure would provide pressure to consolidate the composite as the resin was warmed to its normal flow temperature. Steel pins driven through the prepreg lay-up provided holes for bolted attachments

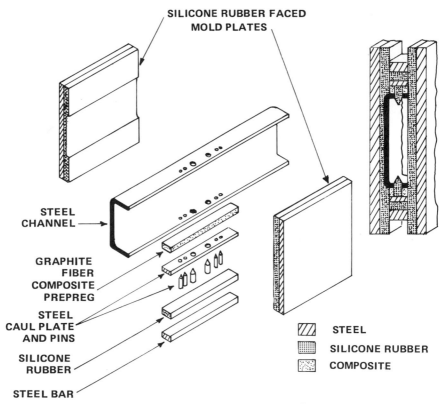

FIG. 4.59   Parts produced by press molding using elastomeric tooling.[25]

so that no subsequent drilling or punching of the reinforcements was necessary. The weight of the channel's reinforcement was reduced by a factor of 10 compared with a steel reinforcement.

Other typical press molded parts include:

Avionics shelves laid up and cured between two flat patterns

Bulkheads using cast elastomeric rubber and inserts used to

form return flanges

Cargo floors made by a multicomponent processing technique (co-curing and cobonding)

Landing-gear frames using elastomeric tooling blanks

A typical press-molding sequence used to fabricate several of these components started with the preparation of materials. Flat patterns of composite prepreg were cut from broad goods by a high-speed reciprocating knife or roller presses using steel-rule dies. Segment kits developed using these flat patterns were subsequently laid up directly into the mold.

Figure 4.60 shows a closed-loop conveyor system to carry the molds for a landing-gear bulkhead through a series of work stations, i.e., lay-up, tooling insertion, press molding, and unloading. The press-mold dies for each appropriate part are individually prepared for molding. The lower portion of the die set is routed by the conveyor system to the first work-station, lay-up. At the lay-up station precut flat patterns and three-dimensional woven preforms are transferred from the kit boxes to the mold cavity. The mold is then transferred by the conveyor system to the tooling station, where thermal-expansion inserts and cauls are properly positioned. The laid-up assembly is now ready for molding. The press-molding machine itself is an integral part of the conveyor system, virtually eliminating any need to handle or transfer molds. The press is preheated and ready to accept the mold upon transfer via the conveyor system. Following the molding operation the finished part is transferred to an unloading station, where it is removed from the mold. The mold cavity and associated tooling are cleaned and transferred to the lay-up station for another press-mold cycle.

## Matched-Cavity Press Molding

Matched-cavity molding is used mainly for built-up shaped assemblies and sandwich structures. Kv–Ep has been successfully used in fabricating covers in matched hot dies using plugs and shaped preforms.

## THERMOPLASTIC FORMING (THERMOFORMING)

Thermoforming consists of heating a thermoplastic material to its softening point and forcing it against the contour of a mold. The process, which is analogous to stamping, is readily adaptable for fabrication of large secondary-structure airframe parts and economical because of the short cycle time, good material utilization, and elimination of labor intensiveness.

Fiber-reinforced thermoplastic resins such as ABS, nylons, polycarbonates, and polysulfones and thermoplastic esters such as polybutylene terephthalate, and PET are particularly

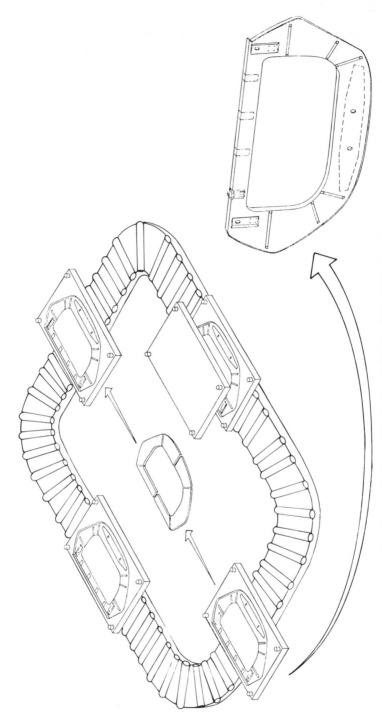

**FIG. 4.60** Press-molding flow plan for landing-gear bulkhead. (*Sikorsky Aircraft, Division of United Technologies.*)

well suited to this method since they exhibit good melt-flow characteristics even with fiber reinforcements and filler contents of 40% or more. No degradation of properties occurs in forming as long as the processes are performed below the thermal-degradation temperature of the fibers and matrix resins. Since thermoplastics have no sharp melting points, an increase of temperature causes gradual softening of the sheet. At set temperature the thermoformed piece can be removed from the mold without warping (Table 4.8).

**Table 4.8**   Thermoforming Temperatures for Different Thermoplastics

| Material | Set temperature | | Forming temperature | |
|---|---|---|---|---|
| | °F | °C | °F | °C |
| Acetate | 160 | 71 | 260 | 127 |
| Styrene | 185 | 85 | 260 | 127 |
| Polycarbonate | 280 | 138 | 335 | 168 |
| PVC | 150 | 66 | 200 | 93 |
| ABS | 185 | 85 | 260 | 127 |
| Polybutylene terephthalate (estimated) | 380 | 193 | 450 | 232 |

Parts are formed by heating the sheet of reinforced thermoplastic to a temperature above the glass transition temperature of the resin and then mechanically forming the sheet into a mold or over a mandrel. The formed parts cool to a temperature below the plastic range and are removed from the mold. At this point the part is still hot and is placed on a fixture to cool to room temperature.

Cycle times do not depend upon chemical reactions and are typically short (seconds or minutes). The reinforced-thermoplastic material blanks are precut to size and heated to the forming temperature before being placed in the mold. Inexpensive fixturing prevents distortion while the part cools outside the mold. Mold temperature is usually maintained below the glass transition temperature of the resin. For unusually complex shapes, the mold temperature can be raised to a higher temperature and then lowered at a controlled rate to minimize internal stresses in the part before it is removed from the mold.

## Thermoforming Methods

There are at least a dozen methods of thermoforming: vacuum-assisted forming, pressure forming, plug-assisted forming, drape forming, matched-mold forming, slip forming, and free forming. The most promising for airframe components is pressure or vacuum forming into a female mold and adaptations of matched-mold forming and pressure-plug-assisted drape forming. Simple drape forming is not suitable since the high loading of continuous-fiber reinforcement severely hinders forming by this method, especially the forming of complex compound curvatures.

Matched-mold forming and pressure-assist forming differ only in the method of applying pressure. In the latter the rigid male mold plug is replaced by an elastomeric plug or pressure bag, depending upon the shape and geometry of the part. When an elastomeric plug is used, it is oversized in one dimension so that it will deform to fill the mold cavity and exert pressure on the part when the mold closes. The techniques of thermoforming are illustrated in Fig. 4.61.

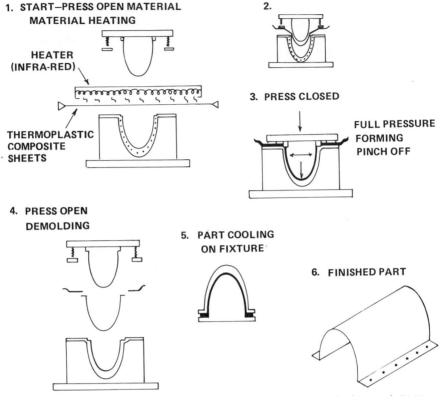

1. START—PRESS OPEN MATERIAL
   MATERIAL HEATING

HEATER
(INFRA-RED)

THERMOPLASTIC
COMPOSITE
SHEETS

2.

3. PRESS CLOSED

FULL PRESSURE
FORMING
PINCH OFF

4. PRESS OPEN
   DEMOLDING

5. PART COOLING
   ON FIXTURE

6. FINISHED PART

**FIG. 4.61**  Sequence of thermoforming reinforced-thermoplastic composite. *(Sikorsky Aircraft, Division of United Technologies.)*

## Thermoforming Molds

Thermoforming molds can be made of most common materials. Cast aluminum is favored, although cast, filled polymers of the phenolic, polyurethane, epoxy and polyester types may also be used. Gypsum, wood, and machined-steel molds have been used successfully. Female molds provide easier release and produce thicker and stronger edges or rims. They also give sharper definition of outer surface detail and are less likely to incur gouge or scratch damage on the mold surface. Male molds allow deeper draw than female molds, but good plug assist will offset this advantage.

Thermoforming molds used with vacuum or air pressure must be perforated with holes or channels so that air can be evacuated or pressure buildup minimized. The female mold requires a 2 to 3° (0.035 to 0.052 rad) draft angle for best release; the tendency to shrink away from the mold upon cooling favors thermoforming of parabolic or tapered shapes. Allowable variations in sheet thickness should be no greater than 4 to 8%; otherwise uniform heat-up is harder to attain in the sheet and cycles are longer. Keeping sheet thickness within the tolerance limits helps suppress blowouts or thin spots in the part.

## Thermoforming Process

Parts fabricated by this process are typically of high quality and easy to duplicate. The tooling normally incorporates the removal of flash and any excess material during the forming operations. The following typical secondary helicopter structures can be thermoformed more economically than they can be produced by conventional laminating techniques:

Nose fairing

Main-rotor pylon fairing

Engine cowling and air intake

Engine cowling aft

Tail-pylon tip cap

Tail-rotor gear-box fairing

Tail-cone fairing

Main and nose landing-gear strut fairing

Forward pylon work platform

A typical manufacturing sequence begins with cutting flat patterns from thermoplastic composite sheet stock nominally 0.040 to 0.080 in thick (1 to 2 mm). The cut details are then assembled with the help of templates and joined together temporarily by ultrasonic tack welding. At the molding station the pattern assembly is placed between a preheated mold cavity and a heated platen. The flat platen is preheated just above the thermoplastic heat-distortion temperature in order to make it pliable for molding and to prevent cracking and tearing (Fig. 4.61). When the flat pattern reaches the established temperature, the heating element is removed and the matched metal molds are closed. Pressure is applied, and the part is simultaneously formed to its final shape and trimmed of excess material. Rubber plug punches help expand the part into the cavity where necessary. At the completion of the cycle the mold is opened and the part is removed and transferred to a cooling fixture. The mold remains at temperature and is ready to accept the next flat pattern. When the molded part has cooled sufficiently, it is removed from the cooling fixture and is ready for final fit-out.

The tail-rotor drive-shaft cover (Fig. 4.62) is typical of components fabricated by this process. The part is molded in three identical sections, which are subsequently joined on assembly for lower die cost and better mold utilization. The unstiffened part is lightly loaded, has moderate curing, which is amenable to single-stage forming, and can tolerate the local fiber crimping inherent in the process.

## STAMPING

Parts are being produced from plastic sheet today faster than they can be injection-molded from pellets or compression-molded from wads of SMC. Thermoplastic sheet of several kinds can be stamped virtually like sheetmetal, with hydraulic presses, matched-metal dies, and a little heat. Large, complex parts that require little or no subsequent finishing can be formed, and scrap, when there is any, can often be reused in the production of new sheet. Production rates are 200 truck fender liners per hour, for example. The rate is 2, 3 or more times higher if multiple dies are used.

Plastic stamping competes not only with metal stamping but with injection molding and compression molding of sheet-molding compounds. Overall fender-liner production is faster than with injection molding, and compression molding of SMC is no match in speed since these thermosets are cured in the dies, necessitating longer dwells. Stamped parts can

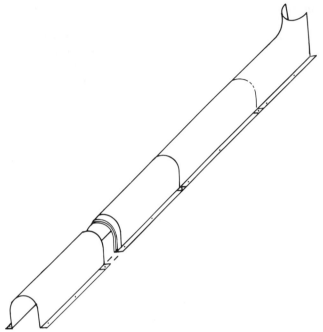

**FIG. 4.62**   Tail-rotor drive-shaft cover. *(Sikorsky Aircraft, Division of United Technologies.)*

be light in weight, strong, rigid, tough, and corrosion-resistant at reasonable cost. These features, combined with their economical production, are leading to many applications, primarily for weight and cost savings, e.g., front-end auto retainers or support-fascia bumpers, shells for hinged bucket seats, and oil pans.

## Stampable Plastic Materials

For engineering applications, several plastics can be stamped, e.g., unreinforced polypropylene copolymer, polypropylene 40% filled with calcium carbonate, glass-reinforced polypropylene, and glass-reinforced nylon. Which is chosen depends on the performance requirements of the part to be produced and material cost. Polypropylene copolymer–polypropylene blended with polyethylene for greater toughness is the least strong, rigid, and heat-resistant, but is quite tough, extremely light, and the lowest in cost. If a tensile strength of 3 kips/in² (21 MPa), a flexural modulus of 105 kips/in² (724 MPa), and a heat-deflection temperature of 115°F (46°C) at 0.27 kips/in² (1.9 MPa) suffice, it is usually the choice.

Filling polypropylene with 40% calcium carbonate does not alter tensile strength but more than doubles flexural modulus [250 kips/in² (1724 MPa)] while increasing heat-deflection temperature to 140°F (60°C). The automotive industry is the major user of stampable polypropylene; auto manufacturers produce their own sheet and stamp most of their parts. Reinforced polypropylene is far stronger, more rigid and heat-resistant, and somewhat tougher. Containing 40% long-glass fibers, it provides a tensile strength of 10.7 kips/in² (74 MPa), a flexural modulus of 800 kips/in² (5516 MPa), and a heat-deflection temperature of 310°F (154°C). Still stronger, more rigid and heat-resistant but heavier and costlier are two grades of nylon 6 sheet. Reinforced with 34% glass, in the form of both long and short fibers, it provides a tensile strength of 19.5 kips/in² (135 MPa), a flexural modulus of 1420 kips/in² (10.1 GPa), and a heat-deflection temperature of 432°F (222°C).

## Stamping Process

Sheets extruded in widths of 80 in (2032 mm) and thicknesses of 0.050 to 0.100 in (1.3 mm to 2.5 mm) are cut to required length and taken to the stamping line (Fig. 4.63). The blanks are heated on platens to 280 to 300°F (138 to 149°C), then automatically placed on the die cavity and formed to shape by a mating die fixed to a single-action ram. Dwell time after bottoming is 10 to 20 s, depending on sheet thickness (the thicker the sheet the longer the dwell). To cool the part during dwell, the dies are cast in aluminum and contain integral passages for water cooling. During forming, the sheet flows and thins out, but thin-out rarely exceeds 15%.

Stamping reinforced plastics is different. Not only are precut blanks heated to above their melting temperature but the blanks are formed completely within the dies; no draw ring is used. The dies incorporate the concept of variable volume (as opposed to the fixed volume of injection-molding dies) and are heeled to protect telescoping shear edges or shutoffs. Stops are used to preclude metal-to-metal contact, ensuring that all available pressure from the press will be applied to the blank rather than to stops or pressure-relief devices. The dies also provide their own guidance on closing and sustain any lateral forces that may arise during stamping. To enable displaced air to escape as the press closes, the dies are vented through grooves on the ejection pins. Air escapes can also be provided through the telescoping shear edges.

Although sharp corners, ribs, bosses, holes, and inserts can be formed, the long-glass fibers that optimize the mechanical properties of the material do not flow easily into long thin ribs; hence thick ribs are preferred.

Glass-reinforced polypropylene has found the broadest use. Some of the most significant automotive applications are the shells for bucket seats, front-end retainers, and front fascia-support panels for soft-fascia bumpers (Figs. 4.64 and 4.65).

## REACTION INJECTION MOLDING AND LIQUID INJECTION MOLDING

The various high-pressure impingement-mixing systems are more widely classified as *reaction-injection molding* (RIM). Compared with low-pressure molding and mechanical mixing, RIM offers the following advantages:

1. Since the mixing head is self-cleaning, no solvent flush is required, as it is in low-pressure molding.
2. Since there is no mechanical mixing, outputs are higher and faster-reacting urethane mixes can be used.

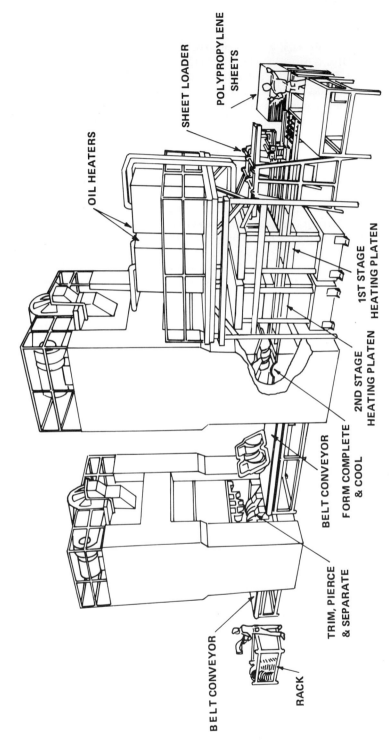

OIL HEATERS

SHEET LOADER

POLYPROPYLENE SHEETS

1ST STAGE HEATING PLATEN

2ND STAGE HEATING PLATEN

BELT CONVEYOR

FORM COMPLETE & COOL

TRIM, PIERCE & SEPARATE

BELT CONVEYOR

RACK

FIG. 4.63  Stamping line for thermoplastics. (*From Am. Mach., June 1981, p. 131.*)

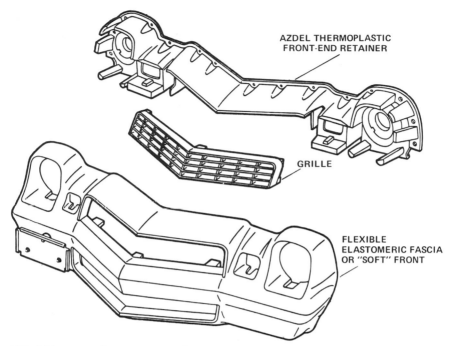

FIG. 4.64    Front-end retainers. *(PPG Industries, Inc.)*

FIG. 4.65    Front fascia-support panels. *(PPG Industries, Inc.)*

3. Since reaction time is faster, mold residence time is reduced because the material cures more rapidly, in turn reducing mold cycle time by as much as 75%.

4. Since high-pressure impingement mixing reduces air entrapment in the reaction mixture, part appearance is improved and surface defects eliminated.

Compared with injection molding for the manufacture of such parts as automotive bumpers, the advantage of RIM lies in the fact that mold pressures are very low and less costly molds can be used. On some runs, in fact, aluminum-filled epoxy tools have been used. On longer, quality runs, however, nickel or chrome-plated steel is generally recommended, but big savings in machinery can be achieved. Whereas injection molding requires high-torque extruders and very-high-tonnage presses to process high-viscosity melts and mold large parts, RIM can achieve these goals with much less expensive equipment.

RIM is suitable for making flexible, semiflexible, or rigid materials and has been used for producing solid, microcellular, or foamed parts. Current RIM applications use urethane exclusively, but better properties are needed. The search has begun for new materials with better heat resistance, higher mechanical properties, and improved impact resistance. Epoxies, polyesters, vinyl esters, nylon 6, and even interpenetrating polymer networks are all in the running. With their versatile chemistry and a wide range of properties, epoxies may be the best bet. The prime advantage of an epoxy system is higher service temperatures than urethanes. Unreinforced epoxy has both a high flexural modulus and a high tensile modulus. The high-modulus polyurethanes are generally lower in modulus than the best epoxy systems.

## RIM Materials

Epoxy RIM materials may be a little less sensitive to mix-ratio variations than polyurethane; mechanical properties are not very sensitive to a mix-ratio variation, unlike the case for some urethanes.

### Reinforcements

Reinforcements are necessary. Since epoxies are thermosets and highly cross-linked, impact strengths are in the range of 0.5 to 1 ft·lb/in (27 to 53 J/m). Reinforcements are necessary for high modulus; certain types of reinforcement, e.g., high-aspect-ratio fibers, are very useful in boosting the impact strength. In most cases, they help boost other mechanical properties as well.

Addition of small particulate fillers with small aspect ratios, e.g., milled glass, wollastonite, and processed mineral fibers, increases the modulus but can lower impact strength. Other advantages of small-aspect-ratio reinforcements include lower shrinkage and thermal expansion coefficient. Other epoxies, in conjunction with glass-mat reinforcements, produce RIM moldings with good flexural and impact properties and good appearance. Two different epoxy resins have been molded in 3-min cycle times at 300°F (149°C).

New developments include combined reinforcements for epoxy RIM molding. Foam cores or combination reinforcement packages are placed in the mold before injection, in a technique called *ultimately reinforced thermoset reaction injection* (Fig. 4.66). Parts with up to 78 continuous fibers have been produced. Reinforcement packages have been designed with E-glass, graphite, and Kevlar fibers and combinations of them, as well as with foam cores. Unidirectional, longitudinal, bidirectional, and/or bias-oriented fibers can be mixed in any proportions.

A good example of the process is an 11-ft-long (3.4-m) propeller blade, two of which make up a 22-ft (6.7-m) propeller. The blade, an easy part to resin-inject with high reinforcement, was originally aluminum. The part is now in production by compression molding. More recently, foam-cored versions have been produced with a resin-injected composite skin, using a two-component epoxy (Fig. 4.67).

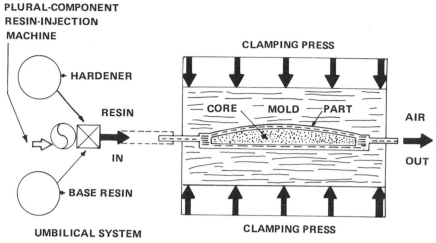

**FIG. 4.66** Reinforced thermoset resin injection with fiber combinations (glass, graphite, and Kevlar).[26]

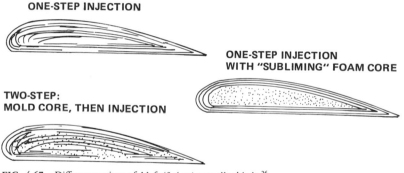

**FIG. 4.67** Different versions of 11-ft (3.4-m) propeller blade.[26]

Parts with or without cores and with transitions from solid laminate to a varying-cross-section core and back to solid laminate have been produced. Parts like these may find applications in the automobile industry for bumper beams, transmission-support beams, and body parts. Other developments include hockey-stick blades; water skis fabricated from a resin-injected glass-wrapped two-component epoxy with a thermoplastic foam core; and skateboards from two-part epoxy.

## Liquid Injection Molding

Two LIM techniques are not quite the same as RIM but accomplish essentially the same end and are used for much the same purpose. The first involves feeding the reactive components into the mixing chamber, where they are acted upon by a mechanical mixer (instead of using high-pressure impingement mixing). Unlike low-pressure mixing, the mixer does not normally need to be flushed since a special feed system automatically dilutes the residue in the mixer with part of the polyol needed for the next shot, thereby keeping the ingredients from reacting. Higher-viscosity materials can be handled in this way. The second departure involves special urethane formulation so that very large, thick-sectioned parts [up to 6 in (152 mm)] can be processed on very short cycles (1 min or less). The end product is neither a foam nor an elastomer but a cross-linked rigid solid urethane plastic.

## BLOW MOLDING

Industrial blow molding is just beginning to approach its potential. Engineering resins are now being blow-molded more often and in interesting applications. Designers have the option of choosing irregular shapes for such components as automobile gas tanks; parts can be manufactured to fit the available space and often need not even look like containers. Small parts, e.g., Christmas tree bulbs of butadiene styrene, are manufactured by injection blow molding, a process used almost exclusively for containers. Extrusion blow molding of polysulfone and other engineering resins has been proposed for such industrial components as electric-motor housings, traditionally made of sheet metal. Use of mineral and fiber reinforcements for strength and impact improvements, already in use for many plastic parts, is in the developmental stage for blow-molded parts. Polyethylene with 55% calcium carbonate has been successfully run. Blow molding of fiber-reinforced pressure vessels deserves exploration.

## KNITTING

Fiber-glass-reinforced materials, originally introduced as a replacement for woven roving, are manufactured by knitting rather than by weaving. Modification of the knitting process has resulted in material which can be produced as either a uniaxial fabric or biaxial fabric and can be cut and handled without falling apart. Since knitting does not crimp the rovings as weaving does, mechanical properties have been increased as much as 50% over similar woven reinforcements, resulting in molded products that weigh less and cost less. Fabrics are now being developed which are custom-tailored for each reinforced-plastic-molding process.

## THREE-DIMENSIONAL WEAVING

Reinforced detail preforms with unique fiber orientations and properties have been fabricated directly by three-dimensional weaving. Structural details and substructural elements for concentrated-load fittings and high-stress applications demand consistent high quality, strength, and uniformity. These fittings can be made from preforms woven three-dimensionally, which make it possible to build composites with fiber reinforcement not only in the $xy$ plane, but through the $z$ direction as well. The process, operating in a semiautomated, almost continuous manner, is not limited by operator skills. Fibers or rovings are woven into controlled orientations, building shaped details at significant savings in weight and volume. The resulting products are exceptional in performance, since the fibers are oriented and arranged for maximum efficiency of reinforcement.

## Three-Dimensional Reinforcement

Fiber alignment, which gives reinforced composites their exceptional mechanical properties, is also responsible for one of their most serious drawbacks—low strength in the direction perpendicular to fiber orientation. In flat panels subjected to bidirectional loads, the solution is relatively easy—cross-ply lay-up, i.e., alternate layers of fibers aligned perpendicular to each other. Parts in actual use, however, seldom see simple bidirectional loads; forces often are more complex. Stresses in the third direction must be absorbed by relatively weak matrix materials. Thus, materials engineers and designers working with composites must take the possibility of delamination into account. However, three-dimensional reinforced composites overcome delamination problems, increasing freedom of design.

A wide range of properties from isotropic to anisotropic is possible with three-dimensional weaving since the types of fibers and number of fiber ends can be varied in each of the $x$, $y$, and $z$ axes. Off-axis yarns can be added obliquely to the $x$, $y$, and $z$ axis if so desired. Similarly preforms for parts requiring long gradual buildups can be fabricated by stepping or layering the weave or by stacking woven preforms. These degrees of freedom can be used individually or in almost any series of combinations to build materials with unique mechanical properties unobtainable by other means (Fig. 4.68).

## Materials and Equipment

Any continuous filament can be used for reinforcement. Three-dimensional composites have been fabricated from glass, boron, and graphite fibers. Matrix materials include epoxy, phenolic, fluorocarbon, polyimides, and other resins with low viscosity at temperatures between 150 and 200°F (66 and 93°C) and reasonably long pot life. Solvent resins can be used, but special procedures are required to eliminate voids caused by evaporated volatiles.

Three-dimensional composites usually are formed by fabricating a three-dimensional structure of reinforcement and impregnating it with the resin matrix, usually by vacuum impregnation. Several techniques have been developed to form three-dimensional reinforced structures. A number of them involve weaving on specially designed looms that interconnect fabric layers in various ways, namely multilayer looms, rapier looms, and special machinery combining braiding, weaving, and winding.

The three-dimensional preform is normally woven from dry fiber for subsequent impregnation, using materials in their least expensive form, fiber and resins. Different weave geometries and shapes, e.g., tapers, are produced readily. The quality of a particular shape is generally limited more by processing principles for fiber placement and equipment than by fiber orientation alone.

### *Textile Forms*

Various types of textile forms have been tried and used for three-dimensional reinforcement of composites:

*Multiple warp:* Warp yarns of each layer are interlocked with filling yarns of adjacent layers. Thicknesses from 0.080 to 0.500 in (2 to 12.7 mm) are available and possibly 1 in (25.4 mm).

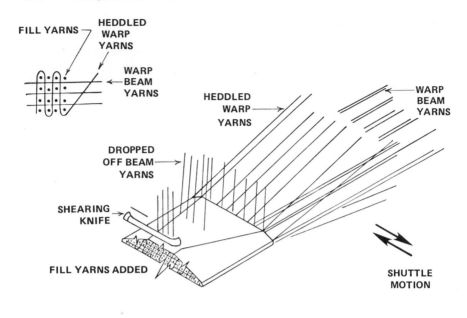

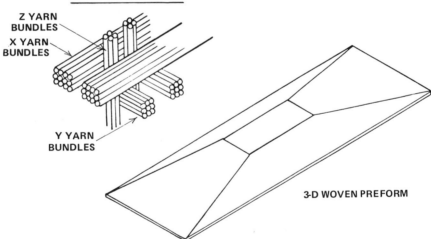

**FIG. 4.68** Detail preforms fabricated directly by three-dimensional weaving. *(Sikorsky Aircraft, Division of United Technologies.)*

*Braiding:* Layers of helically wound and interlocked yarns are woven in a cylindrical shape. Interlocks can be produced at every intersection of the yarns.

*Tufted fabrics:* Loops of fibers are sewn through a woven backing cloth. Available machinery can tuft through 10 to 14 layers of fabric.

*Pile and loop fabrics:* Adjacent fabric layers are held together by interference of piles and loops in each.

*Needle and fabric felts:* Fibers are forced through alternating layers of fabric and mat to link them mechanically. The barbed needle that is required causes damage to the fabric by tearing.

## Weaving vs. No Weaving

Techniques used to fabricate parts involve *no-weaving* methods. Since fibers are straight and continuous in all three directions, there is less damage to them during fabrication and greater control of properties in each direction. The no-weaving process works like this. Stainless-steel tubes are positioned in a fixture in a pattern corresponding to the configuration of the $z$-axis filaments in the finished composite. Filaments are threaded between the tubes to form alternate plies of aligned reinforcement. Fiber orientation usually shifts 90° (1.6 rad) between plies, but other angles are possible. When sufficient cross-plies have been laid down to produce the required thickness, they are compressed and the steel tubes are removed. Bundles of filaments are threaded through the holes left by the tubes to form the $z$-direction reinforcement. Shown in Fig. 4.69 is one of the first specially designed

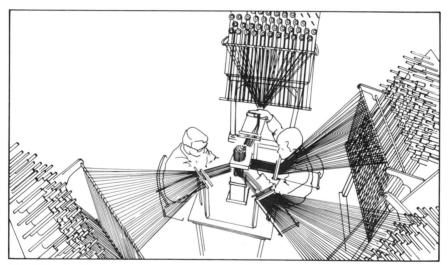

**FIG. 4.69** Specially designed loom for weaving a three-dimensional cylindrical segment.

looms. The resulting shape is a block reinforced in three dimensions. Techniques have also been developed for forming cylindrical structures where the $z$-axis reinforcement is oriented in the radial direction and the $x$- and $y$-axis reinforcements run axially and circumferentially. The process is also used to fabricate cones and truncated cones.

Three-dimensional reinforced blocks have been made with maximized fiber densities as well as cylinders and cones, but there is freedom to choose the proportion of reinforcement in each direction. Thus, the strength and direction characteristics of three-dimensional reinforced parts can be tailored to meet specific multidirectional loading requirements. Moreover, different fibers can be used in each direction, further increasing freedom in selecting directional properties. For example, the direction of heat flow through a three-dimensional reinforced part can be controlled by using graphite, a heat-conducting fiber, in one direction, and glass, a fiber that is a poor conductor, in the others.

## Applications

Three-dimensional composites can be machined by conventional techniques. Blocks can be machined into a wide variety of shapes, including cylinders and cones. However, cylinders machined from blocks lack radial and circumferential reinforcement, and machining rectilinear blocks into cylinders leaves weak, unreinforced points, especially where cylinder walls are thin. In such applications as journal bearings, gears, and deep-submergence pressure vessels radially reinforced three-dimensional composites are used. Joints and closures cause fewer problems in three-dimensional composites than in bidirectional reinforced materials. In fact, end closures for three-dimensional composite cylinders can be screwed in place with threads cut directly into the composite.

For cylindrical journal bearings the radially oriented fibers perpendicular to the bearing surface produce low wear rates. Low-friction fluorocarbon resins have been used as the matrix for self-lubrication, while the fibers provide mechanical support. In operation, fluorocarbon smears over the fiber ends, lubricating them as well. Heat-conducting fibers in the radial direction can reduce heat buildup at the bearing surface.

Three-dimensional composites have been used for rocket nozzles, reentry-vehicle heat shields (Fig. 4.70), and nose-tip applications. The composites combine the functions of an ablative material and a structure to contain it. Thermally conductive filaments give good thermal shock resistance.[8] A washing-machine manufacturer has considered three-dimen-

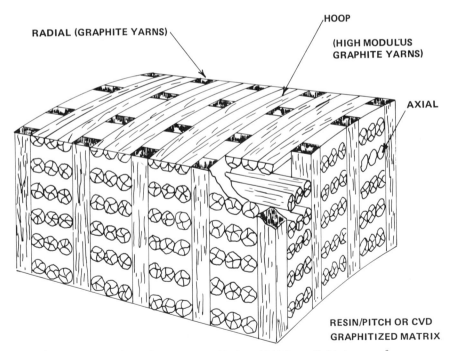

FIG. 4.70 Schematic of three-dimensional graphite heat shield from cylindrical segment.[8]

sional composites to replace bidirectional reinforced laminated gears. The three-dimensional composite would eliminate delamination caused by complex loading. Numerous selected applications have recently been explored for use in a newly designed composite helicopter. The airframe design contained 70 three-dimensional preform fittings made by weaving. By design, a degree of flexibility was incorporated in the weave and preform shape, allowing the shape to be modified somewhat before curing to enable a single type of uncured preform to be used for a family or a variety of parts to fit specific applications such as tail-cone attachment fittings. An example of subsequent forming was bending flat preforms into angle-shaped reinforcements for the keel beams in the helicopter. These preforms were woven on a modified multilayer loom.

## BRAIDING

In the braiding operation, a mandrel is fed through the center of a braiding machine at a uniform rate, and the fibers or yarns from the carriers are braided around the mandrel at a controlled angle. The machine operates like a maypole, the carriers working in pairs to accomplish the over-and-under braiding sequence (Fig. 4.71). Parameters in the braiding

FIG. 4.71   Braiding theory ("dancing the maypole").[9]

operation include strand tension, mandrel feed rate, braider rotational speed, number of strands, strand width, perimeter being braided, and reversing-ring size. Strand tension is set by the carrier springs and must be high enough to work the strands together at the point of formation of the braid but not high enough to break the tape. The axial feed rate of the mandrel can be increased to a point where the braid opens and gaps are left between the strands. Strand width is an important parameter, since the wider the strand the faster the rate of braid production. The widest strand that can be braided with some modifications is 0.35 in (9 mm). Strand width and number of carriers used determine the braid angle for a given part perimeter.[9]

## Materials and Equipment

Fiber glass, Kevlar, carbon, or graphite yarns and tows, either dry or prepreg, have been braided. A 144-carrier braiding machine (Fig. 4.72) was modified to fabricate seamless heat shields by continuous braiding of shingled plies (Fig. 4.73). The shingled plies have a bias filament orientation, simulating the construction of tape-wrapped parts. The braided material is a PAN-precursor carbon tow designed for heat-shield use.

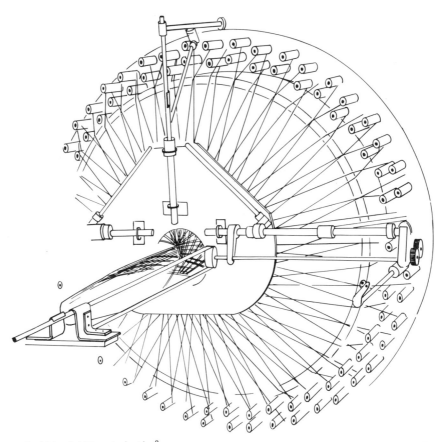

**FIG.** 4.72   A 144-carrier braider.[9]

## Applications

Besides heat shields an automatic braiding machine (Fig. 4.74) winds filaments around a core to produce lightweight ducts for aerospace applications. Coated with Teflon resin, fiber-glass yarns have been designed for use in high-temperature, fluid-sealing components. The coated parts are low-yield texturized yarns with bulk, conformability, and absorption qualities similar to those of asbestos. Although knitting and weaving have been used to form components, braiding has been the most successful in producing packings, sleevings, and rope gaskets. The coated yarns perform at continuous temperatures up to 450°F

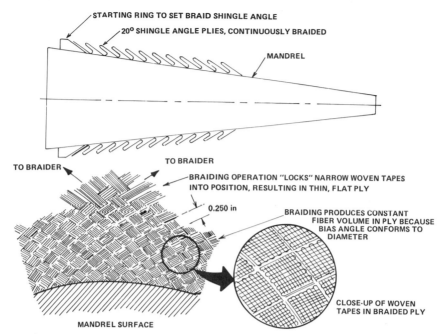

**FIG. 4.73** Schematic of braided heat-shield fabrication and single braided ply viewed from heat-shield forward end (20° = 0.35 rad; 0.250 in = 6.4 mm).[9]

(232°C). By texturization, the glass fibers in the yarns are fluffed up and intertwined randomly, distributing the tensile stress over a series of fiber lengths rather than one continuous length. The texturized yarns are strong and durable, and the added bulk provides resilience, conformability, and absorption.

Tubes of filament-woven fiber glass have proved outstanding as double insulation in hand tools. They are presently being developed for portable tools and certain induction motors (Fig. 4.75). The tube-braided structure lends itself to press fitting; it also develops a very high torque strength in either driven direction, which the normal filament winding will not do. The tube is usually a Gl–Ep structure with an operating capability up to 356°F (180°C). As seen in Fig. 4.75, the tube is slipped or slightly press-fitted into the motor laminations and the armature shaft is subsequently driven into the assembly to lock the laminations and the shaft securely.

## REINFORCED MOLDING COMPOUNDS

Reinforced molding compounds include two materials commonly identified in the plastics industry as premix or bulk-molding compound (BMC) and the closely related but slightly different product sheet-molding compound (SMC).[10,11] BMC has been defined as "a fiber reinforced thermoset molding compound not requiring advancement of cure, drying of volatiles, or other processing after mixing to make it ready for use at the molding press."[1] To this might be added "and which can be molded without reaction by-products under only enough pressure to flow and compact the material." If the word "mixing" is changed

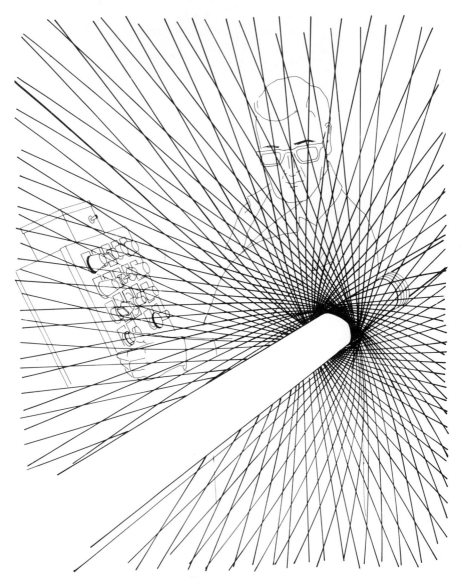

**FIG. 4.74** Automatic braiding machine.[28]

to "manufacture," the definition can apply equally well to SMC. The principal differences between BMC and SMC are in the manufacturing methods and the form in which they reach the molder. Each can be formulated from the same basic ingredients.

BMC is made by combining all the ingredients in an intensive mixing process. It emerges from the mixer in a fibrous putty form and can be used directly, the operator merely weighing out charges. Some formulations can be compacted and extruded into bars or "logs" of simple cross section to facilitate handling. SMC, as the name implies, reaches

FIG. 4.75   Braided tubes for induction motors. *(Polygon Company.)*

the molder in the form of a thin, semitacky sheet, which is cut and plied to suit the article being molded. It is made by impregnating a chopped-strand mat with previously blended resin, filler, and other ingredients.

## BMC (Premix)

BMC first became a practical possibility when glass-fiber rovings came on the market. Earlier some reinforced polyester molding compound was made by chopping fiber-glass fabric prepreg into small pieces. In spite of disadvantages of high cost, residual solvents, no internal release agent, no fillers, etc., some useful products offering properties available in no other known material were successfully molded.

The earliest BMCs were probably made about 1950, employing a process of impregnating roving strands with the blend of resin, filler, etc., and chopping them to length in the wet stage. Since wetting glass fibers with a resin containing much filler is difficult and slow, these premixes had a high glass content. Because no compounders of conventional thermoset compounds were initially interested in BMCs, such simple features as internal mold release were a long time coming. The first move to high volume occurred with the development of a BMC based on sisal fibers and the molding of automobile-heater housings. While the easy-molding, low-cost sisal BMC continued to dominate the high-volume automobile market, the development of a resin coating for glass-fiber strands that preserved their integrity, i.e., kept them intact as bundles of fibers through the mixing process, made possible large-area moldings with strength, chemical resistance, electrical-insulation values, and other desirable properties. Consequently heavy, large-area electrical and chemical items as well as a few commercial products such as internal parts of appliances, where surface quality was a secondary consideration, became commonplace. Surface waviness and coloring problems limited product applications to parts not normally exposed to the consumer's view.

The applications of glass-fiber BMC was seriously limited by irregular waviness of the molded surface. Developments in Europe about 1960 and use of chemical thickening agents and thermoplastic additives markedly reduced cure shrinkage, resulting in improvement in the quality of the surface as well as limitation of other distortions. These improvements unfortunately required some compromises, and efforts to produce a distortion-free BMC without sacrifice of other properties continues.

## Materials

BMCs are high-performance engineering thermoset molding materials, based on unsaturated polyester resins and fibrous reinforcements. Because they are moderate in cost, penetration of existing markets is not always easy where phenolics, melamines, ureas, epoxies, ceramics, and die castings may have a price-gap advantage and have been used in established applications for a long time. Therefore, switching such applications to BMC is not always feasible.

When functional performance is critical, the material preference will be for BMCs since they feature such characteristics as nontracking, arc quenching, arc resistance, and current-interrupting capacities as well as greater flexural modulus, higher impact strength and mold-in color. There is now a complete range of fiber-glass-reinforced hybrid BMCs. Each compound differs in reinforcement content, performance, shrinkage, and color to meet specific commercial requirements. Other characteristics such as wear resistance, low density, energy absorption, Underwriters' Laboratory approval, conductivity for shielding and electrostatic discharge, and transparency to microwave energy can also be built into the BMC. Hybrid BMCs have recently been used for residential circuit breakers with a fourfold increase in current-interruption rating.

BMC has the consistency of putty of molding clay and is often extruded into logs or ropes for easier handling and mold-charge preparation. Usually polyester is reinforced with chopped-glass strands from 0.125 to 1.250 in long (3.2 to 32 mm) but typically 0.125 to 0.50 in (3.2 to 12.7 mm). Glass loading is usually 10 to 30%. The catalysts, fillers, release agents, thickeners, and thermoplastic low-shrink–low-profile additives are very similar to those for SMC. Epoxy or vinyl ester resins are sometimes used instead of polyester in specialty applications.

BMC is usually molded by compression molding in heated matched steel dies, but it can also be transfer-molded or injection-molded on both plunger and screw machines. The BMC must be force-fed into the injection-molding machine by a *stuffer*. BMC can also be formulated in pellet or granular form for easier and faster injection molding. A pelletized thermosetting polyester molding compound with glass reinforcement which can be molded in automatic systems is solid polyester molding compound (SPMC). Ordinary BMC can be injection- and transfer-molded. BMC must be fed into the injection unit by a pressurized hopper. By contrast, SPMC is free-flowing and can be used in regular injection, compression, and transfer machines. It can be conveyed with conventional vacuum equipment.

SPMC is a true thermosetting polyester, reinforced with up to 20% glass fiber; it has a shelf life of about 6 months.

## SMC

Most large-area parts have been produced by the preform and mat die-molding process, in which the mat or preform and resin filler are brought together at the press and the resin is distributed with varying degrees of uniformity throughout the reinforcement by the force exerted as the mold closes. The development of a chemical thickening process provided a solventless low-viscosity resin system that permits impregnation of mats having a minimum of soluble binder. The resin, wet in the initial stage, readily impregnates the mat. In time (3 to 48 h) it becomes dry or slightly tacky without further processing except being rolled up in a polyethylene separating film. While SMC has not replaced conventional mat and preform molding in products with demanding mechanical properties or lent itself to a high order of mechanization, it has nevertheless found a place in the market between the areas served by BMC and preform.

### *Materials*

SMC is the basic composite material from which a number of other variations have been developed. SMC is made of thermosetting resins, reinforcing fiber, thickeners, and other fillers deposited on a carrier film to form a sheet of material. The cured sheet is compression-molded under heat and pressure into the final part. SMC is commonly made of polyester resin with glass-fiber reinforcement, chopped to less than 2 in (51 mm) in length and randomly oriented; approximately 1 in (25 mm) is the usual length. Epoxy and vinyl ester SMCs are also available. Other elements of the SMC compound are catalysts to cure the laminate; thermoplastic additives to reduce shrinkage during the cure; thickeners to increase resin viscosity so that it is tack-free and can be molded; fillers to lower cost, improve the surface, or add certain properties; and mold-release agents. The reinforcement can be chopped short fibers, deposited randomly (most common and generally what one thinks of as SMC), longer fibers oriented in a single direction, continuous fibers unidirectionally oriented, or combinations of these. Typical glass loading is 20 to 35%.

**Hybrids:** Recent years have seen rapid changes in the use of carbon- and glass-fiber hybrids processed in SMC equipment. Continuous carbon fiber is added to the normal chopped-glass fiber, resulting in a high modulus in the machine direction, provided by the continuous carbon fiber, and good transverse properties, provided by the chopped random-oriented glass fiber. The technique allows for rapid processing of large volumes of fiber in a convenient unsaturated-polyester molding compound. Table 4.9 shows typical properties for another hybrid approach of all-continuous fibers with a unidirectional alignment.

For injection-molding applications, a significant upgrading of some of the important engineering properties of many glass-filled thermoplastic resin systems can be accomplished by hybridizing with chopped carbon fibers. The individual carbon fibers are approximately 0.015 in (0.38 mm) long and blend easily with all matrix materials used in injection-molding compounds. Typical properties include lower coefficient of friction, greater strength, higher stiffness, increased thermal and electrical conductivity, greater fatigue life, and longer wear life. Table 4.10 presents typical properties of a nylon 6/6 injection-mold-

**Table 4.9** Effect of Carbon-Glass Fiber Ratio on Typical Properties of Hybrid Composites†[33]

| Carbon-glass ratio | Tensile strength | | Tensile modulus | | Flexural strength | | Flexural modulus | | Interlaminar shear strength | | Density | |
|---|---|---|---|---|---|---|---|---|---|---|---|---|
| | ksi | MPa | $10^6$ psi | GPa | ksi | MPa | $10^6$ psi | GPa | ksi | MPa | lb/in$^3$ | g/cm$^3$ |
| 0:1 | 88 | 607 | 5.3 | 36.6 | 135 | 931 | 5.1 | 35.2 | 9.5 | 66 | 0.069 | 1.91 |
| 1:3 | 93 | 641 | 9.0 | 62 | 155 | 1069 | 9.1 | 62.7 | 10.8 | 74 | 0.067 | 1.85 |
| 1:1 | 98 | 676 | 12.4 | 85.5 | 175 | 1207 | 11.3 | 77.9 | 11 | 76 | 0.065 | 1.80 |
| 3:1 | 115 | 793 | 16.3 | 112.4 | 185 | 1276 | 16 | 110.4 | 12 | 83 | 0.060 | 1.66 |

†Polyester resin matrix, fiber volume = 65%.

**Table 4.10** Typical Properties of Nylon 6/6†-Carbon-Fiber-Grade Hybrid[33]

| Reinforcement carbon-glass wt % | sp gr | Ultimate elongation, % | Ultimate tensile strength | | Tensile modulus | | Flexural strength | | Flexural modulus | | Compressive strength | |
|---|---|---|---|---|---|---|---|---|---|---|---|---|
| | | | ksi | MPa | $10^6$ psi | GPa | ksi | MPa | $10^6$ psi | GPa | ksi | MPa |
| 20:25 | | 3.9 | 18.1 | 124.8 | | | 34.3 | 236.5 | 1.65 | 11.4 | | |
| 25:20‡ | 1.42 | 3.8 | 16.5 | 113.8 | 1.71 | 11.8 | 34.0 | 234 | 1.92 | 13.2 | 17.0 | 117 |
| 30:15‡ | 1.41 | 3.5 | 14.7 | 101.4 | | | 31.0 | 213.8 | 1.89 | 13.0 | 15.0 | 103 |

†Injection-molding compound.    ‡0.25-in (6.4-mm) chopped glass fibers.

ing compound blended with chopped carbon fiber and 0.25-in (6.4-mm) chopped glass fibers.

Although compression molding of SMC has used glass fibers as an almost exclusive reinforcing fiber, development of hybrids of glass and carbon fibers in SMC is active. Adding carbon fibers to glass enhances dimensional stability and provides electrical conductivity, increased stiffness, and increased strength. A thin veil mat of carbon fiber molded into the surface of SMC has produced an electrically conductive surface that permits electrostatic painting of a plastic part without using a conductive primer. The molded-in carbon fiber also provides a convenient method of suppressing radio-frequency or electromagnetic interference.

## Equipment and Process

SMC is made on a machine that deposits a layer of the resin compound on a polyethylene film, lays down the glass reinforcement, and places another resin-carrying film on top to form the composite. The compound is kneaded between compaction rolls to wet the glass fibers and is wound under controlled tension into package-size rolls. Machines used to produce SMC material are shown in Figs. 4.76 and 4.77. SMC can be manufactured on site by the molder or purchased in rolls.

Final thickening or curing can take from 1 to 5 days, after which the SMC can be molded. This time can be shortened to almost instant maturation by modifications in chemical technology and by using accelerated heating techniques (conventional or radio-frequency). To mold a part, SMC is precut and layered to build up a charge for a standard compression molding press. Heated matched steel dies are used. Molding pressure is typically in the range of 0.8 to 1 kip/in$^2$ (5.52 to 6.9 MPa). Low-pressure molding compounds permit molding at greatly reduced pressures or are used for larger parts at the higher pressure ranges.

## Properties and Design

SMC and BMC offer an extraordinary range of desirable properties. The nearly infinite variability of resin, filler, and reinforcement types and contents would seem to permit tailoring a compound for any design or performance requirement. Within limitations this is true and is probably the most outstanding single quality of these materials. It also makes definition of detailed properties difficult. Nevertheless, the important attainable characteristics, if the compound is properly formulated, are as follows:

1. Excellent electrical performance, especially resistance to tracking
2. High strength, particularly impact strength
3. Heat and flame resistance
4. Dimensional stability
5. Chemical resistance
6. Rapid cure
7. Low molding shrinkage
8. Low molding pressure
9. Ability to mold thick, thin, and variable thicknesses
10. Low cost

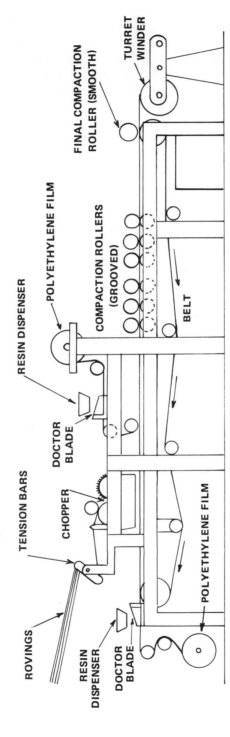

FIG. 4.76 Belt-type SMC machine.[32]

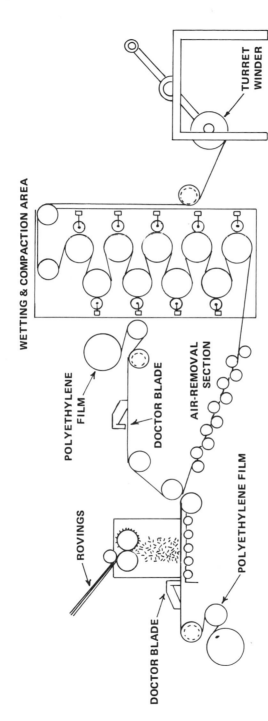

**WETTING & COMPACTION AREA**

**TURRET WINDER**

**POLYETHYLENE FILM**

**DOCTOR BLADE**

**AIR-REMOVAL SECTION**

**ROVINGS**

**POLYETHYLENE FILM**

**DOCTOR BLADE**

FIG. 4.77   Stack-roll SMC machine.[24]

General principles of design for any kind of molding apply to reinforced molding compounds. Almost any shape can be molded if the cost and complexity of the mold is not a factor. Some special design consideration must be given to reinforced compounds because of their low elongation or extensibility. In terms of metal characteristics their yield and ultimate strengths are identical. This means that a comparatively small distortion (even at a high stress level) will result in breakage of a part. The general answer to the problem is to make all sections of any BMC or SMC part thick enough or to contour them in such a way that the probable applied local forces will not result in stresses above the ultimate. Large flat areas should be divided up with convolutions or ribs. Edges of parts should be flanged or increased in thickness. If a part has a mounting flange or extending feet with bolt holes, generous gussets should be provided extending from the adjacent wall to beyond the bolt holes. It should be pointed out that increases in thickness cause very little more than an increased material cost, and sometimes the material added in one place can be removed from another less critical section of the part.

## Applications

Some representative successful applications are described below to illustrate what reinforced molding compounds are good for and where they should be used. The sheer volume of material consumed makes automotive applications of prime interest. Among them are components of heating and ventilating systems, hoods, radiator support (Fig. 4.78), and heater housings. Getting more plastics into an automobile takes all the ingenuity and perseverance designers can muster. It is difficult to trade plastic off against its chief rivals, lightweight steel and aluminum. A 65%-glass-reinforced SMC has been used for a new auto bumper system consisting of a face bar and two corner braces. The resulting assembly weight was reduced from 18 to 8 lb (8.2 to 3.6 kg).

Production of the first SMC door for a truck is the major step taken toward the pro-

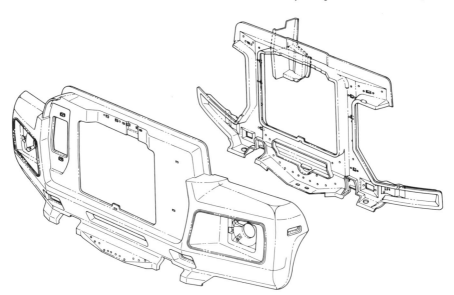

**FIG. 4.78** Fiber-reinforced plastic radiator support. (*Owens-Corning Fiberglas Corp.*)

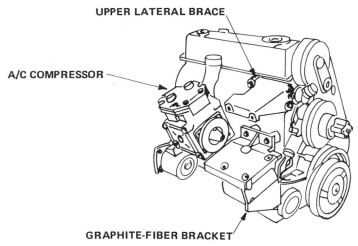

UPPER LATERAL BRACE

A/C COMPRESSOR

GRAPHITE-FIBER BRACKET

FIG. 4.79 Graphite-fiber-reinforced bracket for auto air compressor. (*Armco Composites.*)

duction of an all fiber-reinforced-plastic cab. Another advance is production of the first automotive application of a graphite-fiber-reinforced air-compressor mounting bracket (Fig. 4.79). The part has twice the fatigue resistance of steel and is 70% lighter. An all-fiber-glass-reinforced truck door is said to be the first mechanically stressed vehicle body component using no metal supports or stiffeners. The door, which saves 40 lb (18.1 kg) of weight, is molded in three pieces. All three are compression-molded in matched metal dies from glass-reinforced SMC. The outer and inner panels are 30%-glass-reinforced, and the internal structural frame is reinforced with 30% continuous and 20% chopped glass fiber.

A proposed split-bench seat has successfully passed rigid auto tests. The compression-molded parts are made of vinyl ester high-strength SMC with 65% random glass fibers (Fig. 4.80). Another proposed part is a single-leaf fiber-glass-reinforced spring, expected to

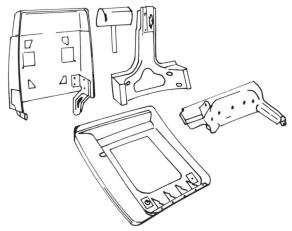

FIG. 4.80 Proposed split-bench seat of high-strength SMC.[29]

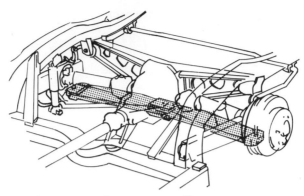

FIG. 4.81    Single-leaf Gl–Ep rear spring.[29]

replace a 10-leaf steel spring. The 7.5-lb (3.4-kg) Gl–Ep spring replaces a steel spring weighing 43 lb (19.5 kg) and is the third major substitution of reinforced-plastic structural components for steel in 3 years. In 1979 a stamped fiber-glass-reinforced plastic seat sheet was introduced, and in 1980 a lightweight bumper backup system.

The transverse rear spring is a taper-leaf design, thicker at the center and tapered toward the end for correct compliance. It also has a constant-area cross section (constant stress) for maximum load-carrying capacity, so that it is narrower at the center than at the ends, something like a bow tie. The spring is 49 in (1245 mm) long (Fig. 4.81). The Gl–Ep spring material used has a specific strain energy 6 times that of the SAE5160 steel currently used; graphite composites are even better, with a specific strain energy 6.5 times that of steel, but are not cost-effective. Other applications include compression molded highly flexible SMC auto rear decks and jet-ski handles (Table 4.11).

## Other SMCs

SMCs with improved physical properties and new additives, capable of being molded under reduced pressures, have developed into a number of new materials.

**HMC and XMC:**   High-strength molding compound (HMC) is a SMC containing 65% chopped glass fiber instead of the usual 20 to 35%. It is about twice as strong as short-fiber-reinforced SMC with no sacrifice of mold-flow characteristics. The chopped glass-fiber strands are 0.5 to 1.5 in long (12.7 to 38 mm), randomly oriented. Instead of the fillers associated with SMC, HMC uses additives such as catalysts, low-shrink–low-profile compounds, mold-release compounds, and thickeners. HMC can be made on standard belt and stack-roll SMC machines. HMC composite offers isotropic strength properties approaching twice those of conventional SMC by employing high glass-fiber content with little or no filler. Engineering properties of HMC also can be enhanced by hybridizing with carbon fibers. Fiber lengths from 1 to 2 in (25.4 to 50.8 mm) have been used. A hybrid SMC of glass and carbon fibers provides a high-strength dimensionally stable composite with good flow characteristics and the ability to form bosses and ribs. Because of the lower volume of reinforcement, SMC properties are lower than those of HMC.

Directionally reinforced molding compound (XMC) is a directionally oriented moldable resin–glass-fiber sheet containing 65 to 75% continuous reinforcement. It is made on stan-

**Table 4.11** Properties of High-Impact SMC†

| Type | Tensile strength | | Tensile modulus | | Flexural strength | | Flexural modulus | |
|---|---|---|---|---|---|---|---|---|
| | ksi | MPa | $10^6$ psi | GPa | ksi | MPa | $10^6$ psi | GPa |
| Standard | 10 | 68.9 | 1.7 | 11.7 | 27 | 186 | 1.5 | 10.3 |
| Flexible | 10.2 | 70.3 | 1.3 | 8.9 | 23.4 | 161 | 1.4 | 9.6 |

†From *Mater. Eng.*, April 1980, p. 59.

**Table 4.12** Typical Physical Properties of Hybrid Moldable Sheet[33]

| Type | Glass-carbon, wt % | | Tensile strength | | Tensile modulus | | Flexural strength | | Flexural modulus | |
|---|---|---|---|---|---|---|---|---|---|---|
| | | | ksi | MPa | $10^6$ psi | GPa | ksi | MPa | $10^6$ psi | GPa |
| XMC | 70 | 0 | 70 | 483 | ...... | ...... | 135 | 931 | 5 | 35 |
| | 26 | 35 | 100 | 690 | ...... | ...... | 165 | 1138 | 13.5 | 93.1 |
| HMC | 60–65 | 0 | 35 | 241 | 1.7–2.1 | 11.7–14.5 | 55 | 379 | | |
| | 28 | 30 | ...... | ...... | 3.5 | 24 | | | | |
| SMC | 30 | 0 | 12–15 | 83–103 | ...... | ...... | 24–30 | 165–207 | 1.5–1.9 | 10.3–13 |
| | 0 | 35 | 15 | 103 | 3.9 | 27 | | | | |

dard filament-winding equipment and offers strength properties 5 times greater than SMC in the prime direction of reinforcement. Continuous fiber-glass strands wetted with polyester compound are wound under tension on the mandrel of a filament-winding machine. The winding angle is small, 7.5° (0.13 rad), so the reinforcement is essentially unidirectional (slight X or diamond pattern). Individual sheets are cut and stripped from the mandrel after winding. Strength is highly directional along the line of glass-fiber orientation.

The polyester compound content is 25%. Hybrid epoxies and vinyl esters have also been molded. Thermoplastic monomers, fillers, and thickeners have been used in the compound. Magnesium oxide or magnesium hydroxide is added to SMC, BMC, XMC, and HMC composites to give proper molding viscosity and prevent separation of the resin from the fillers and reinforcements. Two types of XMC have been developed. XMC-2 has continuous unidirectional reinforcement, and XMC-3 has two-thirds continuous and one-third chopped random reinforcement. The strength of XMC is 2 to 3 times that of HMC in the direction of reinforcement but lower than HMC perpendicular to the continuous reinforcement. XMC is compression-molded, but the material does not flow in the direction of continuous reinforcement. Since flow is obtained only in the perpendicular direction, the mold charge must include XMC layers in perpendicular directions or a combination of XMC with HMC. Combining XMC with HMC also yields a balance of isotropic and anisotropic properties in the final part. XMC composite is intended for use in beam-type applications or in combination with HMC composite to mold a wide variety of structural shapes. Some typical physical properties of XMC, HMC, and SMC moldable sheet are given in Table 4.12.

**UMC and LMC:**   Unidirectional molding compound (UMC) is a system of chopped and continuous fibers produced on a modified SMC machine. Both the chopped and continuous fibers can be glass, aramid, high-modulus graphite, or combinations of them. The use of long continuous fibers means higher tensile strength in the parallel direction since the lengths and types of fibers used for reinforcement determine the strength of the molded material. One graphite-reinforcement combination, for instance, is 28% continuous graphite with 43% chopped glass fiber. An aramid formulation is 21% Kevlar and 33% fiber glass. A production application of UMC is an air-conditioning mounting bracket for an automobile. The 2-lb (0.91-kg) part replaces a 7-lb (3.2-kg) cast-iron version. The compression-molded bracket consists of 30% polyester resin, 50% of 1-in (25.4-mm) chopped glass fibers, and 20% continuous graphite by weight. Low-pressure molding compound (LMC) is SMC made with polyester resins formulated to allow molding at greatly reduced pressure to produce longer parts and/or use less expensive presses and molds.

**SMC-C, SMC-D, and SMC-R:**   This family of SMCs consists of continuous-fiber SMC, directional-fiber SMC, and random-fiber SMC. The combinations like C/R describe SMC composites containing different kinds of reinforcements, and appended numbers (SMC-C30/R20) indicate the percentage of the reinforcement type in the final composite sheet. SMC-C is a continuous reinforced fiber deposited in a unidirectional fashion. A belt-type machine (Fig. 4.82) can be used to produce SMC-C or SMC-C/R, although modifications are necessary. Reinforcement content ranges up to 70 wt %. Continuous reinforcement may be from glass rovings or a unidirectional mat. The random fibers in a SMC-C/R construction may be graphite, aramid, or a different type of glass. SMC-D contains discontinuous fibers 4 in (102 mm) or longer deposited unidirectionally. Two methods have been used to manufacture SMC-D composites. In the on-line approach, a long-fiber cutter is placed

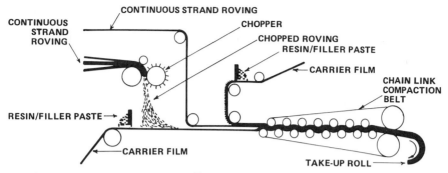

FIG. 4.82 Machine for making SMC-C/R.[24]

on, and synchronized with, the SMC machine. The other method is to make SMC-C/R, and then use the long-fiber cutter to pierce the carrier film and cut the continuous rovings, converting SMC-C/R into SMC-D/R (Fig. 4.83). SMC-R is equivalent to SMC, described earlier. Glass-fiber lengths are less than 3 in (76 mm), commonly 1 in (25.4 mm). The fiber orientation is random, and nominal glass content ranges from 30 to 70%, most compositions being molded with 50% or higher glass content.

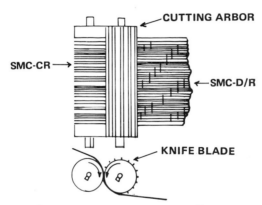

FIG. 4.83 Machine for making SMC-D/R.[24]

**TMC:** Thick molding compound (TMC), one of the newest fiber-reinforced plastics, is suited for compression, injection, and transfer molding and can be processed on the same equipment as SMC and BMC materials. TMC composites can be produced up to 2 in thick (50.8 mm), in contrast to the 0.19 in (4.8 mm) maximum thickness of SMC. Glass-fiber length in TMC is 0.25 to 2 in (6.4 to 50.8 mm). During fabrication complete wet-out of resins, fillers, and reinforcement fibers is achieved, resulting in improved mechanical properties and reduced porosity. Low porosity means that TMC affords better surface qualities on molded products than either SMC or BMC, which also means that less rework is needed on TMC end products. Superior impregnation of the reinforcing fibers in TMC also minimizes some other problems inherent in SMC, e.g., uneven fiber distribution and variations in resin coating from surface to center.

The equipment used to fabricate TMCs (Fig. 4.84) provides instantaneous wet-out of the glass fibers, eliminating the degradation of fibers caused by more intensive methods of

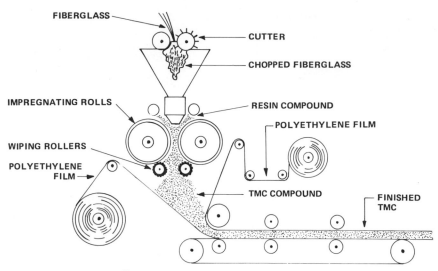

FIBERGLASS

CUTTER

CHOPPED FIBERGLASS

IMPREGNATING ROLLS

RESIN COMPOUND

POLYETHYLENE FILM

WIPING ROLLERS

POLYETHYLENE FILM

TMC COMPOUND

FINISHED TMC

**FIG. 4.84**   TMC machine.[30]

mixing and allowing the use of highly filled high-viscosity paste systems. BMCs are normally mixed in blade mixers, which inherently degrade the glass fibers, and SMCs are limited to formulations having low initial viscosities to ensure proper glass wet-out.

The combination of TMC's thicker sheet and expanded compounding versatility (by allowing much higher filler content) provides many molding and molded-part advantages:

Reduced cost of compound

Lower handling cost at the press

Improved part properties with regard to flame retardance, electrical insulation, and corrosion resistance

Wider choice of reinforcements

Improved flow and improved appearance

Better strength in extreme flow regions

Resin paste and reinforcing fibers are fed directly to the nip of the impregnating rolls. Wiper rolls throw the impregnated fiber onto the moving polyethylene carrier film. Since carrier-film speed is independent of the impregnated-fiber deposition rate, sheet thickness is adjustable. The final composite is guillotine-cut and boxed.

Because of its many advantages in properties and its versatility in processing, TMC is being used in business-machine housings, appliance components, machine bases, automotive grille opening panels, circuit breakers, auto-engine covers, and bus-bar supports.

## Resin-Transfer Molding

Filling a gap between hand manufacturing lay-up or spray-up of parts and compression molding of SMC and BMC in matched metal molds is resin-transfer molding (RTM), also called resin-injection molding. The closed-mold process has higher production rates, more

**Table 4.13**  Advantages of RTM†

| Compared with lay-up and spray-up | Compared with matched-metal-die compression molding |
| --- | --- |
| Two finished surfaces on parts | Lower-cost molds and molding equipment |
| Highly reproducible thickness | Use of stiffeners, ribs, inserts, etc., possible |
| Low monomer loss (closed process) | Shorter lead times for molds |
| Higher output (labor and material savings) | |

†From *Mater. Eng.*, May 1981, p.50.

consistent parts, reduced styrene emissions, and both material and labor savings compared with lay-up and spray-up (Table 4.13). Tooling costs of RTM are typically twice those of lay-up and spray-up methods.

The key to RTM is low pressure in the mold. Reactive polyester or epoxy is injected into a closed mold (usually made of reinforced plastic) containing reinforcing fiber or mat. The molds are backed up with epoxy, gravel, or even concrete, frequently with embedded steel frames for reinforcement and mold handling. The mold is closed and clamped, and catalyzed resin is pumped in. Air is vented from the extreme edges of the mold, and resin displaces the air to fill the mold. Resin pressure must be kept very low to allow low-cost tooling to be used without damaging the mold.

Growth of this process in the future will depend on tooling costs. Since at present the life of RTM molds (about 3000 parts) is much shorter than that of matched-metal dies (about 100,000 parts), tooling-cost per unit is higher for higher production runs. Molds of electroformed nickel must be considered since they could increase mold life 10 times. Also needed are improvements in mold handling, especially for gel-coated molds.

RTM has been used in several marine applications because of its excellent corrosion resistance, e.g., boat hatch covers, a radar arch supporting a 150-lb (68-kg) radar unit, 5.5- and 7-lb (2.5- and 3.2-kg) fertilizer hoppers, bathtubs, water tanks, tractor fan shroud, entire boat hulls, and computer housing.

## Elastic-Reservoir Molding (ERM)

The ERM molding system is similar to SMC in concept but radically different in materials. The ERM structure is made as a continuous sandwich composed of a layer of flexible foam with outer skins of reinforcement. The resin may be epoxy, polyester, or vinyl ester, and the reservoir is open-celled polyurethane foam. The outer-layer facings are Gl–Ep and/or Gr–Ep. At the start the sandwich is up to 2 in thick (50.8 mm). Overall glass content is about 30%, but the surfaces have glass contents of about 70%. Unidirectional glass cloth can be used for the surface skin.

### Processing and Tooling

Low pressure in producing ERM composites offers several benefits, namely, producing much larger sections with the same presses and lighter and cheaper molds. Simple cast-epoxy or cast and machined aluminum are recommended for low-pressure molding and are inexpensive. The cost of molds and presses is approximately one-tenth that of high-pressure equipment, and the time to construct the molds is approximately one-fifth that for standard molds. The time of molding is approximately the same. Because of the lower pressures, damage and wear to the mold are considerably less; the final properties and appearance are essentially the same as those of high-pressure composites.

The process (Fig. 4.85) used a flexible open-cell urethane foam as a "reservoir" to carry

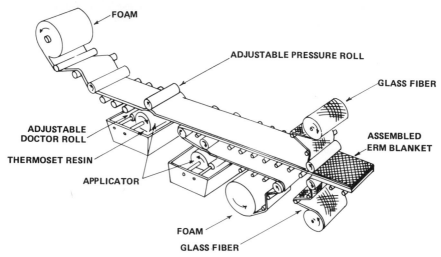

**FIG. 4.85** ERM process. *(Automation, December 1975, pp. 62, 63.)*

the resin binder into the process. After the urethane has been impregnated with the binder, sheets of dry fiber-glass or graphite cloth are added on the outside, making a four-layer sandwich. The sandwich is put in a mold, compressed at low pressure, and cured at moderate temperatures for a few minutes, giving composite sections that are strong, stiff, and rigid. Final laminates can be as thin as 0.125 in (3.2 mm), and thickness can be varied in different areas of a part. Molding ERM is similar to other thermoset processes, where both heat and pressure are used, but molding pressures of about 0.1 kip/in$^2$ (0.69 MPa) are low, compared with SMC [0.75 to 1 kip/in$^2$ (5.2 to 6.9 MPa)] and mat or preform [0.3 to 0.5 kip/in$^2$ (2.1 to 3.5 MPa)]. Curing temperatures depend on the time available for curing. Extremely high temperatures cannot be used, but temperatures of 200 to 300°F (93 to 149°C) produce curing times of 1 to 2 min. Because of the low pressure, large-surface-area parts up to 100 ft$^2$ (93 m$^2$) are said to be feasible. Two ERM sandwiches can be molded around a central core of rigid urethane or styrene foam to form a structure 6 in (152 mm) or more thick. Surface skins of thin metal or flexible vinyl can be integrally molded on.

## Materials and Applications

Reinforcements can consist of chopped strand, continuous strand, unidirectional or woven rovings, or combinations. Composites can also be produced with plywood, balsa, and rigid urethane cores. There are two main limitations for the process: ERM is not suitable for molding ribs and bosses and is limited to fairly simple shapes. Applications being looked into include a gull-wing automobile door and hoods, deck lids, and front fascia panels for sports cars. Epoxy ERM is being considered for oil pans to meet high-temperature and fatigue-resistance requirements. Development work is being done on seats and seat backs. ERM is cost-effective for the fabrication of durable, impact-resistant secondary airframe structures. Table 4.14 shows the potential weight and cost savings for an ERM fabricated door vs. a corresponding metal structure.

**Table 4.14**  Weight and Cost Comparison for Access-Door Designs†

| Design configuration | Weight lb | Weight kg | Weight savings, % | Estimated cost (1981) | Cost savings, % |
|---|---|---|---|---|---|
| Baseline aluminum beaded and fastened | 5.7 | 2.6 | .. | $1200 | |
| Unidirectional Gr–Ep and woven Gl–Ep hybrid | 3.9 | 1.8 | 31 | 840 | 30 |
| Woven Gr–Ep cloth | 4.7 | 2.1 | 18 | 975 | 19 |

†Data from Grumman Aerospace Corp.

## Vibrational Microlamination (VIM)

Using a combination of heat and vibration, this process can produce parts of practically any size with superior strength and integrity from any thermoplastic material. Without subjecting the plastic to thermal degradation the special techniques involved in the process produce parts with no stresses or flow patterns. Most VIM-produced components have been relatively large, boxlike shapes, e.g., bins, hoppers, tanks, racks, hoods, and trays from polyethylene, polypropylene, and polyvinylidene fluoride. The process can also mold parts that are impossible by conventional molding methods, e.g., cylindrical parts several feet long with uniform wall thickness and no draft or parting line.

### Processing

The VIM process (Fig. 4.86) uses a single-surface mold, male or female, mounted on a frame or platen. The heated mold is placed in an environmentally controlled chamber, and

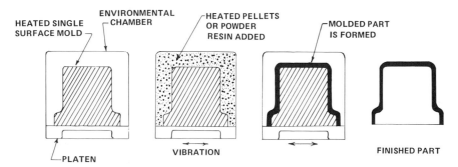

FIG. 4.86   VIM process.[11]

warm thermoplastic material in pellet or powder form is fed rapidly onto it. Processing temperature is considerably below that used in extruding or injection molding but is sufficient to cause the plastic particles to fuse under the additional influence of vibrational energy. Temperature control is critical because the resins must not reach the molten state. This control is particularly important with crystalline materials, whose softening and melting temperatures are close together. When the desired thickness has been reached, the mold is removed from the chamber, another mold goes in, and the finished part is removed from the first mold after a brief cooling, usually by cold air or water.

## Materials and Tooling

The nature of the VIM process enables a broad range of composite components to be made by simply removing the mold and its plastic coating from the chamber, placing or spraying a reinforcement or other embedment on it, and reinserting the mold in the chamber where more plastic is added to the structure. Typical laminates include woven fabric or strands of aramid, glass, or carbon fiber. Shorter fibers of glass, carbon, or other materials also can be incorporated into molded parts and can be applied, either in layers or along with the resin pellets or powder, to form a fiber-reinforced structure in a single operation. The fibers are distributed randomly, not aligned as they would be in injection molding with uniform strength in all directions. Since the VIM process is carried out at atmospheric pressure, molds are lightweight and much less expensive than those required for injection molding. Most molds are fabricated from aluminum sheet; mold surfaces must be mirror-smooth so that parts can be demolded easily. For female molds, where part release is not a problem (since the part shrinks away from the mold during cooling), electrode forming of a nickel shell, backed up with a glass-fiber-reinforced layer, is an acceptable mold-making technique.

## Reinforced RIM (RRIM)

RRIM urethane elastomer is a new material whose use should increase rapidly during the next 5 years with growing automotive weight reduction. The urethane elastomer made by the reaction-injection-molding process (RIM) is modified by the addition of 20 wt % milled glass fiber 0.062 in (1.6 mm) long. The resultant material reduces the thermal coefficient of expansion considerably and increases the flexural modulus to more than twice that of unfilled material. This is reinforced RIM (RRIM) urethane elastomer. The material is not intended to replace SMC and BMC parts, where rigidity is essential. RRIM parts for automobiles will be serious candidates where lower stiffness is acceptable or desirable, e.g., for fenders and door panels. If RRIM fenders replaced steel fenders, the weight saving would be about 6.3 lb (2.9 kg) per fender. Some experts claim that 5% of the body surface area of all cars and trucks will be converted to RRIM urethane elastomer by the 1985 model year and that substituting a RRIM urethane part for steel results in a 60% weight reduction.

Materials for exterior auto-body panels must have certain physical properties, e.g., withstanding temperatures of 325°F (163°C) since primers and paints cure at that temperature. In attaching RRIM parts to steel or aluminum components, thermal-expansion coefficients must match as closely as possible. Both these requirements can be met with the efficient use and selection of fibers. Adding fibers to polyurethane materials results in a lower impact strength compared with SMC, but preliminary impact values obtained from RRIM-produced liquid epoxy specimens are equal to, or better than, the SMC system at both room temperature and −20°F (−29°C).

## METAL-MATRIX-REINFORCED COMPOSITES

MMCs are at approximately the same stage of development as the filament- and/or fiber-reinforced organics were in the 1960s. A major barrier to growth has been the high cost of reinforcing materials. Boron, Borsic, and SiC are expensive but ultimately will cost less. MMCs are used in highly specialized applications where performance is critical. Common matrix metals are aluminum, titanium, and magnesium, but some attention has been paid to lead and copper. Even silver, gold, and superalloys have been reinforced with silicon carbide, molybdenum, and tungsten fibers.

Reinforcing fibers range from the ordinary to the exotic (Table 4.15). Graphite fibers

**Table 4.15**   Representative Metal-Matrix Composite Materials†

| Fiber | Matrix | Potential applications |
|---|---|---|
| Graphite | Aluminum | Satellite, missile, and helicopter structures |
| | Magnesium | Space and satellite structures |
| | Lead | Storage-battery plates |
| | Copper | Electrical contacts and bearings |
| Boron | Aluminum | Compressor blades and structural supports |
| | Magnesium | Antenna structures |
| | Titanium | Jet-engine fan blades |
| Borsic | Aluminum | Jet-engine fan blades |
| | Titanium | High-temperature structures and fan blades |
| Alumina | Aluminum | Superconductor restraints in fusion power reactors |
| | Lead | Storage-battery plates |
| | Magnesium | Helicopter transmission structures |
| Silicon carbide | Aluminum, titanium | High-temperature structures |
| | Superalloy (cobalt-based) | High-temperature engine components |
| Molybdenum, tungsten | Superalloy | High-temperature engine components |

†From *Iron Age*, July 2, 1979, p. 51.

are the most widely used, especially in aluminum matrixes. After fiber and metal-matrix material are produced, the material is fabricated into a variety of tapes and windings. For example, Gr–Al panels have been prepared by sandwiching an array of Gr–Al wires between aluminum face sheets and consolidating by hot pressing. Another process takes ultrahigh-modulus graphite fibers and by using liquid-metal infiltration produces composite "wires" with high specific strength and high specific modulus.[12] In the infiltration process multifilament graphite fiber rows are intimately coated with a fine layer of Ti–B by the reduction of $TiCl_4$ and $BCl_3$ with zinc vapor. The Ti–B coating activates the surface of the fibers and promotes wetting and infiltration by the molten aluminum, providing intimate bonding of the fibers to the aluminum. The process works equally well with magnesium.

Boron reinforcement has fathered a whole family of fiber variations. Boron filaments are produced by the chemical vapor deposition of boron on a substrate carbon or tungsten filament. In Borsic modification a surface coating of SiC is deposited over the boron to provide a higher-temperature capability. The boron is used primarily in an aluminum matrix and has a fiber loading of 45 to 50%.

Developments continue, one of the newest materials being BC-coated boron in a titanium matrix. This experimental material can be processed at relatively low consolidation pressures and may prove suitable for use in structures exposed to temperatures greater than 1000°F (538°C). Another new material is SiC deposited on a tungsten or carbon core and subsequently put into a 2024 or 7075 aluminum or titanium matrix.

## Fabrication Methods

The key to the use of MMCs and their bright future for weight-sensitive applications is fabrication. Realistic predictions of mechanical properties help determine which of several MMC fabrication methods should be chosen. End properties depend to a large degree on how the composite is made and fabricated into shape. In a typical composite, where filaments have an average strength of 450 kips/in$^2$ (3102 MPa) in the virgin state, strength drops to 402 kips/in$^2$ (2722 MPa), or 11% in the fabricated condition. Thus, the filaments in MMC parts can contribute their full residual strength to the composite. This means that predicted strength is attainable with state-of-the-art fabricating methods.

It is now possible to fabricate MMCs with a reinforcement, e.g., highly reactive boron filaments, keeping filament degradation to a minimum and obtaining an interfacial bond fully utilizing residual filament strength. Predicted strength and modulus values are attainable, and results indicate significant improvements in fatigue and stress rupture properties as well.

Fabricating methods include hot-pressure bonding, liquid-metal infiltration, electro-deposition, plasma spraying, and powder metallurgy. In any method the objectives are to incorporate the reinforcing filaments without breakage, with minimum degradation of filament reaction, with maximum filament loading, and with an interfacial bond strong enough to transmit loads from the matrix to the filaments. The reactive nature of boron requires a lot of emphasis on reducing filament degradation from reactions with the matrix and has led to a concentration on hot-pressure bonding.

### Hot-Pressure Bonding

This process has been used successfully to fabricate aluminum, magnesium, and titanium MMC panels. As shown in Fig. 4.87, filaments are spaced between foil layers of the matrix metal on a drum and sprayed with a binder (subsequently burned off) to hold the composite in place. The green composite is then removed from the drum, cut to size, and stacked in a hot-press die[13] (Fig. 4.87, bottom).

Light pressure is kept on the stack, and, as bonding temperature is reached, full bonding pressure is applied for the required time. Pressure is then released and the assembly cooled slowly to minimize residual stresses from mismatches in thermal expansion.

Composites made by hot-pressure bonding have several notable advantages:

High absolute strength

Strength-to-density ratios 2 to 4 times higher than those of conventional materials

Excellent strength retention and good stress-rupture properties at elevated temperatures

Outstanding fatigue strength

Full compliance with rule-of-mixture predictions

### Hot Isostatic Pressing

A unique approach to MMCs involves coating boron or graphite filaments with metal and then bonding the coated filaments into composite shapes by hot isostatic pressing. The coating on the filament becomes the matrix in the finished composite. Filament spacing is controlled by coating thickness. Hot isostatic pressing was chosen because temperatures are

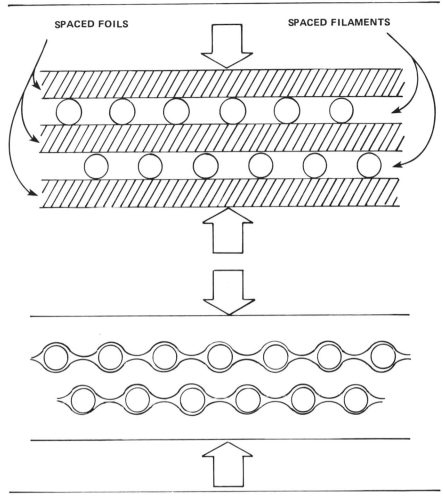

**FIG. 4.87** Cross section of filaments and foil matrix.[31]

much lower than those for sintering or hot pressing and gas pressure is distributed uniformly throughout the part, resulting in minimum distortion. The method permits fabrication of B–Ti composites well below 1600°F (871°C), the temperature at which boron and titanium react. Spacing of 0.0004 in (0.01 mm) between boron fibers does not cause voids or extensive damage to the fibers, suggesting that thin powder coatings would be sufficiently ductile under isostatic pressing to permit fabrication of dense structures.

Graphite-filament-reinforced metals have also been produced by electroless deposition of nickel or cobalt on pretreated graphite fibers, densification of fiber bundles by rotary swaging, sealing the bundles in mild-steel tubes, and hot isostatic pressing to consolidate the composite. The composites show good filament-to-matrix bonding, high density, and good separation of fibers.

### Liquid-Metal Infiltration

This method can produce structural shapes, such as rods, tubes, and I beams, with maximum properties in a uniaxial direction, but it is difficult because of the high temperatures involved and the relative instability of some filament materials. For example, at liquid-metal temperatures, boron filaments are stable only in structural-grade magnesium. Similarly, SiC is stable to liquid aluminum. The basic process involves passing a bundle of filaments through a liquid-metal bath so that the individual filaments are wet as they enter the bath and wiped of excess metal as they are drawn through an orifice at the bottom of the crucible (Fig. 4.88). The compressive strength of B–Mg composite produced by this technique varies with the filament volume that can be achieved:

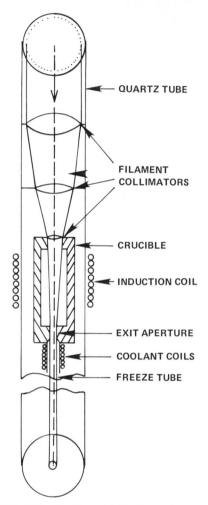

|  | Compressive strength | |
|---|---|---|
| Percent | ksi | MPa |
| 69.0 | 457 | 3151 |
| 46.7 | 236 | 1627 |

Tensile strength for the 69% filament composite exceeds 200 kips/in$^2$ (1380 MPa).

### Electrodeposition

This was one of the first methods used with reactive filaments because the moderate temperatures involved do not damage the filaments. Figure 4.89 shows a schematic of a typical setup for electrodepositing the matrix. Although multilayer circumferentially wound reinforcing structures and monolayer tapes have been formed by electrodeposition, isolated voids may occur between filaments in monolayer tapes or between filaments and filament layers in

QUARTZ TUBE

FILAMENT COLLIMATORS

CRUCIBLE

INDUCTION COIL

EXIT APERTURE

COOLANT COILS

FREEZE TUBE

**FIG. 4.88** Liquid-metal infiltration-casting apparatus.[31]

multilayer structures. Voids are increasingly difficult to eliminate with higher filament loadings. Nevertheless, monolayer tapes with up to 45% filament loading can be made without voids, and multilayer circumferentially wound composites have shown tensile insensitivity to hoop-oriented voids. Test data of B–Al rings fabricated by electrodeposition shows transverse strength only 20 to 30% of matrix strength because of voids. Electrodeposition processing can produce continuous monolayer composite tape in widths (limited only by the engineering ability to collimate thousands of circumferentially wound filament structures) several feet in diameter.

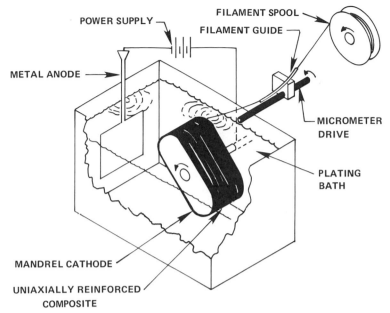

FIG. 4.89 Schematic of electrodeposition.[31]

## Plasma-Spray Deposition

Plasma-spray and chemical-vapor deposition have been used to form tapes and surface-of-revolution forms. In plasma spraying the metal matrix is sprayed onto the filaments to the desired thickness. Plasma arc spraying has been used with aluminum matrix on boron fibers; tensile properties proved superior to those of high-strength titanium alloys. Plasma spraying has the advantages of:

Simplifying fabrication of complex shapes

Allowing variety of fiber spacing and orientation

Reducing fiber-matrix reaction

Since droplets of plasma-sprayed matrix quench rapidly on contact with fiber, temperatures high enough to cause an adverse reaction are minimized.

B–Al broad goods have been produced by attaching boron filaments to aluminum-alloy foils by plasma-spraying aluminum. The fiber count, fiber diameter, and quantity of aluminum used are controlled to yield the proper fiber volume in the laminate. The selection of the type of fiber and aluminum matrix depends on the metallurgical process used to form the individual plies in the consolidation process. Since conventional boron fibers are severely degraded by contact with molten aluminum, most parts are made by diffusion bonding. Borsic fibers are required for consolidation processes such as brazing, in which the matrix is melted. The lower operating pressures in the consolidation process are traded off against the higher fiber costs. Since the use of B–Al is limited, broad goods are produced by drum wrapping rather than by continuous-tape machines.[14]

In the plasma-spray process, a thin sheet of aluminum foil is placed on the drum. The continuous boron fiber is then wound onto the drum over the foil to the desired spacing. After winding is complete, a uniform layer of plasma-sprayed aluminum is applied while the drum is rotating. The plasma-spray process is controlled to minimize both fiber degradation when the molten aluminum touches the fiber and oxidation of the aluminum. As an alternative, a mixture of metal powders can be sprayed onto an array of Borsic fiber and aluminum foil in the proper proportion to produce a low-melting eutectic master alloy. The master-alloy material can be consolidated at significantly lower temperatures and pressures than diffusion bonding requires. Ti–Bsc composite tapes have not been successfully plasma-sprayed but have been made with fugitive binders or as diffusion-bonded monolayer sheet sandwiches.

In a typical chemical-vapor deposition process filaments are wound on a mandrel and then heated to the pyrolysis temperature of the chemical vapor supplying the metal. This method can produce a fully dense, metal-matrix monolayer tape and, with repetition, a multilayer circumferentially wound composite similar to that made by electrodeposition.

## Powder Metallurgy

B–Al composites have been made by powder-metallurgy techniques involving cold pressing and sintering or hot pressing. Cold pressing and sintering is more difficult because the high pressures needed to press the aluminum powders to the required density can break the boron fibers or degrade fibers during sintering. Composites have been made by hot-pressing pure aluminum powder and boron-fiber mats. Powder is vibrated into stacked boron mats; after induction heating, pressure is applied gradually when the final temperature is reached. Some deformation can be expected by this method, particularly if there is good fiber-matrix bonding, but stresses can be annealed out.

## Other Methods

Other ways of producing MMCs include explosive consolidation, electroforming, rolling, diffusion bonding, and using fugitive organic resin binders.[15] The basic resin approach attaches boron fiber to metal foils with an acrylic or polystyrene binder. The process normally starts by placing the foil on the drum. When the fiber is wound, it is coated with the organic binder, causing the fiber to adhere to the foil. After winding, a second foil is placed over the surface and held in place with the binder. The product is called *green tape*.

A continuous casting process has been used to produce structural shapes of B–Al and B–Mg. Rods, tubes, I beams, and angles have been produced in continuous long lengths. Selective placement of filaments in the I beam puts the strength where it is really needed. In the B–Al beam (Fig. 4.90), the flange contains 65 vol % boron filament, while the web contains only 10% boron.[16]

Structural forms of constant cross section have been produced in AZ92 and EZ33 aluminum and magnesium alloys reinforced with boron filaments up to 65 vol %. Rods 0.020 and 0.045 in (0.51 and 1.14 mm) in diameter have been produced in lengths up to 5000 ft (1524 m) in the same alloys or in 2024 aluminum alloy reinforced with Borsic filaments. The rods, with 65 to 70 vol % filament, can be used as composite preforms in diffusion bonding or can be used directly for such structures as antennas.

Other direct casting of boron and graphite fibers has been attempted. Boron fibers were pultrusion-coated with magnesium and formed into a bundle; graphite fibers were nickel-coated by electrolytic and nonelectric plating processes. These material forms were then recast by conventional techniques into magnesium rods and bars.

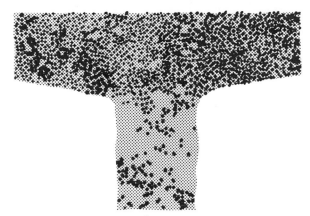

FIG. 4.90   B–Al cast I beam.[16]

The newly developed aluminum oxide FP has the necessary chemical compatibility and thermal stability to meet the requirements of cast fabrication. Developmental work demonstrated that excellent fiber-wetting and void-free plates, rods, and bars could be fabricated using $Al_2O_3$ fibers and molten-metal vacuum-infiltration techniques. The steps involved in using the fiber FP are shown in Fig. 4.91. The process is straightforward and direct, using virgin fibers and metal-alloy block as starting materials. The steps to producing fiber preform are analogous to those involved in producing an organic-resin-bonded

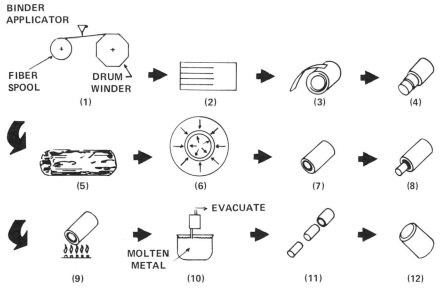

FIG. 4.91   Current FP technology for metal-matrix tubing: (1) prepare fiber sheet, (2) slit sheet to size, (3) lay up on male form, (4) install steel casting shell and remove form tool, (5) install vacuum bag, (6) pressure-compact at 100 lb/in² and 350°F (690 kPa and 77°C), (7) remove vacuum bag, (8) install inner steel tube and weld ends, (9) burn out binder at 1200°F (649°C), (10) infiltrate fiber, (11) remove inner and outer shells, (12) metal-matrix tube.[17]

component with the addition of high-temperature binder burnout and preheat followed by molten-metal infusion.[17]

## Consolidation-Diffusion Bonding

One of the most practical techniques today is to produce B–Al preform sheets from B–Al composite sheets and plates consisting of filaments and a special acrylic resin. The preform sheet shown in Fig. 4.92a consists of a layer of boron filaments bonded to the aluminum

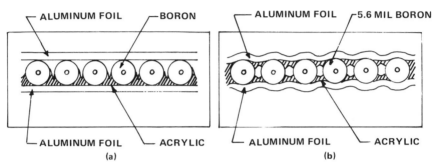

FIG. 4.92 (a) B–Al acrylic and (b) roll-bonded preform.[32]

foil by a specially formulated acrylic resin, which combines flexibility for handling with absence of residue in the subsequent diffusion bonding. A further refinement to improve handleability, particularly for fan-blade fabrication, is to roll the preform, deforming the aluminum foils slightly (Fig. 4.92b) to provide a mechanical lock in addition to the acrylic bond; this holds the filaments in place for subsequent cutting, forming, and diffusion bonding. The process is also suited to $B_4C$–B and SiC filaments. The sheets have been fabricated into MMC tape, sheets, and structural components by vacuum diffusion bonding. The acrylic resin has been formulated to disappear with complete volatilization during the preheating cycle. Figure 4.93 is a flowchart of a typical fabrication cycle. Sheet sizes up to 3 by 10 ft (0.9 by 3 m) have been produced with monolayer and plates containing anywhere from 50 to 100 plies. It should be noted in the figure that alternate layers of matrix foil and oriented fibers are laid down until the necessary amount of material for the final thickness has been assembled. Then by a combination of heat, pressure, and time in a vacuum [typical is 900 to 1000°F (482 to 538°C) for 1 h at 2 to 10 kips/in$^2$ (14 to 69 MPa) in a vacuum of $10^{-3}$ to $10^{-4}$ torr (0.133 to 0.0133 Pa)] the matrix is caused to flow around the fibers, bond to the next layer of matrix, and grip the reinforcing fibers slightly.

Successful aluminum-alloy matrix materials include:

| | | |
|---|---|---|
| 1100 | 5052 | 7039 |
| 3003 | 5456 | 7075 |
| 2024 | 6061 | 7178 |

Many metal-matrix parts are diffusion-bonded without vacuum using an air-bonding process. If bagging is required, 0.020 to 0.030 in (0.51 to 0.76 mm) steel retorts are typically used. Steel corrugations are placed at either end of the lay-up to provide for vac-

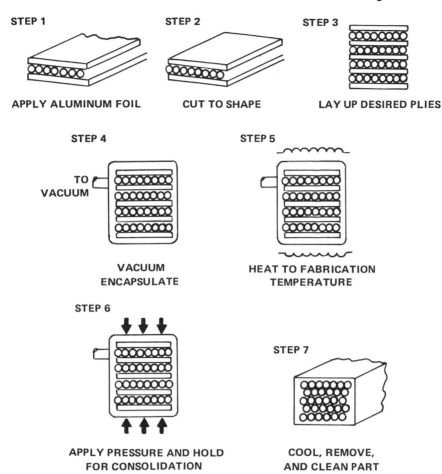

FIG. 4.93  Flowchart of B–Al composite fabrication.[32]

uum flow, and the two steel sheets forming the bag are seam-welded. Stainless-steel tubing for vacuum introduction is fitted into gaps in the welds, and the joints are made by brazing. A typical metal-matrix lay-up-tooling-bag assembly is shown in Fig. 4.5.

Closely akin to diffusion bonding is diffusion brazing, which is effective in fabricating B–Al components from composite monolayer.[18] The process relies on the diffusion of a thin surface film of copper into the aluminum matrix to form a liquid phase when heated above the Cu–Al eutectic temperature of 1018°F (548°C). A nitrogen environment is satisfactory for monolayer storage with permissible storage times of 80 h for the cleaned monolayer before vapor deposition and 800 h after copper coating. Techniques for producing 20-$\mu$in (508-$\mu$m) adherent physical-vapor-deposition copper layer on the monolayer have been developed, and complex B–Al structures have been fabricated with structural shapes utilizing a mechanical forming technique to pressure-form all monolayers simultaneously before bonding (Fig. 4.94). These structures demonstrated strengths in excess of design ultimate loads at 600°F (316°C). An integrally stiffened B–Al panel was fabricated

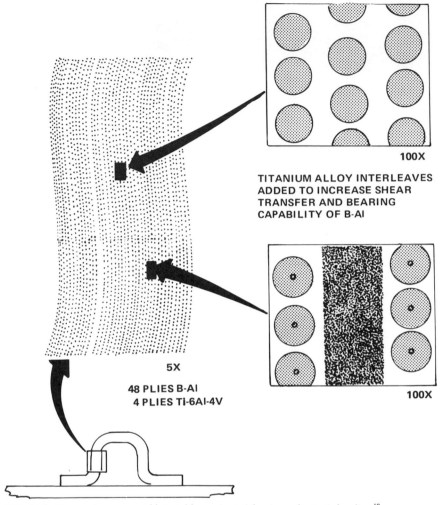

100X

**TITANIUM ALLOY INTERLEAVES
ADDED TO INCREASE SHEAR
TRANSFER AND BEARING
CAPABILITY OF B-Al**

5X

**48 PLIES B-Al
4 PLIES Ti-6Al-4V**

100X

**FIG. 4.94**  Structural component fabricated by mechanical forming and eutectic bonding.[18]

utilizing simple reusable tooling which can be cost-competitive with titanium machinings
(Fig. 4.95).

### Processing for Diffusion Bonding

**Ply Cutting:**   The laminating process for metal-matrix parts begins by cutting individual
plies from sheets of fiber-foil broad goods or preconsolidated monolayer material. The
extreme hardness of boron or Borsic fiber would cause severe tool-wear problems if the
fibers were actually cut. Fortunately, a simple shearing action, such as that provided by
hand shears, produces clean cuts with little wear because the individual fibers snap as soon
as the shearing force is applied. The only actual cutting action occurs in the metal matrix.

Hand cutting of plies is extremely slow and expensive; ply cutting can be done econom-

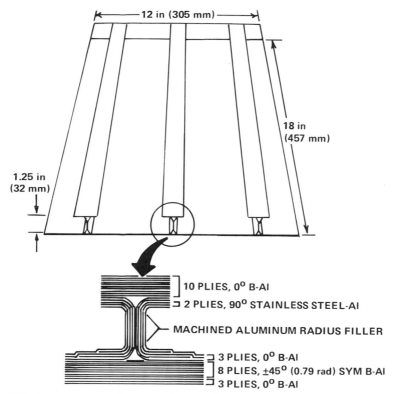

**FIG. 4.95**  Integrally stiffened B–Al skin panel.[18]

ically and effectively by automatic or semiautomatic methods. One simple method involves the use of a *roll cutter,* designed to cut 43 plies for a fan blade in one operation. It consists of 0.5-in-thick (12.7-mm) steel plate with 0.125-in (3.2-mm) steel mandrels used as cutters duplicating the 43 different ply shapes. Double-stick tape was used to attach the cutters to the backing plate, facilitating quick removal and reattachment when a ply shape is changed.

When a full set of plies is cut, a plasma-sprayed Bsc–Al tape is placed on the 43 cutters and over it a 0.5-in (12.7-mm) sheet of cork. The sandwich is passed through sheetmetal rollers; the pressure forces the cork down over the cutters' sharp edges, breaking the fibers and cutting the metal matrix. The results are more uniform shapes and cleaner edges than by hand-cutting methods, and the process is fast. Steel-rule, clicker, or blanking dies can be substituted, and the rolling motion could be replaced by a pressing motion to do the cutting. Within the next several years the process will be fully automated. One manufacturer now uses a fully automatic ply-cutting machine in which the cutting tool is a pin that nibbles the material away along the cut line by progressively punching small holes in the tape at a rapid rate.

**Preforming:**  In processing highly contoured parts, e.g., engine blades, a preform is used to minimize fiber movement during consolidation of the part. In *cold preform* the ply is made to conform to a contoured surface and tack-welded to its adjacent ply or plies. The contoured tool is copper or some other high-electrical-conductivity material to facilitate tack welding. *Hot preform* tools, which normally creep-form the ply or plies, are machined from stainless-steel or nickel-alloy plate stock, but the lower pressures involved permit the use of lower-strength steels or even titanium.

**Consolidating:** Diffusion-bonding parameters depend on the matrix of the composite and are shown in Table 4.16. In consolidation one of the most critical factors is the die

**Table 4.16** Diffusion-Bonding Parameters[14]

|  | Pressure | | Temperature | |
|---|---|---|---|---|
| Process | ksi | MPa | °F | °C |
| Consolidation | 3.5–12 | 24–83 | 900–1800 | 482–982 |
| Brazing | 0.1–1 | 0.69–6.9 | 800–1200 | 427–649 |

design: its size and contour must allow adequate transfer of the applied pressure to the preform. Depending on the matrix material, dies are made from stainless steel, nickel alloy, or aluminum alloy.

For parts made from monolayer material, consolidation without applied vacuum is possible; if fugitive binder materials are used, the pressing must be done under vacuum. Parts that are consolidated under vacuum are encapsulated in a hermetically sealed steel or titanium bag, as determined by the requirements of service temperature, cost, weldability, and formability. Stainless-steel or titanium tubing attached to the bag connects to the vacuum source.

Blades and vanes are normally fabricated in a closed die. The die design shown in Fig. 4.96 has been used successfully to hot-press Bsc–Al at 1100°F (593°C) and 5 kips/in$^2$ (35

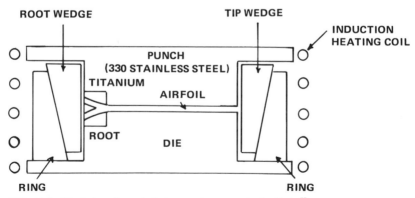

**FIG. 4.96** Die-design schematic for hot-pressing composite fan blade.[27]

MPa) in 1.5h. A manufacturing development program for vanes demonstrated that diffusion bonding can be successfully applied to B–Al vanes (Fig. 3.58).

SiC filaments, being relatively new in the market, are still being evaluated for consolidation by molding and laminating. The SiC filaments are being hot-molded into 6061 aluminum laminates at 1100°F (593°C), pressures of 0.4 to 0.8 kip/in$^2$ (2.8 to 5.5 MPa), and 48% fiber volume. An SiC–Al composite [with the SiC continuous fibers laid in a 0°/ 90° (0/1.6 rad) crosswise pattern] has a strength of over 120 kips/in² (827 MPa) and a modulus of over 20 $\times$ 10⁶ lb/in² (138 GPa). Low-cost hot molding of aluminum composites is a fabrication method being developed for such aircraft structural components as skin panels, stringers, and beams so that they can be molded directly to shape without having to be machined to profile. Laminates of SiC are also being hot-pressed into 6 Al–4V titanium at 1700°F (927°C) and 6 kips/in² (41 MPa) for 30 min and 35% fiber volume. SiC–Ti composites have higher mechanical properties, due to the stronger matrix, and maintain usable properties up to exposed temperatures of 1200°F (650°C). Hot-pressed SiC–Ti engine components are being tested as blades, vanes, and shafts.

A promising turbine composite material is tungsten-fiber-reinforced superalloy. Fiber-reinforced superalloy composites could increase operating temperatures in hot-section gas-turbine components by 400 Fahrenheit (222 Celsius) degrees over those currently possible with superalloys.[19] This class of composites embeds metal or ceramic fibers in a superalloy matrix. The fibers provide high-temperature strength, creep-rupture resistance, and low- and high-cycle fatigue strength. The superalloy protects the fibers from the engine environment.

A promising first-generation composite uses tungsten wire in a superalloy matrix. Tungsten-fiber-reinforced superalloy (TFRS) has much greater strength than conventional superalloys used for turbines at temperatures above 1800°F (928°C). For example, composites using W–Re–Hf–C will have over 4 times the strength of conventional superalloys at 2000°F (1093°C). TFRS also has an advantage over conventional superalloys in thermal conductivity and thermal expansion. The substantially higher conductivity of TFRS compared with that of superalloys, coupled with greater strength, can reduce coolant-flow requirements in impingement-cooled moderate-strength TFRS blades, in contrast to directionally solidified superalloy blades of similar configuration. By reducing thermal gradients in a blade high thermal conductivity can also reduce thermal strains that might otherwise cause low-cycle fatigue damage. The lower thermal expansion of TFRS (about half that of most superalloys) also reduces thermal stresses and strains. Ceramic thermal-barrier coatings are likely to prove more durable on TFRS because TFRS expansion nearly matches the expansion of thermal-barrier coatings.

At first glance, making TFRS composite components appears to be complex and expensive, but this is not necessarily so. The procedure parallels that for making other composites. Fibers and matrix foil are pressed at temperatures to form composite sheets that are cut, stacked, and bonded together to form components. Hollow components are produced using cores removed chemically after final pressing. Both solid and hollow first-stage turbine blades have been made to demonstrate the feasibility of this method. The future for TFRS lies in determining the benefits that can be achieved by applying the material to several different turbine-engine components.

# REFERENCES

1. Lubin, G.: "Handbook of Fibreglass and Plastic Composite," Van Nostrand Reinhold, New York, 1969.

2. Shockey, F. D.: Advanced Composite F-5 Fuselage Component, *5th Natl. SAMPE Tech. Conf., Kiamesha Lake, N.Y., Oct. 9–11, 1973,* pp. 423–431.

3. Sanders, H., J. A. Munyak, and L. Poveromo: Effective Methods for Fabrication of Large Polyi-

mide-Matrix Aircraft Components, *5th Natl. SAMPE Tech. Conf., Kiamesha Lake, N.Y., Oct. 9–11, 1973,* pp. 627–642.

4. Poveromo, L. M., C. Paez, and R. Sorraffe: Radome Design/Fabrication Criteria for Supersonic E. W. Aircraft, *10th Natl. SAMPE Tech. Conf., Kiamesha Lake, N.Y., Oct. 17–19, 1978,* pp. 166–186.

5. Improved Advanced Composite Bag Molding Processes, AFML-TR-73-161, Contr. F-33615-71-C-1705, May 1973.

6. Warner, G.: Design Considerations for Composite Leaf-Springs, *SME Compos. Conf., Los Angeles, June 10–12, 1980,* EMBO-420.

7. Mitchell, S.: T700 Composite Engine Inlet Particle Separator Swirl Frame, *MTAG Non-Met./Compos. Progr. Rev., Orlando, Feb. 23–25, 1981,* pp. R1–R29.

8. McAllister, L. E., and A. R. Taverna: Reinforced Carbon-Carbon Composites for Advanced Reentry Vehicle Applications, *AVCO Systems Div. Tech. Rep.* AFML-TR-71-57, Contr. F33615-69-C-1758, Wright-Patterson AFB, Ohio, May–November 1970.

9. Seibold, R. W., and E. F. Disser: High-Speed Braiding: An Approach for Fabrication of Reentry Vehicle Heatshields with Seamless, Shingled Construction, *10th Nat. SAMPE Tech. Conf., Kiamesha Lake, N.Y., Oct. 17–19, 1978,* pp. 227–237.

10. Winship, J.: Plastics: Your Future Feedstock? *Amer. Mach. Spec. Rep.* 677, May 15, 1975, pp. 53–68.

11. Dreger, D. R.: Processes That Produce Massive Plastic Parts, *Mach. Des.,* Jan. 24, 1980, pp. 58–64.

12. Banker, J. G.: Metal Matrix Composite Fabrication by Liquid Infiltration, *SAMPE Q.* 5(2):39–46 (January 1974).

13. Krukonis, V. J., and T. Schoenberg: Manufacturing Methods for Low-Cost Metal Matrix Composite Materials (Continuous Green Tape), *AVCO Corp. Systems Div. Final Tech. Rep., May 1974–April 1975,* AFML TR-75-126, Contr. F33615-74-C-5123, January 1976.

14. Debski, R. T.: Boron/Aluminum Fan Blade Development, *Pratt & Whitney Govt. Products Div., Final Rep.* AFWAL-TR-80-4202, *Oct. 18, 1976–July 31, 1980,* Contr. F3316-76-C-5318, Wright-Patterson AFB, Ohio.

15. Doble, G. S., and I. J. Toth: Roll Diffusion Bonding of Boron Aluminum Composites, *Proc. 1975 Int. Conf. Compos. Mater.,* vol. 2, pp. 775–788, Metallurgical Society of AIME, New York, 1976.

16. Shaver, R. G.: Metal/Matrix Compositing by Continuous Casting, pp. 232–241 in Composite Materials Engineering Design, *Proc. 6th Symp. Compos. Mater. Eng. Des., St. Louis, May 11–12, 1973.*

17. Pinckney, R. L., and A. K. Dhingra: Design and Fabrication of High-Modulus Fiber-Stabilized Magnesium Transmission Cases, *4th Conf. Fibr. Compos. Struct. Des., San Diego, November 1978.*

18. Niemann, J. T., and R. A. Garrett: Eutectic Bonding of Boron-Aluminum Structural Components, II: Development and Application of the Process, *Weld. J.,* 53(8):351s–360s (August 1974).

19. Petrasek, D. W., and R. A. Signorelli: Tungsten Fiber Reinforced Superalloys Status Review, *Lewis Res. Cent. NASA Tech. Mem.* 82590, January 1981.

20. DOD/NASA Structural Composites Fabrication Guide, 2d ed., vol. II, May 1979, Contr. F33615-77-C-5256, Lockheed Company, Manufacturing Technology Division, Air Force Material Laboratory, Wright-Patterson AFB, Ohio.

21. Schwartz, M., and G. Jacaruso: A Giant Step toward Composite Helicopters, *Am. Mach.,* March 1982, pp. 133–140.

22. Lost, C. T.: Pultrusion: Composites Road to Mass Production, *Iron Age,* 221(5):27–29 (January 1978).

23. Mayfield, J.: Hornets Fly on Composite Wings, *Am. Mach.*, November 1978, pp. 107–110.

24. Stedfield, R.: The Molding-Compound Alphabet Soups, *Mater. Eng.*, September 1978, pp. 50–55.

25. Faddoul, J. R.: Preliminary Evaluation of Fiber Composite Reinforcement of Truck Frame Rails, Rep. NASA TM X-73582, Lewis Research Center, *SAE Congr. Expos., Detroit, Feb. 28–Mar. 4, 1977.*

26. *Mater Eng.*, June 1980, p. 43.

27. Advanced Composites Design Guide, 3d ed., Air Force Materials Laboratory, Wright-Patterson AFB, Ohio, 1973.

28. Cornsweet, T. M.: Manufacturing Methods for Advanced Composites, *SAMPE Q.* 3(2):28–33 (January 1972).

29. Stedfield, R.: Composites Come On Strong, *Mater. Eng.*, April 1980, pp. 57–60.

30. *Prod. Eng.*, May 1980, pp. 60–65.

31. Alexander, J. A.: Five Ways to Fabricate Metal Matrix Composite Parts, *Mater. Eng.*, July 1968, pp. 58–63.

32. Joseph, E., V. Krukonis, and A. W. Hauze: Exploratory Development and Evaluation of Low-Cost Boron Aluminum Composites, *Avco Corporation, Systems Division, 3d Q. Rep.*, Contr. F33615-74-C-5082, ASD, November 1974.

33. Saylor, D. K.: Hybrids, *Mod. Plast.*, Encyclopedia Issue 1978–1979, December 1979, pp. 182–184.

# BIBLIOGRAPHY

Ault, G. M., and J. C. Freche: Status of Composites for Aeropropulsion Applications, *AIAA Mon. J.*, Aug. 8, 1979.

Bania, P.: ATDE Metal Matrix Shaft, Avco Lycoming Division, Contr. F33615-80-C-5176, Wright-Patterson AFB, Ohio.

Bonassar, M. J., and J. J. Lucas: Fiber Reinforced Plastic Helicopter Tail Rotor Assembly (Pultruded Spar), *Sikorsky Aircraft Final Tech. Rep. August 1975–October 1978*, Avradcom TR-79-45, Cont. DAAJ02-76-C-0001, ATL, Ft. Eustis, Va.

Brayden, T., Jr.: Laminate Conductivity and Events Occurring during the Cure Cycle, *12th Natl. SAMPE Tech. Conf., Seattle, Oct. 7–9, 1980.*

Brown, R. L. E.: "Design and Manufacture of Plastic Parts," Wiley, New York, 1980.

Cahuzac, G., and Y. Grenie: The Automatic Weaving of 3D Contoured Preforms, *12th Natl. SAMPE Tech. Conf., Seattle, Oct. 7–9, 1980.*

Champion, A. F., and H. K. Street: Fabrication of Boron-Reinforced Magnesium Composites by Diffusion Bonding of Plasma-Sprayed Monolayer Tapes, *Sandia Lab. Summ. Rep.* SC-DR-720677, September 1972.

Christian, J. L.: Fabrication Methods and Evaluation: Boron/Aluminum Composites, *Proc. 1975 Int. Conf. Compos. Mater.*, vol. 2, pp. 706–736, The Metallurgical Society of AIME, New York, 1976.

Dharan, C.: Pultruded Braided Hybrid Composites, *12th Nat. SAMPE Tech. Conf., Seattle, Oct. 7–9, 1980.*

Edmonson, R. E., and R. W. Harrison: Low-Cost Process for Boron/Aluminum Tape, *General Electric Co., Aircraft Engine Group, Interim Proc. Rep.* IR 284-1 (V), *June 1971–January 1973*, Contr. F33615-71-C-1646 (February 1973).

Fackler, M. B.: Design and Materials: A Partnership, *6th Major SME Conf. Struct. Compos. Manuf. Appl. Los Angeles, June 10–12, 1980*, EM80-423.

Gigerenzer, H., and G. C. Strempek: Fabrication of Discontinuous Graphite-Aluminum Composites via Pultrusion, *Fiber Materials, Inc. Final Rep.*, DAAG46-76-C-0068, AMMRC-CTR-77-8,

Biddeford, Me., February 1977; Fabrication of Graphite-Aluminum Composites via Pultrusion, *Fiber Materials, Inc. Final Rep.* DAAG-46-77-C-0036, AMMRC TR-78-16, Biddeford, Me., March 1978.

Gray, D., and R. Beck: Titanium Metal-Matrix Composite Shafts, Teledyne CAE, PR4, Contr. F33615-80-C-5016, Wright-Patterson AFB, Ohio, May 1981–August 1981.

Hill, S. G., and J. T. Hoggatt: Development of Hybrid Thermoplastic Composites, *Boeing Aerospace Rep.* D180-18752-3, March 1977.

Hoffstedt, D. J., L. C. Ritter, and D. J. Toto: Low-Cost Forming Influence on Reinforced Thermoplastic Mechanical Properties, *Boeing Vertol Final Tech. Rep.* AMMRC-TR81-36, *January 1980– December 1980,* Contr. DAAG46-79-C-0092, Army Materials and Mechanics Research Center.

Hordon, M. J.: Tensile Properties of Boron Carbide Ribbon in Titanium Matrix, *J. Compos. Mater.* 7:521–524 (October 1973).

Prewo, K. M.: The Fabrication of Boron Fiber Reinforced Aluminum Matrix Composites, *Proc, 1975 Int. Conf. Compos. Mater.* vol. 2, pp. 816–838, The Metallurgical Society of AIME, New York, 1976.

Slate, P. M. B.: Explosive Fabrication of Composite Materials, *Proc. 1975 Int. Conf. Compos. Mater.,* vol. 2, pp. 743–757, The Metallurgical Society of AIME, New York, 1976.

Smith, C. W.: Development of Manufacturing Methods for Joining Thermoplastic Composites, General Dynamics, Convair Division, Contr. N00019-76-C-0227 Mod P00024, Department of the Navy, Navair, Washington, 1977.

Steinhagen, C. A., and M. W. Stanley: Boron/Aluminum Compressor Blades, *General Electric Co., Aircraft Engine Group, Final Rep. November 1971–November 1973,* AFML TR-73-285, Contr. F33615-71-C1230, Air Force Materials Laboratory, Wright-Patterson AFB, Ohio, October 1973.

Tallbacka, D. W.: TMC Uses and Applications, *SPI Reinf. Plast./Compos. 35th Ann. Tech. Conf., Los Angeles, May 1980.*

White, M.: Braiding of Helicopter Main Rotor Blade Spars, *26th Natl. SAMPE Tech. Conf., Los Angeles, April 28–30, 1981; Avradcom Rep.* TR81-F-9, Kaman Aerospace Final Report, Contr. DAAG46-78-C-0070.

Yamada, M., T. Iwai, K. Matsumoto, and J. Walton: TMC-Combining SMC-BMC Compounding with New Impregnating Efficiency and Economy, *SPI Reinf. Plast./Compos. 33d Annu. Tech. Conf., Washington, Feb. 7–10, 1978.*

Zeitz, R. R.: Filament Winding Precision Resin Impregnation System, *Avradcom* TR-79-15, Bell Helicopter Textron, Final Technical Report, Contr. DAAJ01-77-C-0777, U.S. Army Aviation Research and Development Command, St. Louis, Mo., August 1977–March 1979.

Gr-Ep prepreg before automatic nesting and cutting on a Gerber reciprocating-knife machine. *(Sikorsky Aircraft, Division of United Technologies.)*

# Automated Fabrication Methods

More attention must be paid to production-line methods to automate producing, cutting, applying, and handling composite materials. Research and development efforts already have made some of the structures amenable to computer-aided design and numerical-control equipment. About 70% of the total cost of boron and graphite filamentary composites is spent on fabricating into end shapes. Automation could reduce this cost by 50%. This does not mean that older methods, like hand lay-ups and spray-ups, built-up composites, and reinforced molding compounds, will disappear overnight. They will be with us for some time to come along with the sophisticated SMCs, BMCs, and RRIMs. All are true composite "structures," built to develop optimum properties.

Each of the continuous processing techniques discussed in Chap. 4 was developed around a specific end product, and until recently applications were limited to these areas. Now, however, these processes are being recognized for what they really are, the equivalent of the plate-sheet, and structural-profile rolling mill in the steel industry or the extrusion plant in the aluminum industry.

Future payoffs of composites will achieve their potential only if companies can speed up the making of parts, which are largely handmade. Many companies hope to solve that problem by automating the fabrication of composite parts. Proposed systems include a "factory of the future" with a robot which moves 4-

ft-wide (1.2-m) bolts of graphite broad goods to cutting tables, where shapes are cut by a reciprocating knife and then arranged for curing. The sequence will take a mere 15 s compared with 30 min for present manual methods. A company building composite stabilizers on a partly automated assembly line is now putting the entire system under computer control. Automated processing of tapes has been widely investigated to reduce lay-up costs. One approach was using automated facilities at the material supplier's plant to handle alternative materials, thus reducing handling in the processor's facility. These concepts include wide tapes [up to 48 in (1.2 m)], uniweave fabric, and fabrics of conventional construction which are unrolled, automatically cut into full or partial plies, sorted into kits, and then laid up onto a tool.

Machines for tape lay-up have been under development since 1967. At first they used numerical control to lay up plies of simple geometry as determined by simple blade cutters. A head able to cut the tapes sequentially into arcs is under development to resolve this problem, which is specific to ply-on-ply layups. The output of tape-lay-up machines is significantly higher than that for manual lay-up, but lay-up also typically includes ply trimming, stacking, and inspection before the part is ready to prepare for cure. Hence the automated lay-up machines or wide-goods dispensers are now being incorporated into various modular systems which can convert prepreg into fully laid-up parts. The modules include prepreg trimming units, robotic or flip-table stacking units, devices to seat plies on complex curvature molds, TV cameras for lay-up inspection, and modules to form flat multiple-ply lay-ups in substructure components. These automated systems are essential for cost-effective, large-scale production and represent the future technology. A computerized quality-control system is also being developed to monitor the conversion of prepreg into a fully inspected part.

Automation efforts vary from company to company as well as in various industries from computer-aided design (CAD), to computer-aided manufacturing (CAM), to computer instructions to assist in preparation of manufacturing instructions for fabricating tools.

The next step for composite tools is into the paperless factory, where automation is being developed and applied to include handling, storage, and retrieval of frozen materials through automated cutting, knitting, and lay-up. Assembled parts would be carried automatically to autoclaves for curing on an overhead monorail or other conveyor. This chapter disscusses the accelerated use of CAD and CAM and the tooling and factory automation being generated in composites fabrication.

## COMPUTER-AIDED DESIGN AND MANUFACTURING

With CAD and CAM, design changes from metal to composite can be made with minimal risk, since the design can be evaluated as it develops. Much of the trial and error in the design process is eliminated, and by proving the design in the computer before the prototype is built, development time and prototype expense can be cut down. The immediate goal for many companies is to extend the basic computer-graphic and numerical-control technology into tool design and tool building. By computerizing the entire body-tooling cycle from design to production tooling, maximum lead-time and dollar savings can be achieved. A generalized CAD-CAM network of systems will not only encompass composite design, engineering, tooling and mold design, and mold machining but will also help maximize utilization of equipment and material.

Along with this design-and-build technology is the need to improve productivity of basic production machines, materials, and processes. To produce materials at competitive costs will require new machines to compound the material and manufacture the parts. CAD and CAM will make it possible to develop computerized techniques to measure and mix the materials and control the machinery. In addition, by marrying process computers to sensing devices that can determine position, motion, and shape, manufactured pieces will be inspected automatically, making it possible to have machines that refuse to put out unacceptable parts.

Several parts of CAD-CAM integration have already been implemented. For example, a unique application of CAD-CAM has applied the principles of integrated manufacturing to a composite fuselage section. From the beginning all functions in designing and producing the fuselage worked together using a common data base, which included loft data, three-dimensional layout, stress analysis, detail drafting, tooling, numerical-control programming, manufacturing, and quality assurance. Using loft data, the designer creates a three-dimensional skeleton model of the section (Fig. 5.1) showing shape and all basic

**FIG. 5.1** Computer-aided design-interactive system depicting fuselage section. *(Sikorsky Aircraft, Division of United Technologies.)*

geometry, including section cuts and accurate locations of hat sections and ribs and their intersection with the fuselage surface. Stress analysts modify the design model to create a structural finite-element model. Then they produce a computer run to obtain stress, loads, and deflection data. Next comes detail design. From the stress-analysis results and aerodynamic and design considerations, the number of plies, ply patterns, and ply orientation is obtained. The designer determines each individual ply pattern and the overall ply-installation design. Now the CAD system performs the following functions:

1. Flat patterns of the individual plies are developed, accurately representing the three-dimensional geometry. There are only 50 different ply shapes (Fig. 5.2).

**FIG. 5.2**   Flat-pattern nesting. *(Sikorsky Aircraft, Division of United Technologies.)*

2. Tooling holes are coordinated and incorporated in all drawings, and section cuts are made automatically for tooling and substructure definition.

Now the CAM system takes the above three-dimensional work with the following results:

1. Trim lines, tooling holes, and the panel and ply geometry shape are accurately depicted.
2. An optimized cutter path for each ply is developed to cut the composite epoxy cloth plies (broad goods). This includes *nesting,* an arrangement of the parts to be cut from each sheet that minimizes scrap.
3. In the production shop, individual plies are placed in the bonding fixture in proper sequence to secure the desired contours and orientation of the composite fibers (Fig. 5.3).
4. After the part is cured, quality assurance refers to CAD-CAM-generated points that define the outer skin shape. These points are compared with measured dimensions to check the finished part accurately and quickly.

Estimates from several companies provide the following results from the implementation of CAD-CAM systems:

50% reduction in design costs

30% reduction in tooling costs

80% elimination of rework costs resulting from design, details, and tooling

Improved use of materials, reduced waste, and 10% reduction of overall manufacturing costs

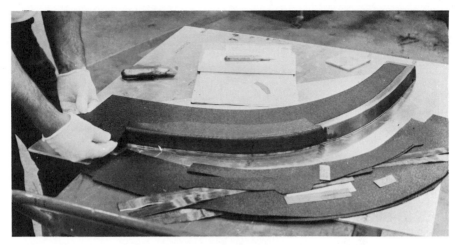

**FIG. 5.3** Graphite prepreg plies being oriented on steel tool. *(Sikorsky Aircraft, Division of United Technologies.)*

## TOOLING

Of all the elements required for successful composite fabrication, tooling developments (materials and processing) rank at the top of the list. Innumerable tooling programs have been concerned with establishing the tooling technology required to fabricate large and complex advanced-composite structures. Experience has shown certain tooling characteristics to be essential for tools used to cure large and complex composite parts:

1. Compatible thermal expansion between the tool and the part
2. Rapid heat transfer
3. Tool stability
4. Tool surfaced to produce a smooth part
5. Long service life
6. Minimum maintenance

Most standard tooling materials will not meet all these requirements, especially compatible thermal expansion between tool and composite part, which is often difficult to achieve unless the tool for making the part is made of the same material and has the same ply orientation as the part. This is not to say that steel, aluminum, or fiber glass should not be used for making composite-curing tools. Tools made of these materials are excellent for making flat or slightly contoured parts, where thermal expansion is not a problem; if shrinkage factors are calculated and incorporated into the tool for the particular prepreg material being used, adequate complex composite parts can be made (Figs. 5.4 and 5.5).

### Tooling Materials

Besides the conventional steel, aluminum, and electroformed nickel (Chap. 4) other materials have been successfully used. A Gr–Ep system developed for large shell-type tools used

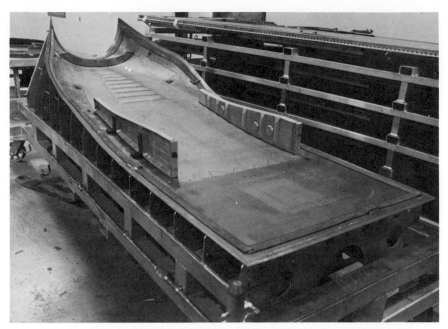

FIG. 5.4    Steel tool for compound contoured parts. *(Sikorsky Aircraft, Division of United Technologies.)*

**FIG. 5.5**    Hybrid composite (Kv–Gr–Ep) produced in tool shown in Fig. 5.4. *(Sikorsky Aircraft, Division of United Technologies.)*

epoxy-impregnated graphite unidirectional tape and was thermally stable up to 365°F (185°C).

## Fiber Glass

The two basic methods of fabricating high-temperature plastic tools use laminated Ep–Gl or cast epoxy. The basic methods and variations of these are shown in Table 5.1 and Fig. 5.6, which can be used to select the most cost-effective materials and fabrication procedure. Direct comparisons of labor and materials cost, shrinkage allowances, and tolerance devia-

**Table 5.1** Methods of Tool Fabrication

| Method | Resin curing temperature | | Description |
| --- | --- | --- | --- |
| | °F | °C | |
| Laminated tools | | | |
| Plastic-faced plaster | 275 | 135 | Low-cost short-flow method of fabricating tooling for prototype, test, or short-run production parts (up to 3 parts can be made); maximum service temperature 275°F (135°C) |
| Room-temperature cure; high-temperature epoxy laminating | 400 | 204 | Wet-lay-up method; tool is made directly on assembly model, vacuum-bagged, and gelled for 16 h at room temperature then removed from model and postcured in self-supporting condition |
| Intermediate high-temperature epoxy laminating | 400 | 204 | Wet-lay-up method; tool made on plastic-faced plaster transfer from model; tool is laminated in two stages, vacuum-bagged, and oven-postcured while still vacuum-bagged. |
| High-temperature epoxy laminating | 500 | 260 | Wet-lay-up method similar to plastic-faced plaster but requiring 500°F (260°C) resin system |
| Cast tools | | | |
| Intermediate high-temperature epoxy casting | 400 | 260 | Uses metal-filled-epoxy casting compound, which is poured into plastic-faced plaster mold, gelled for 16 h at room temperature, and oven-postcured in the mold |
| Intermediate high-temperature epoxy-cored casting | 400 | 260 | Similar to above except that the casting is cored out to reduce weight and material cost and improve heat-up and cool-down rates |

tions for various high-temperature tool-fabrication methods should be available to the designer.

**Guidelines for Selection of Fabrication Methods:**  In addition to the data presented in Table 5.1 and Fig. 5.6 other factors must be considered in selecting a tool-fabrication method.

*Production Process and Required Service Temperature:*  A cast high-temperature tool may cost less to fabricate than a laminated tool, but the cast-tool heat-up and cool-down characteristics may not be within the cure envelope for that particular process. Having less mass, a large laminated tool will heat up and cool down much faster than a cast tool. Moreover, a laminated tool is less prone to damage by thermal shock than a large cast tool.

*Tool Configuration:*  A tool with joggles or intricate detail requires a high level of skill on the part of the tool fabricator to produce a high-quality tool by laminating. Consideration should always be given to casting tools of this type.

*Tool Size:*  The overall dimensions of a tool normally dictate whether a tool should be laminated or cast. As a general rule, it is not practical to cast a high-temperature plastic tool with over 10 ft² (0.9 m²) of tooling surface.

*Tool-Handling Requirements:*  Tool-handling requirements should be an important consideration in selecting the fabrication method. A tool having more than 10 ft² (0.9 m²) of tooling surface should be laminated because a lighter-weight tool is easier to handle and transport in production. Since it has better durability, the laminated tool is particularly suited for manufacturing operations that require transportation of a tool through several

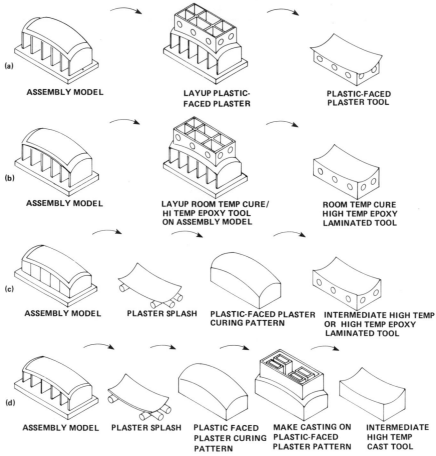

**FIG. 5.6** Fabrication sequence for high-temperature plastic tools: lamination at (*a*) 275°F (135°C), (*b*) 400°F (204°C), (*c*) 400 and 500°F (204 and 260°C) and (*d*) casting at 400°F.

work areas for fabrication of a single part. A cast tool is more prone to impact damage and is more suited, for example, to a vacuum-forming operation, which requires but one setup to make many parts.

**Metal-Faced Fiber Glass:** A cost-effective method for molding dimensionally accurate and structurally sound composite parts has been developed using metal-faced fiber-glass tools. Coatings such as zinc spray, steel spray, and nickel-zinc spray are successful coatings using the setup shown in Fig. 5.7 and produced several prototype stabilizer fins for the E-2C and JA-37 aircraft.

## *Graphite*

Solid carbon has been used by several companies as tooling to produce hat sections and conical and cylindrical parts. These tools have been used for Gr–Ep, Kv–Ep, and Gr–PI with good success. The parts produced from Gr–PI were cured at 800°F (427°C) and 0.2

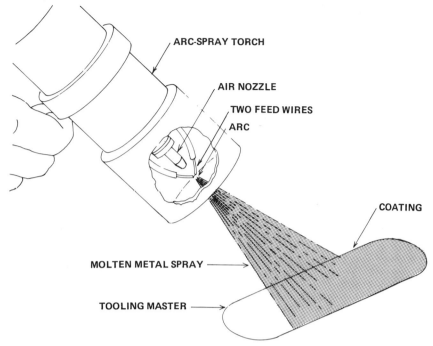

ARC-SPRAY TORCH

AIR NOZZLE

TWO FEED WIRES

ARC

COATING

MOLTEN METAL SPRAY

TOOLING MASTER

**FIG. 5.7**   Schematic representation of arc-spray equipment.[8]

kips/in$^2$ (1.4 MPa) in an autoclave. Before use the graphite tool is fired in a furnace with an inert atmosphere to burn out any contamination; it is then coated with a parting agent, which is sprayed or brushed on in layers. This application continues until it is impossible to rub or smear any graphite from the tool surface. To prevent shop personnel from dirtying their hands when handling the graphite tools all surfaces of the tool should be coated with a parting agent. Once the tool has been coated and initial parts cured, they require no subsequent coating.

There are several advantages in using this type of tooling material:

1. They can be cemented or sections can be bolted together; threading graphite presents no problem.
2. Heaters can be installed inside a tool and cemented in place to assist in draping prepreg material by preheating before cure.
3. Graphite tooling offers flexibility in curing at 350°F (177°C) or at 700 to 800°F (371 to 427°C), depending on the material.
4. The coefficients of expansion of Gr–Ep and Kv–Ep materials with graphite tooling are practically the same.

### Ceramics

Ceramic tooling can eliminate molding stresses because the coefficients of thermal expansion of tooling and composite parts can be closely matched to produce quality parts (differences in thermal expansion cause loss of dimensional accuracy and distort or damage composites). In addition to being thermally and dimensionally stable, ceramics are low-cost, easily cast, and resistant to thermal shock. Unlike fiber-glass tooling, ceramics do not undergo shrinkage during thermal cycling at temperature below the sintering temperature of 2000°F (1093°C).

One candidate material, castable fused silica, has been tested sucessfully and has shown high resistance to thermal shock. Tests are performed on ceramics both with and without reinforcement because ceramics alone are characteristically brittle. Crack resistance is improved significantly with high-tensile-strength reinforcements. Fiber-reinforced ceramics have shown a threefold increase in flexural strength and a sixfold increase in deflection over unreinforced ceramics. The fabrication of durable thin shell-type tooling is possible with reinforced ceramics. Development is still required for reinforcement and coatings for ceramic tooling.

## FABRICATION EQUIPMENT

The various types of equipment required to support the fabrication of composite structures are under continuous development, change, upgrading, modernization, and automation. The special features of each type of equipment have been developed either by commercial machine and tool builders or by a company with particular needs, the main incentive being increased productivity and elimination of manual operations.

## Filament Winding

The increased emphasis on automating filament-winding equipment, impregnating directly or applying prepreg in the winding operation, and developing the technology to fabricate complex hybrids is subject to certain limitations:

Control of fiber fraction, void content, and porosity

Inconsistent surface finish of parts

Difficulty of making concave surfaces and undercuts

### Helical-Winding Machines

Helical-winding machines with multiaxis motion are quite versatile. The spindle axis, which can rotate clockwise or counterclockwise, is the independent axis (Fig. 5.8). While the mandrel rotates, the carriage alongside traverses parallel to this axis. A rotatable cross-feed traverse provides greater versatility in placing the fibers. Oscillating mandrels, which traverse as they rotate, are also used (Fig. 5.9).

Mechanical or Computerized Programming: Filament-winding machines are programmed in three ways: by hand calculations for each setup (*mechanical*) or by stored memory for various setups (*comput-*

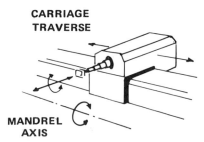

**CARRIAGE TRAVERSE**

**MANDREL AXIS**

FIG. 5.8 Multiaxis motion. *(From Am. Mach., May 1981, p. 127.)*

FIG. 5.9 Carriage and cross-feed traverse in microcomputer-controlled multiaxis winder. *(McClean Anderson.)*

*erized)*, which can be triggered at a keyboard *(microcomputer-controlled)*. Programming consists of coordinating the motions of the spindle and the material-delivery system or fiber-payout eye (Fig. 4.36). For mechanically programmed helical winders this involves selecting a drive-gear ratio between spindle and carriage and determining the size of the carriage-drive sprockets and the length of chain.

Which type of programming is best depends largely on production volume, the number of different parts to be made, and, to some extent, the complexity of the part. Mechanical machines are much less expensive but require more time and effort to program. On the other hand repair is easier, making users less dependent on equipment manufacturers in the event of breakdown.

Programmable calculators, which can be furnished with appropriate programs as machine accessories, have greatly simplified mechanical programming. Mechanically programmed helical winders can produce almost any symmetrical component that does not require varying the winding angle.

It would be a misapplication to assign a computer-based machine to every job. If only circumferential winding is required, special or even general-purpose mechanical machines are more economical; nor are computerized machines generally necessary for polar winding. Mechanical controls permit simple and rapid programming of polar winders. Replacing the simple but versatile mechanical system of a polar machine with a computerized one would be hard to justify in almost every case. On the other hand, computerized machines are much more versatile and easier to program since changes in the program can be accomplished almost simultaneously at the control console. Thus cost effectiveness increases with the number of different components to be produced. Computerized machines costing about $1 million are usually the only choice for winding complex configurations, e.g., rocket-motor cases (Fig. 5.10), light-truck drive shafts (Fig. 5.11), rocket-launcher tubes of varying diameter, and unsymmetrical helicopter- or wind-turbine blades with varying cross section.

The wind-turbine blades, 127 ft (38.7 m) long and weighing 13 tons (11,793 kg) were designed to survive hurricane winds and drive multimegawatt electric generators. The blades are formed one at a time on a massive winding machine (Fig. 5.12), the major elements of which are a mandrel, mandrel drive, filament carriage, and data-management center. As the mandrel is rotated, the rail-mounted carriage moves along its length, wrapping it in epoxy-coated fiber glass. The computer, which controls the machine's five axes of motion requires 5 million bytes of memory to wrap a single blade. In most filament-winding operations the control program consists of a series of repeated patterns, but because of the blade's size and shape, the winding pattern does not repeat.

Fiber glass is paid off carriage-mounted spools and through eyelets that form 52 rovings. The rovings are combined to form a "tape," approximately 6 in (152 mm) wide, that is drawn through an epoxy bath and then passes through the feed eye and onto the rotating mandrel. An operator on the carriage monitors the computer-controlled process.

The microcomputer-controlled machines store programming parameters on board or on

**FIG. 5.10** Winding a rocket-motor case on a mandrel. *(McClean Anderson.)*

**FIG. 5.11** Winding drive shafts at the rate of one per minute. *(McClean Anderson.)*

digital tapes. The machine operator programs the winding parameters at the computer keyboard. Once such factors as bandwidth, winding angle, number of layers, length of winding, and even end diameter are keyed in, the computer calculates proper machine setup and positions the fiber-payout eye at a preprogrammed mandrel position. The operator then keys a run command, and the winding sequence begins.

Machines are programmable in both linear and nonlinear modes; i.e., the relationship between spindle rotation and carriage traverse may be represented by linear or nonlinear functions. The fiber placement is greatly aided by the use of a programmable, servo-operated cross-feed and the rotating fiber-delivery eye (Fig. 5.11).

## Applications

Filament-wound glass-reinforced-epoxy pipe has recently been successfully made for geothermal wells. Two 7-in-diameter (178-mm) pipes, fitted with molded threads, were used to reach a water level at 5833 ft (1778 m). In 20 years, 400 geothermal dual wells are expected to be drilled and the use of filament-wound pipe will naturally increase.

Filament winding is proving more cost-effective than brick or steel construction for large chimney liners at industrial plants. One example is the two liners [0.5- to 1.375-in (12.7- to 34.9-mm) wall thickness, 19-ft (5.8-m) diameter] being assembled for an 805-ft (245-

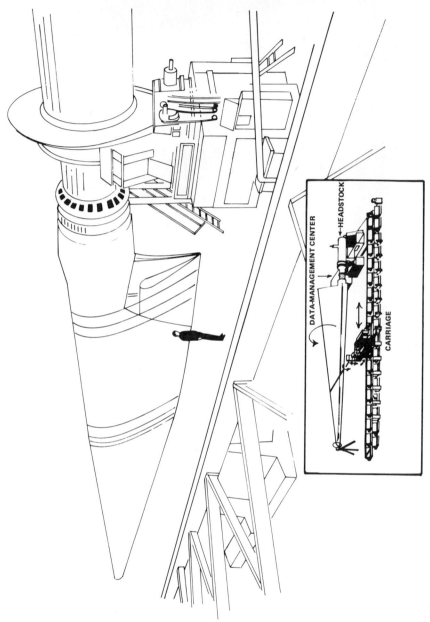

**FIG. 5.12** A five-axis machine. During blade winding the computer controls five machine axes: carriage travel, mandrel rotation, feed-arm extension, feed-arm elevation, and feed-eye rotation.[9]

m) chimney at an energy-generating station. The liners are assembled from 22 sections, each of which is bound with steel rings to maintain form and provide additional strength.

The sections are wound from glass-reinforced polyester at helix angles varying from 8 to 30° (0.14 to 0.52 rad). Chopped-glass fibers are sprayed between individual layers during winding. Graphite, mixed in with the first layer of resin, makes the inside surface conductive. A copper wire running down the length provides a ground for electrostatic electricity.

Filament winding is now an established method for manufacturing metal-lined pressurized-gas tanks used by fire-fighters, mountain climbers, and scuba divers. The thin aluminum liner serves as a barrier to gas permeation and as the mandrel for filament winding. The aluminum-lined tanks are subsequently wrapped with glass- or aramid-reinforced epoxy; they are 50% lighter than all-metal tanks and designed to resist rupturing on impact.

The government is considering producing submarines with a multiwall hull construction filled with syntactic foam and produced by computerized winding and foam molding. The approach would be similar to that used years ago to wind insulated cargo containers. The advantages over a metal hull include greater ability to withstand pressure, a tenfold improvement in strength-to-weight ratio, corrosion resistance, and much lower manufacturing costs. Similar projections at major aircraft manufacturers foresee filament-winding structural components for future generations of passenger aircraft, including entire fuselages. These giant parts would be wound under computer control and vacuum bagged to impart a good surface finish.

## Specialized Winding Machines

Behind the new interest in computerized winding is growing demand for continuous-fiber-placement equipment able to mass-produce complex components with extremely high strength-to-weight ratios. Conventional gear-driven machinery often cannot fill the bill, largely because of the amount of time needed to go from one winding pattern to another. Now the microprocessor is enabling equipment builders to come up with winders that can control each axis of motion for fiber placement in relation to mandrel rotation, allowing precise fiber patterns to be wound automatically with great accuracy and speed. For applications that call for capabilities beyond even those of the most sophisticated winding techniques equipment manufacturers are turning to hybrids, machinery systems that combine winding with other techniques.

**Filament Molding:**   Automated filament molding is an approach developed for volume production of such parts as structural steering-wheel inserts, suspension springs, transmission supports, door beams, aircraft structural members, and robot arms. The system involves a programmable multiaxis winder, special mandrels and molds, and a carousel of compression presses synchronized to operate in sequence with the winder (Fig. 5.13). Compression molding was selected because it improves mechanical properties, imparts a configuration with tighter tolerance, keeps void content low, and provides a smooth surface finish. Output of the system can range from one part every 20 s to one part per minute, depending on the number of presses. The resulting material is continuous resin-impregnated fiber-glass-hybrid molding compound, described in Chap. 4 as XMC. These composite sheet goods offer exceptional strength (as a result of continuously wound filaments rather than random chop), can be made with a variety of resins, and feature a hybrid

WINDING HEAD

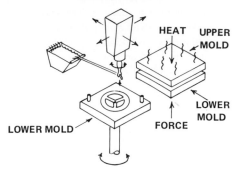

FIG. 5.13 Production of a hybrid steering-wheel insert involves two steps synchronized to operate in sequence: (*a*) the part is filament-wound; (*b*) the mandrel mold is indexed to the compression press for completion.[9]

makeup of two or more fibers. For example, using both carbon and glass fibers, the machine can be programmed to place the high strength, high-cost carbon fibers only in those portions of the product where great strength is required and to use the much cheaper glass fibers in the less critical portions.

*Dual-Mandrel Process:* The process works as follows: A steel mandrel approximately 4 ft (1.2 m) long and 8 ft (2.4 m) in circumference is mounted in a conventional filament-winding machine (Fig. 5.14). The mandrel is slotted to facilitate removal of the finished sheet. Before winding begins, a plastic film is applied to the mandrel to serve as a protective surface covering for the XMC. Using the microcomputer controls on the filament winder, the engineer can program each layer of the composite sheet for the desired winding angle. No matter how many layers of laminate are needed, each can be individually programmed for any sequence of fiber angles from 10 to 90° (0.17 to 1.6 rad) and any pattern of continuous-fiber reinforcement. Conventionally, the fibers are wet-wound onto the drum. If molding and strength requirements dictate, some chopped reinforcement can be added just under the fiber band being applied. Once the composite sheet has been wound around the mandrel, the fiber band is severed and another plastic film is applied to the sheet surface. The molding sheet is removed by cutting all the fiber bands bridging the slot in the mandrel. The finished sheet (Fig. 5.15) is then unloaded onto a flat storage rack or table.

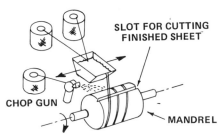

FIG. 5.14   Winding the sheet.[10]

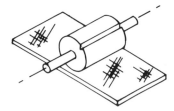

FIG. 5.15   Continuously wound laminate removed from the mandrel.[10]

*Curing and Molding:* The plastic film applied to top and bottom surfaces serves to make handling easier, provide protection, and help curing. The resin system used in processing is designed to cure to the proper viscosity for molding. During the maturation period, which can last from hours to days, the wet resin cures to a leathery consistency. When cured and ready for molding, the composite sheet is cut into mold charges with dies or conventional cutting tools (Fig. 5.16). Before molding, the plastic film is peeled from both surfaces. Finally, the mold charge is placed in the mold cavity (Fig. 5.17), the press is closed, and heat is used to finish cure the composite sheet.

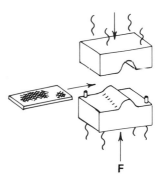

F

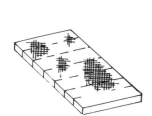

FIG. 5.16  Cutting molding charges.[10]          FIG. 5.17  Charge in mold cavity.[10]

The difference between the XMC part and one made of SMC is the continuous-fiber makeup of XMC. Figure 5.18 shows a proposed computer-controlled filament winder fitted with ancillary equipment to produce XMC material. Molded composite for passenger cars, trucks, and aircraft wheels now being produced from 50% XMC for higher strength and modulus (rigidity) and 50% HMC (a high-glass-content SMC, see Chap. 4) are in the prototype-testing stage. The wheels are said to reduce weight by 40 to 50% over steel wheels and weigh about the same as aluminum.

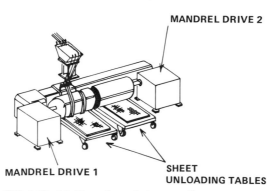

MANDREL DRIVE 2

MANDREL DRIVE 1          SHEET
                         UNLOADING TABLES

FIG. 5.18  Machinery for producing XMC.[10]

**Ring Winder:** The opposite of the oscillating mandrel shown in Fig. 5.11, the ring winder has a ring-shaped winding system that traverses and orbits the fiber supply around a stationary fixed mandrel for helical or circumferential winding patterns. An advantage of this arrangement is that it avoids the complicated mandrel support and laydown control techniques required to allow for mandrel sag and whip when conventional rotating mandrel machines wind long, limber parts such as helicopter blades, airplane-wing spar boxes, or small-diameter pipe.

**Automated Tape Lay-Up System (ATLAS):** A new federally financed tape-laying machine has been developed to meet the needs of the intricate shapes being considered for production. Added sophistication in the parts being produced and the means of producing them spotlighted the need for thorough knowledge of the tape-placement characteristics and capabilities of a machine. Further, the ability to place and compact the tape accurately beyond current limits should enhance a part's internal structural composition with an attendant increase in strength or reduction in materials. The program was also intended to help open up new markets for preimpregnated composite tapes.

*Machine Capability and Operation:* Although the numerically controlled machine wrapped geometrically complex shapes experimentally and produced prototype hardware (rotor blades for helicopters), it has never been fully used in a production environment. Some of its unique features are worth describing, however. Originally the closed-D design of the spar of the helicopter blade was fabricated using matched metal dies, integrally heated and cooled under computer control. When the ATLAS machine was used to wrap the spar, the design called for fiber and epoxy filaments with a styrofoam core. Epoxy composites were chosen because they could be formed into more aerodynamically efficient shapes and were lighter and stiffer than metal.

Looking like a conventional gantry-type skin mill, or planer, the ATLAS featured six axes of numerical control traversing the gantry at 60 ft/min. (30.48 cm/s) to lay 3-in-wide (76-mm) filament tape over an entire 34.3-ft (10.5-m) rotor blade. The six-axis machine was designed to rotate both the tool and the part as the tape was laid (Fig. 5.19). The prototype machine used a head-mounted infrared device to heat the tape just before laydown, making the surface tacky so that it would adhere and compact better. All six axes are driven by independent motors. Programs for running the machine are on punched

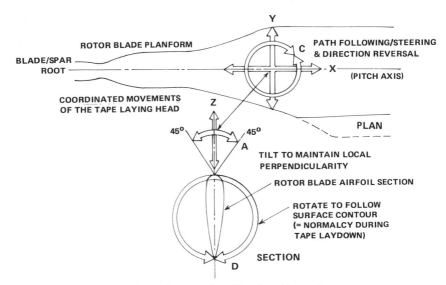

**FIG. 5.19** Diagrammatic sketch of six axes of coordinated machine motion.

paper tape; an electrooptical device enables the machine to generate its own programs. A null-seeking line follower is put on the machine in place of a laydown-roller head. Adhesive-backed paper tape with a black line in the center is then placed by hand in the pattern designed for the epoxy-impregnated tape. During programming runs, encoders report to the computer the position and movement rate of each axis. This information is processed, and the program is produced.

Automatic tape slitting is another innovation incorporated in the machine. It is virtually impossible to get a wide tape to adhere to a compound-curve surface at an angle since the tape tends to slip and form a tight spiral. To counteract this, blades on the machine are set to slit the tape into narrow strips, 0.27 in (6.8 mm) wide, permitting coverage and correct adhesion to the surface. This prototype machine was the first tape-laying machine capable of placing tape onto and over compound contours of primary aircraft structures such as rotor blades and writing new control programs directly.

## Automated Broad Goods and Tape Lay-Up

The automation of lay-up of composite broad goods and tape is very design-dependent. When engineers design a structure, they must know in advance whether it is for automated manufacture. Research and development have developed processes and equipment that industry may be able to use in production 3 to 5 years from now. Figures 5.20 and 5.21 show concepts for automated tape-laying machines capable of laying prepreg tapes from 1 to 12 in (25.4 to 305 mm) wide at speeds up to 600 in/min (254 mm/s) for parts up to 15 ft (4.6 m) wide and 50 ft (15 m) long. Broad-goods laminating and dispensing equipment can produce feedstocks 8 ft (2.4 m) wide and theoretically of unlimited length.

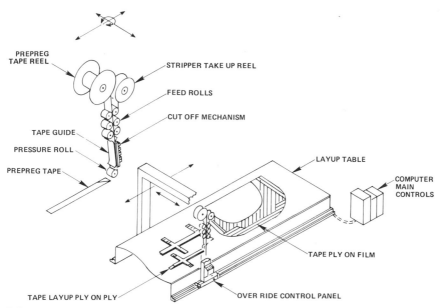

**FIG. 5.20** Automated tape lay-up. *(Sikorsky Aircraft, Division of United Technologies.)*

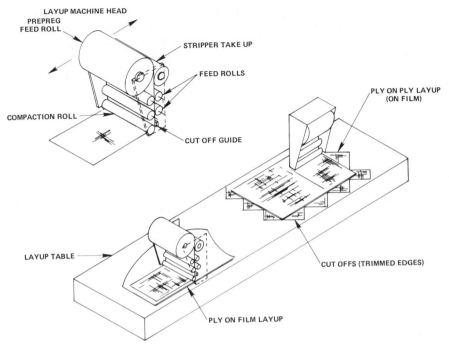

**FIG. 5.21** Automated broad-goods lay-up. *(Sikorsky Aircraft, Division of United Technologies.)*

Several advantages are gained by broad-goods or tape materials lay-up. The process can be automated, hybrid materials can be layered, lay-up can be performed on the tool and preforms, and ply-on-ply and ply-on-film can be utilized. With these advantages come limitations:

Difficulty of making compound curvatures

Difficulty of achieving prepreg conformability

Time-consuming material transfer and handling

Size factor in handling large complex parts

Automated tape-lay-up systems use tapes consisting of unidirectional, cross-plied, or woven fabric in various widths. Provisions in the machine for automated movement in vertical, transverse, and longitudinal modes permit accurate tape positioning. Cutting or shearing preliminary laminate compaction are provided to permit lay-up of multilayered panels in a continuous operation. Use of a specific width of tape and the ability to shear off at an angle make it possible to lower scrap and to lay up parts closer to their net size. The tape-lay-up system offers a high degree of flexibility thanks to the number of degrees of freedom available, e.g., prepreg width, ply angle, and angle trimming of tapes by machine during lay-up.

The automated broad-goods-lay-up system cuts part segments from machine-placed broad-goods prepreg in various widths. The broad goods are transported to the mold on a transfer film although the system can be used in either the ply-on-ply or ply-on-film mode.

Broad-goods lay-up is used primarily for large patterns, while tape lay-up is used for smaller and more intricate shapes.

## Manual and Semiautomatic Tape Application

In the early stages of development tape lay-up was largely manual. Individual plies were generally laid up on plastic templates; following mating of the plies the template was removed after each ply was laid up. For actual production, lay-ups have to be made ply-on-ply, omitting the templates, but in hand lay-up it is difficult to control the ply position without templates for indexing. For this reason studies were undertaken to develop semiautomatic or automatic equipment to control lay-up indexing, which would increase lay-up quantity per worker-hour and improve production rates. Early semiautomated lay-up relied on several methods. (1) A large plastic belt positioned over two rollers with a transversing payoff head allowed large sheets of unidirectional broad goods to be laid up quickly. Detail plies were then cut from those sheets using templates and subsequently laid up. (2) A movable payoff head was mounted on an indexing transverse movable area similar to a drafting machine. The lay-up made on templates was transferred to the previously laid-up ply. Other semiautomatic lay-up machines have been devised to achieve the same results. Most used tape 3 in (76 mm) wide both for boron and for graphite, but tapes up to 12 in (305 mm) have been used.

## Automated Tape Lay-Up

A leading exponent of automated prepreg tape lay-up is General Dynamics. The mechanization of composite parts manufactured by this company has been slow to evolve simply because production rates have been low. When production of composite vertical and horizontal stabilizers increased to 40 per month and tests proved the feasibility and economics of several small tape-laying machines, an automated tape layer was built and installed (Fig. 5.22). The numerically controlled machine, which can lay up 6-in (152 mm) Gr–Ep tape

**FIG. 5.22**   Tape laying with six numerically controlled axes.[3] *(General Dynamics, Ft. Worth Division.)*

along any angle orientation, consists of a flat stationary bed on which to lay the plies of tape, a movable two-column gantry providing 3-ft (0.9-m) x-axis motion, a saddle mounted on the gantry for 10-ft (3-m) y-axis motion, and 6 in (152 mm) of z-axis saddle ram travel.[2,3] To lay the tape, a rotating dispensing head mounted on the saddle contains the tape roll and drive, a backing-paper windup spool and drive, a rotating guillotine shear for cutting the tape, and a roller for pressing tape (Fig. 5.23). Three additional numerical-

**FIG. 5.23** Automated tape-layer head.[3] (*General Dynamics, Ft. Worth Division.*)

control axes are involved in the head: the c axis for head rotation, d axis for shear rotation, and u axis for tape positioning. The machine can lay up 12 lb (5.4 kg) per worker-hour on an aircraft vertical stabilizer with only 5% waste.

Another company using tape is the Vought Corporation. Figure 5.24 shows a tape-laying machine (TLM) designed to lay 3-, 6-, or 12-in-wide (76-, 152-, or 305-mm) glass, graphite, or Kevlar composite tapes for part sizes up to 15 by 50 ft (4.6 by 15 m) at speeds up to 600 in/min (254 mm/s). This machine is capable of laying a composite skin or panel directly on the mold, ply-on-ply, or producing tailored broad goods. Another tape-laying machine installed at Rockwell International is capable of making Gr–Ep composite parts at production costs 10 to 20% below those of metal components.

An aircraft door may have eight layers of tape, but some components may have hundreds, e.g., one B-1 structural component with a finished thickness of 2.2 in (55.9 mm). Each layer, after curing, contributes about 0.005 in (0.13 mm) to the total thickness.

FIG. 5.24   Tape-laying machine. *(Vought Corp.)*

The TLM can lay 12-in-wide (305-mm) tape at speeds up to 720 in/min (305 mm/s). This is about the maximum speed for 3-in (76-mm) tape on other machines. The head of the machine also strips the paper backing off the prepreg and can be adjusted to accommodate narrower widths. Tape is pulled from a roll, through a shear mechanism, and is applied by a Teflon shoe or bar that is used in place of a roller to press the material down.

The machine is electrically and pneumatically operated with a manual push-button control. The tape head is suspended from an overhead beam and lays tape in only one direction along the $x$ axis (there was not sufficient economic incentive to modify the head for tape laying in all directions).

Grumman has developed a first-generation modular manufacturing system, the integrated laminating center (ILC) (Fig. 5.25) that can be readily expanded and can handle a wide variety of aircraft skins and material. The ILC automates tape laying or broad-goods dispensing, nesting and trimming, inspection, and ply handling and stacking without the costly Mylar templates used in manual operations. The continuous operation works as follows. At the tape-laying station (Fig. 5.26) prepregged boron or graphite tape is laid down automatically on a carrier material and is cut to length in a predetermined pattern. The tape head itself has four axes of rotation for flexibility in configuring ply details. After all the tape strips have been laid and cut for the first ply, the transfer gantry prepares for the second ply. The gantry which has a vacuum head to hold the ply, simultaneously transfers the first ply to the laser trimming station (Fig. 5.27). As the second ply is being prepared by the tape head, the laser trimmer cuts the first ply to net shape.

After cutting, the trimmed part is visually inspected and the data recorded with a video monitor. If the trimmed part meets specifications, it is transferred via a vacuum flip table to a mold form, where additional plies are added as the process continues. The total system keeps cycling (rather like a computer program loop) until the total number of plies is

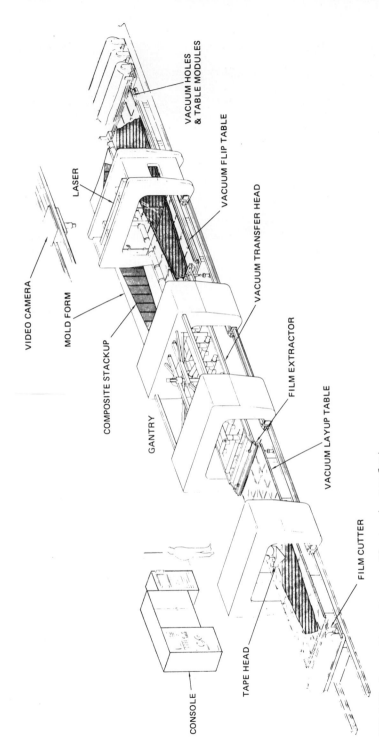

**FIG. 5.25** Integrated laminating center. (*Grumman Aerospace Corp.*)

VACUUM HOLES & TABLE MODULES

VACUUM FLIP TABLE

VACUUM TRANSFER HEAD

FILM EXTRACTOR

VACUUM LAYUP TABLE

FILM CUTTER

TAPE HEAD

CONSOLE

GANTRY

COMPOSITE STACKUP

MOLD FORM

VIDEO CAMERA

LASER

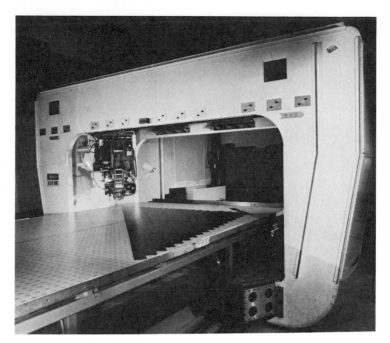

**FIG. 5.26** Tape-laying station of ILC. *(Grumman Aerospace Corp.)*

**FIG. 5.27** Vacuum-transfer head of ILC. *(Grumman Aerospace Corp.)*

**FIG. 5.28**   Horizontal stabilizer of F-14 aircraft. *(Grumman Aerospace Corp.)*

reached. Plies are built up according to the required design and can vary in size, shape, and even material.

Figure 5.28 shows the F-14 production horizontal stabilizer which consists of four B–Ep covers, each containing 100 individual plies. The manual operation required about 65 worker-hours per cover. The ILC system has reduced that figure by more than 50% to roughly 30 worker-hours. More than 500 ship sets have been produced.

**Innovative Tape and Broad-Goods Systems:**   Several companies have also developed innovative broad-goods systems in conjunction with their tape system.

*AIMS:*   One is the advanced integrated manufacturing system (AIMS), developed in conjunction with the ILC for production-rate fabrication. The system involves stiffened, complex-contoured skins with a mix of tape and broad-goods material forms and the use of translaminar reinforcement by stitching (Fig. 5.29). The first add-on AIMS module is the broad-goods dispenser (Fig. 5.30). In operation it automatically pays out Kv–Ep, Gr–Ep, or Gl–Ep prepreg material in widths up to 80 in (2 m). The dispenser interfaces with the ILC via the transfer gantry.

Broad goods are unreeled from the dispenser by the transfer gantry on the ILC; then the laser trim gantry moves over the length of prepreg material, cross-cuts it to the necessary length, and cuts computer-generated nested pieces that will later be laid up to form stiffeners or skin plies. The module also features quick-lead chucks, rollaway freezer storage, and optional dewrinkler roll action.

A second module is a contour ply handler which locates, drapes, and forms large composite plies which have been transferred to it by the ILC's vacuum flip table onto a contoured plastic-covered tool. This module interfaces with the ILC flip table so that it can automatically receive large-area plies prepared on deformable carrier film. Plies are formed to contour by rotary brush action in a programmable cycle that controls tool rise, split-

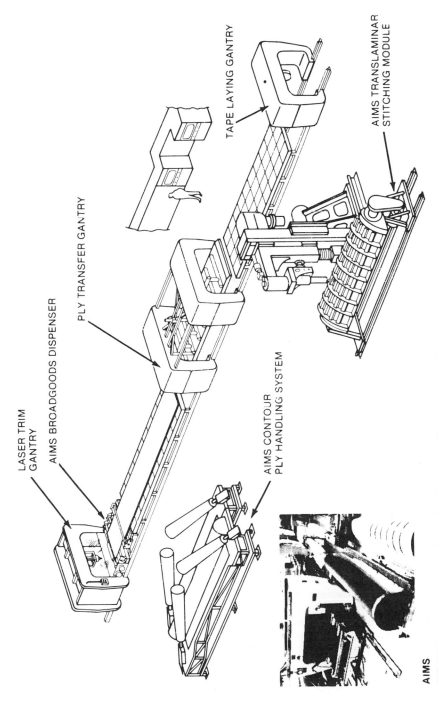

LASER TRIM GANTRY

AIMS BROADGOODS DISPENSER

PLY TRANSFER GANTRY

TAPE LAYING GANTRY

AIMS TRANSLAMINAR STITCHING MODULE

AIMS CONTOUR PLY HANDLING SYSTEM

AIMS

FIG. 5.29   Automated integrated manufacturing system. (*Grumman Aerospace Corp.*)

**5.27**

**FIG. 5.30** AIMS broad-goods dispenser installed in an ILC system.[4] *(Grumman Aerospace Corp.)*

table separation, and brush-arbor motion. The tacky plies are separated from each other by a thin polyethylene backing that is removed before the plies are laminated. This automated system eliminates manual tailoring of strips to contour and permits direct use of automatically prepared flat plies for contoured configurations (Fig. 5.31).

The third module, the translaminar stitcher, is a self-digitizing five-axis module capable of stitching along straight, bowed, twisted, and highly contoured (circumferential) paths. Its computer interpolates between selected point inputs and inserts the required stitch pitch (Fig. 5.32). After a short vacuum-compaction cycle in an autoclave to remove some of the resin from its plies, a skin is draped over a slotted stitching rack, which allows the presser foot of the stitcher to reach up from below the skin and complete the machine's chain stitch pattern at the rate of 20 in/min (508 mm/min). The stitcher has been used to sew hat stiffeners and flanges with Kevlar thread to the skin built up by the contour ply handler. Stitching an engine duct with Kevlar thread eliminated 1600 mechanical fasteners during fabrication[4] (Fig. 5.33).

*IPS:* The integrated process system (IPS), developed by Vought Corporation, was constructed and tailored specifically for the component or family of components to be produced (Fig. 5.34). The IPS consists of the TLM, described earlier (Fig. 5.24); a rapid ply cutter, which uses a high-speed reciprocating knife for cutting detail preforms from single and laminated plies of composite materials; and a laminating station (Fig. 5.35), which laminates up to four plies of broad goods up to 8 ft (2.4 m) wide to supply the laminator-spreader. The latter is mounted on the rapid ply cutter ways to laminate and spread broad

**FIG. 5.31** Ply handler.[4] *(Grumman Aerospace Corp.)*

goods up to 8 ft (2.4 m) wide tailored to individual family requirements. Another component of the IPS is the overhead material dispenser (Fig. 5.36), which lays out fabric broad goods for parts too large to be kitted by the normal system. The material is automatically dispensed and draped by a semiautomatic control system. The head indexes, rotates 180° (3.2 rad), and traverses across the mold at the selected angle in steps initiated by the operators. The machine also incorporates a bagging station.

## Automated Broad-Goods Lay-up

The Northrop Corporation began with a different approach, but their purpose was to reduce the cost of composite manufacturing by automation. The system features an automatic dispensing unit whereby composite broad goods 5.5 ft (1.72 m) wide by up to 40 ft (12.2 m) long are dispensed onto a table through a material-dispensing unit (Fig. 5.37). The material is dispensed down the length of the table, and a programmed automatic machine cuts the plies. After anywhere from 12 to 100 different patterns in different orientations have been cut and deposited on a table 6 by 48 ft (1.8 by 14.6 m), a robot for automated lay-up moves in (Fig. 5.38). It picks up one ply at a time and deposits it in the tool until the laminate is complete. Since a part may comprise 50 or more plies, the reduction in lay-up time is considerable.

**FIG. 5.32** Translaminar stitcher.[4] *(Grumman Aerospace Corp.)*

Material forms have been developed that are aimed for automation and low cost. At first only 3- to 4-in wide (76- to 102-mm) prepreg was available, but now there is 5-ft-wide (1.5-m) interwoven broad goods, with the capability of up to 10-ft (3-m) widths. Northrop engineers have examined three-ply broad goods, uniwoven (all plies running continuously in one direction with a cross-stitching of fiber glass), bidirectional woven, and unitape, a uniwoven with bonded cross-stitching. Besides the robots for ply transfer and stacking the system includes a monorail to convey the composite-laden table top to a work

**FIG. 5.33** Completed duct.[4] *(Grumman Aerospace Corp.)*

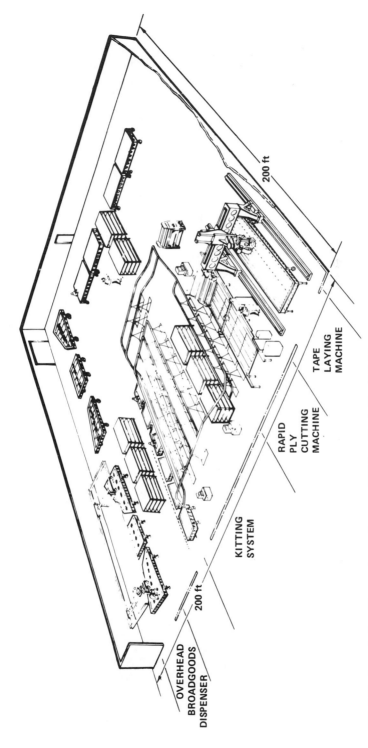

FIG. 5.34  Isometric view of integrated processing system (200 ft = 61 m). (*Vought Corp.*)

**OVERHEAD BROADGOODS DISPENSER**

200 ft

**KITTING SYSTEM**

**RAPID PLY CUTTING MACHINE**

**TAPE LAYING MACHINE**

200 ft

**FIG. 5.35** Laminating station. *(Vought Corp.)*

**FIG. 5.36** Overhead material dispenser. *(Vought Corp.)*

**FIG. 5.37** Material-dispensing unit laying out broad goods 5 ft (1.5 m) wide.[5] *(Northrop Corp.)*

**FIG. 5.38** Robotic transfer of graphite plies to laydown curing tool.[5] *(Northrop Corp.)*

station. Video cameras direct the operation of the transfer unit. Plies can be positioned so that fibers lie at +45° (0.79 rad) in one layer, −45° in another, and 90° (1.6 rad) to each other in the next. The plies are laid up in a contoured tool which is taken to an autoclave or oven for curing to obtain the final part. It is possible for the table top to hold graphite sections on one side and Kevlar or fiber glass on the other, so that the transfer head would combine them in the tool as desired; or the same head could operate from two tables, one on each side.

The culmination is Northrop's automated system, an integrated flexible automation center (IFAC) (Fig. 5.39), which works as follows. An automatic dispensing unit dispenses broad goods 5 to 6 ft (1.5 to 1.8 m) wide at 350 ft/min (107 m/min), makes a perpendicular cut, and returns to storage position. The cutter then goes in and cuts, being picked up by a computer signal collected by the monorail and a *flying carpet*. Standard cutting-table bases were modified to accept the segmented flying-carpet table tops (Fig. 5.39), which enable the electrically driven table-lift devices to move the table tops to the monorail for subsequent transfer. The carpet moves to various robot areas, e.g., the wing station, where it triggers descent of the robot. The programmed robot picks up the graphite. When it has performed its assigned task, another signal is picked up and a prepreg scrap remover is brought in to move back the transfer head and scrape the prepreg off it.

At the robot station one of three video cameras checks the identity of the plies and the cut. If everything is correct, the robot picks it up; if not, the robot is shut off automatically. In addition, a sensor table examines the ply on the robot as it comes across. It checks the ply orientation (on the head) and for damage that may have occurred between the cutting table and the robot.[5] Other automated broad-goods lay-up systems are in their initial development stages.[6]

## Automation of Cutting and Kitting

Each of the automated composite systems described above contains some form of cutting and kitting machine. For example, one company's kitting system includes a mold-transfer line, a transfer and storage system, core storage, and a computer with multiple terminals and accessories (Fig. 5.40). The mold-transfer line is parallel to the unloading work station. The line itself is also the hand-lay-up work station, which includes a bagging system. The overhead handling system has two work stations for loading and unloading. The loading station is at workbench height and parallel to the rapid ply-cutter bed for the easy transfer of details to hanging trays (Fig. 5.34).

Another system is shown in Fig. 5.41. When the prepreg materials are received at the high-speed cutting and kitting operations, computer-developed flat patterns for prepreg details, nested by computer for the most effective use of materials, are cut by appropriate high-speed techniques. While they are being cut, the individual details are marked and identified by a high-speed marking system. The identified cut prepreg details are removed from the cutting machines and assembled into kits. Computerized, automated takeoff and materials-transfer equipment accomplishes this efficiently and can position and prestack details for building preforms. Assembled kits containing prepregs, adhesives, preforms, and core details are inspected, serialized, and tagged with visual and optical or magnetic coded labels. The kits are then routed by mobile kit-storage rack or conveyor to the next operation or to refrigerated storage, as required.

At the lay-up station the labels on the kits are scanned and the information fed to the computer. Tooling and support materials are prepared. Precut prepreg details and film

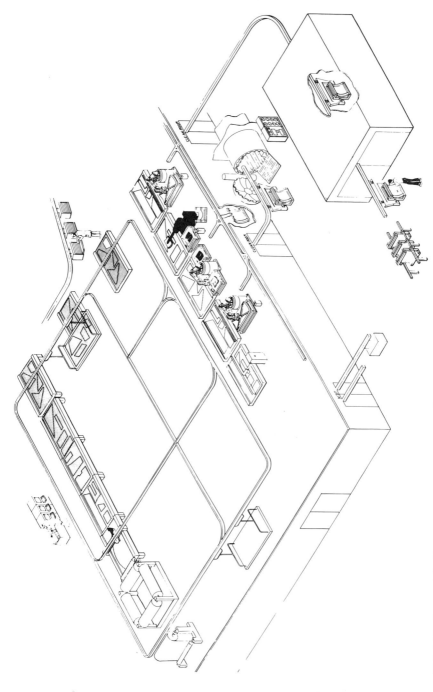

**FIG. 5.39** Integrated flexible automation center with monorail.[5] (*Northrop Corp.*)

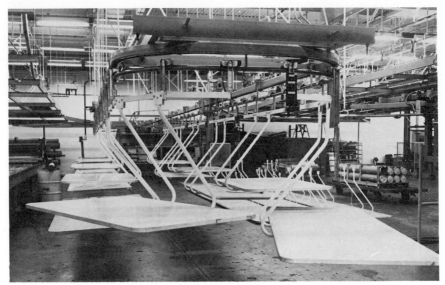

**FIG. 5.40**   Computer-aided kitting system. (*Vought Corp.*)

adhesives, core details, and preforms are removed from the kits and sequentially loaded directly into the molds for lay-up and curing.

Kitting for automated lay-up can be used in a slightly different form. Prepreg and adhesive details prepared by automated direct tape lay-up machines and other high-speed cutting techniques are sequenced and accurately oriented and positioned on stable film tapes. The tapes bearing the details are then rolled up and packaged in cassettes for use in high-speed placement and direct-lay-up equipment. By thus minimizing machine-head motion, indexing, and rotation, rates of material lay-up are maximized. Figure 5.42 depicts a composite factory of the future.

## Automation in Processing SMC and RIM

The manufacture of plastics must be sophisticated to compete with sheetmetal and other weight-saving materials. Development of chopped-fiber composites is essentially complete, and they are ready for the commercial market. Continuous-fiber composites still require basic development work. New applications, especially in the automotive industry, will involve fiber-reinforced plastics for body panels and structured components. The key is to convert exotic aerospace composite-manufacturing methods to high-volume automotive production. The near-term focus is on SMC and RIM.

### *Presses for SMC*

The presses of the future must be able to work with robots and loaders for cycle times measured in seconds rather than minutes. An experimental press for compression molding of polyester–chopped-fiber-glass systems (Fig. 5.43) is aimed toward better control of:

Parallelism during press closing

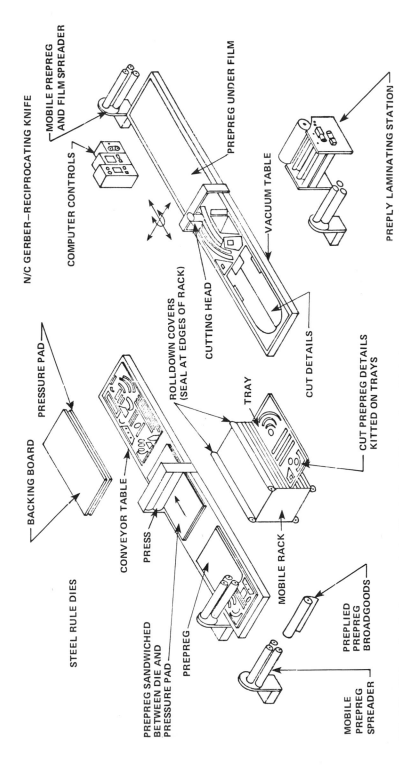

MOBILE PREPREG AND FILM SPREADER

N/C GERBER–RECIPROCATING KNIFE

COMPUTER CONTROLS

PREPREG UNDER FILM

VACUUM TABLE

PREPLY LAMINATING STATION

CUTTING HEAD

ROLLDOWN COVERS (SEAL AT EDGES OF RACK)

TRAY

CUT DETAILS

CUT PREPREG DETAILS KITTED ON TRAYS

PRESSURE PAD

BACKING BOARD

STEEL RULE DIES

CONVEYOR TABLE

PRESS

PREPREG SANDWICHED BETWEEN DIE AND PRESSURE PAD

PREPREG

MOBILE RACK

PREPLIED PREPREG BROADGOODS

MOBILE PREPREG SPREADER

FIG. 5.41   High-speed cutting by two methods. (*Sikorsky Aircraft, Division of United Technologies.*)

5.37

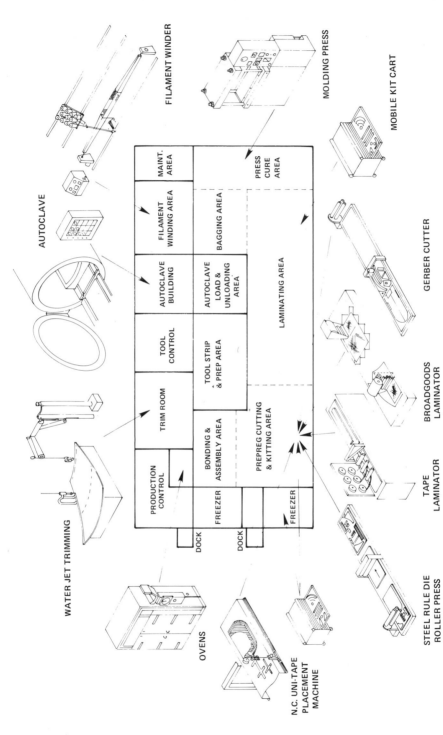

**FIG. 5.42** Composite factory of the future. *(Sikorsky Aircraft, Division of United Technologies.)*

FILAMENT WINDER

MOLDING PRESS

MOBILE KIT CART

AUTOCLAVE

GERBER CUTTER

WATER JET TRIMMING

BROADGOODS LAMINATOR

TAPE LAMINATOR

STEEL RULE DIE ROLLER PRESS

OVENS

N.C. UNI-TAPE PLACEMENT MACHINE

MAINT. AREA

FILAMENT WINDING AREA

AUTOCLAVE BUILDING

TOOL CONTROL

TRIM ROOM

PRODUCTION CONTROL

PRESS CURE AREA

BAGGING AREA

AUTOCLAVE LOAD & UNLOADING AREA

TOOL STRIP & PREP AREA

LAMINATING AREA

BONDING & ASSEMBLY AREA

PREPREG CUTTING & KITTING AREA

FREEZER

FREEZER

DOCK

DOCK

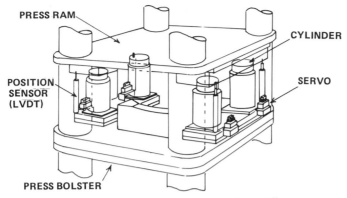

**FIG. 5.43** Laboratory compression-molding machine for SMC.[11]

Closing velocity
Mold pressure and temperature

Controlling these critical variables is important. For instance, when press platens move 0.100 in (2.5 mm) out of parallel during closing, the chemical reactor or mold is out of control, causing partially formed molecules to degrade as they are forced to reflow. Automatic load and unload equipment feeds and positions the molding charge and moves away in less than 5 s with the assistance of servo controls and a minicomputer. The minicomputer is programmed for closing rate, pressure, and temperature control. Temperature and pressure input (from sensors) allows the press to compensate automatically for variations in material and mechanical devices. This fast new system allows a multistage catalyst system to be used, reducing cycle times from today's 2 to 3 min to under 1 min, which will make plastics competitive with sheetmetal stampings for automobile chassis parts. An experimental SMC automobile tail gate has been produced in a modified version of the experimental press and is being evaluated.

## RIM Automation

Although RIM does not have the control problems of compression-molded SMC, it has a special problem in that mold release must be applied between all parts. Robotics and automated spraying speed up mold release, and coating is more uniform, requiring less frequent cleaning than manual application. Combining fast chemistry, an automatic loader, and automatic mold-release system will triple RIM productivity.

## SMC Hybrids

Structural automobile parts including seat shells, bumper bars, door beams, and chassis components are excellent candidates for a hybrid SMC (continuous and chopped-fiber reinforcements). A more sophisticated type of compounding machine is required to produce these SMC hybrids. An experimental SMC machine can produce standard SMC and up to a nine-layer composite. The machine, which features a gravimetric paste-dispensing system, encloses the paste, dispensing it directly onto chopped fibers of mat. Paste-layer consistency is monitored and controlled; a glass compaction belt forces out gross air. The SMC unit also uses a series of microprocessors to integrate all mixing and SMC functions (Fig. 5.44).

The hybrid SMC is a promising material system for the future. Currently, automobile front ends of SMC use 27 wt % chopped-glass fibers, but for structural applications, SMC can contain up to 70% total reinforcements. Adding continuous glass fiber to random chopped-glass fiber significantly boosts strength; compression molding provides low-cost processing and high throughput. For even greater stiffness in the fiber direction, several fiber manufacturers are hybridizing carbon and glass fibers in SMC. Carbon hybrids not only increase mechanical properties but also decrease the weight of the laminate for stiffness-critical applications. The end result of the program is to optimize the carbon-glass ratio in structural SMC, using carbon fiber to improve properties.

**Processing** The hybrid-SMC process is similar to existing practice. First random chopped fibers are added to the paste, then continuous-strand glass rovings, and then a polyester prepreg of unidirectional carbon fiber. The most dramatic increase in mechanical properties over those of an all-glass structure occurs with just a small addition of carbon fibers. Putting carbon fibers over a continuous-glass–random-glass system rather than an all-random SMC material works well. Continuous glass fibers form a good transition between the high-elongation random-fiber composite and low-elongation carbon-fiber skins. Increased strain to failure and load sharing occur between the two carbon and glass-fiber plies. With all-random glass fibers and carbon, failure occurs at much lower strain levels. The hybrids are aimed for initial-load-carrying applications, e.g., structural frame members, beams, bumper backups, and wheels.

*SMC Compression Molding and Pulmolding:* Compression molding of hybrid SMC and pulmolding have been used to produce highly stressed structural parts such as frames and leaf springs. Both processes get their strength from continuous fibers. Hybrid SMC molding uses the same molding equipment as conventional SMC molding. The difference lies in the process of building the charge. Mold charges must cover 90% or more of the mold area, causing resin to flow perpendicular to fiber direction. When part geometry is deep, both detailed preform patterns and a preform lay-up procedure are required. High production rates require automatic equipment: a programmable pattern cutter, a pattern-assembly machine that makes two-dimensional sheets into a three-dimensional shape, and a method of preheating the three-dimensional charge and loading it into the mold.

Pulmolding, a cross between pultrusion and compression molding, is suited for parts with a uniform cross section, e.g., leaf springs. Continuous fibers (under tension) are pultruded through resin, preheated, gathered by a preform tool, and then drawn into a moving, heated compression mold. Table 5.2 shows the processes for the next decade.

## New Hybrid Processes

A process developed in Japan combines extrusion, pultrusion, and filament winding all in one continuous production line. As shown in the schematic (Fig. 5.45), the first stage is the extrusion of a thermoplastic inner liner like PVC. Next, polyester-resin-impregnated glass rovings oriented in the axial direction are formed around the liner by pultrusion. Next several strands of glass are filament-wound over the axial strands to provide hoop strength. The thermoset resin is cured by a combination of ultraviolet radiation and heat, and, finally, an outer, impermeable skin of thermoplastic is extruded over the whole. The key

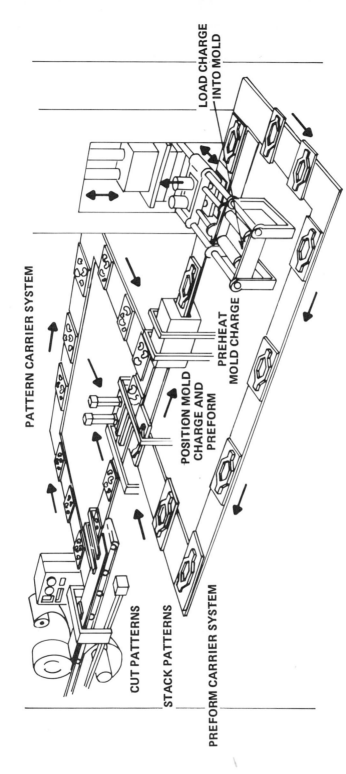

FIG. 5.44 Future automated hybrid-SMC system.[11]

**Table 5.2** Processes for Body Panels and Structural Components[11]

| Short-term | Mid-1980s | 1990 |
|---|---|---|
| RIM, RRIM, and SMC | Compression molding of hybrid SMC: chopped- or continuous-fiber-reinforced | Reaction molding of epoxies, nylons, and polyesters |
| | Reaction injection molding with inserts (tubular-metal and disposable foam cores) | Pulforming, pulmolding |
| | | Filament winding |
| | | Automated postmolding operations |

to the process is the special polyester, which is said to gel under ultraviolet exposure in half the time required by other polyesters.

In another intriguing new process reinforcements are woven into continuous three-dimensional shapes that serve as preforms for pultrusion. The process can produce flat panels, I beams, tubes, or square channels using a special loom configured for the shape to be produced. Preforms have been woven of glass, graphite, aramid, nylon, polyester, SiC fibers, and epoxy prepregs. The virtue of such continuously interwoven preforms is that weave density is uniform in all directions, so that there is no weak direction subject to delamination; the ability of the interwoven reinforcement to arrest crack propagation is said to be excellent. Reinforcement effectiveness can be optimized by providing different directional orientations in different parts of a shape. Two or more separate fibers, e.g., glass and graphite, can be interwoven simultaneously, each fiber being retained in a specific part of the piece if desired. Production of preforms is said to be fast and independent of the shape of the preform.

## MMC and New Material Composite Automation Processing

In order for MMC hardware to be competitive with conventional metallic hardware, manufacturing techniques must be streamlined for production quantities. Analysis of most advanced component production shows that the area of major cost is usually handling impregnated material from receipt to lay-up in a mold ready for curing. Although development programs discussed elsewhere in this book have successfully demonstrated cost benefits for integrated fabrication of relatively large aircraft components, little has been done for material handling in smaller applications such as engine components. Manufacturing methods established for aircraft structures in general are not always suitable for engine components, which are among the most sophisticated in terms of performance and environmental and quality-control requirements. Many engine applications require much higher production rates than aircraft structures. Furthermore the size and the types of plies used in engine applications and the types of higher-temperature composite materials require different methods of automated ply transfer than those used for typical aircraft component applications.

### Automatic TLM

One of the few automatic TLMs in use for MMC has a head designed to be used for small parts such as blades and vanes. The machine can be programmed to perform three functions: (1) cut the ply shapes by using a laser beam, (2) pick up and position the plies, and

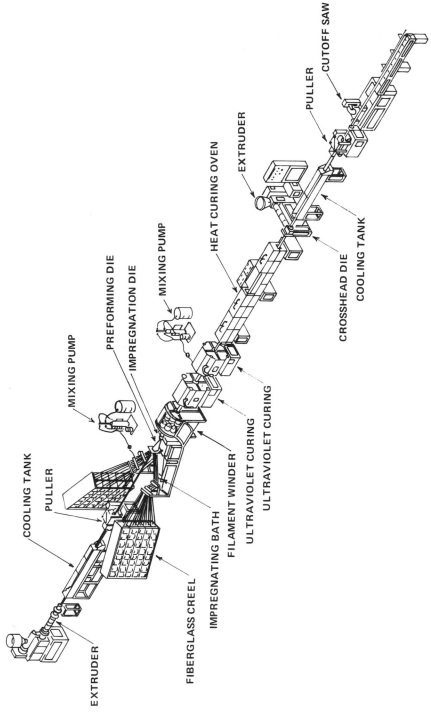

**FIG. 5.45** Production line of thermoplastic extrusion, thermoset pultrusion, and filament winding. *(From Plast. Technol., December 1981, p. 24.)*

**5.43**

(3) tack-weld them in place. This last operation is required for MMCs since, unlike resin matrix materials they are neither pliable nor tacky.

Consolidation of MMC is accomplished by applying heat and pressure, either in an inert atmosphere or in air. A sophisticated piece of equipment developed to combine these operations is a hydraulic press. One of the largest is rated at 1500 tons (1361 metric tons) (Fig. 5.46) and is equipped with automated control of pressure and temperature cycles. Bonding cycles as short as 10 min for MMCs have been developed.

**FIG. 5.46** Bonding press. *(Hamilton Standard, Division of United Technologies.)*

### Ply Cutting, Pick Up, and Stacking

A Gr–PI flap has been considered for production processing with an automated ply-stacking and cutting system. The key elements of the system are (1) a steel-rule die for ply cutting, (2) a roller to provide the ply-cutting force, and (3) a pickup head and stacking foot for stacking cut plies[7] (Fig. 5.47). The polyurethane-coated roller is mounted on the table track and attached to the carriage. Gr–PI is placed over the steel-rule die, and the carriage is translated, pulling the roller over the die. The Gr–PI is cut successfully when the graphite fibers are paralleled to the steel rule.

Up and down motion of the pickup head is accomplished by a pneumatic cylinder; rotation is provided manually, another pneumatic cylinder locking the head at the correct angular position. Vacuum density is directly related to the ability to pick up plies consis-

**FIG. 5.47** Automated flap ply-stacking system.[7] *(Hamilton Standard, Division of United Technologies.)*

tently with no droop. Stacking is accomplished by rotating the pickup head and stacking foot to a common position. Tacks on the stacking foot for holding the position of the plies after stacking work well, especially when the stacking foot is coated with Teflon.

High-volume production equipment for composite engine structures is being developed. Streamlining these multifaceted manufacturing operations will make advanced-composite hardware competitive with conventional metallic hardware.

## REFERENCES

1. Swazey, E. H.: "Manufacturing Methods for Low Cost Tooling for Advanced Composite Shell Type Structures, *General Dynamics Ft. Worth Div.,* AFML-TR-75-112, *Final Rep. June 1973–June 1975,* Contr. F33615-73-C-5119.

2. Davis G.: Composites Production Integration, *General Dynamics Ft. Worth Div.,* AFWAL-TR-80-4018, *Final Rep.,* Contr. F33615-77-C-5018.

3. Davis, G: Composites Manufacturing Operations Production Integration, *General Dynamics Ft. Worth Div.,* IR-448-8(VIII), *Final Rep.,* September 1980, Contr. F33615-78-C-5217.

4. Flesher, A. L.: Composites Manufacturing Operations Production Integration (Composites Contour Part Laminating), Grumman Aerospace Corporation, Bethpage, N.Y., IR-419-8-(VIII), Contr. F33615-78-C-5080, May 1980–July 1980.

5. Stansbarger, D. L., et al.: Composites Manufacturing Operations Production Integration (Flexible Composites Automation), Northrop Corporation, Aircraft Group, IR-447-8-(VIII), *8th Interim Rep.,* Contr. F33615-78-C-5215, Hawthorne, Calif. August 1980.

6. Miller, M. F., et al.: F-16 Technology Modernization Semiannual Report, TM45 (IV), General Dynamics, Ft. Worth Division, Contr. F33657-80-G-0007, July 1981.

7. Memmott, J. V. W.: Composites Manufacturing Operations Production Integration, *Hamilton Standard Div. United Technologies Corp.,* IR-449-8-(VII), *7th Interim Rep.,* Contr. F33615-78-C-5218, Windsor Locks, Conn.

8. Poveromo, L.: *MTAG Meet., Orlando, Fla., February 1981,* vol. 1, p. C-5.

9. Wood, A. S.: Filament Winding Goes Computer, Sheds Design Limitations, *Mod. Plast.,* May 1981, pp. 51–56.

10. *The Composite* (McClean Anderson Co.), January 1981, p. 4.

11. *Mater. Eng.,* January 1981, pp. 55–62.


# BIBLIOGRAPHY

London, A., G. Lubin, and S. Dastin: Improved Manufacturing of the F-14A Composite Horizontal Stabilizer, *5th Nat. SAMPE Tech. Conf. Kiamesha Lake, N.Y., Oct. 9–11, 1973,* pp. 80–90.

Moquet, G., and J. Lamalle: Filament Winding Products in Aerospatiale Aquitaine, *Joint SAE ASME, Conf. Colorado Springs, July 27, 1981,* ATAA 81-1462.

Tape Laying with NC, *Amer. Mach.,* August 1981, pp. 120, 121.

Thorpe, M., and J. Mingle: Spray Metal Composite Tooling, *26th Natl. SAMPE Tech. Conf. Apr. 28–30, 1981, Los Angeles,* Sess. 4A.

Yurenka, S., and J. T. Parks: An Automatic Fabrication Process for Composite Aircraft Structures, *14th Natl. SAMPE Tech. Conf. Cocoa Beach, Nov. 5–7, 1968* (II-IA-3).

Trimming, drilling, routing, and fastening used to fabricate a composite structure.
*(Sikorsky Aircraft, Division of United Technologies.)*

# Cutting, Machining, and Joining Composites

Advanced composites such as boron, aramid, and graphite fibers in epoxy and metal matrixes are now fully qualified and accepted materials for safety-of-flight components. The inherent structural efficiency of these materials has led to selective application of advanced composites in current aircraft designs and promises of even greater use in future advanced-technology vehicles. However, these new materials are not necessarily adapted to the same technological methods as their metallic predecessors. Some of the new techniques and processes identified with composite manufacturing are

Methods of cutting cured and uncured composites

Machining technology for routing, trimming, beveling, countersinking, and counterboring

Drilling technology

Mechanical fastening

Joining technology, which includes welding, adhesive bonding, brazing, and diffusion welding

Coatings, e.g., paint and special coatings for protection against lightning and electromagnetic interference

The cost of machining B–Ep, Gr–Ep, and B–Ep and Gr–Ep hybrids has been reduced, and new machining and cutting techniques offer possible further cost reductions. The same is true for the other manufacturing processes previously described, which will be discussed in detail in this chapter.

## CUTTING

### Uncured Composites

The conventional approach for cutting uncured composite materials involves manual cutting with a carbide disk cutter, scissors, or power shears. Since manual trimming for prepreg plies using knives, carbide wheel "pizza" cutters, and similar tools is slow, may require tooling, and is subject to human error, this manual operation should be replaced by an automated one wherever possible.

### *Scissors and Power Shears*

Yarns, rovings, fabrics, woven rovings, and prepregs of Kevlar can be cut and trimmed using sharp, clean tools. Scissors must have low tolerances between cutting surfaces. A serrated scissors (Fig. 6.1) cutting Gr–Ep works equally as well with Kevlar. The serration

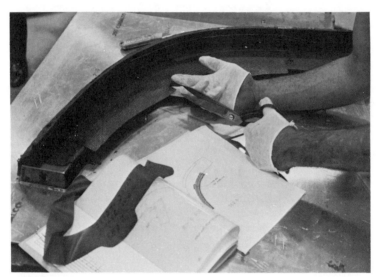

**FIG. 6.1**   Hand scissors cutting Gr–Ep prepreg. *(Sikorsky Aircraft, Division of United Technologies.)*

prevents the material from slipping out from between the cutting surfaces. In cutting prepregs, the tool must be frequently cleaned with acetone to prevent resin buildup and loss of cutting action. Electric and pneumatic power shears can cut up to 32 layers of Kevlar fabric at once. Offset cutting shears can cut Kevlar laminates up to 0.060 in (1.5 mm) thick and allow the cut product to slide past the cutting blades without interference.

### *Power Cutters and Saw Blades*

Rotary power cutters can effectively cut one layer of Kevlar fabric. The blades are self-sharpening and have extended cutting life. The unit is easily held and guided with one hand. Multiple plies of fabric or prepreg can be cut with abrasive grit, toothless bandsaw, or saber-saw blades. Both saw blades will fit conventional equipment. The blades will cut up to 50 plies of Kevlar.

## Laser Beam, Water-Jet, Reciprocating-Knife, and Steel-Rule Blanking

Alternative techniques, many of which have been automated and therefore are potentially more economical, are substitutes for the manual cutting. Each method produces high cutting rates of single or multiple plies and reproducibility of ply configuration and is applicable to the various types of composite prepregs. The water jet uses an ultra high-pressure focused stream of water to cut the workpiece. It therefore requires a method to collect the fluid; the possibility of water absorption must not be overlooked. The laser beam may damage the resin in the areas of the cut and may score the work stand. The reciprocating knife does not contaminate the work but penetrates into the closely packed bristles which form the surface of the work table.

**Laser-Beam Cutting:** Continuous-wave 250-W $CO_2$ lasers have produced cutting speeds up to 400 in/min (169 mm/s) in single-ply Kevlar prepregs and up to 300 in/min (127 mm/s) in single-ply B–Ep. Although B–Ep laminates up to four plies thick were also effectively laser-cut at a feed rate of 60 in/min (25.4 mm/s), the slow feed rate resulted in a marginal cut with a significant amount of resin retreat. The best cuts have occurred in two-ply B–Ep laminates cut at a feed rate of 120 in/min (51 mm/s).

Feed rates of 300 in/min (127 mm/s) on single-ply Gr–Ep tape is the most effective cutting rate; for two-ply laminates 150 in/min (63.5 mm/s) is recommended. Although Gr–Ep tape laminates up to three plies thick have been successfully laser-cut, an excessive edge bead of resin develops at 60 in/min (25.4 mm/s).

In uncured Gl–Ep laminates laser cutting has been accomplished at 300 in/min (127 mm/s) for single-ply thicknesses and at 90 in/min (38 mm/s) for thicknesses up to three plies. Higher feed rates reduced cut quality, apparently because the resin could not vaporize quickly enough.

Woven graphite broad goods are readily cut at 300 in/min (127 mm/s) with nitrogen-assist gas pressure. No evidence of heat damage to the fibers is visible. A Gr–PS with its thermoplastic matrix exhibits the greatest matrix damage. In tests as much as 0.025 to 0.036 in (0.6 to 0.9 mm) of the matrix material has been removed. Cutting rates up to 270 in/min (114.3 mm/s) are easily obtained with the laser.

**Water-Jet Cutting:** The water-jet process severs material by forcing water through a small-diameter jet at high velocities. As the water jet impinges on the surface, it cuts by inducing localized stress failure and eroding the material. Typical cutting conditions are a water-stream diameter of 0.010 in (0.25 mm) and water-jet velocities up to 2900 ft/s (884 m/s). A schematic representation of commercially available water-jet cutting equipment is shown in Fig. 6.2. The cutting performance of the water-jet is affected by:

Jet pressure

Nozzle orifice diameter

Traverse speed

Type and thickness of material cut

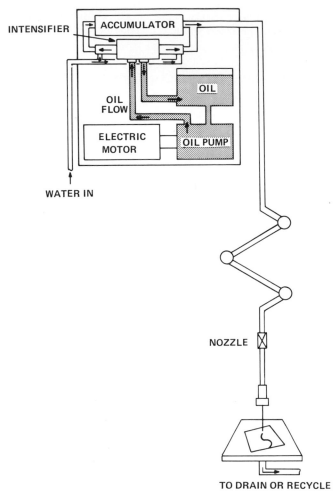

**FIG. 6.2**  Schematic representation of high-pressure water-jet cutting system.[1]

Tests conducted by several aerospace companies have found that water-jet cutting of prepregs of Gr–Ep, B–Ep, Kv–Ep and Gl–Ep produce good, clean edge finishes with no serious detrimental curing or strength effects due to the water absorbency.[1] When the cutting head is mounted on automated indexing and positioning equipment, the water-jet method becomes very practical for cutting prepregs.

**Reciprocating-Knife Cutting:**  There are two reciprocating-knife cutting machines, both of which incorporate high-speed reciprocating knives that are driven through the material to be cut by a minicomputer-controlled, *xyzc* positioning system. In one system the cutting knife penetrates through the material into closely packed plastic bristles that constitute the surface of the cutting table. The surface is nondegradable and does not require periodic renewal.[2] The cutting knife ranges in width from 0.050 in (1.3 mm) for the diamond cutter up to 0.175 in (4.4 mm) for the carbide cutter. The system cuts in a chopping mode; i.e., the knife rises above and plunges through the material onto the table.

The second system[1] can cut desired patterns in a continuous line at high speed. Curves, sharp corners, and notches can also be cut without lifting the knife from the material. The knife can be lifted, as required, to start new cutting lines, to pass over sections without cutting, or to cut holes of any diameter. The system uses a blade 0.250 in (6.4 mm) wide and cuts either by chopping or slicing. In the slicing mode, the knife remains buried in the material after the first stroke [each stroke is 0.75 in (19 mm)] and is always at least 0.125 in (3.2 mm) below the material being cut. Computer-controlled rotation of the knife about the $c$ axis keeps the blade properly positioned at all times.

Cutting test results for both systems summarized in Table 6.1 indicate that the slicing and chopping system can cut a greater number of Gl–Ep and Gr–Ep plies at twice the feed rate of the chopping system. Visually the quality of the edges of laminates cut by both systems is about equivalent.

**Steel-Rule-Die Blanking:**  Up to the present time, die life is not a major problem in the aircraft industry because of the relatively small quantities produced (compared with the appliance or automotive industries, for example). If a die does require reconditioning, only a minor expenditure is involved. The steel-rule dies (Fig. 6.3) are positioned above a flat mild-steel plate to permit blanking on the downstroke. Each die consists of 0.118-in-thick (3-mm) one-side-beveled, hardened-steel strap embedded in a wooden base with a cork stripper plate. This cutting-edge configuration gives a higher quality than other standard configurations. Single-ply laminates of all composite materials except Kevlar cut cleanly and easily, requiring only minor die-position and pressure changes. Kv–Ep requires more buildup with paper and metal (Fig. 6.4) for additional pressure and a better impression than the other materials to achieve a clean separation of the blanked configuration. Multiple cutting is clean and easily performed for all materials (Fig. 6.5) except Kevlar prepreg. The criterion used for selecting the maximum number of plies is squareness of the cut. As the number of plies increases, edge squareness decreases. Kv–Ep requires significantly higher pressures since the fibers are difficult to sever, limiting the maximum number of plies cut (Table 6.2).

## Cutting of Cured Composites

The primary candidates for production implementation are high-pressure water-jet cutting and laser cutting. Since they exhibit different operations and performance characteristics, it is necessary to establish some criteria for evaluation purposes. Areas of concern include:

*Part contamination:*  A serious problem in the fabrication of laminated parts, contamination can take the form of cutting fluids, lubricants, or solids derived from the base material, backing material, or from the machine itself.

*Quality of the cut:*  In this case quality means smoothness and absence of fiber fraying or induced ply separation.

*Stress on workpiece:*  Induction of stress onto the workpiece causes ply separation and movement on the cutting surface, resulting in mislocation or miscutting.

*Localized-heating effects:*  Problems include heat distortion.

*Resolution:*  This is the ability to cut small-radius curves with minimal error.

*Freedom from secondary (postcut) operations:*  Trim on cleanup operations increases both parts cost and the likelihood of later rejection.

**Table 6.1** Comparison of Recipro-Cutting Tests of Uncured Composites[1]

Chopping system

| Plies | Orientation, deg[a] | Cutter[b] | Feed rate in/min | Feed rate mm/s | Cutter strokes per min | Remarks |
|---|---|---|---|---|---|---|
| | | | | Graphite-epoxy | | |
| 1 | 0[c] | Carbide | 600 | 254 | 5000 | Good cut |
| | 0[d] | Carbide | 600 | 254 | 5000 | Some fibers not cut |
| 5 | zzfzz | Carbide | 600 | 254 | 5000 | Fuzzy |
| | | HSS | 600 | 254 | 5000 | Good cut |
| 8 | zzfnnfzz | HSS | 600 | 254 | 5000 | Good cut |
| | | | 900 | 384 | 6000 | Good cut |
| 13 | zzfnnfzzzzfzz | HSS | 600 | 254 | 5500 | Acceptable, slightly fuzzy |
| 21 | zzfnnfzzzzfnnfzzzzfzz | HSS | 600 | 254 | 5500 | Fuzzy |
| | | | | Glass-epoxy | | |
| 1 | 0 | Carbide | 600 | 254 | 5300 | Best, slightly fuzzy on exit |
| | | | 300 | 127 | 4200 | Good cut |
| | | | 900 | 384 | 6000 | Good cut |
| | | HSS | 600 | 254 | 5300 | Not good, wide cut |
| 4 | 0 | Carbide | 600 | 254 | 5300 | Not good, fuzzy |
| | | | | | 6000 | Not good, worse |
| | | | | | 3600 | Better |
| | | | | | 3100 | Best, slightly fuzzy on exit |
| 6 | zznnzz | Carbide | 450 | 191 | 3700 | Not good |
| | | | | | 4500 | Worse |
| 8 | zzznnzzz | Carbide | 450 | 191 | 4400 | Not good, fuzzy |
| | | | | Kevlar-epoxy | | |
| 1 | 0 | Carbide | 600 | 254 | 3700 | Not good |
| 4 | 0 | Carbide | 600 | 254 | 4500 | Best, some fuzz on exit side |
| 6 | zznnzz | Carbide | 600 | 254 | 5000 | A little fuzzy |

Slicing system

| Plies | Orientation, deg[a] | Cutter[e] deg[f] | Depth in | Depth mm | Feed rate in/min | Feed rate mm/s | Comment |
|---|---|---|---|---|---|---|---|
| | | | | Graphite-epoxy | | | |
| 1 | 0 | 15 | 0.031 | 0.8 | 300 | 127 | Best cut |
| 4 | znnz | 15 | 0.053 | 1.4 | 300 | 127 | Good cut |
| 8 | zzfnnfzz | 15 | 0.091 | 2.3 | 300 | 127 | Best cut |
| | | | | Glass-epoxy | | | |
| 1 | 0 | 25 | 0.035 | 0.89 | 600 | 254 | Good cut |
| 4 | 0 | 25 | 0.076 | 1.9 | 600 | 254 | Good cut |
| | | | | Kevlar-epoxy | | | |
| 1 | 0 | 25 | 0.045 | 1.1 | 600 | 254 | Best cut |
| 4 | 0 | 25 | 0.085 | 2.2 | 600 | 254 | Rather fuzzy |

[a]Except as noted cut direction is 0°. Code for orientation: z = 0° (0 rad), f = 45° (0.79 rad), and n = 90° (1.6 rad), so that zzfzz stands for 0°/0°/45°/0°/0°.

[b]Cutter width is 0.25 in (6.4 mm). All carbide cutters operate in the chopping mode and all HSS (high-speed steel) in the slicing mode.

[c]Cut direction 90°.    [d]Cut direction 10° skew.    [e]Cut direction 0°, carbide cutter, width 0.125 in (3.2 mm).

[f]15° = 0.26 rad; 25° = 0.44 rad.

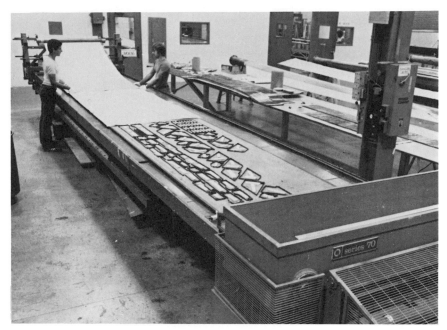

**FIG. 6.3** Steel-rule-die blanking of Gr–Ep prepreg. *(Sikorsky Aircraft, Division of United Technologies.)*

**FIG. 6.4** Buildup of paper and metal in cutting. *(Sikorsky Aircraft, Division of United Technologies.)*

**6.7**

**FIG. 6.5** Removal of multiple-ply cut material. *(Sikorsky Aircraft, Division of United Technologies.)*

**Table 6.2** Summary of Steel-Rule-Die Blanking of Uncured Laminates†[1]

| Material | Maximum number of plies | Configuration | Cut quality |
|---|---|---|---|
| Gr–Ep | 18 | Circles: 2.5 in (64 mm) diameter with 0.44-in (11-mm) diameter holes | Excellent |
| | | Triangles: 3 in (76 mm) with 0.68-in (17-mm) diameter holes | Excellent |
| B–Ep | 18 | Triangles: 3 in (76 mm) | Excellent |
| | | Same with 0.68-in (17-mm) diameter holes | Excellent |
| Kv–Ep | 12 | Triangles: 3 in (76 mm) | Good |
| Gl–Ep | 27 | Triangles: 3 in (76 mm) | Excellent |

†All blanking was done with top and bottom polyethylene cover sheets.

*Freedom from wear:* Cutting techniques which are susceptible to wear reduce machine output by creating downtime and raise cost by adding to inspection requirements and increasing rejects.

*Cutting flexibility:* This is the ability to change directions without interrupting the cut.

*Versatility:* This is the ability to cut different materials without reduction of effectiveness.

The two processes are evaluated in Table 6.3.

## Water-Jet and Laser-Beam Cutting

**Water-Jet Cutting:** In the past cured composite parts were trimmed by bandsawing, hand routing, shearing, nibbling, and sanding. These processes were labor-intensive and noisy and produced critical dust problems. Water-jet cutting has reduced the dust and noise problem and produced clean trimmed edges that eliminate hand sanding. Cured materials follow the rule of thumb; i.e., the cut quality improves with increasing nozzle pressure,

**Table 6.3** Evaluation of Water-Jet and
Laser Cutting

| Evaluation criterion | Water-jet | Laser |
|---|---|---|
| Contamination | Superior | Superior |
| Quality of cut | Superior | Superior |
| Workpiece stress | Excellent | Superior |
| Localized heating | Superior | Poor |
| Radius cutting | Superior | Superior |
| Postcut operations | Excellent | Excellent |
| Freedom from wear | Superior | Superior |
| Versatility | Superior | Superior |

increasing nozzle orifice diameter, decreasing traverse speed, and decreasing material thickness and hardness. Softer materials cut with a better edge quality than the harder ones. The order of decreasing hardness is B–Ep, Gr–Ep, Gl–Ep, and Kv–Ep, with hybrid materials occupying positions midway between the parent materials. Table 6.4 reflects some results of production water-jet cutting.

**Table 6.4** Production Water-Jet Cutting of Kevlar and Hybrid Materials

Jewel size 0.012 in (0.3 mm), pressure 50 kips/in$^2$ (345 MPa)

| Material | Thickness | | Cutting speed | | Edge quality |
|---|---|---|---|---|---|
| | in | mm | in/min | mm/s | |
| Woven cured Kv–Ep | 0.079 | 2 | 42 | 17.8 | Clean |
| | 0.160 | 4 | 42 | 17.8 | Clean |
| | 0.2 | 5 | 42 | 17.8 | Clean |
| | 0.275 | 7 | 42 | 17.8 | Clean |
| Filament-woven Kevlar | 0.160 | 4 | 42 | 17.8 | Clean cross-fiber cut |
| | 0.44 | 11 | 7.5 | 3.2 | Clean cross-fiber cut |
| | 0.49 | 12 | 14 | 5.9 | Clean cross-fiber cut |
| Prepreg Kevlar | 0.625 | 16 | 17 | 7.2 | Clean |
| Gr–Kv–Ep | 0.160 | 4 | 42 | 17.8 | Clean |

The process typically operates as follows: at 60 kips/in$^2$ (414 MPa) [absolute maximum capacity; usually the unit runs at 50 kips/in$^2$ (345 MPa) in conventional operation] the water is 12% compressible; for the first 12% of the stroke of the plunger, nothing will come out. To keep the jet running, compressed water is stored in the accumulator when the pump is making its initial compression (like a storage tank on an air compressor) (Figs. 6.6 to 6.8). The water moves through high-pressure tubing and coil assemblies. Hose is not available for operating pressures above 30 kips/in$^2$ (207 MPa), but connecting the rigid tubing lengths to high-pressure swivels provides flexibility in the movements of the cutting nozzle. The water moves to the nozzle, where it passes through a sapphire orifice and emerges as a coherent cutting stream. At this point the jet is a tight core surrounded by a shroud of light mist. Sapphire is used for the nozzle because it is an erosion-resistant material that will maintain its geometry and is fairly easy to fabricate. Cutting power rises as the nozzle diameter is increased; the tradeoff is that the cost of the pump also rises as the nozzle diameter gets larger. Common diameters for cutting applications range between 0.003 and 0.012 in (0.08 and 0.30 mm).

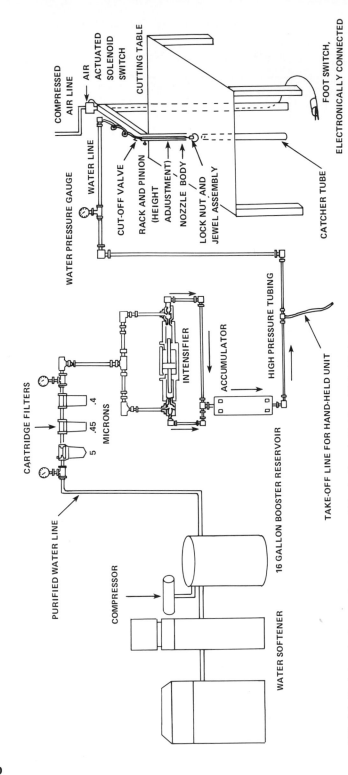

**FIG. 6.6** Typical production water-jet cutting system. (*Sikorsky Aircraft, Division of United Technologies.*)

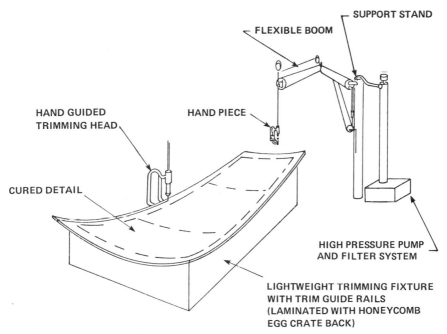

**FIG. 6.7**  Schematic closeup of water-jet system.

**FIG. 6.8**  Five water-jet cutters in operation. *(Sikorsky Aircraft, Division of United Technologies.)*

**Laser-Beam Cutting:** Kv–Ep is readily cut with a laser beam at thicknesses up to 0.3 in (7.6 mm) with up to 2 kW of laser power. At 1 kW Kv–Ep up to 0.2 in (5 mm) thick can be effectively cut; 0.3-in-thick (7.6-mm) Kv–Ep also can be cut at 1 kW, but very heavy charring occurs. Charring can be reduced in thinner materials [up to 0.09 in (2.3 mm)] by increasing speed. Material thicknesses greater than 0.09 in cannot be cut at speeds great enough to reduce significant charring with the laser. Using helium or $CO_2$ in a jet-assisted laser instead of air does not reduce charring significantly although helium results in lower maximum cutting speeds at constant power. Discoloration occurring on the top surface of laser-cut Kv–Ep can easily be removed with a rubber eraser lightly applied. Loose char on laser-cut edges is easily rubbed off, but the edges remain black (Fig. 6.9). Speeds

FIG. 6.9  Kv–Ep cut by laser beam, showing charring. *(Sikorsky Aircraft, Division of United Technologies.)*

of 10 to 300 in/min (4.2 to 127 mm/s) have been evaluated for the thicknesses mentioned above; 30 in/min (13 mm/s) offers the best results for 0.062-in-thick (1.6-mm) Gr–Kv–Ep hybrid and 150 in/min (64 mm/s) for 0.035-in-thick (0.89-mm) Kv–Ep.

Table 6.5 summarizes further tests conducted with laser-beam power levels up to 11

**Table 6.5**  Summary of High-Power Laser-Beam Cutting of Cured Composites[1]

Assist gas, $N_2$ at 150 kips/in$^2$ (1 GPa); jet assist, copper nozzle located about 0.25 in (6.4 mm) above workpiece at 45° (0.79-rad) angle

| Material | Amount of Gr–Ep, % | Material thickness in | Material thickness mm | Power, kW | Speed in/min | Speed mm/s | Extent of thermal damage to edge in | Extent of thermal damage to edge mm |
|---|---|---|---|---|---|---|---|---|
| Kv–Ep | 0 | 0.120 | 3.05 | 3 | 120 | 51 | 0.050 | 1.3 |
| Gr–Ep | 100 | 0.066 | 1.68 | 8 | 120 | 51 | 0.060 | 1.5 |
|  |  | 0.197 | 5.00 | 8 | 30 | 12.7 | 0.281 | 7.1 |
| Gr–B–Ep | 60 | 0.266 | 6.76 | 8 | 30 | 12.7 | 0.156 | 3.9 |
|  | 90 | 0.357 | 9.07 | 11 | 25 | 10.6 | 0.156 | 3.9 |
| Gr–Gl–Ep | 50 | 0.260 | 6.60 | 8 | 40 | 16.9 | 0.050 | 1.3 |

kW. In general, these results show that Gr–Ep and its hybrids require minimum power levels of 8 kW, cutting speed decreases with thickness, and thermal damage is an inverse function of cutting speed.

## MACHINING

Quality machining of composite materials must meet certain basic criteria, i.e., no fraying or delamination of cured-composite edges. The information below applies to machining generic resin matrix composites (Gr–Ep, Kv–Ep, Gr–Gl–Ep, and B–Ep). Standard machining equipment can be used with modifications. In general, spindle speeds and feeds depend upon the thickness of the laminate being machined and the type of cutting method. The cutting tools required to perform the machining operations include countersink, cutoff wheels, router bits, bandsaw blades, high-speed steel drills, and reamers. For long tool life, drills and routers should be made from tungsten-carbide; circular, band, and saber saws and countersinks should be made from M2 high-speed steel.

It is particularly important to keep tools sharp in order to provide quality cuts and minimize the possibility of delamination. Proper backup support of the work is required to eliminate delamination, along with cooling methods to control resin buildup on the tool caused by excessive frictional heat. Water and water-soluble coolants in a mist or flood application are satisfactory, as well as hydrocarbon fluids for B–Ep. If water-soluble coolants are used, the machined part must be rinsed thoroughly with water to remove excess coolant. When no liquid coolant is used with diamond cutting tools, the dust from machining operations must be collected in a vacuum system and the operator must wear a respirator.

Some general rules follow:

1. Use diamond tools on all machining operations which require cutting B–Ep.
2. Use solid backup for the workpiece, e.g., Masonite, to reduce chipping and delamination. Aluminum is a good backup material for drilling. It is advisable to have the backup cover the tool exit surface of the workpiece in the entire area to be cut. The backup must be held in intimate contact with the workpiece by clamping or otherwise.
3. Extreme caution should be exercised to avoid overheating the resin when epoxy-resin-matrix composites are being machined. Overheating destroys the part locally, and damage can be repaired only by removing the overheated area and filling the void with another material. Overheating occurs when the machined surface turns brownish-black.

### Routing, Trimming, and Beveling

These operations are essentially equivalent, involving the use of hand routers and mechanical Marwin machine routers and Roto-Recipro machines. Diamond-cut carbide and four-fluted milling cutters have been used to machine Gr–Ep and Gl–Ep laminates, while carbide opposed-helical router bits have been used to machine Kv–Ep and Kv–Gr–Ep hybrids. Diamond-coated router bits have been used with the Roto-Recipro machine to rout and trim B–Ep and B–Gr–Ep hybrids. Speeds for these operations range from 3600 to 45,000 r/min.

## Routing

In manual routing a higher-torque but lower-speed router (Buckeye) works best. Diamond-cut carbide bits attain higher feed rates at less operator effort than a six-flute configuration.

Using a coolant tends to extend tool life and increase the cutting force (probably due to sludge formation) with no effect on cut-edge quality. Based on operator effort, the maximum diametral tool wear would be about 0.0015 in (0.04 mm) before tool change would be required.

An opposed-helix carbide cutter (Fig. 6.10) has successfully routed Kv–Ep cured laminates. Since routing with a conventional aircraft tool-steel router can be difficult, the router bit in Fig. 6.10 was developed on the principle of shearing the outermost fibers toward the interior of the composite. The opposed-helix cutter is especially useful for severing laminates into sections, cutting slots and notches, and trimming honeycomb-sandwich panels. The portable version of the cutter has performed better than diamond-cut carbide cutters with Kv–Ep and Kv–Gr–Ep hybrid laminates. Although cut quality is equivalent for both types of cutters, only the opposed-helix, carbide cutter can trim 0.25-in-thick (6.4-mm) material.

The principal limitations of manual routing and trimming are that high cutting forces are required and productivity is closely related to operator skill. Use of a Marwin profiler in composite machining operations will avoid these limitations because feeds can be controlled while maintaining constant speed. Normally slower feed rates tend to give better cut quality. Tool wear, feed rates, and cut quality are equivalent to those obtained with the Buckeye router, which requires a great deal of effort to trim composite materials thicker than 0.125 in (3.2 mm). The Marwin router is limited to use with flat parts and simple formed parts.

**FIG. 6.10** Opposed-helix router bit for trimming Kv–Ep. *(Du Pont Co.)*

For routing and trimming B–Ep and B–Ep and Gr–Ep hybrids diamond-plated router bits are used. Spray mist or flood cooling is required for all power-feed routing of B–Ep; spray mist or air-blast cooling is recommended for hand routing. Power feed is recommended in routing B–Ep thicker than 0.050 in (1.3 mm).

## Trimming and Beveling

Manual trimming and beveling with a Buckeye router against a guide has been successfully used on Gr–Ep, Gl–Ep, Kv–Ep, and Gr–Gl–Ep hybrid. The router bits were diamond-cut carbide types and depths of cuts were 0.06 or 0.13 in (1.5 or 3.3 mm) and 45° (0.79 rad) by 0.125 or 0.25 in (3.2 or 6.4 mm). Good cuts have been made on thicknesses up to 0.27 in (6.9 mm). The quality of Kv–Ep trimmed edges has been improved using the router shown in Fig. 6.11. Kv–Ep trimmed edges can also be improved by sanding.

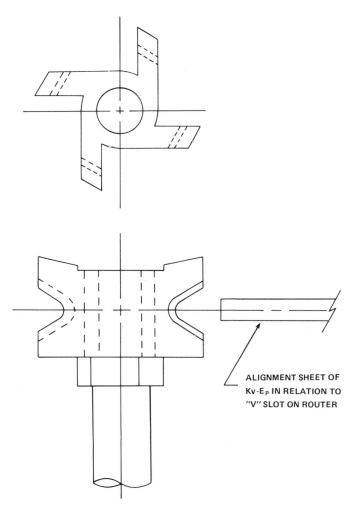

ALIGNMENT SHEET OF
Kv-E$_P$ IN RELATION TO
"V" SLOT ON ROUTER

**FIG. 6.11** Router for trimming Kv–Ep. *(Du Pont Co.)*

## Sawing and Sanding

Several critical points must be remembered in sanding composite laminates:

1. For fit or trim
   a. Use tool correctly (Fig. 6.12)
   b. Use right-angle sander of 20,000 r/min or greater

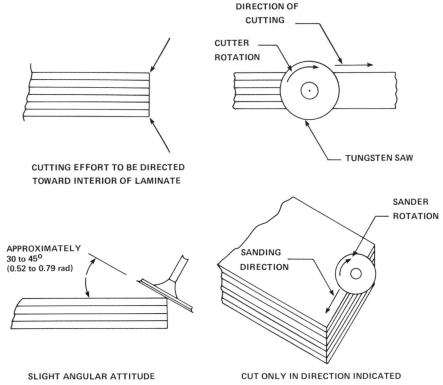

**FIG. 6.12** Sanding composite laminates.

    *c.* Use 2-in (50.8-mm) 80-grit aluminum oxide disks for bulk removal (dry)

    *d.* Use 2-in (50.8-mm) 240- to 320-grit SiC for edge deburring (water)

    *e.* Address edges at proper angle

    *f.* Do not try to remove any defects on surface of parts by sanding

    *g.* Do not oversand beyond trim areas

    *h.* Do not deburr holes; countersink holes with sanders

2. For bonding

    *a.* Lay out the area to be sanded

    *b.* Use 240-grit

    *c.* Use tools of 20,000 r/min or greater

    Sanding Kv–Ep laminates requires wet paper for cooling and to prevent buildup of waste between abrasive particles. The preferred grit sizes are 120 and 240.

    Sawing of cured laminates has been successfully performed with band, circular, or saber saws. The composite must be clamped to eliminate vibration, which can cause delamination. Cutting edges should be checked frequently to maintain sharpness.

## Bandsawing

Normally a fine offset high-strength-steel stagger-tooth blade is used [14 to 20 teeth per inch (25.4 mm)] and a surface speed of 4500 to 6500 ft/min (22.9 to 33 m/s) with 6000 ft/min (30.5 m/s) preferred. This cutting method uses the heel rather than the hook of the cutting-tooth blade for cleaner cuts. The cutting blade should always be sharpened before cutting. Blade sharpening consists of placing a 300- to 400-grit alumina-SiC honing stone in contact with the blade and exerting pressure on the points of the cutting teeth. The best feed rates are 6 to 12 in/min (3 to 5 mm/s), and as a general rule the height of three teeth should match laminate thickness.

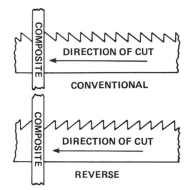

FIG. 6.13 Reverse-bandsawing Kv–Ep cured laminates. *(Du Pont Co.)*

Bandsaw cutting of Kv–Ep is usually performed with a fine-tooth blade, preferably using water as a coolant. The band should be run in reverse, so that the heel of the tooth enters the composite first (Fig. 6.13). B–Ep and B–Gl–Ep hybrids are bandsaw-cut with diamond-coated blades, 40 to 60 grit, and speeds of 2000 to 6000 ft/min (10.2 to 30.5 m/s). Additional tool life has been obtained using carbide-coated blades, which operate at speeds of 1400 and 3000 ft/min (7 and 15 m/s) and feed rates at 20 to 40 in/min (8.5 to 16.9 mm/s).

## Circular Sawing

By using a metal slitting saw Kv–Ep has been successfully cut with circular saws, but diamond blades are preferred for most composite materials. Cutting speed for composites normally varies from 2000 to 10,000 ft/min (10.2 to 51 m/s); 0.25-in-thick (6.4-mm) Gr–Ep is best cut with a carbide-tipped blade and speed between 2000 and 4000 ft/min (10.2 and 20 m/s). Figure 6.14 shows that 3 in/min (1.3 mm/s) maximum feed rate is ideal for all parts up to 0.1 in (2.5 mm) thick. Thicker parts are cut at proportionately lower feed rates.

## Saber Sawing

Saber sawing normally cuts Kv–Ep with the blade shown in Fig. 6.15, which cuts the outermost fibers on both sides of the laminate toward the interior. This blade has five alternating teeth in opposed directions; blade speeds of 2500 strokes per minute are recommended, but blade speed and feed rates may vary with material thickness.

## Countersinking

Conventional countersinking tools from 1750 to 6000 r/min have been used successfully on Gr–Ep, Gl–Ep, Kv–Ep, and Gr–Gl–Ep hybrids. The schematic in Fig. 6.16 illustrates

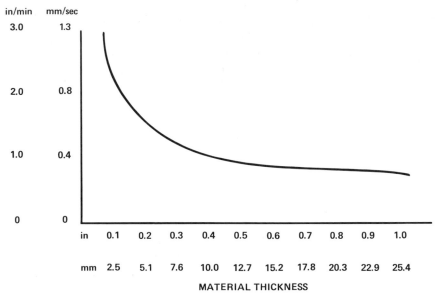

**FIG. 6.14** Feed rate vs. thickness of composite materials.[13]

a new countersink used on Kv–Ep composite. Other butterfly countersinking tools modified with serrations have also been used. Carbide, high-strength steel, and diamond-plated and sintered countersinks have been used with results dependent on the composite. Carbide countersinks have shown good results with Gr–Ep and Gl–Ep with optimum combinations of speed and relief angles.[1]

In countersinking B–Ep and Gr–Ep hybrids the feed rate has a pronounced effect on diamond tool life. Low feed rates are recommended. Plated countersinks become totally worn when the exposed diamonds become worn flat; cutting is then accomplished with a great deal of effort. Ultrasonic countersinking tests on B–Ep and hybrids showed that angular wear on sintered diamond tools becomes excessive after 100 holes have been countersunk. Although the amount of wear on the countersink itself is slight, the angular change approaches the maximum allowable countersink-angle tolerance. The pilot on both types of countersinks wears rapidly when boron fibers are cut. The pilot in the plated tool is replaceable; the pilot in the sintered tool can be refurbished by nickel plating. Although plated tools wear more rapidly than sintered tools, their lower cost makes plated countersinks more cost-effective. Because of the relatively short life of sintered countersinks, it is recommended that a rough countersink be made first with a 40- to 60-grit tool and a secondary finishing operation be made with a 60- to 80-grit tool.

## Counterboring

The most efficient cutting-tool materials for B–Ep and hybrids is diamond; carbide is best for graphite, Kevlar, and fiber glass composites.[1] In general, only a few counterbores can be made per tool, since the entire cutting edge bears on the workpiece material during the entire operation. In fact, torque and thrust forces are 3 times that required for countersinking Gr–Ep. Although diamond tools provide excellent counterbores with B–Ep and hybrids, the high thrust loads limit the number of counterbores that can be obtained.

## Milling

In general, to produce a surface cut that will extend past the edge of a composite laminate the edge of the laminate should be milled first to prevent delamination (Fig. 6.17). The use of fluorocarbon coolant is recommended because of its cooling efficiency. The coolant is applied as a spray mist during machining; the distance between spray applicator and cutter is adjusted so that frost forms on the cutter.

In milling Kv–Ep conventional fluted cutters are satisfactorily operated at speeds of 80 ft/min (0.4 m/s) and feed rates of 1 ft/min (5 mm/s). In milling Gr–Ep high-speed-steel end mills or carbide cutters can be used provided they are multifluted. Four-flute end mills are recommended for efficiency and to reduce the cutting forces to a point where there is less chance of delamination. All cutters should be sharp; dull ones cause delamination.

High-speed-steel cutters should always be of the four-flute positive-rake type; carbide milling cutters should also be of a positive-rake type with chip loads of 0.004 to 0.006 in (0.10 to 0.15 mm) per tooth. Radiused-end mills last longer than square-cornered ones.

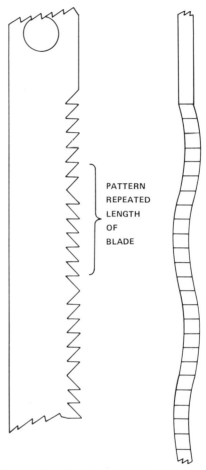

**FIG. 6.15** Alternating-tooth saber-saw blade. (*Du Pont Co.*)

Plunge-cut milling is not recommended unless there is sufficient backup support to prevent delamination. A Gr–Ep ribbed structure is shown in Fig. 6.18 immediately after milling and before bonding.

At first both high-speed-steel and carbide end mills were used on B–Al composites, but edge machining resulted in almost immediate tool failure and produced a large burr on the workpiece. This led to development of a valve-stem cutter plated with 60-grit diamond (the tool was dipped in a diamond-grit slurry and coated by electroless nickel plate), which makes as many as 80 cuts with very little wear.

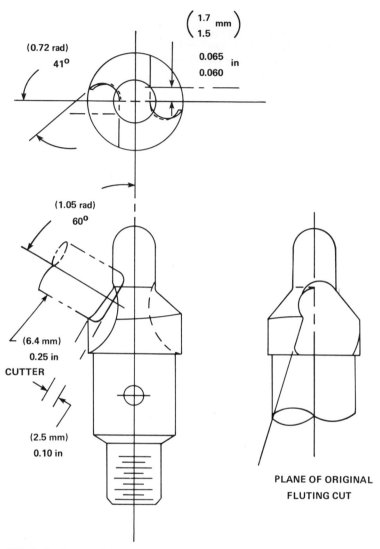

$$\left(\begin{array}{c} 1.7 \\ 1.5 \end{array} \text{mm}\right)$$

(0.72 rad)
41°

0.065
0.060 in

(1.05 rad)
60°

(6.4 mm)
0.25 in
CUTTER

(2.5 mm)
0.10 in

PLANE OF ORIGINAL
FLUTING CUT

**FIG. 6.16**   Countersink for Kv–Ep composite. *(Du Pont Co.)*

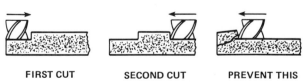

FIRST CUT          SECOND CUT          PREVENT THIS

**FIG. 6.17**   Proper milling technique.[13]

FIG. 6.18   Gr–Ep milled structure before bonding.

## Grinding

Kv–Ep composites can be ground or chamfered by strict adherence to specific conditions. A glazed work surface with a minimum of uncut fibers is obtained. Coolant is necessary to prevent wheel loading. $Al_2O_3$ or SiC grinding wheels are satisfactory; the SiC wheel flooded with soluble oil has been used to grind B–Al composites. Wheel wear is not excessive, and the ground edge of the composite is satisfactory.

## Shearing and Punching

Shearing can be performed, but edge will be coarse or ragged. Recommended practice is to shear 0.010 to 0.020 in (0.25 to 0.50 mm) outside of the final trim line and then sand or mill to the net dimension. Fiber tearing also occurs in punching composite materials, so that a reaming operation has been developed to minimize ragged edges. Punching B–Ep is discussed below in the section on drilling. In most types of machining operations it is good shop practice for an operator to wear a face-mask respirator to prevent inhaling the exceedingly fine but stiff particles that are given off.

## DRILLING

One of the first questions asked when attention began to focus on composite materials was why it takes so long to drill holes in them. The answer, of course, is that the available tooling was not designed for cutting composites. Gl–Ep and Gr–Ep are so abrasive that only tungsten carbide tooling can drill through them. Drill tips were designed for metalworking, the tip heating the metal to provide the plastic flow needed for efficient cutting. Since composites cannot tolerate this heat, production must be slowed down to keep the heat as low as possible. Drill designers had to abandon cutting tips with neutral and negative rakes and wide chisel points because a drill with a neutral rake scrapes the material and causes it to resist penetration by the drill tip. The operator must exert pressure to drill the hole, and pressure causes the heat buildup.

Neutral rake also tends to push the reinforcing fibers out in front, requiring a great deal of pressure to penetrate the piece. This pressure causes the fibers to bend, resulting in furry, undersized holes. The pressure also produces excessive heat, which causes galling and chip clogging in the resin. The release of pressure as the tool bit breaks through the part causes a sudden and momentary increase in feed rate. As the tool plunges through the last few fibers, the cutter shaft, not the cutting edge, removes the remaining material. The result is chipping and cracking.

The best way to analyze a drilling operation is to examine the chips. Ideal chip form for composites is a dry, easily moved chip that looks like confectioner's sugar. If the speed of the cutting tool is too high, heat will make the resin sticky and produce a lumpy chip; if the cutting edge is scraping and not cutting the plastic, the chips will be large and flaky. Either type will eventually clog any evacuation system.

Good tool geometry for both resin and MMCs starts with positive rake. The reinforcing fibers are pulled into the workpiece and sheared or broken between the cutting edge and the uncut material. Positive rake on the cutting edge removes more material per unit of time and per unit of pressure than negative rake, but the more positive the rake, the more sensitive and fragile the cutting edge becomes. A small chisel edge, the second element of good tool geometry, improves the penetration rate, which translates into more pieces per hour. The optimum chisel edge for composites is as close to a point as possible. Good geometry also means a cutting-tool shape that facilitates chip handling, so that the chips are produced and then removed immediately above the entrance of the hole. From this point a properly designed vacuum system can dispose of the chips in conformance with safety and environmental standards.

In the last few years tooling has been developed that has greatly improved drilling operations in Gr–Ep, Kv–Ep, Gl–Ep, B–Ep, and their hybrids as well as metal-matrix compositions. Some tool bits are made of particles of tungsten carbide smaller than 1 $\mu$m; three models are shown in Fig. 6.19. The solid-shank style and the twist drill are used in automatic drilling equipment. The drill-guide system is designed to be used with an air or electric drilling motor. Fitted with a socket adapter, it can drive all drill sizes from 0.118 to 1.0 in (3 to 25.4 mm). The internal compression-control spring regulator withdraws the tool after drilling. Its pressure control compensates for the breakthrough lurch,

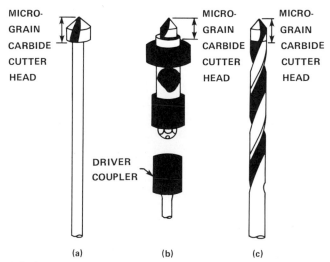

**FIG. 6.19** Cutting heads used in drilling composites: (*a*) solid-shank drill, (*b*) drill guide system, (*c*) fluted twist drill.[14]

or sudden increase in feed rate as the tool bit breaks through the last few fibers. It also eliminates partially drilled holes, because the operator must depress the unit completely each time since no change in pressure is felt as the hole nears completion. In recommended operations these drills have run at speeds of 300 to 600 ft/min (1.5 to 3 m/s).

## Kv–Ep Drilling

Spade drills are best for drilling Kv–Ep, leaving very little fuzz and fraying on hole edges. Like carbide drills, these drills have a tendency to burn if they are used too long. Conventional drills can also be used, but firm sacrificial backing must be provided at the exit surface. Twist and flat-ended high-speed-steel drills perform quite well on Kv–Ep, especially with a firm backup of the composite to eliminate fuzzing and delamination at the hole exit. Backup can be obtained using Micarta or brass or by leaving the glass peel ply on the composite until after drilling. The 0.003-in (0.08-mm) layer of fiber glass on the top and bottom surfaces of the Kv–Ep composite produces the best holes, leaving clean entrance and exit surfaces. Drill speeds range from 25,000 to 35,000 r/min. The high-speed-steel drill is shaped like a twist drill fluted on the end. In a production environment 45 to 50 holes were made before resharpening was required although signs of wear have been found after drilling five holes. The best lubricant is water.

## B–Ep, Gl–Ep, and Hybrid Drilling

Drills with titanium diboride coating improve the life of drills by 887% when used on Gl–Ep.[4] The best way to drill B–Ep and B–Gl–Ep hybrid is with diamond-impregnated core drills and an ultrasonic machine as a drilling-assist tool; however the cores have a tendency to get stuck in the drills. The best combination is diamond-impregnated core and reamer drills and countersink tools. Although the cores sometimes get stuck, like ultrasonically powered drills, they can be removed by stopping the drill spindle and leaving the ultrasonic unit on—unfortunately a slow process for production applications. The best speed for drilling B–Gl–Ep laminates is 2000 to 3000 r/min. The best machine spindle feed rates for drilling B–Gl–Ep laminates are 0.33 to 0.20 in/min (0.14 to 0.08 mm/s). A 5% commercial surfactant in water is the best coolant when drilling B–Ep and B–Gl–Ep hybrids. Piercing (discussed later in this chapter) is a viable production technique.

## Gr–Ep Drilling

Tungsten carbide drills have been successfully used in drilling Gr–Ep using a backing plate. The drill is shaped like a standard twist drill and can be used for 50 to 60 holes before it needs resharpening (standard drills only last 5 to 6 holes). Various coolants have been tried to prevent drills from breaking, dulling, or burning up. A lubricant called Boelube has worked well, but water is the best. To reduce surface delamination during drilling a layer of woven glass 0.003 in (0.08 mm) on both sides of the laminate is recommended.

Solid carbide Daggar drills have also proved quite successful in drilling Gr–Ep, producing a clean hole with little or no breakout. Drilling has been accomplished without using backup material. Since the Daggar drill will maintain hole tolerance, no reaming is necessary and the drill is adaptable as a hand-held unit (Fig. 6.20) or in a permanent fixture using air feed. The best results have been achieved at 900 r/min with the hand-fed units and 2100 r/min with a 3 in/min (1.3 mm/s) feed with the fixtured air-feed units.

A polycrystalline-diamond-tipped drill has a tungsten backing on both surfaces to facilitate silver-soldering it into a slot in the tungsten shank. Two straight flutes and tip geometry are ground into the drill. These tools have drilled 2000 holes of 0.191 in (4.9 mm) diameter in 0.125-in-thick (3.2-mm) Gr–Ep without resharpening. One advantage is that these drills can be reground 4 or 5 times because they maintain size better than the carbide tools.

Drilling Gr–Ep and metal simultaneously has given high-quality holes and countersinks. Solid carbide drills and countersinks at speeds of 21,000 and 5500 r/min were successfully used on aluminum-perforated foil-coated and aluminum-pressed-powder-bonded Gr–Ep laminates 0.118 and 0.135 in (3 and 3.4 mm) thick. A survey[5] disclosed that Gr–Ep and aluminum, Gr–Ep and steel, Gr–Ep and titanium, and Gr–Ep and Invar have been successfully drilled at the same time. The drills were straight flute and twist, and the tooling material was carbide and high-speed steel.

**FIG. 6.20** Daggar drill in hand-feed unit.

## Ultrasonic Drilling

This technique has been used to drill and countersink Gr–Ep and B–Gr–Ep holes in prototype aircraft structures such as the B-1 horizontal stabilizer (Fig. 6.21). The drills are water-cooled. Tools used are all-diamond types, sintered or coated. The machine is versatile in that it can drill, countersink, ream, and counterbore. The ultrasonic drill-countersink is fitted with a sintered-diamond core drill plus an electroplated nickel or diamond sizing band behind the tip to maintain size and concentricity. The countersink surface is also plated because it can be stripped chemically and replated at low cost to extend tool life. This application used specially designed, core-drill–countersink combination tools ranging in diameter from 0.190 to 0.5 in (4.8 to 12.7 mm) (Fig. 6.22).

The application of ultrasonic energy to diamond core drills when drilling either B–Gl–Ep or B–Ep hybrids increases drill life. For example, 50 holes were drilled by a portable drill and diamond-core drill without ultrasonics in 0.4-in-thick (10.2-mm) Gr–B–Ep hybrid material. When ultrasonic energy was applied 100% increase occurred [200 holes in 0.22-in-thick (5.6-mm) hybrid].

## Special Techniques

As engineers are confronted by complex designs, new materials, and obstacles to productivity, special techniques evolve for making holes.

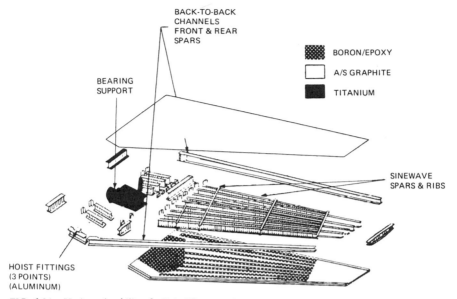

BACK-TO-BACK
CHANNELS
FRONT & REAR
SPARS

BORON/EPOXY

A/S GRAPHITE

TITANIUM

BEARING
SUPPORT

SINEWAVE
SPARS & RIBS

HOIST FITTINGS
(3 POINTS)
(ALUMINUM)

**FIG. 6.21**  Horizontal stabilizer for B-1. *(Grumman Aerospace Corporation.)*

**FIG. 6.22**  Ultrasonic drilling machine. *(Sikorsky Aircraft, Division of United Technologies.)*

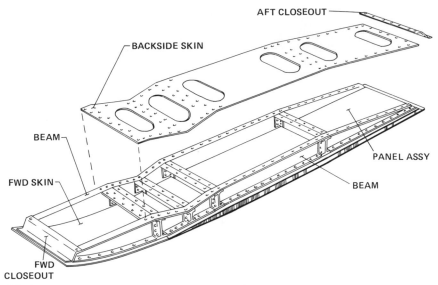

**FIG. 6.23**   Demonstration component: engine inlet door.[6] (*Rohr Industries.*)

## Integrally Formed Holes (Molding)

Figure 6.23 shows an aircraft component used to demonstrate a method of integrally forming fastener holes during the processing of composite structures.[6] One fabrication method used female tooling to control the contour of the outside surface of the structure. This approach was made feasible by using the stripper plate as the mold shell or cure form (Fig. 6.24). Pins were then designed to be extracted from the cured laminate and mold. The female tool was also used to fabricate detail beams for both the drilled and formed hole components.

The basic tools (cure forms, pin plates, and caul plates) were Gr–Ep in order to minimize thermal distortion between the tool and part. The low thermal contraction of the tool material also allowed the forming pins to maintain their position. When the cured mating

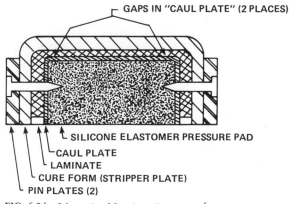

**FIG. 6.24**   Schematic of female tooling system.[6]

parts were brought together on assembly, the fasteners could be installed. The caul plates were used to provide a smooth surface on the inner surface of the channel beams.

This program has several features for consideration:

1. Hole-preparation time reduced by 59% in Gr–Ep laminates up to 0.28 in (7.1 mm) thick
2. Formed holes in quasi-isotropic Gr–Ep laminates achieved 92 to 107% of the bearing strength of drilled holes
3. Best-suited production application for the formed-hole method would have:
    *a.* A gentle contour
    *b.* A large number of fasteners
    *c.* Multiple rows of fasteners

## *Piercing*

Figure 6.25 shows a B–Gl–Ep hybrid strap 0.090 in (2.3 mm) thick in which 194 holes were pierced in 10 s. This compares with the previous method, which required 6 min per hole in drilling with diamond-core drills and ultrasonic assist. A series of tests established that

1. Pierced holes are essentially equivalent to drilled and reamed holes
2. Punch material is crucible steel CPM 10V, used for its exceptional wear resistance and

**FIG. 6.25** B–Gl–Ep hybrid strap with pierced holes.[15] *(Sikorsky Aircraft, Division of United Technologies.)*

**FIG. 6.26** Press and die used in production.[15] *(Sikorsky Aircraft, Division of United Technologies.)*

good toughness. Carbide was rejected because it is prone to breaking and chipping and expensive to sharpen and maintain.

3. Flat punch-point configuration produces the best hole, e.g., minimal delamination and desired hole size.

4. Punch life with Teflon lubricant is 2250 hits. Figure 6.26 shows the Verson brake press with the die in place; it has produced over 400 parts without rejection.

## MECHANICAL FASTENING AND ADHESIVE BONDING

Currently the two most successful methods for joining composite structures are mechanical fastening and adhesive bonding. Eliminating as many joints as possible is one of the most important goals in designing composite structures. Joining of resin-matrix composites has been largely limited to adhesive bonding and mechanical fastening. Several welding techniques have been used but not extensively; see the section on joining, below.

### Adhesive Bonding

Adhesive bonding is preferred for most composites, and mechanical fasteners are used only where access into a structure is required. The main problems with fasteners in composites include low bearing strength and stress concentration at the fastener hole. Bonded metal inserts or edge members are often used so that the fastener is in the metal. One disadvantage of this approach is cost. Glass-fiber softening strips have been used to minimize stress-concentration effects.

Adhesives are limited by low shear strength, peel strength, and environmental resistance. Tapered and overlapped step-lapped joints and careful design to avoid eccentricities and discontinuities, which result in unfavorable loading, can overcome some of the disadvantages of adhesives in bonding composites. Chopped fibers added to adhesives give them a higher strain-energy capability providing substantial improvement in joint integrity. As the heat resistance of each type of adhesive is improved, there is every reason to believe that adhesives will soon become available to break the 600°F (316°C) continual-service-temperature barrier. Polyimide adhesives provide these improvements. For example, one family of condensed polyimides provides lap shear strengths of 4 kips/in$^2$ (28 MPa) at room temperature and 2 kips/in$^2$ (14 MPa) at 482°F (250°C). The adhesive is usable up to about 575°F (302°C) and has found application in joining B–Al to itself and to titanium. Another addition polyimide essentially free of volatiles has provided bond strengths of 2.5 to 3 kips/in$^2$ (17 to 21 MPa) at room temperature and up to 482°F (250°C). Like Gr–PI, this adhesive is being used in metal-matrix honeycomb structural panels.

## Mechanical Fastening

This joining process has been successful in attaching fiber glass, boron, and Kevlar but poses problems for graphite. When metallic fasteners come in contact with graphite composite structures, corrosion often results. Composite fasteners may represent an alternative. Aluminum alloys are least compatible with Gr–Ep structures; stainless steels are somewhat better. Compatibility of nickel-base and titanium alloys is excellent with graphite. Using

FIG. 6.27   Kv–Ep facing with patch of titanium fasteners. *(Sikorsky Aircraft, Division of United Technologies.)*

0.003-in-thick (0.08-mm) fiber glass or adhesive with scrim will insulate graphite and aluminum or steel and avoid corrosion.

Corrosion problems are of particular concern because use of composite materials is growing in primary and secondary aircraft structures. The low environmental resistance of most metallic fastening systems and the cost of titanium fasteners have created an interest in fasteners made of high-performance filamentary-reinforced composites. Gl–Ep or Gr–Ep do not corrode galvanically.

Weight savings are another important consideration. A Gr–Ep fastener is 5 times lighter than a stainless-steel fastener of equivalent size, and a glass fastener is 4 times lighter. The key to the use of metal fasteners in spite of the cost will be decided by the design; the loads and/or stress in the structure may dictate titanium rivets (Fig. 6.27) rather than steel, aluminum, or composite.

Trade-offs in fastener selection for a composite component for a helicopter are shown in Fig. 6.28. Joints produced by mechanical fastening, adhesive bonding, and co-curing of Kv–Ep are compared in Table 6.6. The three forms of joining various components of a commercial helicopter were designed and constructed as shown in Figs. 6.29 to 6.31.

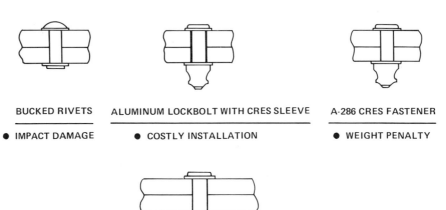

**FIG. 6.28**   Trade-offs in fastener selection.[16] *(Sikorsky Aircraft, Division of United Technologies.)*

**Table 6.6**   Comparison of Three Joining Methods

Best = 1

| Criterion | Co-cure | Bonding | Mechanical fasteners |
|---|---|---|---|
| Production cost | 1 | 2 | 3 |
| Ability to accommodate manufacturing tolerances and component complexity | 3 | 2 | 1 |
| Facility and tooling requirements | 3 | 2 | 1 |
| Reliability | 1 | 2 | 3 |
| Repairability | 3 | 2 | 1 |

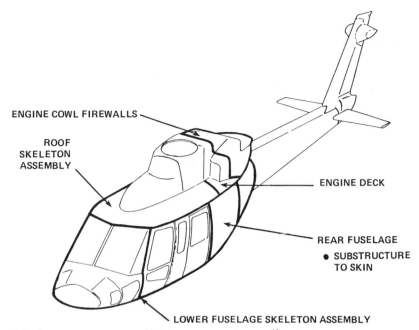

ENGINE COWL FIREWALLS

ROOF
SKELETON
ASSEMBLY

ENGINE DECK

REAR FUSELAGE
● SUBSTRUCTURE
TO SKIN

LOWER FUSELAGE SKELETON ASSEMBLY

**FIG. 6.29** Components assembled by mechanical fastening.[16] *(Sikorsky Aircraft, Division of United Technologies.)*

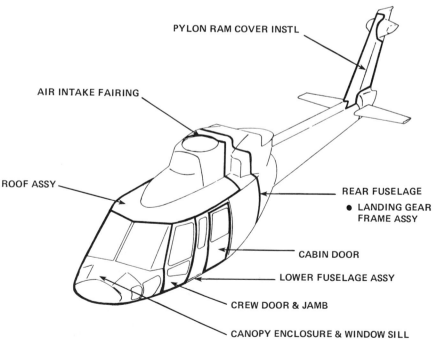

PYLON RAM COVER INSTL

AIR INTAKE FAIRING

ROOF ASSY

REAR FUSELAGE
● LANDING GEAR
FRAME ASSY

CABIN DOOR

LOWER FUSELAGE ASSY

CREW DOOR & JAMB

CANOPY ENCLOSURE & WINDOW SILL

**FIG. 6.30** Components assembled by adhesive bonding.[16] *(Sikorsky Aircraft, Division of United Technologies.)*

**6.31**

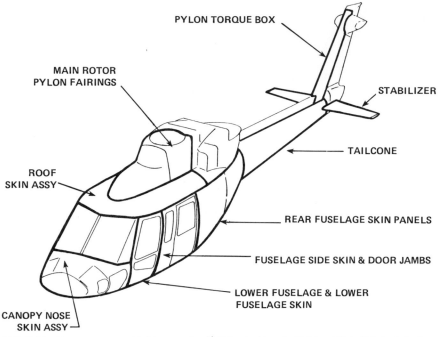

PYLON TORQUE BOX

MAIN ROTOR
PYLON FAIRINGS

STABILIZER

TAILCONE

ROOF
SKIN ASSY

REAR FUSELAGE SKIN PANELS

FUSELAGE SIDE SKIN & DOOR JAMBS

LOWER FUSELAGE & LOWER
FUSELAGE SKIN

CANOPY NOSE
SKIN ASSY

**FIG. 6.31** Components assembled by co-curing.[16] *(Sikorsky Aircraft, Division of United Technologies.)*

To overcome the problems of graphite with metal fasteners Gl–Ep composite fasteners were developed.[7] Besides the manufacturing process development and an exhaustive element and coupon test program three full-scale test articles, representing a section of a helicopter tail boom, were fabricated (Fig. 6.32). The resultant tests indicated that

1. Composite structures joined with glass-composite fasteners are structurally sound for light- and medium-loaded helicopter components.
2. Some detrimental effects from humidity conditioning occur, as evidenced by comparing wet and dry test results of the fastener material.
3. Application of composite-fastener technology to certain helicopter composite components is cost-effective.

## JOINING

Two common processes used to join metals are welding and brazing, but welding is seldom a first choice for joining composite matrixes, whether resin or metal. Work by governmental agencies and industrial concerns has attempted to use some of the familiar welding and brazing processes and develop suitable variations and innovations where necessary.

### Induction Heating

Figure 6.33 shows a self-contained portable welding system used to join composites by induction heating. The device can be used in any atmosphere or in a vacuum, and the

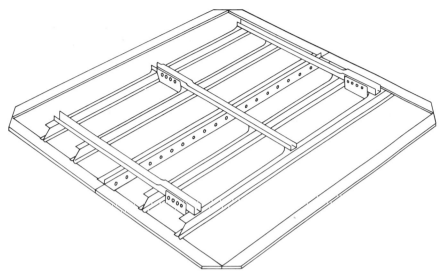

**FIG. 6.32** Demonstration panel of tail boom.[7]

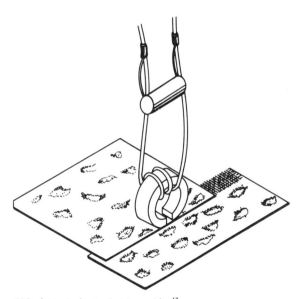

**FIG. 6.33** Inductive-heating welder.[17]

plastic components can be joined in situ. Induction heating offers the following potential advantages over conventional methods of joining thermoplastics and composites, i.e., adhesives, fusing, and mechanical fasteners:

Little or no deformation at the joints

Few component parts

Suitability to almost any type of thermoplastic

A modified wound toroidal inductor core is used to transfer magnetic flux through the thermoplastic to a carbon-steel screen. The air gap cut into the toroid diverts the path of the magnetic flux from the toroid to the screen (Fig. 6.33). The metal screen is typically cut into long strips 0.25 in (6.4 mm) wide and sandwiched between sheets of plastic at a joint to join one sheet of plastic to another or a plastic sheet to a structural beam. The air gap of the toroid is placed on one of the plastic surfaces directly above the screen. When the toroid is energized, the alternating current produces inductive heating in the screen, causing the plastic surfaces on either side of the screen to melt and flow into the screen and form the joint. The temperature of the screen is determined by such factors as the input power, number of coil windings, width of the air gap, and frequency of the alternating current. The toroid is moved along the seam or joint at a controlled speed to produce optimum joining. The low power required (25 to 100 W) permits use of battery or solar power. Various configurations of the plastic welder can be used in the aerospace, automobile, furniture, and construction industries.

## Fusion Bonding

Several fusion-bond approaches have been developed for joining fiber-reinforced thermoplastic structures.[8] One approach placed resistance wires at the bond interface and applied a potential across the joint. The heated wires softened the fiber-reinforced polysulfone, which fused and formed the joint. Typically 0.0025-in-thick (0.06-mm) polysulfone adhesive film was used. Lap shear values in the range of 1.4 kips/in$^2$ (9.7 MPa) were obtained with specimens made from this joint, but the interface in this joint had only 25% of the area fused since the wire did not generate enough uniform heat. It seems likely that if total area fusion were obtained, shear strengths of 4 to 5 kips/in$^2$ (28 to 35 MPa) should be possible.

In another approach stainless-steel screen (80-mesh) was used as a resistance heater. The joint was formed in 90 s using a pressure of 10 kips/in$^2$ (69 MPa), and lap shear values of 3.8 kips/in$^2$ (26 MPa) were obtained. All failures occurred in the adhesive-wire mesh interface. Continued research is warranted since the resistance-wire heating has proved very promising.

In a third promising technique, electromagnetic bonding, a lap-shear value of 3.75 kips/in$^2$ (26 MPa) has been obtained. Electromagnetic bonding uses induction heating between two abutting thermoplastic surfaces to fusion temperature by means of a heat-activated electromagnetic adhesive layer. The electromagnetic material at the bonding interface consists of a dispersion of finely divided metal particles in a thermoplastic matrix. When the interface is subjected to a high-frequency alternating current, fusion temperature is instantly achieved. Under slight pressure a bond is formed. Table 6.7 compares results of several other fusion-bonding techniques and lap-shear tests with those for electromagnetic bonding.

## Ultrasonic Welding

One of the tasks to be performed inside the Space Shuttle is the actual manufacture of lightweight composite thermoplastic (Gr–PS) beams.[9] As these beams take shape (Fig. 6.34), they will be joined at critical points by ultrasonic welding. The welding head will consist of a transducer, which converts a 20-kHz power signal into ultrasonic vibrations,

**Table 6.7**  Typical Values for Bonding Methods[8]

| Bonding method | Lap shear strength | |
| --- | --- | --- |
| | kips/in² | MPa |
| Fusion bond | 8.0 | 55.2 |
| Resistance-heated bond | 3.8 | 26 |
| Electromagnetic fusion | 3.75 | 25.9 |
| Epoxy adhesive | 4.2 | 29 |
| Ultrasonic bonding | 1.4 | 10 |

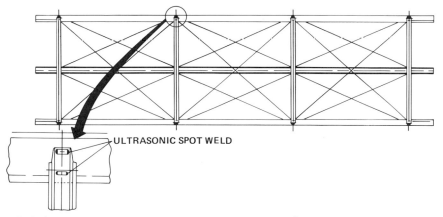

ULTRASONIC SPOT WELD

FIG. 6.34  Schematic of three-bay prototype triangular truss segment.[9]

and a metal horn, whose half-wave-resonant frequency is tuned to match the transducer. The horn will be equipped with weld tips, to impart vibration in the spots to be joined. The vibration will quickly heat the thermoplastic resin in the element-mating surfaces to the plastic state. The transducer power will be turned off and the parts clamped together until the thermoplastic resolidifies, thus creating a fused bond in the weld zones. A typical weld will require about 1 s of excitation and 0.5 s for cooling.

The data base for joining Gr–PS composites by ultrasonic welding in vacuum produced several notable results:

1. There are no identifiable effects on strength or resin-flow characteristics.
2. Conditioning material properly to remove moisture and volatiles prevents outgassing.
3. A properly evacuated piezoelectric ultrasonic transducer operates in vacuum.
4. In vacuum ultrasonic welding produces no loss of efficiency of the welder and no significant heating effects on the weld horn or transducer.
5. Weld heat is conducted away from the weld primarily in the direction of the graphite fibers; the rate of cooling is not significantly different between welds produced in air and in vacuum.
6. Welder performance is affected by gravity but not by zero gravity.
7. The weld power characteristic during welding is sensitive to material thickness and surface conditions, and the amount of weld energy applied correlates well with weld strength.

## Gas Tungsten Arc Welding

Fusion welding, especially gas tungsten arc welding, is an important but problematical joining technique for MMC structures. Its influence on matrix, fiber, and matrix-fiber compatibility becomes a significant factor in practical applications.

### Ti–W and Ti–Gr

Two titanium composite materials were evaluated to obtain weldability information on simple model composite systems and to study fiber-matrix interfacial reactions by exposing the composites to the highly dynamic thermal conditions of the molten weld. Better understanding of the factors controlling the interfacial region and its properties are essential in controlling and predicting engineering properties of composites. Weld specimens for both materials tested were square-groove butt and simple bead on sheet; welds were made both with and without weld-filler metal additions. Bead-on-sheet welds were both transverse and longitudinal to the fiber direction. Welding was both manual and mechanized. Radiographic techniques to inspect weld soundness and fiber orientation after fusion disclosed several problems requiring continued work. Weld porosity was not excessive despite a strong tendency for void formation in close proximity to the tungsten-wire matrix, attributable in part to evolution of gas from the diffusion-welded interfaces and its partial entrapment near the reinforcing wires. In the Ti–W system a dissolution effect was observed in which a concentration gradient indicated interdiffusion and solid-solution formation in the interfacial region.

The reaction zone in the Ti–Gr fiber system showed that a carbide formed around each graphite filament as a result of fusion. Tensile tests on both systems indicate that there is an enhancement in strength even at low filament volumes. These improvements are apparently realized in part through the interfacial regions and their contributions to composite strength.

### B–Al

Welding studies were conducted on B–Al composite to observe the effects on the boron reinforcing filaments and aluminum matrix. Exposure of boron filaments to molten aluminum poses problems of chemical reactivity and its effect on filament properties. B–Al interactions depend on time and temperature and may be sluggish in the solid state or rapid in the presence of a superheated liquid-aluminum matrix. Thermal treatments such as diffusion welding, casting, and arc welding may induce interfacial reactions detrimental to filament strength and composite structural efficiency.

Test results[10] show that thin-sheet B–Al composites can be subjected to the thermal conditions of gas tungsten arc welding without severely damaging the boron filaments. Filler metal can be added and intermixed through the matrix to alter its chemical composition significantly. These results indicate that gas tungsten arc welding of B–Al may be possible if the input of welding energy is controlled. To consider the weldability of B–Al and its potential as a future joining technique requires

Identification of the subsequent fusion-reaction products

Knowledge of the effects of those products on the mechanical properties of the composite

Knowledge of reaction growth-rate kinetics

Means of controlling reaction products during welding

Manual plasma-arc and electron-beam welding are relatively unattractive processes for joining B–Al composites because excessive metallurgical reactions between the aluminum and boron yielded low joint strengths.[11]

## Al–Gr

Since experimental gas tungsten arc welding of Al–Gr composites was conducted with hand-held torches, heat input to the weld was difficult to control. The resulting temperatures caused formation of aluminum carbide on the surfaces of the fibers. Using an automated process would provide adequate temperature control and prevent formation of aluminum carbide. The hand-held torch deposited 4043 aluminum filler alloy on a composite with a matrix of Al–7% Zn. Welds between this composite and 6061 aluminum-alloy sheet were made.

## Resistance Welding

Variations of resistance welding have been developed, especially for the B–Al composite.

## B–Al

**Resistance Spot Welding:**  This is the most fully developed of the resistance joining processes, having been used to fabricate large structural test components employing build-ups from 0.010 and 0.010 in (0.25 and 0.25 mm) up to 0.080, 0.040, and 0.060 in (2, 1, and 1.5 mm). Weld schedules have been developed for B–Al-to-B–Al joints with heat settings somewhat lower than those used for aluminum but approximately twice the electrode pressure. The high electrode pressure is required to prevent expulsion. Schedules have also been developed for B–Al-to-aluminum joints.

**Resistance Seam Welding:**  Seam-welding schedules do not use the very high pressures of spot welding because filament breakage would result. Limited development work has produced metallurgically sound joints that fail in lap shear at the edge of the weld nugget.

**Resistance Brazing:**  Since most brazing processes would expose the B–Al composite above 800°F (427°C) for more than approximately 2 min, which would degrade the strength of the composite material (Table 6.8), a resistance-welding machine was evaluated as a quick heating medium for brazing. Using flat electrodes and standard aluminum brazing foil, brazements were made in a few seconds which developed strengths sufficient to cause failure in the composite with less than 1 in (25.4 mm) overlap.

**Table 6.8**  Strength Reduction with Increased Time at Elevated Temperatures as Reflected in Tensile Tests[11]

| Time to peak, s | Time between 800 and 1050°F (427 and 566°C), s | Tensile strength | |
|---|---|---|---|
| | | kips/in$^2$ | Mpa |
| 2.8 | 8 | 184 | 1269 |
| 9.0 | 8 | 191 | 1317 |
| 30.5 | 147 | 178 | 1227 |

**Resistance Spot Joining:** This process uses standard resistance-welding equipment and commercially available electrodes to join 0.080-in (2-mm) sheet B–Al to 0.040-in (1-mm) sheet titanium. The heat is concentrated in the titanium, not the joint interface, but is sufficient to raise the interface temperature above the melting point of the aluminum. Upon resolidification the molten aluminum attaches itself to titanium without any significant Al–Ti interaction. Joint shear strengths of approximately 10 kips/in$^2$ (69 MPa) have been obtained.

**Resistance Spot Brazing:** This technique was developed to improve joint quality of resistance spot joining. B–Al was copper-plated 0.001 to 0.002 in (0.03 to 0.05 mm) thick, and the titanium was copper-plated to the same thickness over a thin electroless nickel cladding. The nickel is required to get good copper adhesion to the titanium. The joint was heated in a resistance welding machine, under pressure, until a copper–copper joint was formed—less than 10 s. The brazed joint had shear strength equivalent to the spot-joined joint, but the joint quality was excellent for all joints produced (Table 6.9).

**Table 6.9** Typical B–Al Lap-Joint Properties for Resistance Joining[18]

| Process | Matl type[a] | Thickness mils | Thickness mm | Spot diameter in | Spot diameter mm | Load to failure lb | Load to failure kg | Joint efficiency,[b] % | Failure mode |
|---|---|---|---|---|---|---|---|---|---|
| Spot | UD | 40 | 1 | 0.37 | 9.4 | 2650 | 1202 | 100 | Net tension |
| welding | CP | 35 | 0.89 | 0.32 | 8.1 | 800 | 363 | 93 | at edge of |
|  | | 35[c] | 0.89[c] | 0.32[c] | 8.1 | 980 | 445 | 100 | weld |
| Seam | UD | 35 | 0.89 | 0.37[d] | 9.4[d] | e | e | f | Net tension |
| welding | | | | | | | | | |
| Brazing | CP | 40 | 1 | 0.72 | 18.3 | 1750 | 794 | 75 | Joint shear |
|  | | 40[c] | 1[c] | 0.70 | 17.8 | 2070 | 939 | 82 | |
| Spot joining | UD | 25 | 0.6 | 0.30 | 7.6 | 435 | 197 | 55 | Pulled nugget |
|  | CP | 60 | 1.5 | 0.37 | 9.4 | 1070 | 485 | 64 | Joint shear |
| Spot brazing | UD | 40 | 1 | 0.37 | 9.4 | 1090 | 494 | 46 | B–Al shear |
|  | Ti | 40 | 1 | 0.37 | 9.4 | 1400 | 635 | . . . | Joint shear |

[a]UD = unidirectionally reinforced 50 vol % B–Al; CP = 0°/90° (0/1.6-rad) cross-ply reinforced 45 vol % B–Al; Ti = titanium.
[b]Based upon the diameter of the joint and the following composite properties: UD = 160 kips/in$^2$ (1103 MPa), CP = 80 kips/in$^2$ (552 MPa), and CP heat-treated = 90 kips/in$^2$ (621 MPa).
[c]Postweld heat-treated at 980°F (527°C) for 30 min followed by water quench at 350°F (177°C) for 8 h.
[d]Width. [e]2800 lb/in (501 kg/cm). [f]69.2 kips/in$^2$ (477 MPa).

**Weldbonding:** Since weldbonding of aluminum has been applied to several structures in production aircraft, its application to the joining of B–Al composites seems logical. The rough surface finish of the composite increases the surface area of bonding, thereby increasing the joint strength. Several test specimens showed joint strengths so high that failure took place in the parent material. The strengths shown in Fig. 6.35 therefore represent conservative values for this type of joint. The strength was at least 4 times higher than obtained in a spot-welded B–Al joint. The orientation was 0° for these test specimens. For other tension-shear test specimens with a ±45° (0.79 rad) orientation, the weldbonded joints were 1.7 times stronger than the spot-welded joints.[11]

## Gr–PS

**Resistance Welding:** The graphite conducts sufficient current to heat and melt the thermoplastic in the vicinity of the joint and welds the composite material. With a spring-loaded electrode pressure is applied as the thermoplastic softens, and the electrode follows

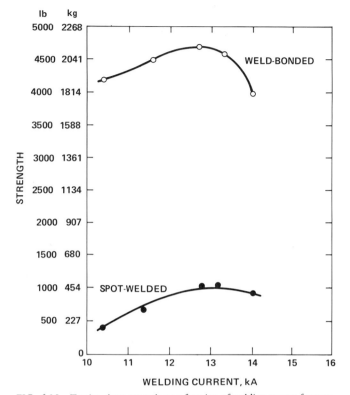

FIG. 6.35  Tension-shear strength as a function of welding current for spot-welded and weldbonded B–Al composite sheet.[11]

the softening material to maintain contact. For a Gr–PS at 36 wt % polysulfone the measured joint tensile strengths were above 1.2 kips/in$^2$ (8.3 MPa). Resistance welding can be done in several configurations. Besides the transverse arrangement of Fig. 6.36 the joint can be heated longitudinally by connecting both power-source electrodes to the center rod and passing current until the thermoplastic softens. Since the joint interfaces offer higher resistance than the bulk, they reach softening temperature first (Fig. 6.36). Microwave heating has also been considered. So far, little effort has been devoted to varying the joint surface

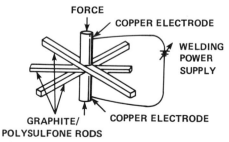

FIG. 6.36  Application of heat transverse to joint to soften the thermoplastic.[17]

condition or its size and shape, but it is expected that optimization of these parameters will improve the joint strength further.

## Gr–Al

Gr–Al composites have been spot-welded to each other and to sheets of 2219 aluminum alloy. Using a 0.003-in-thick (0.08-mm) foil of no. 718 braze filler material (88% Al–12% Si) between the interface of the joints improved weldability. Using this foil filler material eliminated oxide layers in the interfaces and resulted in wetting of the composite surface in the all-composite and the composite–2219 aluminum joints.

## Brazing

Composite brazing has been confined to the metal-matrix types and especially vacuum. Vacuum-brazing B–Al composite material using Al–Si braze filler metals is more critical than soldering because of the high temperatures involved, typically 1070 to 1140°F (577 to 616°C), where it is necessary to protect the boron filaments by coating them with a SiC layer to prevent B–Al filament interaction and subsequent strength degradation.

Vacuum-brazing Bsc–Al composite materials by eliminating diffusion of braze-filler-metal constituents into the aluminum matrix[12] has led to the development of a hybrid composite which combines high-strength Bsc–Al and ductile titanium to form a titanium-clad Bsc–Al. The titanium foil provides the Bsc–Al with a durable outer surface and serves as a diffusion barrier to alleviate fiber and matrix degradation during brazing. Titanium-clad Bsc–Al skin panels have been joined to titanium-clad Bsc–Al stringers by brazing with 4047 or 718 aluminum brazing foil 0.010 in (0.25 mm) thick. Brazing was accomplished at 1100°F (593°C) for 5 min at temperature. Tests showed that titanium-clad Bsc–Al unidirectional sheets have longitudinal and transverse tensile strength respectively 0.9 and 2.5 times that of unclad Bsc–Al unidirectional sheet. Specific buckling strengths are higher for the titanium-clad Bsc–Al skin-stringer panels than for similar unclad Bsc–Al panels. The fiber orientation of the cap and skin material is shown in Fig. 6.37. Subsequently panels considered primary structure have been subjected to flight-service evaluation; results show no deleterious effect on panel properties.

Gr–Al composites with matrixes of both pure aluminum and Al–7% Zn have been successfully brazed to 6061 aluminum alloy. Brazing was performed in a flowing-argon atmosphere without a flux. The braze filler metal was no. 718, and brazing was accomplished in 5 to 10 min at 1094°F (590°C). The results showed that the presence of both magnesium and silicon in the braze filler metal is necessary for satisfactory composite-to-composite brazing and that brazing is achievable when a binary Al–Mg braze filler is used.

## Diffusion Welding

Diffusion welding, discussed earlier as a means of producing composites, is also an excellent joining method. B–Al tape having a plasma-sprayed layer of brazing material (713Al) has been diffusion-welded at pressures below 0.2 kips/in² (1.4 MPa), compared with 10 kips/in² (69 MPa) for welding 6061 aluminum-matrix composites. Aircraft turbine-engine fan

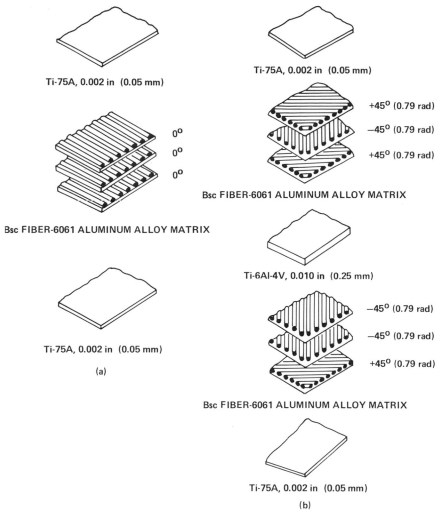

Ti-75A, 0.002 in (0.05 mm)

Ti-75A, 0.002 in (0.05 mm)

0° 0° 0°

+45° (0.79 rad)
−45° (0.79 rad)
+45° (0.79 rad)

Bsc FIBER-6061 ALUMINUM ALLOY MATRIX

Bsc FIBER-6061 ALUMINUM ALLOY MATRIX

Ti-6Al-4V, 0.010 in (0.25 mm)

Ti-75A, 0.002 in (0.05 mm)

(a)

−45° (0.79 rad)
−45° (0.79 rad)
+45° (0.79 rad)

Bsc FIBER-6061 ALUMINUM ALLOY MATRIX

Ti-75A, 0.002 in (0.05 mm)

(b)

**FIG. 6.37**   Fiber orientation in titanium-clad Bsc–Al (*a*) cap and (*b*) skin material.[12]

blades of Ti–6Al–4V alloy have been locally stiffened by diffusion welding with an inlay of Ti–6Al–4V–50B composite in recesses in the blade surface. Various dynamic tests, including engine tests, of blades have resulted in no failures at the composite inlay. The advantage of the composite inlay in this case is a significant improvement in blade vibration stability because of the high elastic modulus of the composite.

Table 6.10 illustrates several blade configurations that have been joined using two variations of the diffusion-welding process using diffusion-aid materials. Other variations of the process with diffusion aids were described in Figs. 4.94 and 4.95. Future applications of diffusion welding include composite structures (filled, laminar, cellular, and metal and/or ceramic) and hybrid structures.

**Table 6.10** Application of Joining Process to Composite Blade Configurations

| Composite configuration | | Configuration cross section | Process applicability | |
|---|---|---|---|---|
| | | | Activated diffusion brazing | Gas-pressure welding |
| Finned shell to strut | Fins at midchord region | CHORDWISE FINS / MIDCHORD REGION / LEADING EDGE REGION | Very good | Either is good, but shell needs support in nonfinned region; with both combined, results are very good |
| | Fins at leading edge | | Not applicable unless gas pressure used | |
| Nonfinned shell to strut webs | Solid shell | | Very good | Either is good, but shell needs internal support |
| | Porous shell | | Not applicable because braze clogs pores | |
| Shell segments with spanwise joints | | MATERIAL B / MATERIAL A | Good | Not applicable |

## Painting and Coating

Surface-coating and finishing composite materials for various applications have been developed, and many are in production.

### Coating

Electric-arc metal spraying is one method for applying metallic coatings to aircraft exteriors made of composite materials. The coating can protect aircraft against laser weapons and reduces the cost of exterior maintenance. Other development work involves an ablative (heat-dissipating) coating that can protect a polished-aluminum surface and flake off or evaporate on contact with a laser beam. The coating might be applied by electric-arc spray or atomization of a solid wire in a flame of burning gas. These processes would require close control to avoid damage to the substrate surface. Other possibilities are electrolytic plating and vapor deposition. Metal coatings on compression-molded SMC are a newly developed automotive application for glass-reinforced resin composites. Two metallizing techniques, sputter coating and electroplating, show promise for SMC components. Structural applications for the composite material include cross-members, radius arms, bumpers and wheels.

The sputter-coating technique takes advantage of high-performance magnetron sputtering, which permits increased deposition rates, thereby reducing substrate heating and deposition over large substrate areas. An advantage of sputtering is the higher kinetic energy of the metal ions, estimated at 10 to 100 times that of conventional vacuum metallizing, which means higher impact energy and thus a denser and brighter coating. To minimize the tendency of metallic sputter coatings to show up cracks, iridescence, and surface defects of the SMC, two organic coatings are applied, a base and a top coat.

Since each resin demands its own electroplating process, suppliers no longer can rely on selling products but must sell processes. The plater, on the other hand, must have a basic understanding of the resin system and the total composition of the SMC. The best etch process for SMC is the Ronason gas process, a proprietary, nonchromic etch system. A high-modulus carbon-fiber paper can be used by itself as an isotropic structural material, or it can be hybridized with carbon, glass, or other fibers in unidirectional and woven composites. Applications include use for electromagnetic-interference shielding and surface veils to improve surface smoothness. In unidirectional woven and nonwoven fabric composites the carbon-fiber paper can also greatly improve surface cosmetics. For example, as a surface veil it produces a smooth outer finish for better aerodynamics. In addition, electrical properties allow electrostatic painting without further surface preparation. The carbon-fiber paper can also be used for static-charge dissipation.

### Finishing and Painting

Finishing composite parts has been a labor-intensive operation, but improvements in manufacturing techniques and use of material have reduced costs. Such improvements include better adhesion of sealers, paints, and finishes, which are highly dependent on surface conditions and preparation of composite materials, especially graphite and Kevlar. The standard method in the past has been to sand or abrade, solvent-wipe, apply fillers and pinhole sealing compounds, sand or abrade, solvent-wipe clean, apply primer-sealer, then paint. This is a laborious and time-consuming effort, especially when extensive filling or surface repairs are required to remove manufacturing-induced surface damage. New methods and new materials have improved the finishing process. Improved separation films and release agents avoid transferring contaminants that previously caused adhesion problems. Peel-ply and bleeder materials with better absorption characteristics facilitate removal of excess surface resin. Incorporation of a fine-textured lightweight veil or outer layer of fiber glass on class A Kevlar parts significantly minimizes the sanding required; note Fig. 6.38.

**FIG. 6.38** Helicopter production part, showing as-molded finish of forward pylon fairing and engine inlet fairing in rear. (*Sikorsky Aircraft, Division of United Technologies.*)

# REFERENCES

1. Marx, W., and S. Trink: "Manufacturing Methods for Cutting, Machining and Drilling Composites," vol. I, "Tests and Results," vol. II, "Final Report August 1976–August 1978," Grumman Aerospace Corp., Bethpage, N.Y., AFML-TR-78-103, Contr. F33615-76-C-5280.

2. More, E. R.: Manufacturing Methods for Composite Fan Blades, *Hamilton Standard Div. United Technologies Corp.*, AFML-TR-76-138, *Final Rep. April 1974–February 1976*, Windsor Locks, Conn., Contr. F33615-74-C-5135.

3. Staebler, C. J., Jr., and B. F. Simpers: Metallic Coatings for Graphite/Epoxy Composites—Phase II, *Grumman Aerospace Corp., Final Rep.*, Contr. N00019-78-C-0602, Bethpage, N.Y., August, 1980.

4. Kellner, J. D., W. J. Croft, and L. A. Shepard: Titanium Diboride Electrodeposited Coatings, *AMMRC Tech. Rep.* 77-17, June, 1977.

5. Study of the Influence of Hole Quality on Composite Materials, Lockheed-California Co., Burbank, Calif., NASA Contr. NAS1-15599, February 1980.

6. Brink, N. O.: Low Cost Composite Fastener Hole Formation, *Rohr Industries Final Rep. August 1977–November 1979,* Chula Vista, Calif., AFWAL-TR-80-4007, Contr. F33615-77-C-5226.

7. Miller, R. L.: Manufacturing Technology for Low-Cost Composite Fasteners, *Vought Corp.* AFWAL-TR-80-4130, *Final Rep. Phase III, March 1979–June 1980,* Dallas Tex., Contr. F33615-77-C-5050, P00003.

8. Hoggatt, J. T., J. Oken, and E. E. House: Advanced Fiber Reinforced Thermoplastic Structures, *Boeing Aerospace Co.,* AFWAL-TR-80-3023, *Final Rep. August 1976–August 1979,* Seattle Wash., Contr. F33615-76-C-3048.

9. Graphite Composite Truss Welding and Cap Section Forming Subsystems, *General Dynamics Convair Div. Final Rep.,* vol. II, "Program Results," San Diego, Calif., Contr. NAS9-15973, Johnson Space Center, Houston, Tex., Oct. 31, 1980.

10. Kennedy, J. R.: Microstructural Observations of Arc Welded Boron-Aluminum Composites, *Weld. J.,* **52**(3):120s–124s (March 1973).

11. Wu, K. C.: Welding of Aluminum-Boron Composites, *5th Natl. SAMPE Tech. Conf., Kiamesha Lake, N.Y., Oct. 9–11, 1973,* pp. 692–702.

12. Royster, D. M., R. R. McWithey, and T. T. Bales: Fabrication and Evaluation of Brazed Titanium-Clad Borsic-Aluminum Compression and Skin-Stringer Panels, *NASA Langley Res. Cent.* NASATP-1573, March 1980 and NASATP-1674, July 1980.

13. Advanced Composite Design Guide, 3d ed., Air Force Materials Laboratory, Wright-Patterson AFB, Ohio, 1973.

14. Mackey, B. A., Jr.: How to Drill Precision Holes in Reinforced Plastics in a Hurry, *Plast. Eng.,* February 1980, pp. 3–5.

15. Schwartz, M., and S. Kosturak: 196 Holes per Shot in Boron-Epoxy, *Am. Mach.,* October 1982, pp. 130–131.

16. Kay, B.: Composite Rear Fuselage, Sikorsky Aircraft, Internal Report, December 1981.

17. NASA Tech. Brief, Summer 1980, LAR-12540, p. 242; MSC-18534, p. 234.

18. Hersh, M. S.: Correlation between Boron/Aluminum Sheet Quality and Resistance Weld Quality and Strength, *Weld, J.,* **50**(12):515s–521s, (December 1971).

# BIBLIOGRAPHY

Browning, D. L.: Space Construction Automated Fabrication Experiment Definition Study (SCAFEDS), pt. III, *Final Rep.,* CASD-ASP78-016, June 29, 1979.

Castle, C. H., P. Melnyk, and W. G. West: Process Development for Boron-Aluminum Fan Blades, *TRW Tech. Mem.* TM-4663, Cleveland, 1972.

Dwyer, J. J., Jr.: Composites, *Amer. Mach. Spec. Rep.* 643, July 13, 1970, pp. 87–96.

Goddard, D. M., R. T. Pepper, J. W. Upp, and E. G. Kendall: Feasibility of Brazing and Welding Aluminum-Graphite Composites (Aerospace Corporation, El Segundo, Calif., Contr. F04701-71-C-0172) *Weld. J.,* 51(4):178s–182s (April 1972).

Grauer, W.: Automated Assembly Fixture Drilling, *Grumman Aerospace Corp., AFML Interim Rep.* 806-5 (I), August 1975, Bethpage, N.Y., Contr. F33615-75-C-5192.

Hauser, D.: Investigation of the Effects of Brazing on the Properties of Fiber-Reinforced, Aluminum-Matrix Composites, Battelle Columbus Lab. *NASA Contr. Rep.* NAS1-14182, CR-145096, Columbus, Ohio, X77-10074, April 1977.

Jacquish, J., C. H. Sheppard, et al.: Graphite Reinforced Thermoplastic Composites, Boeing Aerospace Co., Seattle, Wash., Naval Air Systems Command Contr. N00019-79-C-0203, August 1980.

Kaufman, A., T. F. Berry, and K. E. Meiners: Joining Techniques for Fabrication of Composite Air-Cooled Turbine Blades and Vanes, *Gas Turb. Conf., Houston, Mar. 28–Apr. 1, 1971,* ASME 71-GT-32, pp. 1–9.

Laufenberg, T. L.: An Evaluation of Water Jet Cutting of Composite Materials, *McDonnell Douglas Corp. Intern. IRAD Rep.* MDC-J1790, Long Beach, Calif. December 1978.

Meiners, K. E.: Diffusion Bonding of Specialty Structures, *5th Natl. SAMPE Tech. Conf., Kiamesha Lake, N.Y. Oct. 9–11, 1973,* pp. 703–712.

Metzger, G. E.: Joining of Metal-Matrix Fiber-Reinforced Composite, *WRC Interpretive Bull.* 207, July 1975.

Payer, J. H., and P. G. Sullivan: Corrosion Protection Methods for Graphite Fiber Reinforced Aluminum Alloys, *8th Natl. SAMPE Tech. Conf., Seattle, Oct. 12–14, 1976,* pp. 343–352.

Robertson, A. R., M. F. Miller, and C. R. Mailkish: Soldering and Brazing of Advanced Metal-Matrix Structures, *Weld. J.,* 52(10):446s–453s (October 1973).

Van Cleave, R. A.: Laser Cutting of Kevlar Laminates, Bendix Corporation, Kansas City, BDX-613-1877, September 1977.

Webb, B. A., and J. F. Dolowy, Jr.: Braze Bonding of Borsic/Aluminum Composite Sheet to Titanium, *DWA Composite Specialties Final Rep.* NASA CR-132730, Contr. NAS1-13095, Chatsworth, Calif., June 1975.

Kevlar fibers makes tires stronger.  *(Sikorsky Aircraft, Division of United Technologies.)*

# Applications Development

## COMMERCIAL AIRCRAFT

Application of advanced composites to civil aircraft has generally lagged behind military use because in civil applications cost is a more important consideration, safety is a more critical concern (both to the manufacturer and to government certifying organizations), and a generally conservative outlook obtains because of past experiences with financial penalties from equipment downtime. The major exception is the Lear Fan 2100, the first all-advanced-composite aircraft (Table 7.1). Use of advanced composites in commercial aircraft was preceded by extensive experience with fiber glass in fairings, control surfaces, and other secondary-structure components. Table 7.2 shows the amount of exterior surface area using this material on commercial transports.

Ten years ago, Schjelderup and Purdy[1] proposed several ways to establish the confidence required for significant use of composites in commercial aircraft. They included use in military aircraft primary and secondary structure and in commercial aircraft secondary structure ("secondary" means not flight-critical). Both these approaches have been followed. The many significant military applications will be discussed later. The use of composites in commercial aircraft secondary structure has taken place under two major NASA programs, Flight Service Evaluation and Aircraft Energy Efficiency (ACEE). Although the military programs have helped advance the technology and develop confidence in the materials, it is NASA's contracts with major commercial aircraft builders that has directly stimulated use of composites in transports.

**Table 7.1**  Advanced Composites in Commercial Aircraft and Engines

| Manufacturer and model | Material | Component | Status† |
|---|---|---|---|
| Aerospatiale Concorde | Gr–Ep | Landing-gear well door | 1 |
| Airbus A300 | Gr–Ep | Rudder | 3 |
| | | Outboard spoiler, vertical-fin leading edge, cabin vertical support rods, main landing-gear fairings | 2 |
| | Kv–Ep | Flap-track fairings, fin fairings, trailing-edge fairings, horizontal stabilizer | 1 |
| Boeing, 707-320 | B–Ep | Foreflap | 2 |
| 727 | Gl–Ep | Potable-water tank | 3 |
| | Kv–Ep | Engine cowl | 2 |
| 727-100 | Gl–Gr–Ep | Galley | 2 |
| | Gr–Ep | Seats | 2 |
| 727-200 | Gr–Ep | Elevators | 3 |
| 737 | Gr–Ep | Horizontal stabilizer, spoiler | 3 |
| 747 | Gr–Ep | Outboard aileron | 2 |
| | | Engine inlet outer cowl | 3 |
| | Kv–Ep | Pressure air-storage bottles for escape slides | 3 |
| 747 SP | Gr–Ep | Floor panels | 3 |
| | Kv–Ep | Air-storage bottles | 3 |
| 757-767 | Gr–Ep | Spoiler, nacelle inlet cowl | 3 |
| | Kv–Gr–Ep | Fixed trailing edge | 3 |
| 767 | Gr–Ep | Outboard and inboard aileron, rudder, elevators | 3 |
| | Kv–Ep | Pressure air-storage bottles for escape slides, strut fairing, support fairing, stabilizer tip, wing trailing edge flap | 3 |
| | Kv–Gr–Ep | Landing-gear door, wing-to-body fairing, stabilizer fixed trailing edge | 3 |
| Canadair CL-600 | Gr–Ep | Flooring | 3 |
| | Gr–Ep or Kv–Gr–Ep | Stabilizer, aileron, rudder, elevator, flaps, landing-gear door | 1 |
| | Kv–Ep | Access doors, instrument panels and consoles, fairings, wheel bins, radar dome, air-conditioning duct, leading edges | 3 |
| Cessna: | | | |
| Citation I | Kv–Ep | Radome | 3 |
| Citation II | Gl–Ep | Flap | 3 |
| Citation III | Gr–Ep | Spoiler | 3 |
| | Kv–Ep | Engine nacelle, fairings | 3 |
| | Kv–Gr–Ep | Flap | 3 |
| Citation "Conquest" | Kv-Ep | Seats | 3 |
| DeHavilland: | | | |
| DHC-7 | Kv–Ep | Flooring, window reveals, ceiling panels, lavatories, overhead storage bins, avionics compartment, flaps, wheel bins, fairing fin, tail cone, underwing fairings | 3 |
| DHC-8 | Gr–Ep or Kv–Gr–Ep | Landing-gear door, flaps, aileron, elevator trim tabs | 1 |

**Table 7.1**  Advanced Composites in Commercial Aircraft and Engines (*continued*)

| Manufacturer and model | Material | Component | Status† |
|---|---|---|---|
| Dornier | Gr–Ep | Air brake | 1 |
| Douglas, DC9-80 | Kv–Ep + Gr–Ep | Engine nacelle and cowl | 3 |
| DC-10 | B–Al | Aft pylon | 3 |
|  | Gr–Ep | Upper aft rudder‡ | 2 |
|  |  | Vertical stabilizer | 3 |
| Lear: |  |  |  |
| Lear Fan 2100 | Gr–Ep | Center fuselage, horizontal stabilizer, flaps, vertical stabilizer, wing, tail fin, frame, control surfaces, seats, headliner, interior surfaces, bulkheads | 2 |
|  | Kv–Ep | Propeller blades, nose | 2 |
| Lockheed: |  |  |  |
| L-1011 | Gr–Ep | Vertical fin,§ rudder | 2 |
|  |  | Aileron | 3 |
|  | Kv–Ep | Wing-body and center engine fairings, ceiling panels, window reveals | 3 |
| Rolls-Royce |  |  |  |
| RB211-535 engine | Gr–Ep | Thrust reverser, hinge cowl | 2 |
| United Technologies JT9D-7R4 engine | Kv–Gr–Ep | Nose-cowl outer barrel, fan-wrap cowl, sleeve | 3 |

†1 = experimental, 2 = prototype development, 3 = production.
‡Contract NAS1-14869.  §Contract NAS1-14000.

One of the programs for the L-1011 airplane involved the development and installation of Kv–Ep fairings (Table 7.1). After 7 years of annual inspections the three aircraft revealed damage comparable to that of similar fiber-glass components, and performance was judged satisfactory.[2] With the success of this program Kv–Ep has been selected for numerous components (Fig. 7.1):

Wing-body fairings
Wing fixed leading edge
Wing fixed trailing edge
Rudder
Elevator trailing edge
Diverter fairing

**Table 7.2**  Use of Fiber Glass in Boeing Commercial Transports[37]

| Aircraft | Surface area per plane | |
|---|---|---|
|  | ft² | m² |
| 707 | 200 | 19 |
| 727 | 1,800 | 167 |
| 737 | 3,000 | 279 |
| 747 | 10,000 | 929 |

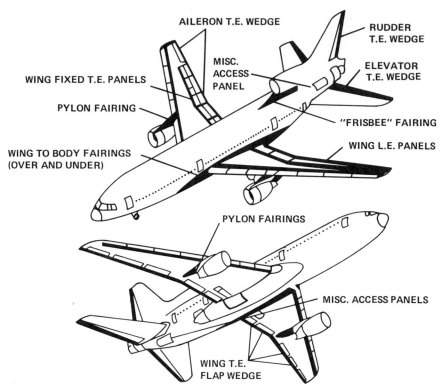

**FIG. 7.1** Use of Kv–Ep composite in the L-1011.[2,3]

In an evaluation of Gr–Ep spoilers on the 737 airplane tests showed little change in strength and no evidence of corrosion. The success of this program led to regular production of these components on 737 aircraft in 1980. The positive results of these NASA Flight Service Evaluation Programs[3], were major factors in the decision to make extensive use of advanced composites in the new 767 and 757 commercial aircraft.[4]

Another NASA ACEE program, started in 1975, greatly expanded the scope of composite applications in commercial aircraft, including three secondary and three primary structures:[5]

1. Secondary structures
   a. 727 elevator
   b. DC-10 rudder
   c. L-1011 aileron
2. Primary structures
   a. 737 horizontal stabilizer
   b. DC-10 vertical fin
   c. L-1011 vertical fin

## Aircraft Components

The 727 elevators use Nomex honeycomb sandwich panels with Gr–Ep face sheets of fabric and tape.[6] The tape is used as the outer layer of the exterior face sheet to provide a smooth, nonporous surface. The outer layer of the inner face sheet is fabric, which is more resistant to fiber breakout during drilling. The use of composites resulted in a direct weight reduction of 26% in the redesigned parts. Five ship sets have been placed in service under the NASA program (Fig. 7.2).

The DC-10 upper aft rudder is also being flight-tested.[5] The Gr–Ep structural box is a two-spar multirib construction with solid skins. The skins are made from unidirectional broad goods and the substructure from fabric. The composite rudder is 30% lighter than its aluminum counterpart. The cost effectiveness of the composite design results from several factors: (1) broad goods and fabrics are used to reduce lay-up time; (2) the structural box is made as a single co-cured unit using high-expansion silicone-rubber inserts in steel molds; (3) curing is done in an oven, which is less expensive to use than an autoclave. Over the past 3 years periodic inspections have shown that the composite rudders easily withstand the daily rigors of commercial service. A lightning strike at the aft edge of one rudder caused minor surface damage, repairable with a simple patch. The DC-10 vertical stabilizer consists of honeycomb sandwich skins with a 4-spar 13-rib substructure.[5,7] The skins have Gr–Ep face sheets over Nomex core, and spar and rib caps were built into the skins, locally replacing honeycomb. The projected weight saving is 27%.

The L-1011 aileron is a two-spar, rib-stiffened structure with sandwich skins having Gr–Ep tape face sheets and syntactic-foam cores.[5] The syntactic foam provides better impact resistance than honeycomb (Figs. 3.69 to 3.71). The L-1011 vertical fin box used

**FIG. 7.2** Composite elevator for a 727, the first installed on a commercial aircraft.[6] *(Courtesy of NASA-Langley Research Center.)*

■ GRAPHITE
▨ KEVLAR
▩ GRAPHITE/KEVLAR

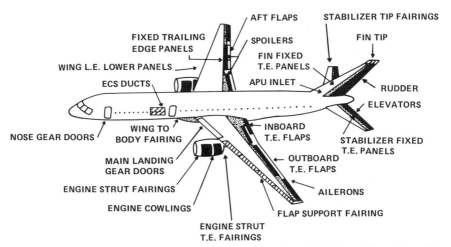

**FIG. 7.3** Planned composite configurations on a 757. *(Iron Age, Mar. 23, 1981, p. 133, and Aug. 12, 1981, p. 61.)*

hat-stiffened solid skins over a substructure consisting of 2 main spars and 17 ribs.[5,7] Unidirectional tape was selected for the skins because of its better mechanical properties and suitability for use with automatic lay-up processes. The three upper ribs were solid Gr–Ep laminates with integrally molded caps and bead stiffeners. The eight lower ribs combined Gr–Ep caps with extruded-aluminum truss webs. Projected weight saving is 28%.

The 737 Gr–Ep horizontal stabilizer box has two spars and eight ribs with solid, I-stiffened skins made from fabric and tape.[5,6] The skin and its integral stiffeners were co-cured. The total weight saving was 29% over the metal structure.

The experience gained from the NASA Flight Service Evaluation and Aircraft Energy Efficiency programs[3,5] has resulted in extensive use of advanced composites in the next generation of commercial transports,[4] DC-9 Super 80, 767 and 757 (Fig. 7.3).

## Miscellaneous Components

Other current aircraft and aerospace applications include escape slides and the pressure bottles to activate them, cabin reinforcement, and drogue chutes for aircraft ejection seats. In addition Gr–Ep and B–Ep strut cylinders and side-brace fittings for landing gears have been introduced into service evaluation, which is continuing.

Carbon composites for aircraft brakes have been qualified for commercial aircraft. The major advantage compared with conventional steel brakes is a weight savings of more than 1300 lb (590 kg), the equivalent of some six additional passengers. This, together with the low wear rate, makes the structural carbon brake a prime candidate for extensive use in commercial and military aircraft. The composite is produced by repeated steps of impregnation and baking of carbon cloth to build up a carbon-reinforced carbon structure. The material is given a special treatment to protect against oxidation, which could affect its properties (Fig. 7.4). The material has been supplied as carbon brake disks for commercial jets such as the Gulfstream II, Challenger CL-600, and the Concorde. The material is still being evaluated for high-performance military jets F-14 and F-15 and the prototype B-1 bomber. When the advantages demonstrated in tests of aircraft brakes led automotive engineers to test brake disks and pads on high-speed tracks, C–C structures outlasted standard racing brakes. The material is tough and absorbs a great deal of heat without glazing, changing shape, or losing braking strength; C–C reduces unsprung weight and improves car handling.

The new DC-9 Super 80 commercial airplane is the first production aircraft to employ 100% advanced composites for the engine pod and inlet cowl. A significant advance in sound suppression has been achieved with the application of an acoustic structure consisting of a fine steel-wire weave bonded to the surface. (This acoustic structure is also being used on the 737 aircraft.) The composite cowling package consists of six different panels per nacelle (Fig. 7.5). The basic concept is a skin and rib design with hybrid Kv–Gr reinforcement in an epoxy matrix. Graphite was selected for its strength and Kevlar for protection from corrosion and impact damage. This hybrid matrix eliminates 30 to 40% of the weight of conventional sheetmetal parts. Each panel has flame retardant and lightning protection applied. This advanced-composite nacelle has achieved a 25% weight savings overall for the total engine nacelle compared with a similar metal nacelle. The first production nacelles have more than 7000 h of flight time.

## MILITARY AIRCRAFT

The initial intensive development effort of composites technology was sponsored and funded by the U.S. Air Force, primarily Wright-Patterson Materials and Manufacturing

**FIG. 7.4** Carbon-reinforced-carbon brake disk.

**FIG. 7.5**  Cowl panel after mechanical assembly. *(Rohr Industries.)*

Technology Laboratories, but the first production application was a B–Ep skin on the horizontal stabilizer box of the F-14, a U.S. Navy aircraft, with full-depth honeycomb and aluminum skins for the top and leading and trailing edges[8] (Fig 5.28). After this beginning airframe engineers were willing to choose the new composite material and design future aircraft components.

The A-7D outer wing (Table 7.3), was selected for a production and service program because it was a primary wing structure that could be installed on, or removed from, the aircraft. The wing featured primarily Gr–Ep construction with boron and Gr–Ep hybrid covers. The wing passed 16,000 h of structural Air Force testing, twice the normal service lifetime of A-7, demonstrating the high fatigue resistance of a properly designed composite structure.

A Gr–Ep spoiler for the S-3A (Table 7.3), was produced in limited quantities (about 35) and put into service. Examination after 6 years shows substantially no deterioration of the structure, thus increasing confidence in the production and service use of composite structures. The S-3A spoiler was fabricated as sandwich construction with Gr–Ep skins and nonmetallic honeycomb core assembled by co-curing the laminate skins and core in one operation.

## Transports

The main purpose of efforts for commercial and military aircraft is the "fly and try" programs on primary and secondary composite structural components to increase the producer's and user's confidence in composite structures, to broaden the base of experience, and to obtain meaningful manufacturing data by producing enough components of a given type to assure economies in composite pilot production. A major deficiency in composites technology is the lack of a data base for predicting life-cycle costs. Data are lacking in the areas of operations, maintenance, reliability, inspectability, and repairability. Appropriate flight-service programs are the best way to obtain the required data, although the return is on a somewhat long-range basis.

**Table 7.3** Advanced Composites in Military Aircraft[9]

| Model | Material | Component | Status† |
|---|---|---|---|
| A-4 | B–Ep and Gr–Ep | Flap | 1 |
| | Gr–Ep | Horizontal stabilizer, speed brakes | 2 |
| A-7 | Gr–Ep | Speed brake | 1 |
| A-7D | Gr–B–Ep | Outer wing panels | 2 |
| A-9E | Gr–Ep | Rudder | 1 |
| A-37B | Gr–Ep | Outer cylinder landing gear, trunnion, landing-gear side brace | 2 |
| AV-8B | Gr–Ep | Wing-box skins, forward fuselage, horizontal stabilizer, elevators, rudder, overwing fairing, ailerons, flaps | 2 |
| B-1 (original) | B–Ep | Torque-box cover skin, longeron | 1 |
| | Hybrid B–Ep and Gr–Ep | Vertical stabilizer, horizontal stabilizer, wing slat | 2 |
| | Gl–Ep | Torque-box cover skin | 1 |
| | Gr–Ep | Torque-box cover skin | 1 |
| | | Secondary airframe structures, leading and trailing edge flaps, weapons bay and avionics doors | 2 |
| C-5A | B–Ep, Gl–Ep, Gr–Ep | Nose radome, wing leading edge, wing trailing-edge flaps, engine nacelles, pylons, cargo doors, landing-gear fairings, troop-compartment floor, aft fuselage panels | 3 |
| | SiC–Al | Wing box | 1 |
| C-130 | B–Ep | Center wing box | 1 |
| C-141 | Hybrid Gr–Ep and Gl–Ep | Aft cargo-door cover (petal door) | 2 |
| E-2A | Gl–Ep | Rotating radome | 3 |
| E-2C | Gl–Ep | Inboard vertical stabilizer fin | 1 |
| F/A-18 | Gr–Ep | Wing skins, horizontal tail box, vertical tail box, wing control surface, tail control surface, leading-edge extension of wings, horizontal actuator cover, dorsal covers, landing-gear doors, rudders, fixed trailing edge, speed brake | 3 |
| F-4 | B–Ep, B–PI, and Gr–PI | Rudder | 1 |
| F-5 | Gr–Ep | Horizontal stabilizer | 2 |
| | | Fuselage component | 3 |
| F-5A | Gr–Ep | Speed-brake door, landing-gear door, wing slat, rudder, horizontal tail, wing leading edge | 2 |
| F-5E | Gr–Ep | Trailing-edge wing flap | 1 |
| F-14 | B–Ep | Horizontal-stabilizer skins | 3 |
| | Gr–Ep | Main landing-gear door, vertical stabilizer | 2 |
| F-15 | Hybrid B–Gr, Gl–Ep, Gr–Ep | Wing | 1 |
| | B–Ep | Vertical tail, horizontal tail, rudder, stabilizer skins | 3 |
| | Gr–Ep | Speed brake | 3 |
| F-16 | Gr–Ep | Empennage skins, vertical fin, fin leading-edge skins, rudder tail skins, horizontal tail skins, forward fuselage | 3 |

**Table 7.3** Advanced Composites in Military Aircraft[9] (*continued*)

| Model | Material | Component | Status† |
|---|---|---|---|
| F-111 | B–Ep | Horizontal tail | 2 |
| | B–Ep and B–Al | Fuselage section | 2 |
| | Gr–Ep | Underwing fairings | 3 |
| | Hybrid Gr–Ep, B–Ep, B–Al, Gl–Ep | Aft fuselage centerbody | 1 |
| F-111B | B–Ep | Wing Box | 1 |
| KC-135 | Gl–Ep | Winglet | 1 |
| QCSEE | Gr–PI | Inner cowl | 2 |
| S-3A | Gr–Ep | Spoilers | 3 |
| T-38 | Gr–Ep | Aileron trailing edge | 2 |
| | | Horizontal stabilizer | 3 |
| T-39 | B–Ep | Wing box | 2 |

†1 = experimental, 2 = prototype development, 3 = production.

One flight-service program considered selective reinforcing of metal structures with composites. A unique opportunity developed when the C-130 transport fleet was retrofitted with strengthened center wing boxes. These aircraft had accumulated fatigue damage in the severe flight environments of Southeast Asia. The standard retrofit involved installation of strengthened aluminum center wing boxes, but a study indicated that about 500 lb (227 kg) of uniaxial B–Ep bonded to the skins and stiffeners of the wing box would reduce the stress levels and increase the fatigue life as much as the aluminum retrofit design but with a 10% weight saving.[10] Some of the details are shown in Fig. 7.6. Laminated strips of uniaxial B–Ep with the required number of plies were bonded to the inner surface of the skin panels under each stringer and to each hat-section stringer on the enclosed crown surface. The nominal area ratio of B–Ep to aluminum of 1:4 was selected on the basis of weight reduction, equivalent ultimate strength, equivalent damage tolerance, and equivalent fatigue endurance.

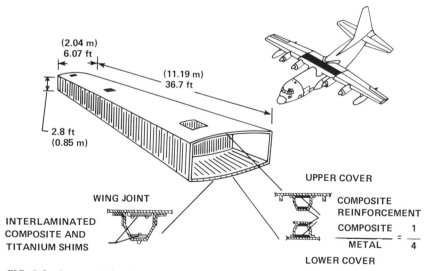

FIG. 7.6 Center wing box for C-130; aluminum weighs 4940 lb (2241 kg), composite-reinforced aluminum weighs 4440 lb (2014 kg), and composite material weighs 350 lb (159 kg).[10]

A significant development was associated with the residual thermal stresses produced when the B–Ep laminates were bonded to the aluminum structure at an elevated temperature due to the difference in the thermal expansion characteristics of aluminum and B–Ep. This led to development of cool-tool bonding, which constrained expansion of the aluminum parts during bonding. The tool is thermally insulated from the parts to be bonded and heat is supplied by an electric blanket rather than an autoclave. As Fig. 7.6 shows, adequate bearing surface was provided in the fastener penetration areas by titanium doublers, which were inserted and integrally bonded into the laminates. The wing box passed all ground testing, which included static tests to limit load, fatigue tests to four lifetimes (40,000 simulated flight hours), and static tests to determine the residual strength.

A newly designed twin fin, called an *afterbody strake,* has been placed on a C-130E for flight-testing. The twin fins, aluminum and Gl–Ep, are attached under the airplane's horizontal stabilizer to eliminate turbulence and smooth out airflow, reducing the aircraft's aerodynamic drag. They also help reduce fuel consumption on the C130 by more than 3% under typical cruise conditions.

On the C-141 aircraft the underside of the aft end of the vehicle contains two aft cargo or petal doors, which form the outer contour of the aircraft. Inside them is a large pressure door, separate from the petal doors, so that the entire petal-door arrangement is outside the fuselage pressure vessel. The lightweight aluminum-core and face-sheet construction is the major reason for the frequent in-service damage experienced by these doors. The composite door cover (Fig. 7.7) designed as a replacement for the existing aluminum honeycomb cover would fully replace the existing cover in form, fit, and function. The door cover consists of a skin made up of a Gr–Ep–epoxy-syntactic-core combination, longitudinal stiffeners and edge members of Gr–Ep, and chordwise frames (at locations matching the existing substructure) of Gl–Ep. All the stiffening members are hat sections, and the frames and edge members are filled with lightweight epoxy potting compound to provide support during fastener installation.[11]

A new program sponsored by the Air Force has been initiated to fabricate a SiC–Al wing-box structure which will satisfy all form, fit, and functional requirements of the new C-5A wing and reduce weight by 20% compared with the present conventional aluminum-alloy structure. Some components will be made of lay-ups of continuous-fiber-reinforced 6061 alloy molded to net shape at pressures of 0.8 to 1 kip/in$^2$ (5.5 to 7 MPa) and temperatures of about 1100°F (593°C). Others will involve consolidation of whisker reinforcements and prealloyed 2124 alloy powder by pressure-molding techniques. Hybrids formed by flame-spraying mixtures of whisker-reinforced 6061 onto alternate layers of fiber-reinforced 6061 and hot-molding to net shape will be evaluated.

## Fighters

The F-15 uses B–Ep for fin, rudder, and stabilator skins (Fig. 7.8), resulting in an estimated 25% weight reduction with respect to an all-metal empennage. A Gr–Ep speed brake uses only 250 individual parts vs. 1200 parts in the aluminum version. Composites constitute about 1% of the structural weight of this aircraft.

Another fighter, the F-16, has made extensive use of Gr–Ep composites. The empennage skins have resulted in an estimated 23% weight reduction over metal designs for the same components (vertical fin box, fin leading edge, rudder, and horizontal tail). The vertical fin has Gr–Ep skins over conventional aluminum rib-and-spar understructure. The fin

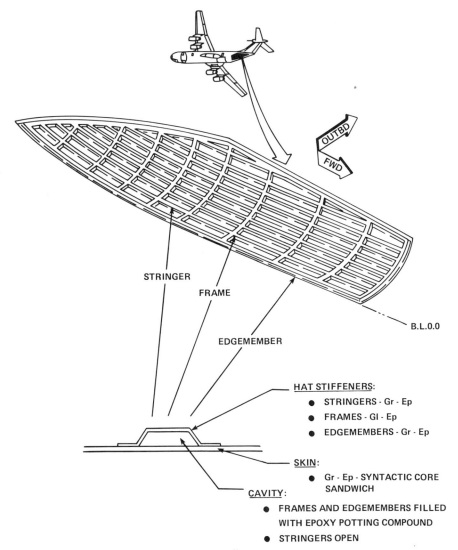

STRINGER

FRAME

EDGEMEMBER

B.L.0.0

OUTBD

FWD

HAT STIFFENERS:
- STRINGERS - Gr - Ep
- FRAMES - Gl - Ep
- EDGEMEMBERS - Gr - Ep

SKIN:
- Gr - Ep - SYNTACTIC CORE SANDWICH

CAVITY:
- FRAMES AND EDGEMEMBERS FILLED WITH EPOXY POTTING COMPOUND
- STRINGERS OPEN

FIG. 7.7  Design concept for composite door cover.[11]

leading edge, rudder, and horizontal tails are all full-depth aluminum-honeycomb sandwich structure with Gr–Ep face sheets. The tail also uses titanium pivot shafts, ribs, and spars. Gr–Ep makes up about 2% of the total structural weight (Fig. 7.9).

As the F/A-18, the next fighter, was being designed, a giant step forward was taken with the use of Gr–Ep for the wing skins, horizontal and vertical tail boxes, wing and tail control surfaces, speed brake, leading-edge extension, and miscellaneous doors.[7] Composites make up about 10% of the structural weight and over 50% of the surface area. Figure 7.10 shows the distribution of composite material in the aircraft.

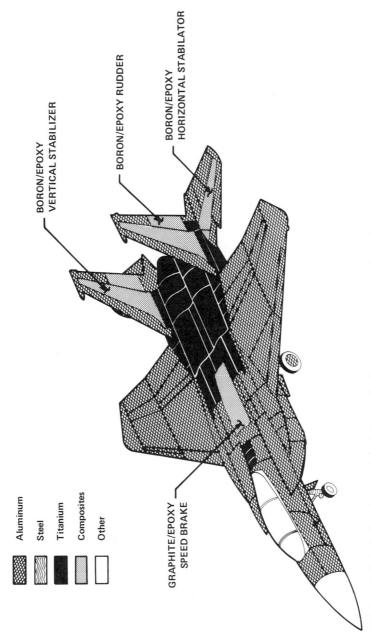

Aluminum

Steel

Titanium

Composites

Other

BORON/EPOXY
VERTICAL STABILIZER

BORON/EPOXY RUDDER

BORON/EPOXY
HORIZONTAL STABILATOR

GRAPHITE/EPOXY
SPEED BRAKE

FIG. 7.8   B–Ep for horizontal and vertical stabilizers and rudder on an F-15. *(McDonnell Douglas, St. Louis.)*

**7.13**

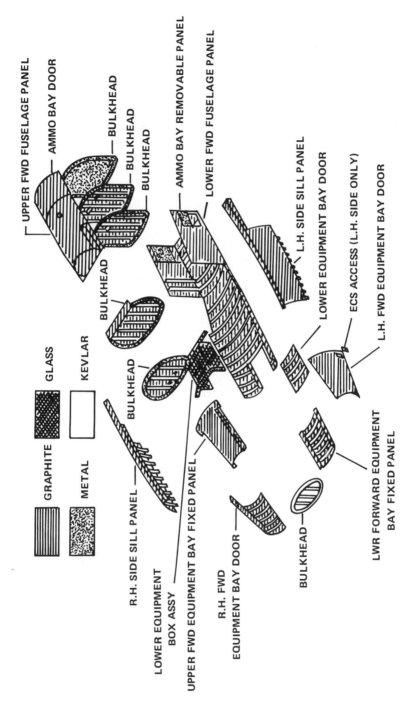

**UPPER FWD FUSELAGE PANEL**

**AMMO BAY DOOR**

**BULKHEAD**

**BULKHEAD**

**BULKHEAD**

**BULKHEAD**

**BULKHEAD**

**AMMO BAY REMOVABLE PANEL**

**LOWER FWD FUSELAGE PANEL**

**L.H. SIDE SILL PANEL**

**LOWER EQUIPMENT BAY DOOR**

**ECS ACCESS (L.H. SIDE ONLY)**

**L.H. FWD EQUIPMENT BAY DOOR**

**BULKHEAD**

**BULKHEAD**

**R.H. SIDE SILL PANEL**

**GRAPHITE**

**GLASS**

**METAL**

**KEVLAR**

**LOWER EQUIPMENT BOX ASSY**

**UPPER FWD EQUIPMENT BAY FIXED PANEL**

**R.H. FWD EQUIPMENT BAY DOOR**

**BULKHEAD**

**LWR FORWARD EQUIPMENT BAY FIXED PANEL**

FIG. 7.9  Major components of composite forward fuselage.[33]

7.14

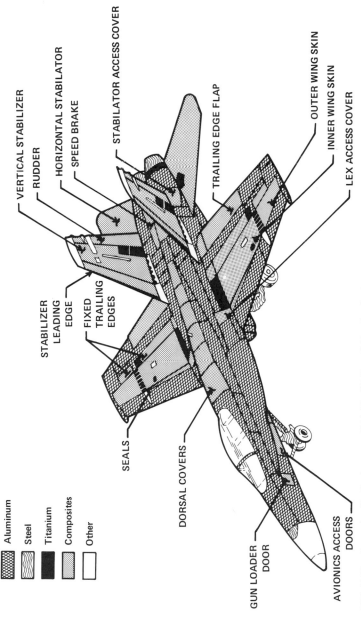

FIG. 7.10  Composite distribution on an F/A-18. (*McDonnell Douglas, St. Louis.*)

VERTICAL STABILIZER

RUDDER

HORIZONTAL STABILATOR

SPEED BRAKE

STABILATOR ACCESS COVER

TRAILING EDGE FLAP

OUTER WING SKIN

INNER WING SKIN

LEX ACCESS COVER

STABILIZER
LEADING
EDGE

FIXED
TRAILING
EDGES

SEALS

DORSAL COVERS

GUN LOADER
DOOR

AVIONICS ACCESS
DOORS

Aluminum

Steel

Titanium

Composites

Other

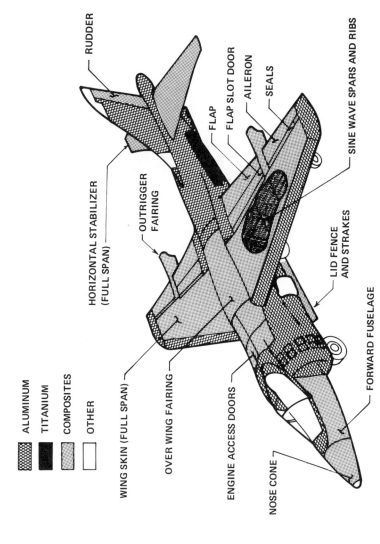

ALUMINUM

TITANIUM

COMPOSITES

OTHER

RUDDER

FLAP

FLAP SLOT DOOR

AILERON

SEALS

SINE WAVE SPARS AND RIBS

HORIZONTAL STABILIZER
(FULL SPAN)

OUTRIGGER
FAIRING

LID FENCE
AND STRAKES

WING SKIN (FULL SPAN)

OVER WING FAIRING

ENGINE ACCESS DOORS

FORWARD FUSELAGE

NOSE CONE

**FIG. 7.11**  Gr–Ep composites on an AV-8B (dark areas). *(McDonnell Douglas, St. Louis.)*

**FIG. 7.12** Gr–Ep wing skin of AV-8B light attack aircraft. *(McDonnell Douglas, St. Louis.)*

A fighter currently under development, the AV-8B, is using Gr–Ep composites extensively. It is said to be the first conventional or V/STOL aircraft to use graphite composites and advanced supercritical wing design. Gr–Ep is used in the wing-box skins and substructure, forward fuselage, horizontal stabilizer, elevators, rudder, overwing fairing, ailerons, and flaps. Gr–Ep composites account for approximately 86% of the aircraft's structural weight [1300 lb (590 kg)] (Fig. 7.11). A significant feature of the AV-8B design is that the wing-box substructure sine-wave spars use Gr–Ep fabric and unidirectional broad goods reinforcements. The multispar wing and low operating stress level give the wing a higher survival rate than an aluminum structure at much lower weight (Fig. 7.12).

Composite weight savings and use in fighter aircraft are summarized in Table 7.4.

## Bombers

Gr–Ep has been used for the horizontal stabilizers of the A-4 attack bomber. The stabilizers (Fig. 7.13) weigh up to 40% less than metal models. Preliminary work has begun on the B-1B or long-range combat aircraft, as the new derivative plane will be called. The original

**Table 7.4** Production Applications of Advance Composites[37]

| Aircraft | Composites portion of structural weight, % | Estimated component weight saving, % |
|---|---|---|
| F-14 | 1 | 19 |
| F-15 | 1 | 23 |
| F-16 | 2 | 23 |
| F-18 | 10 | 35† |
| AV-8B | 26 | 20–25 |

†Overall structure.

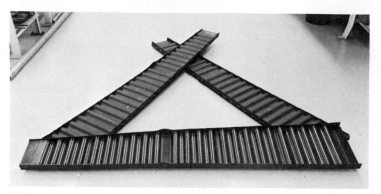

**FIG. 7.13** Gr–Ep ribbed substructure of A-4 horizontal stabilizer. (*McDonnell Douglas, St. Louis.*)

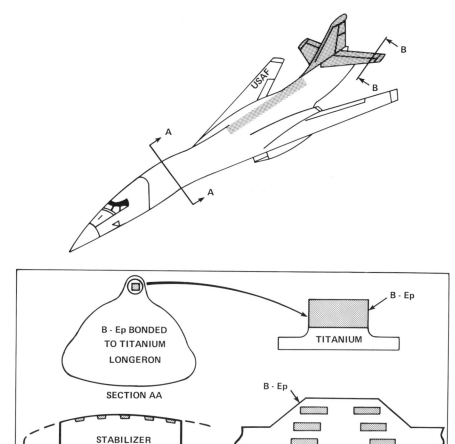

**FIG. 7.14** B–Ep longerons and stabilizers on original B-1. (*AVCO, Specialty Materials Division.*)

B-1 had composites in its design (Fig. 7.14). The B-1 derivative will reflect a number of changes from the original. There will be approximately 3 to 5% composites by weight, compared with about 1% in the original B-1. Composite parts will include aft avionics doors, longerons, weapon-bay doors, slats, and access doors. Overwing fairings will be hybrid, a mixture of Gr–Ep and Gl–Ep. Except for the fairings and the B–Ep longerons, all the parts will be Gr–Ep. It was anticipated that the aluminum tail fin would be replaced by a composite structure in production.

## Miscellaneous Fittings and Aircraft

Several significant development programs on landing-gear doors and drag-link assemblies in the landing gear have been sponsored by various government agencies. One program used a boron composite landing-gear door in the F-5 fighter aircraft. The structure was composed of an aluminum-honeycomb core with an outer covering of bonded layers of boron composite skins, each 0.050 in (1.3 mm) thick. It was 29% lighter than metal doors but had equivalent strength and stiffness. Another program evaluated the use of B–Al for the lower drag-link assembly from the A-7 nose landing gear in lieu of the machined 4340 steel forging heat-treated to 200 kips/in$^2$ (1380 MPa) ultimate strength. The 28-ply B–Al tube was diffusion-bonded to titanium end fittings, and a series of corrosive environmental tests after pebble impact was conducted. There was no reduction in static strength over an as-fabricated specimen, and the composite part survived two lifetimes at 80% design fatigue load levels. This compared favorably with a notched 300M-steel production link, which failed in fatigue after 1.1 lifetimes at 70% design fatigue load levels. Both the B–Al links and the 300M-steel production link were notched and tested in the same sequence.[12]

Concurrent with government efforts to develop long-range bombers was an effort to demonstrate the use of Gr–Ep for rotary launcher-shaft designs to effect weight reduction. The results of the development program[13] showed that the graphite-composite shaft-fabrication process of the 2.125-in-thick (54-mm) wall is adaptable to automated production. The techniques used result in a 40% cost savings, compared with a repetitive ply-by-ply state-of-the-art process.

## Next-Generation Aircraft

It remains to be seen whether there will be an all-composite fighter aircraft, but it is clear that the limit of composites use is far in the future. Figure 7.15 shows that the use of advanced composites for primary aircraft structures has increased from less than 1% to 9.5% in the past decade. Future aircraft are targeted for much more, as suggested in Fig. 7.16. A study sponsored by the U.S. Air Force indicates that using composites for 75 to 80% of the airframe would reduce overall weapon-system weight by 26%. Current studies for subsonic and supersonic V/STOL suggest that 65% composite structures would cut weight by as much as 20%. Further, thanks to an all-composite airframe, a helicopter being evaluated by the U.S. Army promises 22% weight saving, 17% cost saving, and greater resistance to attack and radar detection than metal airframes.

Another area where composites will find wide applications is the forward-swept-wing fighter. Composites can be elastically tailored to counteract the upward-twisting tendency

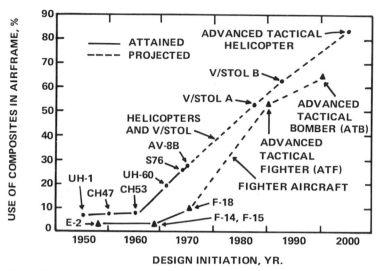

**FIG. 7.15**  Future composite use in aircraft. *(Am. Mach., December 1981, p. 112.)*

of the forward-swept wings, which made aluminum versions of the aerodynamically efficient design too heavy to use. Figure 7.17 is a sketch of the advanced fighter of the next decade and the dominance of composites.

## HELICOPTERS

Better performance is as much a reason for considering composites as weight and cost savings are. Cost effectiveness must be clearly demonstrated before composites are widely used as substitutes for monolithic materials. Composites not only are easier to process but also provide greater versatility in shaping aerodynamic contours. Design studies by various helicopter manufacturers show that composites can be effectively substituted in helicopter airframes in the 1980s and 1990s. From various design concepts and material studies allmolded composite modular panels, which provide integral skin-stringer and frame subassemblies, appear the most promising. The result is a notable reduction in the total number of parts compared with current construction methods.

Helicopters are a fertile field for a variety of composites, including epoxy and MMCs. The first NASA flight-service component for helicopters was reinforcement of an Army CH-54B helicopter tail cone with unidirectional B–Ep. B–Ep strips bonded to the tailcone stringers increased the tail-cone stiffness and reduced the structural mass by 14%.[14,15] Another program currently involves a Kv–Ep cargo-ramp skin on a Marine Corps CH-53 helicopter. The rough service will provide a good comparison with the in-service characteristics of the aluminum ramp currently used. The Kv–Ep skin is 9.2% lighter.

Two NASA programs under way are applied to commercial helicopters. One program on the 206L helicopter has Kv–Ep doors and fairings and Gr–Ep vertical fins which are under a 5- to 10-year service-evaluation period. Components have been fabricated in four different design concepts, and all will be evaluated. The mass saving is about 27.7% over metal parts. The second program will evaluate the durability of components of the S-76

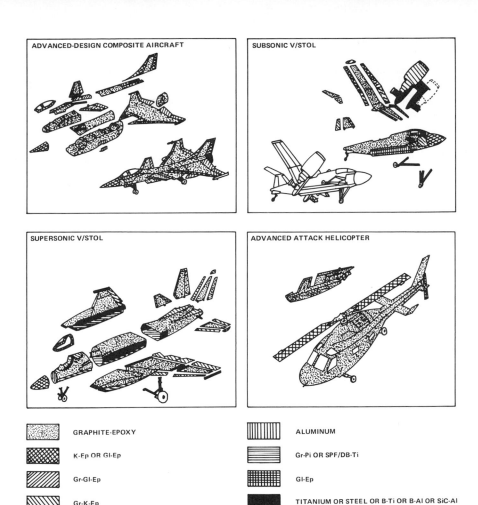

| | | | |
|---|---|---|---|
| GRAPHITE-EPOXY | | ALUMINUM | |
| K-Ep OR GI-Ep | | Gr-Pi OR SPF/DB-Ti | |
| Gr-GI-Ep | | GI-Ep | |
| Gr-K-Ep | | TITANIUM OR STEEL OR B-Ti OR B-AI OR SiC-AI | |

**FIG. 7.16** Four future aircraft and their composite potential. *(Am. Mach., December 1981, p. 113.)*

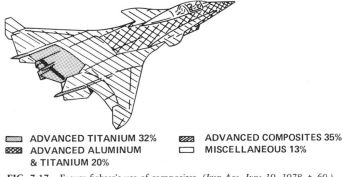

ADVANCED TITANIUM 32%
ADVANCED ALUMINUM & TITANIUM 20%
ADVANCED COMPOSITES 35%
MISCELLANEOUS 13%

**FIG. 7.17** Future fighter's use of composites. *(Iron Age, June 19, 1978, p. 60.)*

helicopter (a tail rotor with a laminated Gr–Ep spar and a Gl–Ep skin and the horizontal stabilizer with a Kv–Ep torque tube with Gr–Ep spar caps, Nomex honeycomb-core, and Kv–Ep skin). Both components will be removed from helicopters after 2 to 10 years of operational service and static and fatigue tests will be compared with base-line certification tests.

Another flight-service program utilizing composite-reinforced metal structures involved the CH-54B. During developmental testing, the original airframe was found to be in resonance for certain combinations of cable sling length and load, creating an undesirable dynamic condition in the tail cone. Although the production fix was to provide thicker top and bottom skins for the aluminum tail cone, a preliminary analysis indicated that uniaxial strips of B–Ep bonded to the tail-cone stiffeners would provide the extra stiffness needed to prevent resonance and result in a 14% weight saving. The general criterion was that the stiffness of the composite-reinforced tail cone be the same as that of the modified production tail cone. Bonding laminates to the aluminum stringers at elevated temperature caused residual stresses and warpage; however, the warpage could easily be removed by applying hand pressure and the residual stresses were not as critical as in the C-130 program because the tail cone is rather lightly loaded. Since in this case composites were used only to meet a stiffness requirement, it was possible to end the composite before reaching joint areas. The composite-reinforced tail cone was installed in a U.S. Army helicopter and has been in service since March 1972. Much of the technology developed under joint NASA and Army funding has been incorporated into the present generation of military helicopters, such as the UH-60A (Fig. 7.18). This helicopter uses an extensive amount of composite prepreg [6275 ft² (583 m²) of Kevlar and 11,279 ft² (1048 m²) of fiber glass] and a nominal amount of graphite and boron (Fig. 7.19).

Advanced composites are also finding their way into commercial helicopters. For example, use of graphite, Kevlar, and fiber glass in the S-76 is extensive. Composites account for 60% of the surface area, their use reduces airframe component weight by about 30%, and 22% of the structural weight is composites. The aircraft has 17,269 ft² (1604 m²) of Kevlar, 2086 ft² (194 m²) of fiber glass, and 300 ft² (28 m²) of graphite prepregs.

Other commercial helicopter manufacturers are using composites in current and future designs. Not only is metal being replaced by composites but sometimes fiber glass as well (Fig. 7.20). A reduction in overall airframe weight of 11% is achievable by such replacement and weight savings of from 20 to 55% on individual helicopter parts.

**FIG. 7.18**   UH-60A Black Hawk helicopter with 17 wt % composites. (*Sikorsky Aircraft, Division of United Technologies.*)

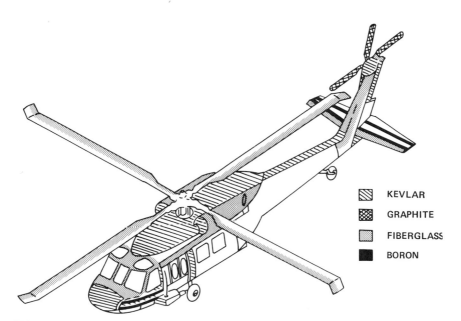

| | |
|---|---|
| ▨ | KEVLAR |
| ▩ | GRAPHITE |
| ▢ | FIBERGLASS |
| ■ | BORON |

**FIG. 7.19**  Composite applications in UH-60A Black Hawk. *(Sikorsky Aircraft Division of United Technologies.)*

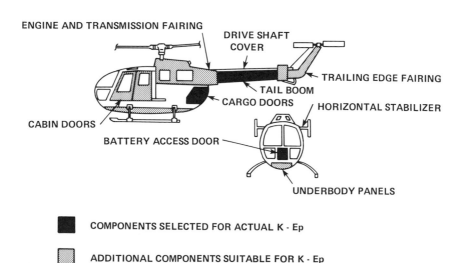

**FIG. 7.20**  Kv–Ep replacing metal and Gl–Ep in a commercial helicopter. *(Prod. Eng., September 1974, p. 50.)*

## Components (Airframe, Fuselage, Landing Gear, Transmission)

A variety of components for military and commercial helicopters have been designed and fabricated from composite materials. Some were experimental and failed or are still being tested; others are in current production. A preliminary design study[14] evaluated helicopter airframes and landing-gear structures for composites, which differ significantly from metals because of the anisotropy of composites. Since these structures normally operate in a moderate temperature environment of −65 to 160°F (−54 to 71°C) and primary design conditions are static, candidate composite materials are evaluated according to the room-temperature static strength of unidirectional laminates. In this study metal matrixes were not considered because they are more expensive and often more difficult to fabricate than resin-matrix materials.

Both B–Ep and Gr–Ep appear to be the prime candidate materials for the major portion of the primary structure, thanks to their high specific strength and modulus in both tension and compression. Kv–Ep, with its density and better modulus than fiber glass, is the prime candidate material for secondary structure. Kv–Ep may also be a candidate for primary structural areas where its high specific tension strength can be used. Kv–Ep combined with Gr–Ep is an excellent hybrid candidate where moderate compression strength will be adequate (Table 7.5).

### Airframe and Fuselage

Although the helicopter industry in the United States is limited to a handful of manufacturers, their product uses more composite materials than the aircraft industry. It should be noted that almost every component fabricated for a helicopter has compound curvature, compared with the flat or mildly contoured components in aircraft wings, etc. Figures 7.21 to 7.24 show several Kv–Ep production components for the commercial S-76 helicopter. The UH-60A helicopter (Fig. 7.18) has approximately 667 composite parts, including the Gr–Gl–Ep hybrid canopy and Kv–Ep fixed covers, work platforms, access doors, fairings, and main-rotor pylon sliding cover (Fig. 7.25).

Other helicopter models have fuel-tank hanger beams with Kv–Ep I beams and Gr–Ep caps, the forward-pylon work platform and engine access doors of Kv–PS, and tail boom and vertical fins of Gl–Ep. Kv–Ep and Gr–Ep have been selected for the nacelle, pylon doors, aft fuselage, and empennage for the YAH-64 helicopter, and Fig. 7.26 illustrates two hybrid components on the YAH-64 helicopter. Composite floors and fuel pods for the 234 commercial helicopter have recently been fabricated and are to be flight-tested. The 15-ft (4.6-m) fuel pods, made of Kv–Ep and Gr–Ep, save 25% in weight over a similar aluminum design. The helicopter also has honeycomb Gl–Ep and Kv–Ep floor panels replacing 30-ft (9-m) extrusions which are riveted and bolted together. The composite floor is 10% lighter and more corrosion- and impact-resistant and transmits less vibration than the metal floor.

Two new commercial production models, 222 and 214 ST, have engine cowlings and fairings made of a honeycomb sandwich with faces of SGl–PI resin and Nomex core. These components were formerly fabricated in titanium since fire resistance was a requirement. The design reduced weight by 22% and eliminated forming problems associated with titanium. Advanced-composite materials have been used in the design of the crew seats for the model 222, especially Kv–Ep for the seat bucket, which was of sandwich construction with aluminum-honeycomb core and woven Kv–Ep faces.

**Table 7.5** Comparison of materials[35]

| Material | Reason for consideration | Density lb/in³ | Density kg/m³ |
|---|---|---|---|
| 7075-T6 aluminum alloy | For comparison | 0.101 | 2795.9 |
| B–Ep | High tensile and exceptionally high compressive stress makes this a good choice for structures designed for reversal of stresses | 0.073 | 2020.6 |
| Gr–Ep, HTS | Similar to that for B–Ep above; smaller fiber size also allows greater flexibility in producing complex shapes | 0.055 | 1522.4 |
| HMS | Moderately high strength coupled with a high modulus makes this a possible first choice for compression-stability-limited structures | 0.058 | 1605.4 |
| AS graphite | Moderate strength and modulus; primary advantage low cost | 0.055 | 1522.4 |
| EGl–Ep | Relatively low-cost with high tensile strength; while lower compression strength is a limiting factor, the material is a good choice for lightly loaded structures | 0.065 | 1799.2 |
| SGl–Ep | A choice that extends the range where EGl-Ep would be used for increased intensity of loadings | 0.070 | 1937.6 |
| Kv–Ep | High tensile strength and very low density offer applications for secondary structures or primary structures designed for tension; increased modulus over Gl–Ep extends the range of usefulness, but material is limited in compression | 0.050 | 1384.0 |

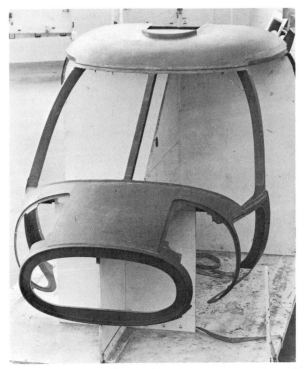

**FIG. 7.21** Canopy of S-76. *(Sikorsky Aircraft, Division of United Technologies.)*

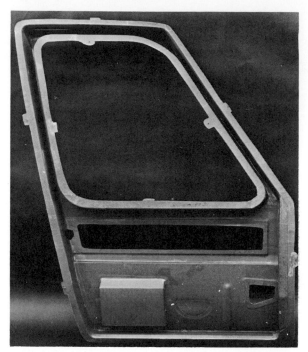

**FIG. 7.22** Crew door of S-76. *(Sikorsky Aircraft, Division of United Technologies.)*

**FIG. 7.23** Aft fairing of S-76. *(Sikorsky Aircraft, Division of United Technologies.)*

Components in the model 206L fabricated from composite materials (to be flight-evaluated in the next 5 years) include the forward fairing of a Kv–PS material, litter door of Kv–Ep with Gr–Ep reinforcement in local areas, a honeycomb-sandwich baggage door with Kv–Ep faces and Nomex core, and the vertical fin and tail bumper (Fig. 7.27) with projected weight savings of 35% over the metal fin. Significant improvements in life-cycle

costs, besides those in weight savings, are expected from use of these materials. Table 7.6 reflects the differences.

Future helicopters will contain airframe, fuselage sections,[16] tail cones, pylons, and stabilizers entirely of composite materials (hybrids of Gr–Ep and Kv–Ep). Figure 7.28 illustrates the design concept for a tail-cone and pylon,[17] and Figs. 7.29 and 7.30 show how these components can be produced economically in the future.

### Landing Gear and Transmission

In studies of landing gear to find where composites may be substituted the rolling gear is normally excluded since it consists of wheels, tires, brakes, and miscellaneous hardware. For the wheels, parts of the rolling gear are not considered replaceable by equivalent parts of composite construction. The remainder of the landing gear usually consists of major steel and aluminum forgings, together with miscellaneous hardware and nonmetallic elements. Structural ele-

FIG. 7.24   Exhaust cover of S-76. *(Sikorsky Aircraft, Division of United Technologies.)*

ments that have been considered for composites are the oleo trunnion, shock strut, drag-strut cylinder and piston, and torque arms. Proposals for the trunnion are shown in Fig. 7.31. Two materials are potential candidates for all-composite construction (Fig. 7.31*a*),

FIG. 7.25   Aft sliding fairing of Kv–Ep for UH-60A Black Hawk. *(Sikorsky Aircraft, Division of United Technologies.)*

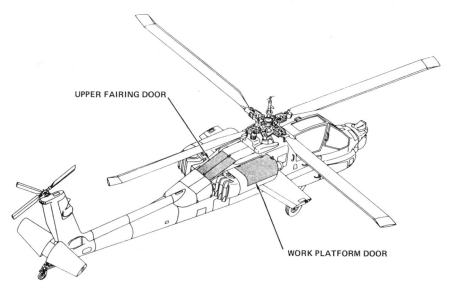

**FIG. 7.26** Two hybrid components of YAH-64 helicopter. *(Hughes Helicopter.)*

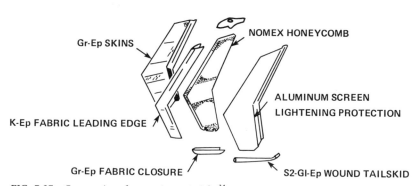

**FIG. 7.27** Construction of composite vertical fin.[34]

**Table 7.6** Composite vs. Metal Components for Model 206L

| Component | Composite weight | | Metal weight | |
|---|---|---|---|---|
| | lb | kg | lb | kg |
| Vertical fin | 10.5 | 4.77 | 14.5 | 6.58 |
| Horizontal stabilizer | 11 | 5 | 13 | 5.90 |
| Upper forward fairing | 7.5 | 3.40 | 9.5 | 4.31 |
| Litter door | 8.5 | 3.86 | 13.5 | 6.12 |
| Baggage door | 2.8 | 1.27 | 3.2 | 1.45 |

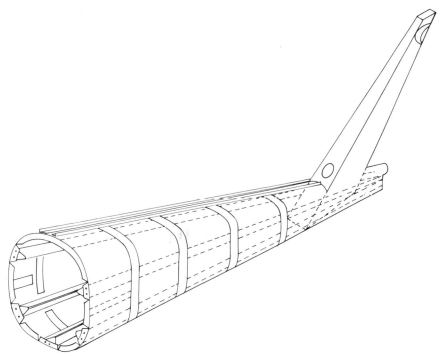

**FIG. 7.28**  Future design of tail cone and pylon in all-composite helicopter.[17]

short Gl–Ep molding or B–Al with brazed connections, but neither is cost-effective. The only proposal judged to have cost-effective potential is selective replacement of the simple elements of the trunnion with cylindrical shapes built up from Gr–Ep laminates (Fig. 7.31*b*). The complex shape and loading of the connections require metallic fittings at these points. For a bonded joint connection of these fittings to the composite cylinders the weight of metal replaced in the inclined and horizontal arms of the trunnion would be very small. The concept with the greatest potential is that in which only the central cylindrical section is replaced by composite material (Fig. 7.31*c*).

For axially loaded members the oleo shock-strut and drag-strut piston and cylinder are similar in their design loading conditions. The idea considered for each of these members is a composite cylinder bonded at each end to steel fittings containing the necessary detailed machined features (Fig. 7.32). Proposals for landing-gear torque arms are shown in Fig. 7.33. The all-composite concept is not considered cost-effective. The alternatives in Fig. 7.33 are similar in that they both contain many parts, with associated high manufacturing and assembly cost. A final design concept for a landing-gear structure is shown in Fig. 7.34. Transmission support structures normally fabricated of aluminum extrusions and mechanically fastened together have been designed as a Gr–Ep composite (Fig. 7.35) and vindicated in Army and NASA tests.[18]

Other components that have been developed include a new type of helicopter drive shaft of advanced-composite materials with only the end fittings of conventional materials. The design marks the first use of B–Ep and Gr–Ep composite materials in mechanical drive trains.

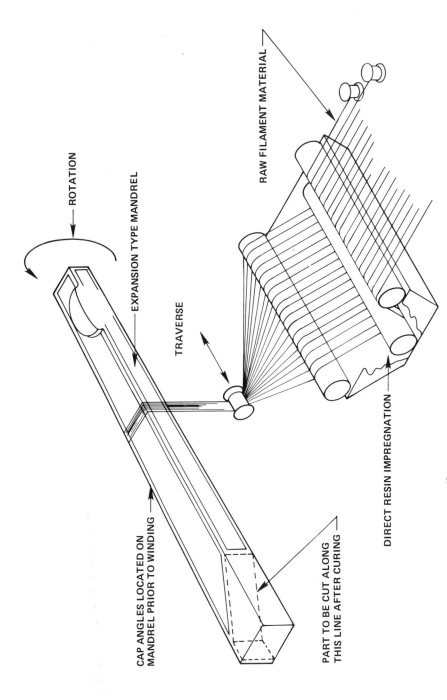

ROTATION

EXPANSION TYPE MANDREL

TRAVERSE

RAW FILAMENT MATERIAL

DIRECT RESIN IMPREGNATION

CAP ANGLES LOCATED ON
MANDREL PRIOR TO WINDING

PART TO BE CUT ALONG
THIS LINE AFTER CURING

FIG. 7.29  Filament-wound vertical pylon.[17]

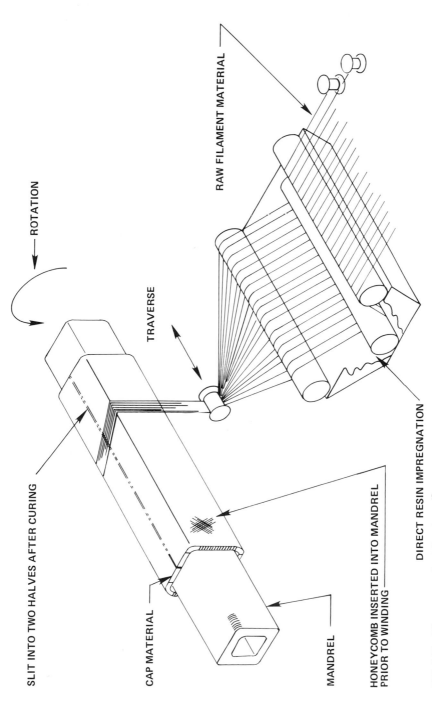

ROTATION

RAW FILAMENT MATERIAL

TRAVERSE

SLIT INTO TWO HALVES AFTER CURING

CAP MATERIAL

HONEYCOMB INSERTED INTO MANDREL
PRIOR TO WINDING

MANDREL

DIRECT RESIN IMPREGNATION

FIG. 7.30   Filament-wound stabilizer beam.[17]

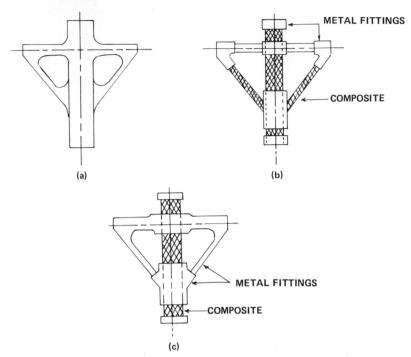

FIG. 7.31   Three concepts considered for landing-gear trunnion.[35]

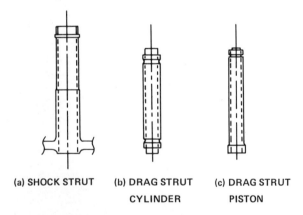

(a) SHOCK STRUT       (b) DRAG STRUT       (c) DRAG STRUT
                         CYLINDER              PISTON

COMPLEX LOCAL DETAILS OF HIGH STRENGTH STEEL PARTS
CANNOT BE PRODUCED WITH COMPOSITE MATERIALS.

(d) TYPICAL SELECTIVE REPLACEMENT CONCEPT
USED FOR ALL AXIAL MEMBERS.

FIG. 7.32   Practical selective-replacement concept for axially loaded gear elements.[35]

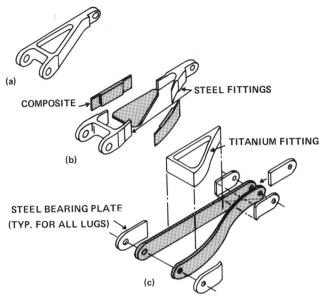

(a)

COMPOSITE →

STEEL FITTINGS

(b)

TITANIUM FITTING

STEEL BEARING PLATE
(TYP. FOR ALL LUGS) ↘

(c)

**FIG. 7.33** Composite landing-gear torque-arm concepts.[35]

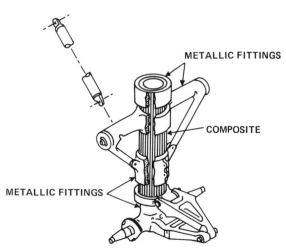

METALLIC FITTINGS

COMPOSITE

METALLIC FITTINGS ↙

**FIG. 7.34** Selective replacement, the most practical concept for landing gear.[35]

FIG. 7.35   Gr–Ep woven fabric in composite transmission-support frame.[18]

A significant weight saving results if the tail-rotor drive shaft can be made of an advanced-composite material rather than steel or aluminum. The first shaft was of fabricated B–Ep. The 97.25-in (2.47-m) length was built up from 10 plies of B–Ep sheet oriented at specific angles (Fig. 7.36); the shaft had a 5-in (127-mm) $OD_2$ and 0.052-in (1.32-mm) wall thickness and weighed 5.3 lb (2.4 kg). Three of these "stovepipes" placed end to end, with fittings and bearing supports between, make up the entire drive shaft system. An aluminum shaft system would require five shorter sections and additional fittings and bearing supports to do the same job. Using B–Ep reduces the overall system weight by 30%; moreover, the composite shaft is twice as stiff as a comparable aluminum shaft. Two Gr–Ep shafts, one using high-strength carbon fibers and the other high-modulus carbon fibers, have subsequently been fabricated and will undergo a rigorous test program.

Gr–Ep is being evaluated as the material for small nose-engine gear boxes, which weigh about 25 lb (11.34 kg) in magnesium; the main-rotor gear boxes weigh approximately 80 lb (36.29 kg). The idea is to fabricate the gear box from graphite fiber in a Kv–PI high-temperature resin matrix; filament winding is the preferred manufacturing method.

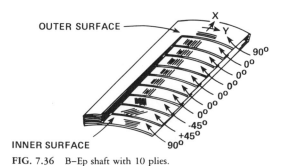

FIG. 7.36   B–Ep shaft with 10 plies.

## Blades

For helicopter manufacturers precision flying machines are only as good as their blades. Pitch, deflection, strength, vibration—all are blade-engineering problems that must be overcome in design and fabrication processes before liftoff on the first flight. Several helicopter manufacturers are now using Gl–Ep for blades since it gives the manufacturer absolute control over final blade manufacturing operations while relieving them of a multitude of material-engineering problems. Preimpregnated glass-fiber composites (prepregs) are tapes, sheets, or strands of partially cured resin with already oriented glass-fiber reinforcements. The manufacturer of the prepreg, not the manufacturer of the blade, controls uniform glass and resin concentrations, thorough saturation of the fiber glass, and other composite specifications. Helicopter manufacturers simply purchase the prepregs for a variety of processes, either forming the blade from tape and sheet and curing it under heat and pressure or winding or wrapping preimpregnated fiber strands before curing.

Gl–Ep blades are being used on the model 214 and the CH-46 and CH-47 military helicopters (Fig. 7.37). These blades have the following advantages:

Glass-fiber composites are stronger and lighter than metal.

Service repairs in the field are quick and relatively simple compared with those for metal blades.

Blades can take a 23-mm high-explosive incendiary hit and still fly back to base.

Glass-fiber blades have demonstrated a service life in excess of 5000 h.

The plastic spar and Nomex core do not corrode or degrade.

The slow, soft nature of glass-fiber failure propagation eliminates the need for on-board safety-monitoring systems.

The blades are insensitive to small manufacturing defects. These blades have the following advantages over metal blades:

Metal-blade spars and aluminum honeycomb suffer from corrosion.

The rapid propagation of metal failure requires on-board blade-safety monitoring systems that malfunction and give metal blades a low mean time between removal.

It is difficult to incorporate varying airfoils and twists on metal blades.

Other blades have been designed to incorporate Gr–Ep into the structure. The SA-360 and 365 helicopters have rotor blades whose blade structure is Nomex honeycomb core with four-ply single-direction Gr–Ep laid at 45° (0.79 rad). This is overlaid with Gl–Ep cloth for protection, and the leading edge is made up of Gl–Ep roving with a stainless-steel cover for protection.

Figure 7.38 shows the machine used to produce the largest helicopter rotor-blade spar. Weighing 570 lb (259 kg) and reaching a length of 41.5 ft. (12.65 m), the spar incorporates a titanium nose cap with a Gl–Ep and Gr–Ep fiber closed-D construction. This type of spar design is used in conjunction with a four-lug all-fiber glass root end. The closed-D spar configuration has the highest torsional stiffness per pound and results in a complete and fully inspectable spar assembly. The component is capable of housing a pneumatic failure-detection system.

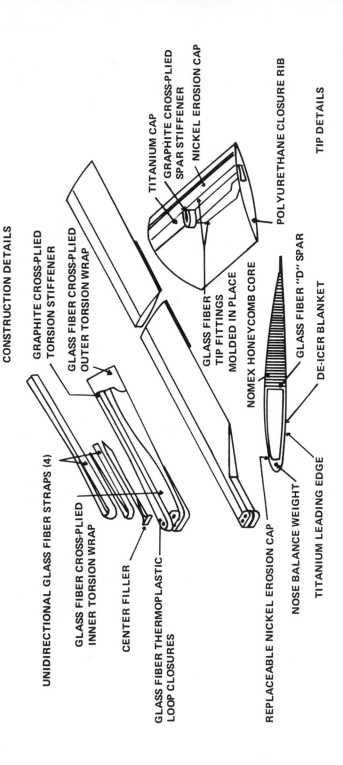

SPAR/ROOT END
CONSTRUCTION DETAILS

GRAPHITE CROSS-PLIED
TORSION STIFFENER

GLASS FIBER CROSS-PLIED
OUTER TORSION WRAP

UNIDIRECTIONAL GLASS FIBER STRAPS (4)

GLASS FIBER CROSS-PLIED
INNER TORSION WRAP

CENTER FILLER

GLASS FIBER THERMOPLASTIC
LOOP CLOSURES

REPLACEABLE NICKEL EROSION CAP

NOSE BALANCE WEIGHT

TITANIUM LEADING EDGE

TITANIUM CAP

GRAPHITE CROSS-PLIED
SPAR STIFFENER

NICKEL EROSION CAP

POLYURETHANE CLOSURE RIB

TIP DETAILS

GLASS FIBER
TIP FITTINGS
MOLDED IN PLACE

NOMEX HONEYCOMB CORE

GLASS FIBER "D" SPAR

DE-ICER BLANKET

TYPICAL OUTBOARD AIRFOIL SECTION

FIG. 7.37  Construction of CH-46 fiber-glass blade. (*Aviat. Week Technol., Mar. 20, 1978, p. 57.*)

**FIG. 7.38** Filament-winding machine for rotor blades. *(Boeing-Vertol Corp.)*

## MISSILES

Composites have demonstrated the wide range of characteristics necessary to satisfy the operational requirements of missiles. High stiffness and strength and minimum weight are the major reasons for the use of graphite composites for critical structural members in ICBMs and other missile systems.

Use of fiber glass to filament-wind rocket-motor cases started in the late 1950s and continued for the next several years. Motor-case programs now in production have used Kv–Ep filament-wound first-, second-, and third-stage rocket motors for the Trident-I (C-4) missile and first- and second-stage Pershing II and MX stage III missiles. The MX missile launch tube, a Gr–Ep structure, will be produced by filament winding. The canister has an 8-ft (2.4-m) diameter and is approximately 40 ft (12 m) long.

A Gr–Ep cloth developed for use in the Trident I missile conforms readily to irregular contours and reduces missile manufacturing lay-up time up to 50% without detracting from the structural characteristics of the material. The new Gr–Ep cloth makes it considerably easier to produce complex geometrical missile shapes; nevertheless, some unidirectional Gr–Ep tape is also used to reinforce the upper and lower cylindrical high-stress surfaces of the missile's equipment-support section. Subsequent improvement in fabrication of the equipment sections has resulted in 20% weight savings over aluminum without sacrificing strength or rigidity. The makeup of the Trident I missile (Fig. 7.39) may reflect the shape of the future for many vehicle structures. As much as 60% of the present Trident structural weight is made up of composites and nonmetallic materials. Studies are expected

**FIG. 7.39**   Trident I missile, containing 60% nonmetallic materials. *(Hercules, Inc.)*

to result in even higher nonmetallic content in later versions, which will probably continue to use a lot of Gr–Ep but with higher-temperature polymers than the original epoxy resin. One possibility is bispolyimides, which could result in a system capable of withstanding 450°F (232°C) temperatures. The present system has limits of 250 to 300°F (121 to 149°C). Besides graphite the missile uses Sitka spruce laminated with Gl–Ep cloth in the nose fairing, Kv–Ep, pyrolytic graphite, and polyphenylene sulfide (PPS). The PPS matrix is reinforced with 40% glass fiber and is used in electronic packaging by injection and compression-molding techniques.

The Tomahawk cruise missile, originally all metal, now uses more composites in its numerous versions. The tail fins and elevons have been changed from aluminum to molded Gr–Ep and compression-molded Gr–Ep, respectively. A program[19,32] has evaluated adhesively bonded thermoplastic wings for the cruise missile (Fig. 7.40). The turbofan-turboshaft engine is being tested with a metal-matrix shaft. Tests are under way to evaluate the spiral-wound shaft made with alternate layers of Ti–6A1–4V and Borsic filament mat with 60 vol % of Borsic filament.

Other composite applications include remote-piloted vehicles. Wings fabricated of Gr–

**FIG. 7.40**   Thermoplastic component of Tomahawk missile wing.[19] *(General Dynamics, Convair Division, San Diego.)*

Ep material for the Firebee II drone reduced wing weight by 54% and cost by 40%. Successful flight tests have proved the flight-worthiness of the composite wing and its ability to match all strength requirements. Gr–Ep weighs half as much and offers twice the load-bearing strength of stainless steel; 22 plies of Gr–Ep over conventional honeycomb core are arranged in various orientations to add strength to the structure. The tail structure is also made of Gr–Ep and has so far been found to be in good normal condition after all test firings.

An important area for composite application is the upper stages of strategic missiles, where weight means a great deal more. Advanced-composite applications include reentry vehicles, deployment modules, and fourth stages. For example, one two-piece Gr–Ep part built as part of a hardware development program for an advanced ballistic reentry vehicle had 21% weight saving, passed all systems test requirements, and promised lower production cost. In a development program for a maneuvering reentry vehicle weight savings were 12% in the forward and 22% in the aft control section. Such vehicles see high temperatures for a longer time; consequently Gr–Ep may not perform as well as graphite in a modified polyimide matrix. In a deployment module, Gr–Ep can save cost and 26% weight over aluminum. The reentry vehicle and deployment module sit on top of the missile upper stage; graphite can save weight in both the shell and internal structure of this stage. Gr–Ep can save weight in satellite systems too. In a global positioning satellite, for instance, graphite can save 25% weight over thin aluminum. Although graphite composites are designed for satellites on a replacement basis for existing systems, the major interest is to demonstrate the technology for use on next-generation systems.

## SPACE HARDWARE

Spacecraft are high on the list of weight-critical structures. Because of the large temperature excursions in space and the need to maintain precise alignment of communication and sensor systems, dimensional stability is frequently a major requirement. Composites reinforced with graphite and Kevlar fibers have high specific strength and modulus and low coefficient of thermal expansion, making them particularly attractive for space vehicles. Because stiffness and low thermal distortion, rather than strength, are frequently the dominant considerations for spacecraft structures, the material systems used tend to differ from those used in aircraft. There is much greater use of ultrahigh-modulus fibers, whose strengths are too low for use in other aerospace applications, and laminates are frequently much thinner. Other important factors include thermal and electrical conductivity, long-term stability under vacuum and space radiation, and low outgassing. The low thermal conductivity of polymer-matrix composites removes them from consideration for some components.

### Background

Most of the applications to date have been antennas, struts, support trusses, and booms, but during the past 20 years applications have included satellites, space vehicles, and others. Composites played an important part in the Courier 1-B satellite, launched in 1960, and typify the use of plastics in satellites. The structure consisted of a Gl–Ep outer spherical shell and two Gl–Ep laminated hemispheres for attachment of solar-cell arrays. The

FIN FACING JOINT BONDLINE

Gr-Ep FACINGS

ALUMINUM INSERT
FOR ATTACHMENT

NOMEX HONEYCOMB CORE

**FIG. 7.41**   Cross section of fin construction.

Explorer I and Orbiting Astronomical Observatory had numerous Gl–Ep composite parts, most of them electrical or heat insulators or conductors (gold-plated), covers, brackets, hinges, etc. The Apollo Lunar Module contained heat-insulator Gl–Ep retainers.

The use of boron and graphite in spacecraft was extremely promising in the late sixties and early seventies, but it was not readily accepted by many designers and engineers primarily because a lack of performance data led to lack of confidence. Since there was no room for error or misapplication in space structures (they could not be sent back for repair and replacement) the tendency was to use known material whenever possible. The potential of advanced composites was so great, however, that some applications for these materials were found, e.g., the use of graphite for a small missile canard or fin, originally of aluminum. The manufacturer decided that this would provide a meaningful demonstration of advanced-composites application because the design requirements would include those typical of aerodynamic fins, control surfaces, wings, and stabilizers. Also, since this particular canard was deflection-limited at the tip, it should provide a meaningful demonstration of the high specific rigidity of graphite composites. Structural graphite was selected because of the difficult contours of the part, which precluded the use of boron, and because of superior handleability and drape in a B-stage preimpregnated state. The final design was a sandwich construction with four ply faces. Each facing was composed of an inner and an outer ply with the fibers oriented parallel to the center chord line. The two inner plies of each facing had fiber orientations of $\pm 45°$ ($\pm 0.79$ rad), respectively, to the center chord line. The fin facings were joined at the center plane of symmetry (Fig. 7.41).

Another application of the high-modulus composites was the experimental use of boron in Lunar Module struts. The original aluminum struts proved to be too flexible and were redesigned in boron, using filament-wound cross plies and longitudinal tape layers. This same construction was seriously considered for the legs of the Lunar Module, which take a severe compressive load upon landing on the moon. The very high compression strength of boron laminates would have made this material ideal for this application, but the project had advanced so far that by the time sufficient reliability information became available on boron, the legs had already been fabricated in aluminum.

## Satellites and Communication Systems

The 14.5-ft (4.4-m) reflector support truss of the Application Technology Satellite (ATS-F) was an early major structural application of advanced composites in spacecraft. High stiffness and low thermal expansion were achieved by using high-modulus Gr–Ep tape. Truss elements were circular tubes consisting of layers of high-modulus graphite tape and an outer hoop of SGl–Ep tape. Hoop layers provide transverse strength and impact resistance and meet transverse coefficient-of-expansion requirements. The glass layer minimizes transverse thermal stress arising from the large temperature excursions in space [$-256$ to $199.4°F$ ($-160$ to $93°C$)]. The use of graphite resulted in a weight saving of 50% over an equivalent aluminum truss. The truss connected the earth-viewing module containing the electronics gear to the 30-ft-diameter (9-m) parabolic antenna dish (Fig. 7.42). The satellites were fabricated successfully, truss and satellite performance has been excellent, and no problems have been encountered in the 5-year operating life. The design of the tubular members consisted of one circumferential ply of Gr–Ep followed by eleven longitudinal plies of Gr–Ep and one circumferential ply of EGl–Ep. Each truss consisted of eight tubes, 2.69 in (66.3 mm) OD by 14.33 ft (4.4 m) long, each weighing less than 8 lb (3.6 kg) (Fig. 7.43).

Composites have become the basic material for spacecraft antennas. The high-gain antenna system of Voyager illustrates some of the unique features of composites. The 12-ft-diameter (3.7-m) Gr–Ep primary reflector is one of the largest composite flight antennas ever fabricated.[20] Another example is the optical bench developed for the High Energy Astronomical Observatory Mission B (HEAO-B) satellite,[21] which required high stiffness and low coefficient of expansion. Dimensional-stability requirements for the HEAO-B optical bench, which is essentially the main body of a telescope, called for a coefficient of thermal expansion of $\pm 0.1 \times 10^{-6}$ per Fahrenheit degree ($0.18 \times 10^{-6}$ per Celsius degree). Gr–Ep was the only structural material that could satisfy this criterion. The successful design and fabrication of a dimensionally stable structure of this size with an overall length of 9.9 ft (3 m) and maximum diameter of 41 in (1041 mm) demonstrates the maturity of the technology. A major problem with the use of epoxy-matrix components for dimensionally stable structures should be noted, however. Epoxy absorbs and desorbs

**FIG. 7.42**    ATS with GR–SGl–Ep truss. *(Hercules, Inc.)*

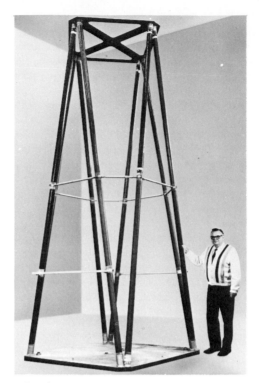

**FIG. 7.43**   ATS support truss. *(Hercules, Inc.)*

moisture, which can cause large dimensional changes. Using moisture barriers and humidity control can be effective solutions in some cases.

A sensor platform consisting of a graphite-composite core-skin sandwich structure with threaded titanium inserts was assembled in a spacecraft used for MAG SAT communications. The entire assembly was bonded, cured, and dried of moisture and then covered with a double layer of 0.0005-in-thick (0.01-mm) aluminum-foil moisture barrier to meet the primary requirements of light weight, high stiffness, and good dimensional stability. A quasi-isotropic lay-up of material was used to give the low coefficient of thermal expansion.

Two major antenna components for commercial advanced telecommunications successfully fabricated from composites and operating in space (Fig. 7.44) consist of the antenna-dish support ribs (eighteen ribs per antenna with two antennas per satellite) and the electronic package and radome support struts (six struts per antenna assembly). The antenna rib is a curved, tapered thin-wall tube about 8 ft (2.4 m) long with bonded titanium end fittings, a bonded clamshell restraint fitting, and an array of saddle reinforcements along its length. The tube consists of a four-ply lay-up of Gr–Ep. The total weight of the graphite structure is under 0.4 lb (181 g) (Fig. 7.45). The radome strut is a straight 1.125-in-diameter (28.6-mm) tube approximately 3 ft (0.9 m) long, including the bonded titanium end fittings. The tube is an eight-ply lay-up of Gr–Ep; maximum graphite weight is 0.28 lb (127 g).

Other types of satellite components have been fabricated from composites:

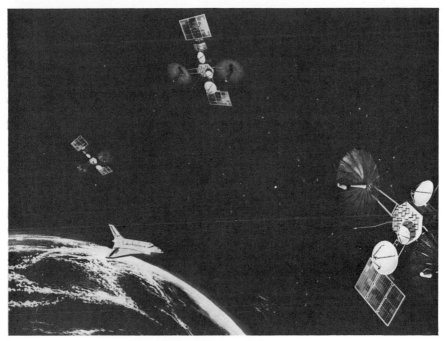

**FIG. 7.44** Tracking and data-relay satellite system (TDRSS). *(Hercules, Inc.)*

**FIG. 7.45** TDRSS antenna partially assembled. *(Hercules, Inc.)*

*Support strut in the Solar Maximum Mission Satellite:* Gr–Ep material was used to fabricate the struts, which are the primary structural members supporting the satellite payload (Fig. 7.46).

*International Ultraviolet Explorer:* Sixteen tubes of Gr–Ep were assembled as part of the optional system for the International Ultraviolet Experiment.

*Camera mount:* Space experiments in star photography used Gl–Ep prepreg for this deflection-critical structure.

*Intelsat I and IV and ANIK communication satellites:* These spacecraft used Gl–Ep composite with a lightweight aluminum-honeycomb-core sandwich for the skins of the shell structure. The shell served as a housing for the mechanical, electrical, and electronic components of the system and as a substrate for mounting solar cells. The Intelsat IV also had an elliptical waveguide constructed of Gl–Ep and aluminum face sheets stiffened with two plies of boron tape.

Space telescopes have used composites in their structures. A Gr–Ep metering cylindrical shell, about 10 ft (3 m) long and 5.5 ft (1.68 m) in diameter, was fabricated to hold the mirror assembly of a space telescope. The scope consisted of a spider assembly of four Gr–Ep spokes and a hub, which was located at the front end of the open cylinder to hold the secondary mirror. The primary mirror was at the other end. Gr–Ep was selected primarily

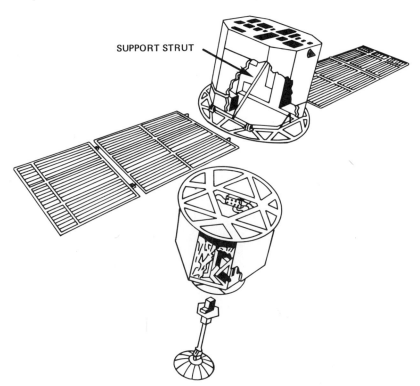

**SUPPORT STRUT**

**FIG. 7.46**   Solar Maximum Mission Satellite, showing strut location. *(Hercules, Inc.)*

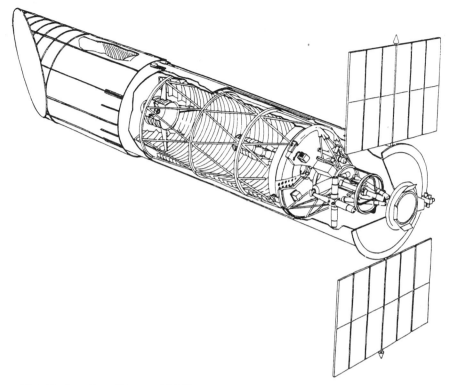

**FIG. 7.47**    Large Space Telescope (LST).[38]

for its low coefficient of thermal expansion, critical for maintaining a stable focal length for the telescope.

The Large Space Telescope used Gr–Ep in a prototype truss. The 11 by 22 ft (3.4 by 6.7 m) model was to hold the primary and secondary mirrors. A distortion-free truss was impossible using conventional metals because a temperature variation of 5 Fahrenheit degrees (2.8 Celsius degrees) would cause expansion or contraction of the truss metal and prevent proper telescope focus. The Gr–Ep was expected to reduce axial expansion in the truss to less than 80 $\mu$in (2 $\mu$m). Gr–Ep was selected because epoxy expands with heat and contracts with cold but graphite does the opposite. Carefully combined, the two materials counteract each other and the structure remains dimensionally impervious to temperature changes (Fig. 7.47).

Other radio telescopes have been launched with antenna arrays 1500 ft (457 m) long. These Explorer satellites contained Gl–Ep and B–Ep inertia booms designed to stabilize the spacecraft during initial orbital operations. The use of composites on spacecraft is growing steadily, as experience and confidence are gained. Satellites flying today contain over 4000 composite parts. Major composite structures made from Gr–Ep and Gr–Kv–Ep hybrid materials include 11 antennas, antenna-support truss structures, and solar arrays. In other satellite applications and spaceborne parabolic antennas where thermal distortion or high specific stiffness was a major consideration Gr–Al and Gr–Mg proved most effective.

Structures studied included a satellite equipment-support shelf, an L-band planar-array antenna, a segmented mirror surface for an optical telescope, a furlable semilenticular antenna rib, antenna-feed support boom, payload-support struts, and a tubular strut support system for a large reflector. Performance benefits evaluated included increased reliability, structural stability in orbit, freedom from contamination, resistance to laser energy, and packing efficiency for shuttle transport. The combination of benefits attributed to Gr–Al and Gr–Mg exceeded those of any other class of materials. In general, Gr–Mg demonstrated a definite advantage over all materials for such space structures.

## Space Shuttle Program

Technological developments were instituted during the early phase of the Space Shuttle program to search out feasible approaches. It was obvious that structural weight fractions approaching those of nonreusable space vehicles had to be achieved for the Shuttle to be a viable system. Composite materials offered a promising new technology, and a number of studies determined potential weight savings and design approaches.

Since the Shuttle is an aircraft as well as a spacecraft, its center of gravity during return from orbit must be far enough forward to provide proper aerodynamic trim. Since the engines and the heavy thrust structure, all located aft, tended to create a problem in this respect, many of the early studies were aimed at reducing weight aft (as well as overall). Examples of composite structures included truss structures representing one planar section of a booster-thrust assembly fabricated from B–Ep tubes and titanium end fittings. Another structural specimen panel was fabricated representing a portion of fuselage structure and fabricated primarily with Gr–Ep composite. Selective composite reinforcement represented another avenue to lower weight. In this case, the basic structure was metal, but full design strength and stiffness were achieved only after bonding composite tape to the metal members. Typical fuselage frame members of titanium were reinforced with B–Ep tape.

The secondary structure was also investigated for composite applications. One was the landing-gear door. A typical door design used Gr–Ep skins and aluminum-honeycomb core. A door sector consisting of two hinges, latches, and an actuator fitting was fabricated and tested under three loading conditions and was failed at 230% of limit load. This program demonstrated the feasibility of meeting Shuttle design conditions for a composite door with an estimated weight saving of 30% over a comparable all-titanium door.

A special feature of Space Shuttle is the use of high-temperature-resistant materials around the exterior of the craft to protect it from the intense heat generated during reentry. The coating consists of reinforced C–C composite on the nose cap and the leading edges of the wings, subjected to temperatures of about 2700°F (1482°C) during reentry. The nose-cap cover is fabricated in one piece and the parts for the leading edges in sections. All these C–C parts are fabricated of resin-impregnated graphite cloth, which is laid up in layers. Each part is heated to pyrolyze the resin, then reimpregnated, this time with alcohol. The pyrolysis–alcohol-impregnation process is repeated several times to drive off the resin, which would otherwise pyrolyze during reentry. The stiff C–C parts are attached to the vehicle's relatively flexible skin by a floating lug-and-pin mechanism. This prevents the graphite from breaking as the wings deflect during flight.

The wings were initially designed as an aluminum structure 30 ft (9 m) long from tip to fuselage and 5.5 ft (1.68 m) thick. In order to bring down weight spars in the torque box, which is the main-load carrying part of the wing, were switched to Gr–Ep. The four aluminum spars in that box originally had a corrugated-web design; now only one is aluminum, and switching the others to Gr–Ep saved 100 lb (45.4 kg).

Advanced thermoplastic composites that offer distinct performance advantages over thermoset composites (notably elongations to break in the 20 to 30% range, against only 0.5 to 7% for thermosets) are being considered for construction in space. Although the project may sound like science fiction, technology now is available. The system being developed is for the automated production of a composite truss beam by a modified pultrusion process. In the Space Shuttle the process will supply beamstock for the construction of large orbiting structures. The system is in fact a blend of two systems, one for producing a complex-geometry C–PS ribbon on earth and another for converting it into beamstock in orbit.

In the first system (Fig. 7.48) prepreg carbon tow is pulled from creels through a carding plate and drawn longitudinally onto a stationary mandrel; then a ring-shaped winding head applies additional carbon fiber, laying it down in a 45° (0.79-rad) pattern around the longitudinal tow "tube." The wound prepreg tube passes through an induction-heating station that melts the polysulfone at around 650°F (343°C). The molten tube is then squashed flat and consolidated into ribbon form by continuous opposed-belt laminating. The finished ribbon, which is approximately 8 in (203 mm) wide and 0.030 in (0.76 mm) thick, is taken up on a storage reel.

In the second orbital-processing system, 8 in (203-mm) ribbon is processed through a *beam builder,* basically a pultrusion system that makes use of three inductively heated heat pipes to raise the temperature of three separate ribbons to 600°F (316°C) at the fold lines for subsequent forming into open-ended, triangular corner members of the beam. Forming and guidance tooling are used to accomplish this last task, after which the unclosed members pass through a final heat die that welds lap seams, closing the triangles. From here the members enter a short chill-die section, where they are radiantly cooled to about 500°F

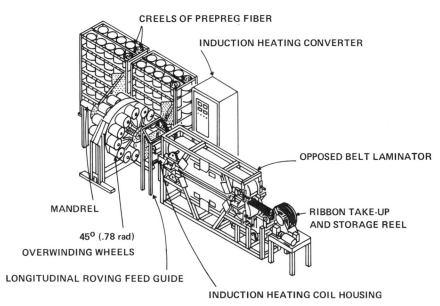

FIG. 7.48   C–PS ribbon-making system. *(Goldsworthy Engineering.)*

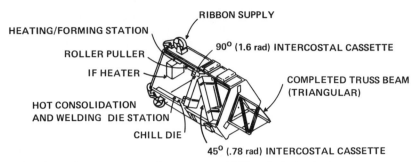

**FIG. 7.49** Orbital-reprocessing pultrusion system. *(Goldsworthy Engineering.)*

(260°C), and solidify into their final structural form. Downstream of this, corner members are joined to make the completed truss beam by intercostal struts, set at 90 and 45° (1.6- and 0.79-rad) angles, which are dispensed from cassettes, like bullets from a clip in the magazine of a gun (Fig. 7.49). Throughput rate of the system is 2 to 6 ft/min (0.61 to 2 m/min), and each triangular corner member is about 2.5 in (64 mm) wide. The triangular member is stronger than an equivalent aluminum structural member, is 30% lighter, and has zero expansion and contraction in the drastic temperature changes of space.

Although the beam builder is an exotic application of advanced thermoplastic composites, it points the way to more mundane uses. Polysulfone won out over other thermosets for the truss beam largely because of its relative ease of processing. Thermosets are tough to control, especially in the vacuum and temperatures of space. The simple heat-chill cycle required to form the polysulfone dispensed with the difficult handling, long cure times, and outgassing constraints in space associated with thermosets. The same rapid and relatively simple processing operation that will take place on board the Space Shuttle can greatly improve the productivity of the composites industry here on earth.

## Space Shuttle

The Space Shuttle itself uses advanced composites in several areas, resulting in a 3600-lb (1633-kg) weight reduction over the initial all-metal design. The components include Gr–Ep skins and Nomex honeycomb on the orbital maneuvering system and titanium I beams and tubes reinforced with B–Ep in the aft thrust structure. Titanium and Inconel tanks overwrapped with Kv–Ep is an advance in pressure-vessel technology; two applications merit special attention, the midfuselage tubular struts and the payload-bay doors.

The Space Shuttle was one of the first production applications of MMC. It has 242 unidirectional B–Al circular tubes, which serve as main-frame and rib-truss struts, frame-stabilizing braces, and nose-landing-gear drag-brace struts, resulting in a 44% weight reduction over aluminum extrusions. The B–Al tubes which help support the fuselage frame of the Space Shuttle (Fig. 7.50) have 92 different configurations and weigh 330 lb (150 kg). In addition to the 320-lb (145-kg) direct weight saving, use of B–Al provides other benefits. The lower thermal conductivity of the composite results in lower heat flow into the equipment and payload bays, reducing thermal-insulation requirements, and the smaller tube diameter permits greater vehicle access.

The other notable composite applications on the Shuttle are the payload-bay doors,

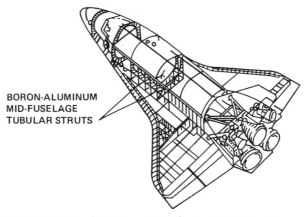

BORON-ALUMINUM
MID-FUSELAGE
TUBULAR STRUTS

**FIG. 7.50** B–Al tubes in Space Shuttle fuselage. *(Mater. Eng., January 1980, p. 66.)*

which are the largest Gr–Ep structures ever built. The right- and left-hand doors are each 60 ft (18.3 m) long. To allow for a large difference in thermal expansion between the graphite doors and aluminum fuselage there are four expansion joints in each door. The door-panel skins are Gr–Ep fabric and tape face sheets over a Nomex honeycomb core. Most of the 3200 lb (1452 kg) of Gr–Ep is contained in the cargo bay's huge double doors; Gr–Ep was chosen because its low coefficient of thermal expansion prevents the doors from warping in orbit due to asymmetrical heating (Fig. 7.51).

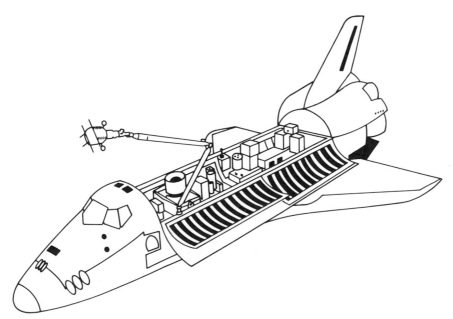

**FIG. 7.51** Space Shuttle cargo-bay doors. *(Aviat. Week Technol., Mar. 1, 1982, p. 54.)*

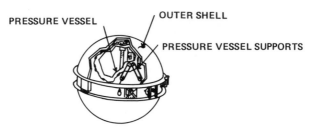

**FIG. 7.52**   Filament-wound composite straps in Space Shuttle Orbiter. *(Mater. Eng., November 1980, p. 46.)*

Other composite applications in the Shuttle include the remote manipulator system, a mechanical arm used to deploy and retrieve payloads from the Shuttle's cargo bay and for servicing in space or after return to earth (Fig. 7.51). The arm is a 905-lb (411-kg) Gr–Ep tubular boom, 50 ft (15 m) long and 15 in (381 mm) diameter, equipped with joints at the shoulder, elbow, and wrist for rotating in the yaw, pitch, and roll modes. It cannot support its own weight on earth, but in orbit it can handle a 65,000-lb (29,484-kg) payload 15 ft (4.6 m) in diameter and 60 ft (18.3 m) long.

The aft body flap for the Shuttle is being built in a development program of Gr–PI and will be extensively tested in ground tests before use in flight. Cryogenic storage tanks for the Shuttle provide oxygen and hydrogen for electric-power generation and oxygen for life support. The inner pressure vessel contains the stored fluid at cryogenic temperatures and supercritical pressures. The outer shell maintains a high vacuum between the two spheres for thermal insulation. The tanks are mounted in the Shuttle from the outer shell. Low-thermal-conductivity straps supporting the pressure vessel from the outer shell are located in the vacuum space between the two spheres (Fig. 7.52). SGl–Ep filament-wound straps were selected because tests proved that this material minimizes heat conduction from outside the tank into the stored cryogen. The low thermal conductivity and high strength of SGl–Ep filament-wound composite material has excellent properties from 250 to −452°F (121 to −269°C).

## Other Space Systems and Components

Recent work sponsored by NASA may be the groundwork for the world's first large platform in space. The program is to determine the structure of an Orbiting Space Station. The slender tubular struts are made of Gr–Ep, 8.5-ft (2.6-m) tapered elements that snap together to form one 17-ft (5.2-m) column.

Large space platforms have been proposed for such applications as communications satellites, megawatt power modules, large antennas, and manufacturing facilities. The structural column members would be carried into orbit by the Space Shuttle and then assembled in gravity-free space, rather like Tinker Toys. Operational space platforms may require half-columns over 33 ft (10 m) long. Column fabrication begins when dry graphite fiber is wound around a hot, tapered steel tube set in a vertical winding machine. Aluminum fittings are mounted on both ends of the tube, and resin is then injected between the tube and an outer sleeve of steel. After, the tube is cured at 240°F (116°C) and allowed to cool, the finished product is removed. The tapered design of the columns allows them to be stacked in the Shuttle's cargo bay like paper cups. The Orbiter could deliver about 5000 of the 33-ft (10-m) columns into space each flight (Fig. 7.53).

Another study is developing the data base for the design and analysis of advanced-composite joints for use at elevated temperatures [550°F (288°C)]. The objectives are to identify and evaluate design concepts for specific joining applications and to identify the

fundamental parameters controlling the static strength characteristics of such joints (Fig. 7.54). The results will provide the data necessary to design and build Gr–PI lightly loaded flight components for advanced space transportation systems and high-speed aircraft[22] (Fig. 7.55).

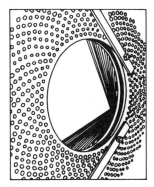

Many advanced spacecraft systems, e.g., Space Lab and Space Tug, proposed as payloads on the Space Shuttle must be designed to resist critical stiffness loads and meet stringent strength and weight criteria. Designs using Gr–Ep material are being proposed to minimize structural weight. A promising concept for spacecraft body struc-

FIG. 7.53   Space columns for erecting platforms in space. (*Mater. Eng., August 1980, p. 30.*)

ture is a grid-stiffened skin with skin laminate configuration and stiffener-grid geometry selected to best suit the design requirements. The basic idea of a grid-stiffened composite panel is that a relatively thin skin is reinforced with a gridwork of stiffeners so that the overall panel can resist design loads without becoming structurally unstable or being overstressed. The advantage of being able to use Gr–Ep in the design of these panels is that the skin laminate can have preferred stiffness and strength directions and the stiffeners can be designed to be structurally efficient by using a high percentage of axially oriented fibers. The designer has many options, including the type of Gr–Ep material, skin-laminate configuration, and stiffener-grid pattern. The particular Gr–Ep orthogrid panel design shown in Fig. 7.56 forms part of a relatively large Gr–Ep shell structure representative of spacecraft body structure which is 157 in (3988 mm) high and 146 in (3708 mm) in diameter. The main feature of the design is that low-cost structural panels will be possible when initial tooling costs can be amortized over a large quantity of panels.

Most space vehicles require storage of gases and fluids in tanks under high pressure and at cryogenic temperatures. The Space Shuttle, for example, uses about 60 of these tanks, several of which have a Kevlar-filament overwrap with a cryoformed 301 stainless-steel liner. These tanks are 25-in-diameter (635-mm) pressure vessels operating at 3 kips/in$^2$ (21 MPa). The tanks save an estimated weight of 438 lb (199 kg).

A little known and less publicized application for composites was part of a lunar surface drill carried to the moon by astronauts during Apollo missions 15 to 17. Drills were connected together to penetrate to a depth of 10 ft (3 m). The tubular drill bore stem was constructed of B–Ep and Gl–Ep combinations, and a tungsten carbide drill bit was used at the end of the stem. A combination drilling and impacting action was used to fracture the soil and rock layers.

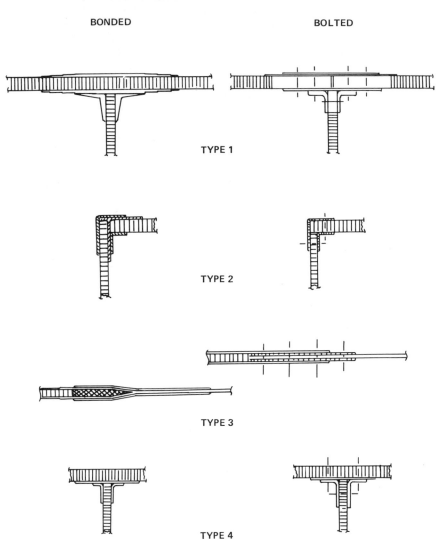

**FIG. 7.54**   Generic joint concepts for four attachment types.[22]

## AEROPROPULSION

Fiber-reinforced composites can improve the performance of both sub- and supersonic air-breathing propulsion systems. A wide variety of turbine-engine components have been designed, fabricated, and tested. The results have been mixed: high-risk, high-payoff, rotating blades demonstrate aerodynamic performance improvements but are constrained by their lack of ability to withstand FOD; static structures such as ducts and frames offer weight savings but are constrained by the interface requirements of the metallic structures being replaced. The use of composites for most of the static structures in a pair of large

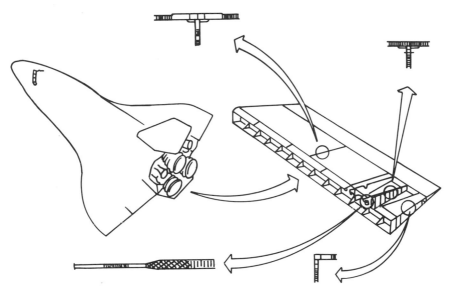

**FIG. 7.55** Component with Gr–PI for future spacecraft.[22]

subsonic high-bypass-ratio energy-efficient turbofan engines has been demonstrated. Both weight and cost savings are possible on external nozzle flaps for a supersonic turbofan engine, and a flight-test evaluation of them is planned. Other programs are pursuing turbofan duct structures for the next generation of engines. Clearly, in the next 10 years, designers of the composite aircraft on the drawing boards should consider composite engines.

To help us review the potential demonstrated by the use of composites in engines to date Table 7.7 divides an engine into the major rotating and nonrotating components.

## Nonrotating Parts

The application of composites to nonrotating parts in some engines goes back 10 to 12 years. Gl–Ep composites have proved their serviceability in components which have logged more than 12 million hours of flight time.

### *Cases, Frames, and Containment*

Fiber-reinforced composite fan frames have been demonstrated for several different engines. The first example was fabrication of a simulated front fan frame for a supersonic

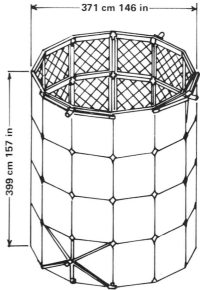

**FIG. 7.56** Lightweight shell structure.[21]

**Table 7.7**  Major Engine Structural Components

| Nonrotating | Rotating |
|---|---|
| Containment cases | Blades |
| Ducts | Disks |
| Vanes | Shafts |
| Nozzle components | |

**Table 7.8**  Components of Composite Engines

| Part | Composition |
|---|---|
| Inlet | Kv–Ep fabric with aluminum honeycomb |
| Fan frame | Gr–Ep tape |
| Outer duct | Gr–Ep tape and Kv–Ep fabric |
| Flaps | Gr–Ep tape and Kv–Ep fabric |
| Inner duct | Gr–PI fabric |
| Blades | Gr-Ep, SGl-Ep, Kv-Ep fabric, and B-Ep |

turbofan engine, followed by two full-size fan frames, part of NASA's quick, clean, short-haul experimental engine (QCSEE) program. Table 7.8 lists the composites used in other components fabricated. The fan frame, which supports the core engine weight, was built with Gr–Ep and is one of the largest, most complex engine components yet fabricated from advanced-composite materials. Designed for a maximum stress level of over 100 kips/in$^2$ (690 MPa), the frame can survive the unbalance forces resulting from loss of five fan blades at one time. In a flight engine, it would weigh only 474 lb (215 kg), about 20% less than an equivalent metal frame. The success of this frame design and its subsequent test program has given designers the freedom to redesign the attachments to the surrounding structures, with the likelihood of even greater advantages in using composite frame structures.

Another application for composites is the containment rings used in transport engines to prevent the fan blades from penetrating the fuselage in the unlikely event that they are severed from the disk or the disk shatters, releasing them. Multilayer Kv–Ep cloth is the most efficient material for containment, and it is possible to integrate containment materials directly into the composite frame.

## Ducts

Ducts are perhaps the most obvious application of composites, since they are roughly cylindrical in shape, closely approximating the geometry of the rocket-motor cases that have for years been made of fiber-reinforced composites. The first major application for ducts was in the augmentor section of supersonic turbofan engines and the inner and outer ducts for production turbofan engines. Figure 7.57 shows the Gr–PI outer duct. The Gr–PI material was developed to withstand temperatures of over 500°F (260°C). Ducts are normally complex parts requiring acoustic treatment and hinges and hardware for attachment to the engine support pylon. Duct skins were processed in an autoclave at 525°F (274°C) and 0.1 kip/in$^2$ (0.7 MPa) with a postcure at 600°F (316°C). In tests Gr–PI ducts have been exposed to engine operation at temperatures exceeding 500°F (260°C) with no deleterious

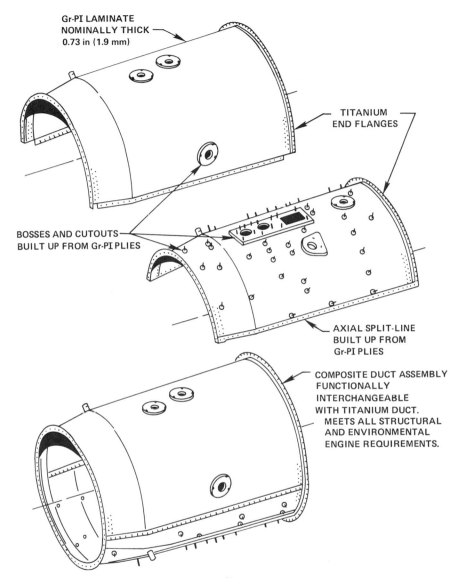

Gr-PI LAMINATE
NOMINALLY THICK
0.73 in (1.9 mm)

TITANIUM
END FLANGES

BOSSES AND CUTOUTS
BUILT UP FROM Gr-PI PLIES

AXIAL SPLIT-LINE
BUILT UP FROM
Gr-PI PLIES

COMPOSITE DUCT ASSEMBLY
FUNCTIONALLY
INTERCHANGEABLE
WITH TITANIUM DUCT.
MEETS ALL STRUCTURAL
AND ENVIRONMENTAL
ENGINE REQUIREMENTS.

**FIG. 7.57** Composite outer duct of F404 engine.[24]

effects. This duct operates at lower temperatures than the augmentor duct but is far more complicated structurally, requiring many bosses and cutouts to interface with engine hardware mounted on, or passing through, the metallic duct. The F404 composite duct uses Gr–PI for the bosses on the duct walls; the augmentor duct mentioned earlier uses titanium bosses. Integration of the fabrication and attachment of these peripheral structures with that of the duct shells obviously provides a significant reduction in the overall cost of duct fabrication.

## Vanes

Several different types of vanes have been developed in the course of applying composites to turbine engine components. Both organic- and metal-matrix vanes have been fabricated. The most successful organic-matrix vanes were produced for the fan-exit guide vanes of a commercial turbofan engine and were actually in production before trailing-edge fatigue problems forced their recall and eventual elimination from the engine. Subsequent redesign efforts developed a suitable modified vane, but the weight savings were not enough to offset the added cost compared with that of aluminum vanes. Significant improvements in the fabrication of organic-matrix composite vanes would have to be coupled with lower-cost materials for them to be cost-competitive today. Gr–Ep material for fan-exit guide vanes for the JT-9D engine appears to be the material for the future. Current metal-matrix vane activities result principally from the fast consolidation technology developed for the fabrication of B–Al blades. The key to production for B–Al vanes is the attachment requirements for vanes and the configuration control needed to join the vanes to the fan structure.[23]

Future utilization of composite vanes will depend upon using low-cost SiC fiber, single-step rapid-consolidation bonding, and $B_4C$–Gr. This new composite material offers an alternative to the brittleness and imperfections of conventional ceramic parts and opens the way for a new generation of gas-turbine materials. A $B_4C$–Gr composite with a chemical-vapor-deposited SiC film on its surface can withstand over 2500°F (1371°C). The SiC film [0.0001 in. (0.0025 mm)] deposited on the composite effectively seals it and protects it against oxidation at high temperatures. Continuous boron carbide filament tapes are stacked in the desired direction to assure maximum strength with the direction of stress. A thermosetting resin binder is converted into the carbon matrix by heating; matrix density can be increased by reimpregnating the binder and repeating the cycle. A key element in integrating composite vanes into the composite engine will be designing the vane attachments to reduce the cost of the composite vanes to the minimum.

## Nozzle Components

Gr–PI external nozzle flaps for a supersonic turbofan engine have been built to solve a fatigue-cracking problem induced by airloads on the titanium flaps. The composite flaps must be fully interchangeable with their metallic counterparts. The nozzle flaps operate in a very severe high-sonic-fatigue, high-dynamic-airload, high-temperature environment [634°F (334°C) maximum service temperature]. Over 28 h of flight and 100 h of testing on flight aircraft have been accumulated, and results to date are encouraging. The use of Gr–PI material in this application results in a cost saving and a 15% weight saving over Ti–6Al–4V.

The flaps are attached to the actuators and hinge points by small titanium hinges. The flaps are both lighter and less costly than their metallic counterparts while providing increased durability. Automation of the fabrication (compression molding) of nozzle flaps is currently under way.

## Rotating Parts

Fan blades have been the focus of attention for the application of advanced composites ever since the early development work sponsored by the government demonstrated the weight savings and improved aerodynamic performance possible with the use of composite blades.

## Blades

The key to composite blades has always been the concern for the lack of FOD resistance. Understanding the effects of FOD is the most critical problem in applying composite technology to engine systems. Major programs under government supervision have been conducted in the areas of organic- and metal-matrix composite blades for both commercial and military applications. The general results have been mixed at best. The problem is the unpredictable behavior of a blade when struck by a foreign object. Blade behavior in tests differed from predicted analytical data and design. This early inability to predict FOD capability of composite blades and the resultant lack of design confidence have brought designers to the point where the only way to establish a new composite blade design with confidence is the classic approach of building it and breaking it.

Organic-matrix fan blades have been only marginally successful. The QCSEE fan-blade design is unique. The hybrid blades, which are made of Gr–Ep, Kv–Ep, glass, and boron fibers, have met both engine-stress and aerodynamic-performance requirements. The primary difficulty has been in developing sufficient resistance to damage from objects such as birds. New Gr–PI materials have been fabricated into blades (Fig. 7.58), and preliminary FOD tests show that they can survive large-bird FOD testing with a minimum loss of blade-tip material.

Metal-matrix blades offer significantly more promise in that some have already passed the FOD requirements for their design point, namely, the Borsic fiber *selectively reinforced* (SERIF) titanium blades for the TF-41 engine. SERIF blades have been approved for accelerated mission testing on a ground-based engine.

Some B–Al blades have passed single-blade rotating tests only to fail multiple-blade multiple-bird-ingestion tests. Other B–Al blades have passed FOD tests and then failed rotating-rig fatigue tests. The JT8D and JT9D turbofan engines [first-stage fan 40 in (1016 mm) in diameter] equipped entirely with B–Al composite blades were successfully tested

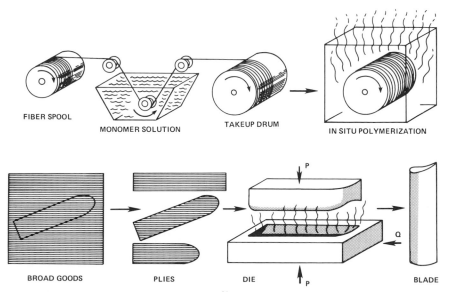

FIBER SPOOL

MONOMER SOLUTION

TAKEUP DRUM

IN SITU POLYMERIZATION

BROAD GOODS      PLIES      DIE      BLADE

**FIG. 7.58** Polyimide process for blade fabrication.[24] *(Courtesy of NASA-Lewis Research Center.)*

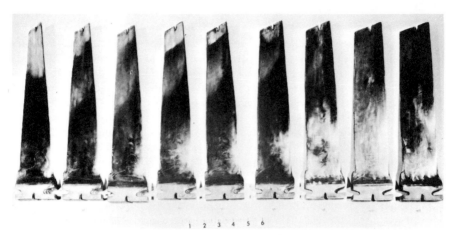

**FIG. 7.59**   B–Al blades before machining. *(General Electric and NASA-Lewis Research Center.)*

for 2 h testing thrust levels during takeoff. The blades were fabricated from tape containing a layer of parallel SiC-coated boron fibers interspersed with plasma-sprayed aluminum and backed with aluminum foil. Fan blades fabricated from this material were flight-tested in the third stage of the TF-30-P9 engine. Other B–Al composites for first-stage compressor blades for the J-79 engine (Fig. 7.59) were fabricated by hot-pressing monotape preforms. This technique uses a protective reproducible outer sacrificial (PROS) sheet process[24] (Fig. 7.60).

Current development programs for metal-matrix blades are using both aluminum and titanium matrix materials, boron carbide–coated boron fibers, and SiC fibers. First-stage fan blades have been fabricated from the Be–Ti composite material. One type of blade was made from a single extrusion of beryllium and Ti–6Al–6V–2Sn alloy powders; the other consisted of multiple extrusions of beryllium rods in wrought Ti–6Al–4V alloy sheaths. The tensile strengths of the beryllium rod–Ti composite blade ranged from 90 to 91 kips/in$^2$ (621 to 627 MPa) compared with 46 to 47 kips/in$^2$ (317 to 324 MPa) for the Be–Ti powder-metal composite blade. The weight of the blades is about 9% less than comparable titanium blades. The material discussed above is similar to the material shown in Figs. 2.44 and 2.45.

Other materials-development programs involve selecting a fiber-reinforced titanium material for high-tip-speed fan or compressor blades for transonic or supersonic engines.[25] Typical fibers include B–SiC (SiC-coated boron), B–B$_4$C (B$_4$C-coated boron), and SiC.

Finally one other significant material system being developed is iron-chromium-aluminum-yttrium (W/Fe–Cr–Al–Y) alloys reinforced with tungsten fibers for turbine blades in aircraft engines. Aircraft engines thrive on heat since higher temperatures mean better performance and greater efficiency. This explains the constant quest for more heat-resistant materials, an effort that since the mid-1960s has brought forth directional solidification, dispersion strengthening, dispersion strengthening combined with precipitation hardening, fiber-reinforced metals, and directionally solidified eutectics as current or potential alternatives to conventional superalloys.

The new W/Fe–Cr–Al–Y alloy is a promising alternative. A first-generation composite of this type seems capable of permitting temperatures at least 90 Fahrenheit degrees (50 Celsius degrees) greater than those now possible with conventional superalloys. Conceiva-

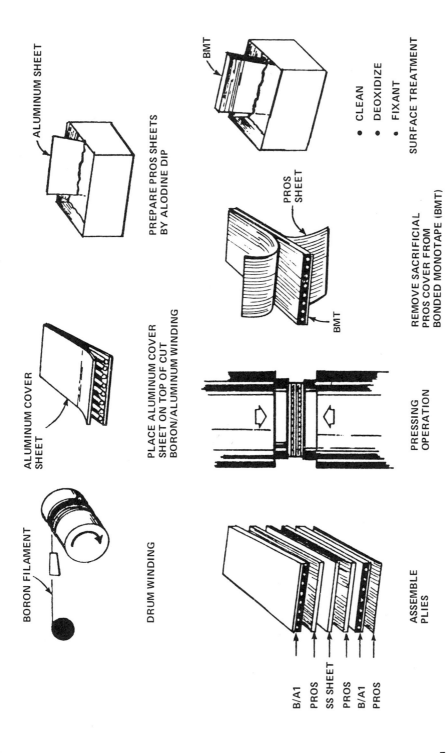

BORON FILAMENT

DRUM WINDING

ALUMINUM COVER SHEET

PLACE ALUMINUM COVER SHEET ON TOP OF CUT BORON/ALUMINUM WINDING

ALUMINUM SHEET

PREPARE PROS SHEETS BY ALODINE DIP

BMT

- CLEAN
- DEOXIDIZE
- FIXANT

SURFACE TREATMENT

PROS SHEET

BMT

REMOVE SACRIFICIAL PROS COVER FROM BONDED MONOTAPE (BMT)

PRESSING OPERATION

ASSEMBLE PLIES

B/Al
PROS
SS SHEET
PROS
B/Al
PROS

FIG. 7.60   Schematic of preparation of B–Al bonded monotapes by the PROS sheet process.[24] (*General Electric and NASA-Lewis Research Center.*)

7.59

bly, temperatures 180 to 270 Fahrenheit degrees (100 to 150 Celsius degrees) greater may be possible when fiber and matrix materials are optimized. The main function of the Fe–Cr–Al–Y matrix is to bind the fibers and protect them from oxidation and hot corrosion in the turbine environment. The matrix must be relatively ductile to transfer applied loads to the fibers. It must also be capable of redistributing local stress concentrations and resisting abrasion and impact from environmental debris. Of course, the matrix material and the fiber material must be chemically compatible and able to coexist without affecting each other in any way that might degrade performance of the composite. Ideally, the material should also have a high melting point; low density; thermal-expansion characteristics compatible with the fiber material; high creep strength; low strain-hardening rate; and resistance to thermal fatigue, oxidation, and corrosion and be easy to fabricate.

Of the many processes tried for converting this unusual material into complex components, such as hollow airfoils for turbine blades, the most promising involves diffusion-bonding of monolayer-composite plies of aligned tungsten fibers sandwiched between layers of the matrix alloy (Fig. 7.61). This approach permits the fibers to be oriented in both uniaxial and off-axis directions, prevents oxidation (precluding the need for protective coatings), minimizes reaction between the fibers and matrix alloy during fabrication (excessive interdiffusion, for example, would degrade the material's structural performance), and permits (when necessary) incorporation of integral passages for cooling and/or weight reduction.

Further, the process can be used to produce complete blades (airfoils and root sections) to near net shape. The only other operations required are machining the root and touch-up grinding. Solid fan and compressor blades have been made by this method, of boron-

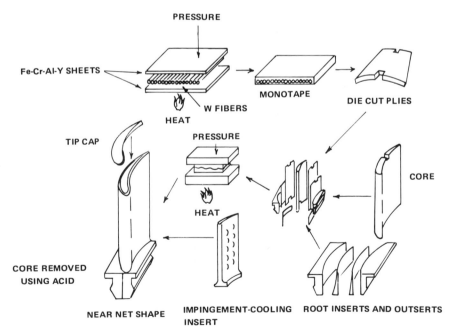

**FIG. 7.61** Schematic of fabrication of monotapes of Fe–Cr–Al–Y sheets and tungsten fibers. *(Am. Mach., November 1981, p. 147.)*

fiber-reinforced aluminum and boron-fiber-reinforced titanium, respectively. A similar approach has been used to form a prototype blade for an advanced gas turbine. The composite blade and the monotape method are much less costly for advanced turbine blades than using directionally solidified eutectic alloys, also being considered for future blades.

### Disks

From the very first development efforts disks seemed to be a natural application for composites. A major problem was associated with geometrical constraint imposed by the requirement to carry blades in the disk, i.e., the problem of designing a composite disk that would provide this load-carrying ability in the same space and with essentially the same attachments as the metal disk being replaced. It was virtually impossible for the outer fibers in the disk rim to transfer any of the load to the inner fibers, which were not in as high a strain field. As a result all work on all fiber-reinforced disks stopped except for some limited applications. In one of these a B–Ep composite replaced titanium in a turbine-engine compressor disk with a weight savings of over 23%. The boron fibers were filament-wound into the outer rim of the disk. The high-modulus, high-strength reinforcement eliminated the bore and web sections of the part, reducing overall weight. Most engineers believe that the prospect of a disk that would be extremely fatigue-resistant and significantly lighter should elicit enough interest to be part of an overall effort to develop an engine composed principally of fiber-reinforced-composite structures.

### Shafts

Manufacturing methods for fiber-reinforced metal-matrix shafts are currently being established for small turbofan and turboshaft engines. Fiber-reinforced shafts have higher specific stiffness than the conventional steel shafts and eliminate the intershaft bearings required with conventional shafting. The type of shaft being considered for MMC (Fig. 7.62) is a thin-wall steel tube with integral splines on both ends and a $B_4C$–B fiber-filament-reinforced titanium-matrix composite cylinder metallurgically bonded to the steel-tube inner diameter. Since the fibers in the titanium matrix run parallel to the axis of the shaft, the composite is used exclusively to improve shaft bending stiffness.

A second fabrication program is currently evaluating the use of Ti–Bsc composite drive shafts. The Borsic filament mat is spirally wrapped in alternate layers with titanium foil around a mandrel to form the MMC shaft assembly, which is subsequently welded to form the full shaft rotor.

Shafts are a relatively new area for composite use, and the next logical step is development of reinforced shafting for large engines. To be most effective, this should be carried out together with the development of at least composite blades and preferably of composite disks as well. Certainly the dynamic interaction of the blade-disk-shaft system will need additional, coordinated development if the use of composites in engines is to be optimized.

## Miscellaneous Engine Components

Composite materials have been substituted for other engine components. Prototypes have been fabricated with limited success while extensive testing still remains.

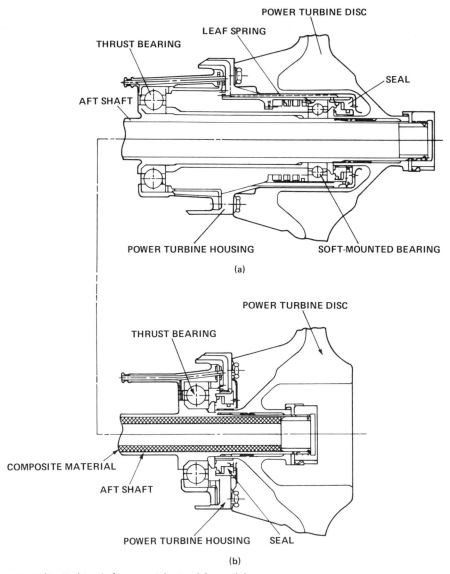

POWER TURBINE DISC

LEAF SPRING

THRUST BEARING

SEAL

AFT SHAFT

POWER TURBINE HOUSING          SOFT-MOUNTED BEARING

(a)

POWER TURBINE DISC

THRUST BEARING

COMPOSITE MATERIAL

AFT SHAFT

POWER TURBINE HOUSING     SEAL

(b)

**FIG. 7.62**   Turbine shaft: present *(above)* and future *(below)*.

## Gear Case

B–Ep was substituted for magnesium and aluminum in the case structure of a turboprop reduction-gear assembly. The composite gear case provided high strength, specific stiffness, light weight, good corrosion resistance, and suitability for molding. Although the conventional magnesium case has reasonable strength and low density, it has a low modulus and resistance to corrosion; and although aluminum has sufficient strength, its specific stiffness (similar to that of magnesium) often results in a heavy case.

At first short boron fibers [1 in (25.4 mm)] were used in the composite, giving essentially an isotropic material because the fibers are randomly oriented. Later a more directional, anisotropic material with controllable properties was chosen. It consisted of collimated continuous 0.0004-in (0.01-mm) boron filaments, glass cloth, and 33 wt % of epoxy. Tests after compression molding showed a 2:1 stiffness improvement in favor of the B–Ep hub over magnesium. Weight was cut 13%, and stresses never exceeded 42% of allowables. Redesign and testing are still required to overcome the stresses produced in some areas of the case, which have limited its operating temperature to 150°F (66°C).

## Seals

As gas-turbine engine temperatures continually rise, the need increases for seal applications to keep ahead. Recent work evaluated the feasibility of SiC composite structures for seals for 2500°F (1371°C) temperature. The material was a Si–SiC composite with sintered SiC substrates, both with attached surface rub layers containing BN as an additive.[26] The SiC composite structures showed good resistance to abrasion, oxidation, gas erosion, thermal shock, and ballistic impact damage for gas-turbine seal applications up to 2200°F (1204°C). For 2500°F (1371°C) applications, further improvements in resistance to hot-gas erosion and abrasion are required. Finally the design data base must be expanded to include

Environmental effects on properties

Sufficient long-time property data

Attachment of silicon-based ceramic components to nickel- and cobalt-based superalloy components

Chemical reactions

A major concern is the chemical reaction between the silicon of the ceramic composite and nickel and cobalt of superalloys, resulting in surface pitting of ceramic components and formation of low-melting silicides.

## Inlet Particle Separator

A new challenge is the interest in applying composites to the inlet particle separator of the GE T700 engine used to power several helicopters. This component is especially difficult because of the high pressures and high temperatures of the anti-icing air under certain engine operating modes. Current plans are to fabricate the particle separator from a combination of composite and metal components so as to take maximum advantage of the material properties of each. Unidirectional graphite will be incorporated to handle the high pressure, and polyimide resin will be used to withstand the 530°F (277°C) maximum temperature condition.

## New Materials Developments

The major limiting factor in the application of composites is the matrix material. Epoxy matrix materials have certain limitations in both processing and properties, two main performance limitations being temperature and moisture. The upper temperature limit for

Gr–Ep is about 300 to 350°F (149 to 177°C), but a temperature capability of 600°F (316°C) would be highly desirable for use around engines and higher-performance missiles. Or at least 400 to 450°F (204 to 232°C), which for many aircraft structures would meet any type of future application. Moisture pickup affects epoxy performance. Properties are lowered, and permanent damage can result from moisture and temperature cycling combined with cycling of loads.

## Polymers

To get the higher temperatures, most attention has focused on the polyimides, which have upper use temperatures at 450 to 600°F (232 to 316°C). Acetylene-terminated polymers promise higher-temperature capabilities. Acetylene-terminated (end capped) sulfone has passed tests at 400 and 450°F (204 and 232°C). Acetylene-terminated quinoxalines are another material under evaluation which could reach the 500 to 550°F (260 to 288°C) range. Polyamide-imide and phthalocyanine are under investigation for 400°F (204°C) use in moisture-sensitive environments.

## Epoxies

Gr–Ep composites are relatively easy to process by standard aerospace lay-up, autoclave, and curing techniques, but this manufacturing method is relatively complex and costly even if automated. It also requires long cure times. Efforts are under way to find materials and processing methods that are simple, easier, and less expensive. Two approaches to an epoxy replacement are being actively pursued, high-vinyl-modified epoxies (HME) and vinyl-modified bisimide (PBBI). Both are moisture-resistant and have inherently better processing capabilities, but both have limited temperature capabilities—about 300°F (149°C) for HME and 350°F (177°C) for PBBI. The big advantage of PBBI is its processability. Since there is no bleeding during the cure cycle, no bleed plies are needed. Therefore thick parts are easily made, and simultaneous curing and bonding of component parts (co-curing) is possible.

## Thermoplastics

Thermoplastic-matrix composites are another possibility. Thermoplastics can be processed in thick sections by standard aerospace lay-up-and-cure techniques or by matched-metal-die molding. The sulfones (polysulfone, polyethersulfone, and polyphenylsulfone) have received the most attention. Other thermoplastics being evaluated include polyimide, polyamide-imide, polyphenylene sulfide, and thermoplastic polyesters (both PET and PBT). For lower temperatures, graphite-reinforced acrylic, polycarbonate, phenoxy, and polyester have been investigated for large space structures.

## Polyimides

Polyimides are the hottest of the high-temperature polymers, since they can handle up to 600°F (316°C). The family of polyimides has two main branches, those which cure by an addition reaction and those which cure by a condensation reaction, but the variations are many. In general, addition-reaction PIs (APIs) are thermosetting and process fairly easily. They have a somewhat lower temperature capability than the condensation-reaction PIs

(CPIs). CPIs are more difficult to process in thick aerospace sections but can handle higher use temperatures. These statements, of course, are sweeping generalities. Obviously API developers are seeking to improve temperature capability and CPI developers to improve processing characteristics.

Addition polymers can be processed by standard autoclave techniques in thick-section parts without voids or by compression molding. Since the boiling point of the solvent is low, volatiles are released easily, avoiding late-cure release, which causes voids in the composite. Up to 76 plies of composite material [0.5 in (12.7 mm) thickness] have been used to fabricate a void-free composite engine fan blade. Cure temperatures for both API and CPI composites are a problem. The addition type must be cured at 600 to 650°F (316 to 343°C). The condensation type requires 700°F (371°C) or higher as a final molding temperature in the autoclave. Since most autoclaves are designed for epoxy laminates, not many have a temperature capability this high. Compression molding in a matched metal die offers another option, since high pressure can be used, but high molding temperatures are still needed.

### Graphite Thermoplastics

Graphite-reinforced thermoplastic composites now offer cost savings in manufacturing operations over conventional Gr–Ep composites without significant losses in mechanical or physical properties. Thermoplastic composites can be processed by matched-metal-die compression molding, press bending, or autoclave molding. They have even been roll-formed successfully.

### Ceramics and Carbon-Carbon

The most recent additions to the reinforcements being used for engine applications are short fiber whiskers and particulates of SiC and continuous alumina fibers. Other reinforced composite materials such as reinforced ceramics and C–C offer an even broader choice of materials systems for application. Engineers are attempting to use the C–C materials in two-dimensional exhaust nozzles, where flat flap liners can provide unusual maneuvering capabilities to aircraft in test engines. Also being considered are afterburners since the material is lightweight and lowers air-cooling requirements. Even though C–C composites must be protected by coatings, there are two types of commercial products: overlay coatings applied after the part is formed and inhibited coatings, i.e., elemental substances (not carbon) that are part of the material. Perhaps even more critical to the use of composites in engines is the designer's freedom to create efficient attachments and peripheral hardware to make optimum use of the composite properties.

## AUTOMOBILES AND TRUCKS

The automotive industry is faced with an unprecedented challenge: it must succeed in meeting the requirements of two uncompromising masters, the customer and the government. One requirement is to produce cars that yield more miles per gallon than the cars of the 1970s. A lighter vehicle, which translates into lighter-weight materials, is one of several ways to increase the mileage from a tankful of fuel.[27]

The Ford Motor Company Light Weight Vehicle (LWV) Program is one of several prototype programs to evaluate the potentials for Gr–Ep composites in automotive applications. The reduction in weight of the prototype car allowed a smaller engine to be used in tests without changing performance based on acceleration. The LWV Gr–Ep composite parts are

| | |
|---|---|
| Hood | Front-seat frame |
| Doors | Accessory drive brackets |
| Hinges | Air-conditioning lateral brace |
| Door guard beam | Air-conditioning compressor bracket |
| Deck (truck) lid | Front and rear bumper components |
| Suspension arms | Apron and radiator support |
| Transmission support | Frame components |
| Drive shaft | |

Another composite experimental vehicle was designed and built of Gr–Ep and tailored around a Ford LTD sedan (Fig. 7.63). The vehicle was 33% lighter than a standard Ford LTD. Designed for a 3750-lb (1700 kg) vehicle, the graphite frame (Fig. 7.64) for the experimental car was only 27% lighter than a standard steel version. If the frame had been designed for the actual composite-car weight of 2500 lb (1134 kg), weight reduction would have been close to 50%. Another prototype car was introduced by Chrysler as the Poly-Car or PXL (Polymeric Extra Light). Five parts were made of Gr–Ep and five of Gl–Ep.

It can be expected that Gr–Ep will be used in production parts in automotive applications when availability and costs of the fibers are acceptable. Gr–Ep in conjunction with chopped fiber glass has been used in the fabrication of engine accessory drive brackets; Gr–Ep with Kevlar fibers was used for engine-valve pushrods, which withstand hot oil in racing cars.

Other components used in Chrysler's Poly-Car include:

*Door hinges:*   Compression-molded Gr–Ep

*Drive shaft:*   Filament-wound Gr–Ep over aluminum shaft

*Rear leaf spring:*   Steel main spring plus Gr–Ep auxiliary leaf spring

**FIG. 7.63**   Gr–Ep lightweight vehicle. (*Milford Fabricating.*)

**FIG. 7.64**  Gr–Ep frame. *(Milford Fabricating.)*

*Bumper energy absorbers:*  Gl–Ep
*Grille opening panel:*  Glass spheres added to SMC and compression-molded
*Road wheels:*  BMC injection-molded and SMC compression-molded Gl–Ep
*Transmission support cross member:*  Compression-molded Gl–Ep
*Door inner panel:*  Compression-molded from SMC made with Gl–Ep

## Materials and Material Forms

Hybrids will no doubt dominate the field of automotive composites. Today, graphite hybrids seem to offer the best approach to maximum weight reduction at a minimum cost. By blending two fibers, graphite and glass, for instance, not only can composite material costs be cut (over all-graphite), but the best properties of each fiber can also be utilized.

A little graphite goes a long way. Replacing 12% of the glass fibers in all-glass-fiber-reinforced resin results in a stiffness increase of 55%. In chopped-glass SMCs, adding 25 vol % of continuous graphite boosts strength by a factor of 4 to 5 and stiffness by a factor of 5 to 6. The combinations are endless and can be tailored to a particular application. Most automotive graphite hybrids use no more than 20% graphite to keep cost to a minimum.

Kevlar and graphite fibers can be hybridized too. For example, by using Kevlar for tensile strength and graphite for compressive strength a Gr–Kv–Ep transmission support was developed. Kevlar hybrids have been developed for drive shafts and bumpers. By using Gr–Gl–Ep hybrids in critical structures with graphite near the surface for maximum strength and stiffness a 26% vehicle-weight saving is possible. Other material combinations include woven Gr–EGl–Ep; hybrid of continuous E and S glass and chopped E glass; chopped-fiber SMC or HMC. The hybrids will be in volume production in the next 10 years.

Epoxy composites will not be acceptable in the auto industry, where fast cycles and high-volume production are required. Likely matrix systems include quick-curing polyesters and vinyl esters and injection-moldable thermoplastics such as nylon, PPS, and polybutylene terephthalate (PBT). Graphite thermoplastic oil pans are under development. Material forms will include compression-molding compounds, SMC, XMC, and UMC, which may contain Kevlar, graphite, glass, and combinations of fibers. Others include injection-molding compounds with Gr–Gl reinforcement, hot-stampable reinforced thermo-plastic sheet, and RRIM materials.

Processes which currently exhibit high-volume production potential are

Compression molding

Pultrusion

Filament winding

The ERM process has been used experimentally to produce prototype hybrid-composite roofs for a sports car. The H-shaped roofs were made by compressing epoxy and foam between Gr–Gl skins. Automobile-door window frames will replace frames presently man-ufactured of steel weldments. A filament-winding procedure using Gr–Ep is currently being developed for this application. Front and rear bumpers fabricated from Gl–Ep HMC composite are undergoing tests. The bumper system consists of a flexible urethane fascia on the outside, supported by a two-piece backup component molded from the HMC composite (Fig. 7.65). To achieve the necessary strength for automotive applications, more fiber glass was added than normally goes into standard SMC. The bumper composition, 65% chopped strand fiber glass and 35% resin, was selected over steel mainly for weight savings, lower tooling costs, and energy-absorbing properties. The design saves more than 40 lb (18 kg) per car.

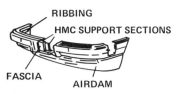

**BUMPER SYSTEM**

RIBBING

HMC SUPPORT SECTIONS

FASCIA

AIRDAM

**FIG. 7.65** HMC composite bumper. *(Iron Age, Sept. 10, 1979, p. 67).*

Air-conditioner compressor brackets have been tested successfully under driving conditions. To achieve production-volume under-the-hood brackets from composites, auto-mated SMC compression molding was used. This part dampens sound and saves fuel through weight savings. Filament-wound Gr–Gl–Ep drive shafts have been undergoing tests for 5 years in light-duty trucks, over-the-road trucks, and ore-mine vehicles. The major advantage of a Gr–Gl–Ep shaft has always been weight. A steel drive-shaft assembly on a light truck, for example, weighs roughly 35 lb (15.9 kg). A comparable hybrid-composite drive-shaft assembly weighs only 17 lb (7.7 kg) and is stiffer and stronger. The hybrid-composite shaft eliminates the need for two-piece construction in selected truck applications, reducing potential maintenance and warranty costs, because no center universal bearing is needed. It is expected that 15,000 light-duty trucks will have composite drive shafts within 2 years.

Many racing cars are using graphite and Kevlar composites to save weight, eliminate environmental and health hazard problems, and give strength and durability. Kevlar has been substituted for asbestos in brake linings, clutch facings, and gaskets for engine heads and manifolds. Other applications of Kevlar are

Starter-motor commutators

Body panels

Radiator and other hoses

Timing and V-belts

Drive chains

Using Gr–Ep in the chassis of a racing car has achieved a 35% reduction over aluminum and is 50% stiffer. With Gr–Ep materials the structure can be narrower than its aluminum counterpart and have larger, venturi-shaped side pods for greater downward forces from ground effects. Testing under actual racing conditions occurred in 1982.

### Composite Engine

Prototype internal parts of auto engines made of Gr–Ep, Gr–Gl-Ep, and Gr–PI have been tested in racing cars. Composite connecting rods have proved their fatigue superiority over aluminum, steel, and cast iron. Push rods, rocker arms, pistons, cylinder heads, and even engine blocks have either been tested or are in development. Advanced composites in engines not only reduce engine weight but also allow engines to turn at higher speed and produce more power. Composite wrist pins made of a steel sleeve internally reinforced with Gr–Ep developed for small-block engines and motorcycles have passed several test racing trials.

## Trucks

An all-fiber-glass door for trucks, with no metal supports or stiffeners, is one of the newest applications. The weight saving is 40 lb (18 kg) per door. The assembly contains an inner panel of 30% glass SMC, an exterior panel of SMC, and a full-length internal structural frame containing 50% glass fiber. The internal frame is the backbone of the door assembly; the continuous glass fiber is oriented along the entire periphery of the frame so that no bowing or deflection under stress or at high speeds can occur. Truck cabs and containers of glass-fiber composite are fast becoming standard on large commercial vehicles. Now light, intermediate, and heavy-duty trucks sport tough composite hoods, front ends, doors, and trailer walls and roofs, which have stood up to the bumps and grinds of cross-country trucking without a hitch (Fig. 7.66). Several light-truck manufacturers have introduced a glass-fiber SMC composite hatch or liftgate. It reduced weight by 33% and eliminated corrosion at the rear door of the truck, one of the most rustprone areas on this type of vehicle.

While structural glass-fiber SMCs have propelled composites into the trucking industry, current advances augur an even broader range of applications to come. Computerized filament-winding and hybrid composites combining Gl–Ep and Gr–Ep, for example, have already made possible structural parts such as axles, frame rails, drive shafts, and leaf springs. These structural components and even an all-glass-fiber composite truck cab are expected in the mid-1980s.

## Conclusions and Recommendations

Predictions for materials use in the car of the future abound. Some sources claim that cars will be made almost entirely from plastics and composites by 1990. To others, only about 12% of the car will be composites. The increase in composite use in cars in the next 10 years will require developments in major load-bearing components. Some of the technological breakthroughs needed include improvements in the use of continuous fibers, especially hybridized Gl–Ep and Gr–Ep, and development of a high-speed production system

**FIG. 7.66** SMC truck door and cab. *(Owens-Corning Fiberglas Corp.)*

for continuous-fiber preforms. Continued development in the automotive and material-supplier industries will produce a typical car of the future weighing approximately 1000 lb (453.6 kg) less than today.

## TRANSPORTATION

Weight savings of up to 80% are possible on various tracked-vehicle components when composites, consisting partially of Gl–Ep fabric and Gr–Ep tape, are used in place of metal. Typical parts evaluated were drive wheel, idler wheel, and road wheel on a combat-tank track and wheels for 5-ton (4536 kg), 8 by 8 cargo trucks. Engineers analyzed each assembly by taking it apart and determining overall stress conditions. A replacement part was assembled by selecting various combinations and orientations of molding compound, Kv–Ep, Gr–Ep tape, or Gl–Ep fabric, to suit the structural requirements (Fig. 7.67). Weight savings were 29% for track-support rollers, 65% for truck wheels, 70% for tank drive wheels, and 65% for truck drive wheels.

A lightweight, molded-plastic structure reinforced by Kv–Ep has formed the gondola of a new lighter-than-air ship. A nose cone and the suspension cables that attach the gondola to the envelope were also made from Kv–Ep and Kevlar fiber. The airship is used for oil-rig and coastal surveillance, fisheries protection, and patrol of territorial waters.

Other areas where composites are making progress include a composite-covered freight car. The first American covered hopper car of Gl–Ep is called the Glasshopper. The car body is made from multilayers of filament wound into a shell and is believed to be one of the largest load-carrying structures that is filament-wound. Because of its light weight the

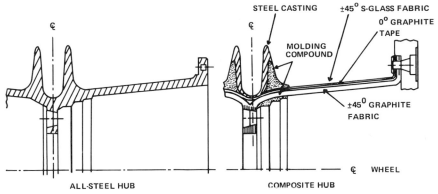

STEEL CASTING    ±45° S-GLASS FABRIC
                              0° GRAPHITE
                                TAPE
         MOLDING
         COMPOUND

                                ±45° GRAPHITE
                                FABRIC

                              WHEEL

ALL-STEEL HUB          COMPOSITE HUB

**FIG. 7.67**    Composite drive-wheel hub in tracked vehicle.[36]

Glasshopper can carry 5.5 tons (4990 kg) more in lading than a conventional steel car at lower fuel costs. The new composite cars weigh 26 tons lb (23,587 kg) compared with jumbo hoppers at 31.5 tons (28,576 kg).

Rapid-transit vehicles use Gl–Ep laminates and wound components in their interior (Fig. 7.68). Candidate designs for the window pillars and roof panels are illustrated in Fig. 7.69.

## Flywheels

Flywheels for mass-transit buses are not far in the future. The system represents an evolution from earlier programs, e.g., the hybrid-flywheel–electric-motor system built for an experimental car. The key to performance is the vehicle's light weight, made possible through the extensive use of Gl–Ep composites. The car's body, from bumper to bumper, was made of Gl–Ep, and so was the chassis, including the frame. The flywheel included an aluminum hub and spokes and nine layers of Gl–Ep filament circumferentially wound

**FIG. 7.68**    Rapid-transit interiors. *(Fiber Science, Inc.)*

**CARBON- FIBER COMPOSITE
SELECTIVELY STIFFENING
"PULTRUDED" FIBERGLASS
PLASTIC BEAM**

**Gr-Ep ALUMINUM-CORE SANDWICH
WITH PLASTIC EDGE MEMBERS**

**GI OR Gr-Ep "PULTRUDED"
MULTIWEB SECTION**

(a)

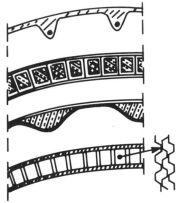

**GI-Ep LAMINATE SELECTIVELY UPGRADED
WITH UNIDIRECTIONAL Gr-Ep COMPOSITE**

**Gr-Ep SANDWICH; ADHESIVE PREPREG
FACINGS; WOUND STRUCTURAL CORE
ON POLYURETHANE FOAM**

**GI-Ep ADHESIVE-PREPREG, EPOXY LAMINATES
WITH PVC HIGH-DENSITY FIRE-RETARDANT
FOAM FILLER**

**GI-Ep FACED SANDWICH WITH SPECIAL
HIGH-STRENGTH CELLULAR CORE
PERMITTING CURVATIVE WITHOUT MACHINING**

(b)

**FIG. 7.69**  Candidate designs for (*a*) window pillars and (*b*) roof panels of future rapid-transit vehicles. *(DMIC Bull., June 5, 1974, p. 2.)*

around the rim. Eight layers of the rim were reinforced with Kevlar fibers. The flywheel package for the bus comprised six composite flywheels mounted on a single aluminum hub. The design used a composite rim made up of multiple rings of Kv–Ep. The rim was attached to a cruciform section made of Gr–Ep. Factors favoring a flywheel include energy savings, low maintenance and absence of noise or air pollution; on the debit side are higher initial cost and higher propulsion system weight. If the demonstration tests scheduled for 1983 are successful, widespread use is predicted within a decade (Fig. 7.70).

## ELECTRICITY AND ELECTRONICS

Demands for improved performance and for lighter-weight products have led to increased use of composites in electrical and electronic applications. A microphone housing that minimizes sound distortion and static interference (problems common in conventional microphones) has been designed using a molded carbon-fiber-reinforced nylon. It is based on the principle that near the reflective surface closest to a sound source there is a zone where sound waves are in phase. Thus, a microphone placed in the zone will pick up one clear tone. Another innovation is the use of glass-fiber-reinforced nylon in a miniature electronic

FIG. 7.70 Artist's concept of a flywheel transit bus, showing flywheel housed in a milk-can container in the rear. *(Ind. Res. Dev., May 1981, p. 99.)*

card holder to demonstrate low mold shrinkage, high dimensional stability, and good electrical properties.

Gl–Ep has been used in strengthening ribs to protect printed-circuit boards. Shock and vibration damage to printed circuits can be prevented by strengthening the boards with reinforcing ribs. The ribs are bonded directly over wiring on completed boards and are made of square or rectangular Gl–Ep tubes that are one-third the weight of solid rods and have high dielectric strength and good rigidity for airborne applications (Fig. 7.71). Slotted Gl–Ep rods are used for bonding to the edges of printed-circuit boards.

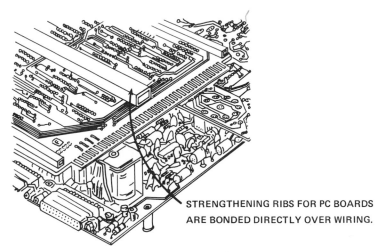

STRENGTHENING RIBS FOR PC BOARDS ARE BONDED DIRECTLY OVER WIRING.

FIG. 7.71 Gl–Ep tubes to protect printed-circuit boards. *(Mach. Des., April 1981, p. 34.)*

A new family of injection-moldable thermoplastic composites has been designed specifically to eliminate electromagnetic and radio-frequency interference in injection-molded enclosures and housings. The composites in two distinct attenuators are based on nylon 6/6, thermoplastic polyester polybutylene terephthalate (PBT), polycarbonate, and C–Gr-reinforced nylon 6/6. The materials are finding use in under-the-hood applications and field housings for microprocessors. Property advantages, especially in the avionics industry, include

| | |
|---|---|
| Low mold shrinkage | Good chemical resistance |
| Low moisture absorption | Good thermal stability |
| Minimal warpage | Minimum weight |
| Maximum attenuation | Flame resistance |

## Antennas

Antennas have been successfully produced in a variety of composite materials and processes. For example, a parabolic antenna was produced from individual molded petals, improving the strength-to-weight ratio and performance with lowered production time and cost. The petals, formed by compression molding of SMC, were cut and shaped to eliminate the need for the backup ring that typically supports conventional dishes.

An antenna built to extremely close tolerances was a key element in a recently completed U.S. Air Force program for a military weather satellite using a microwave sensor to gather vital data about clouds, rain, wind speed, soil moisture, and sea ice. The antenna is a Gr–Ep dish with an aluminum coating for a reflective surface and was contoured to an accuracy of better than 0.001 in (0.03 mm), making it one of the most accurate imaging microwave instruments ever built.

A low-visibility retractable whip antenna has been fabricated of B–Mg composite by a continuous casting process developed to produce the preform shapes of B–Mg. The technique combined continuous pulling through a die and extrusion. The B–Mg rod is stiff enough to permit fabrication of antennas of extremely small diameter that can support themselves in lengths of several feet.

Satellites and space systems offer a broad range of applications. Gr–Al and Gr–Mg composites seem particularly suitable for large spaceborne deployable antennas. Deployable solar panels are another possibility. Qualities which make MMC particularly attractive for space system applications include:

Very high specific stiffness

Excellent thermal conductivity

Good electrical conductivity

Electromagnetic-interference shielding and minimum space charging

Almost zero thermal expansion

No moisture absorption

High-temperature capability

# FILAMENT WINDING AND WEAVING

Filament-winding processes have already been described, but here we discuss some special products which are suited to filament winding because of economies and configuration.

## Windmills

The windmill, a prime source of rural electricity during the first part of this century, is again generating power where it is used, for homes, farms, and small businesses. These low-cost windmills use Gl–Ep composites, making them more efficient and durable. The blades are the most important part of the wind generator. To generate power efficiently at both high and low wind speeds, the blades must catch as much wind as possible. Glass composites produce the most efficient blade shapes to exacting specifications and are economical.

Some blades are molded by hand lay-up in a matched-die plastic mold, but most have been manufactured by filament winding and pultruding (Fig. 7.72). The spline is a glass-polyester pultrusion, and the filament-wound Gl–Ep spar is made of a transverse filament tape, a form of roving used in fabricating wound pipe and similar low-performance structures. Most wind turbines are being field-tested around the United States. Some projects are evaluating groups of wind turbines working together to generate electric power for a utility grid.

Fiber glass is a lot more forgiving than metal in terms of fatigue. To extract energy from the wind, the blades must retard the wind's force, which causes constant bending pressure. There is also gusting, which puts bursts of pressure on the blades. Facing the wind like a weather vane also creates gyroscopic forces, which cause the blades to try to fly out of the rotation plane. Then there are plain tensile loads from centrifugal forces. The material has to stand up to a lot. In metal blades small fractures can propagate quickly and result in an abrupt breakdown, a dangerous condition. If a Gl–Ep wound-composite blade develops a crack, the material's redundant load path prevents the crack from propagating and there is plenty of time to replace the part before it reduces the windmill's efficiency.

## Pipes

In the highly corrosive environments of chemical product processing plants, geothermal-heating applications, and seawater treatment facilities Gl–Ep equipment and structures have been substituted for the traditional steels, giving reduced maintenance, increased durability, and minimal downtime. Traditional construction materials such as steel tend

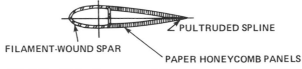

**FIG. 7.72** Gl–Ep filament-wound D-spar for a wind turbine with a 150-ft (46-m) blade. *(Mater. Eng., March 1979, p. 26.)*

**FIG. 7.73**   Gl–Ep blades used in seawater treatment. *(Owens-Corning Fiberglas Corp.)*

to corrode quickly and are subject to extensive repair and maintenance. Gl–Ep pipes and tanks are easy to maintain and simple to install and have exhibited good life-cycle cost benefits in comparison with coated steel and exotic metals.

In extracting magnesium from seawater one of the pretreatment steps uses a tank with 9-ft (2.7-m) Gl–Ep blades (Fig. 7.73). After the injection of additives, the seawater goes up a vertical filament-wound Gl–Ep-composite blend tower, 8 ft (2.4 m) in diameter and 50 ft (15 m) high (Fig. 7.74).

In an unusual composite application, the Bio-Pond, filament-wound Gl–Ep air-intake stacks deliver to compressors air with a high salt content that would tend to clog or corrode most metal and concrete structures. Figure 7.75 shows 30- and 24-in-diameter (762- and 610-mm) filament-wound pipe which distributes the compressed air at 175°F (79°C) over the pond surface to the aeration sites. The fiber-glass composite pipe is required because the severe external conditions make maintenance of steel pipe almost impossible.

Two 7-in-diameter (178-mm) pipes of filament-wound Gl–Ep, fitted with molded threads, were used in geothermal-heating projects in order to reach a suitable water level at 5833 ft (1778 m). This work is taking place in France, where 400 geothermal dual wells are expected to be drilled by 2000. They are expected to save 1 million tons (0.9 metric ton) of oil annually.

## Tanks, Bottles, and Pressure Vessels

Filament-wound Gl–Ep has been successfully applied to the following:

Underwater drones with skin and bulkhead structure
Waste tanks with laminated waste duct (Fig. 7.76)
Potable-water tank for commercial airliner (Fig. 7.77)
Underwater towed body (Fig. 7.78) and lavatory modules (Fig. 7.79)
Helicopter filament-wound fuel tank (Fig. 7.80) holding 650 gal (2461 L)

**FIG. 7.74**  Gl–Ep filament-wound blend tower. *(Owens-Corning Fiberglas Corp.)*

Pressure vessels for industries as divergent as chemical processing and aerospace, once made only of metal, are now being made of composites. Fiber-reinforced thermosets from polyesters to epoxies have become alternatives to steel, aluminum, titanium, and nickel-chromium alloys in pressure vessels. Applications include supply tanks for home water systems, swimming-pool filters, and water-softener tanks. Breathing apparatus with thick metal liners wrapped in a filament-wound Gl–Ep matrix are in use by fire departments and moun-

**FIG. 7.75**  Filament-wound Gl–Ep pipes used at aeration sites. *(Owens-Corning Fiberglas Corp.)*

**FIG. 7.76** Filament-wound waste tanks. *(Fiber Science, Inc.)*

**FIG. 7.77** Filament-wound water tank. *(Fiber Science, Inc.)*

**FIG. 7.78** Towed underwater body. *(Fiber Science, Inc.)*

**FIG. 7.79** Lavatory modules for commercial airliners. *(Fiber Science, Inc.)*

tain climbers. The Kv–Ep pressure vessels carrying helium, oxygen, and nitrogen aboard the Space Shuttle save $10 million in fuel and allow more efficient use of the payload. Kevlar fiber has been chosen over glass as the reinforcement for the thick-walled aluminum pressure tanks which activate the slides and life rafts on several commercial airlines. The need to save weight rather than money influenced the choice. Composite tanks weigh just half as much as the steel tanks they replace.

In an auxiliary power unit an accumulator stores energy in the form of oil under pressure. When released, the oil turns a hydraulic motor, starting a small turbine engine. The accumulator consists of a pressure vessel with a bellows inside. The cylinder is hoop-wound

**FIG. 7.80** Filament-wound helicopter fuel tank. *(Fiber Science, Inc.)*

ACCUMULATOR ASSEMBLY

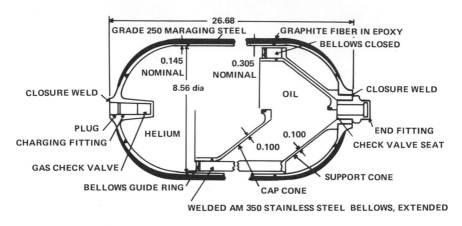

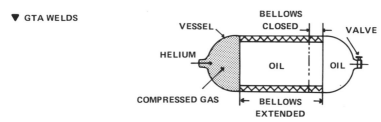

**FIG. 7.81** Accumulator workings simplified: first pump in gas then oil; the oil extends the bellows, which compresses the gas; when the valve is opened, the oil shoots out to turn a hydraulic motor. (*Weld. Des. Fabrication, March 1974, p. 55.*)

with Gr–Ep, and the whole vessel is wound longitudinally with Gl–Ep (Fig. 7.81). Many military aircraft contain this reliable backup power unit.

## Weaving Tubes and Bearings

Filament weaving has become highly versatile. Different interlocks of filaments as they are being woven can accommodate different requirements, e.g., high hoop strength for a rocket tube (Fig. 7.82) or high torque strength for transmission tubes and rods. The process is also used to produce bearings which are ideal for aircraft and aerospace applications with lightweight (less than aluminum), high load-carrying ability, and self-lubrication. There is no lubricant to freeze up in cold or to outgas and fog optical instruments in the vacuum of space (Fig. 7.83).

## MARINE

For over 30 years composites and glass-reinforced plastics have been used for small boat hulls, sonar domes, masts, tanks, decks, etc. In addition to high strength, low weight, and resistance to the marine environment advantages include

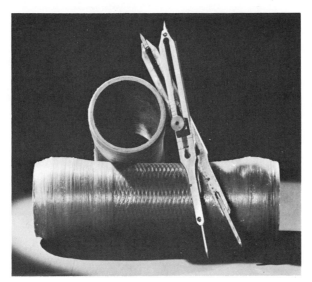

**FIG. 7.82**  Woven Gl–Ep tube used for rockets. *(Polygon Co.)*

**FIG. 7.83**  Filament-woven self-lubricating bearings. *(Polygon Co.)*

Monolithic, seamless contruction, minimizing leakage and assembly problems

Durability and ease of maintenance and repair

High energy absorption

Design flexibility, with the possibility of tailoring the material for a specific application

In some specific designs composite material offered good dielectric properties, absence of magnetic properties, and low thermal conductivity.

## Fishing Boats

Since fishermen are seeking vessels that can store more fish and stay out longer in all climates, the trend is toward larger boats that feature all-weather low-maintenance construction. The favored construction material is Gl–Ep, which has demonstrated excellent strength, durability, and low maintenance under all conditions, including extremely low temperatures. Figure 7.84 shows a typical fishing boat used in Alaska which has needed little maintenance in 2 years of service. If production plans of the major boat builders are any indication, a large part of the American fishing fleet will consist of 68-, 75-, 90-, and even 125-ft. (20.7-, 22.9-, 27- and even 38-m) fiber-glass boats designed to accommodate a variety of fishing techniques.

## Military Ships

The use of reinforced plastic composites in a structural and semistructural marine application is constantly growing as design and service experience is acquired. From their earliest employment in small boats some 40 years ago, use of glass-reinforced plastics, especially fiber glass, has been extended to the construction of commercial and military vessels, a

**FIG. 7.84**   Gl–Ep fishing boat. *(Owens-Corning Fiberglas Corp.)*

**Table 7.9** Estimated Weight of Strut-Foil Systems for a Development Big Hydrofoil[29]

| | Weight of material† | | | | Weight savings, % |
|---|---|---|---|---|---|
| | Titanium | | B–Ep | | |
| Component | tons | t | tons | t | |
| Forward struts | 29.8 | 27.03 | 20.0 | 18.14 | 32.9 |
| Forward foil | 16.8 | 15.24 | 10.3 | 9.34 | 38.7 |
| Aft struts | 51.3 | 46.54 | 35.0 | 31.75 | 31.8 |
| Aft foil | 39.3 | 35.65 | 25.5 | 23.13 | 35.1 |
| Total | 137.2 | 124.46 | 90.8 | 82.36 | |

† In U.S. customary units 1 ton = 2000 lb; 1 metric ton (t) = 1000 kg.

variety of shipboard structures, including fairings, deckhouses, masts, and tanks, as well as other marine structures such as floats and buoys. A summary of fiber-glass developments from the production of the first personnel boat through the work in submarines is available.[28]

Greater interest in high-performance surface ships for the U.S. Navy to achieve higher speed at sea requires a reduction in ship subsystem weights without sacrificing lifetime reliability. Major emphasis is therefore being placed on the reduction of structural weight, largely through the development and use of new high-strength material and refined design techniques. Future marine craft will require high-strength lightweight structures for air-cushion and surface-effect ships, hydrofoils, and planing craft.[29]

Weight savings from the use of composites in high-performance ships will come largely from use in hull structure, topside structures, e.g., as deckhouses, and strut-foil systems of large hydrofoils. Programs have been initiated by the Navy to evaluate advanced composites in three high-payoff areas:

*External decking:* Sandwich of Gr–Ep with a core of aluminum honeycomb with 53% weight savings over existing aluminum structure

*Hull:* Same construction as above with weight savings of 27 to 55% over existing aluminum structure

*Struts and foils:* Gr–Ep with a 54% weight savings

In other hydrofoil designs and studies B–Ep has been considered for struts and foils with significant weight savings (Table 7.9).

## Pressure Vessels

The outlook for Gr–Ep composites improves as the price of the material comes down. Many researchers feel that Gr–Ep offers better promise for deep-submergence pressure hulls than Gl–Ep. Advantages are higher modulus, lighter weight, and exceptional fatigue properties. The material can develop a hoop (circumferential) strength of 130 kips/in$^2$ (896 MPa) and modulus of 14 kips/in$^2$ (97 GPa) in a filament-wound unstiffened cylinder. Combining the cylinder with titanium end closures would result in an efficient hull capable of operating at 20,000 ft (6.1 km) with a stress safety factor of 2.0. This hull configuration could double the net buoyancy of advanced metal hulls while cutting the weight of a manned deepsea vehicle in half.

**FIG. 7.85** Gr–Ep mast and spar. *(Hercules, Inc.)*

## Masts and Booms

In yacht racing materials science is replacing art. The spinnaker pole on the racing yacht *Intrepid,* which weighs only 40 lb (18 kg) vs. 85 lb (38.5 kg) for the earlier all-aluminum design, was made of graphite yarn and B–Ep plus aluminum honeycomb. The boron fibers were included to improve compressive strength. The main boom for the yacht *Valiant* was fabricated of Gr–Ep and Gl–Ep inner and outer skins, aluminum honeycomb, and an extruded-aluminum cap that functions as a sail track. Weight saving over an aluminum boom is 40% (Fig. 7.85).

Submarine performance at periscope depth has traditionally been limited by performance characteristics of periscopes, antennas, and masts, most of which have a circular cross section and produce excessive wake, vibration, and noise. Attempts to streamline them with retractable fiber-glass fairings have met with limited success. Streamlining by different cross-sectional design has been hampered by the designers' use of stainless steel and fiber glass since ensuring an adequate stress level in a mast made from these materials requires a large, bulky structure that cancels the advantage of streamlining. To meet streamlining requirements the naval engineers proposed a titanium mast, which is both expensive and difficult to manufacture. As a result the Navy decided to examine the feasibility of fabricating submarine masts from Gr–Ep composites. The extreme stiffness and high strength-to-weight ratio of this material are highly desirable in mast construction. Implementation of the Gr–Ep mast is expected in 1983.

## Miscellaneous

A corrosion-resistant floating fish cage, built to weather storms and requiring little upkeep, has recently come on the market. Materials used are 90Cu–10Ni expanded-metal mesh for the interior and pultruded Gl–Ep for the exterior. Materials were selected for their resistance to corrosive seawater and resistance to fouling by marine life, which adds weight and reduces buoyancy and water flow through the mesh.

Several MMCs are under study for marine and ship applications. Gr–Al has been studied as a material for strut and foil systems of hydrofoils and for surface-effect ships, where weight reduction is important.

SiC–Al is attractive for stiffened, lightweight superstructures and decking for marine applications and improved-depth-capability pressure hulls for underwater applications. It also holds possibilities as a structural material for torpedoes and mine casings. The SiC–Al has been produced in the 2000, 6000, and 7000 series alloys, and work is in progress to develop a 5000 marine alloy system.

The Gr-Pb composite is being developed and evaluated for plates in nuclear-submarine lead-acid batteries. The composite plates are said to have increased strength and creep resistance, and lead additives such as antimony and calcium may ultimately be unnecessary. The composites may also double the life of the plates.

## RECREATIONAL AND SPORTING EQUIPMENT

Table 7.10 lists sporting goods using advanced composites.

### Bicycles

Given two equally well-trained bike racers, the winner's edge, even though it may be only a tenth of a second, is often a lighter, stiffer bicycle. With the advent of Gr–Ep, the graphite bike was a natural and probably at the time one of the most ambitious applications of advanced composites to sporting goods. Manufacturers are producing bikes with different composites. One bike uses hybrid composite consisting of Gr–Ep wound around aluminum tubing, where the Gr–Ep contributes flexural stiffness and the aluminum torsional stiffness.

**Table 7.10**  Sporting Goods Using Advanced Composites

| Transportation | Outdoors | Indoors |
|---|---|---|
| Racing and touring (Fig. 7.86) | Water skis | Ice-hockey sticks |
| Bicycles | Golf shafts | Racketball rackets |
| Sailboats and kayaks (Fig. 7.87) | Tennis rackets | Squash rackets |
| Canoes and catamarans (Fig. 7.88) | Cricket bats | Ping-pong paddles |
| Racing-car components | Javelins | |
|   (body panels, aerofoils) | Pole-vault poles | |
| Harness-racing sulkies | Fishing rods | |
| Lightweight aircraft | Bowstrings | |
| Oars and paddles | Snow skis | |
| Racing sculls (Fig. 7.89) | Archery bows (Fig. 7.93) | |
| Yacht components (Fig. 7.90) | Bats | |
| Sailplanes | Arrow shafts | |
| Power-boat components (hull and deck) | Arrows | |
| Sailing dinghy | Surf boards | |
| Snowmobiles (Figs. 7.91, 7.92) | Ski poles | |
| | Hang-glider frames | |

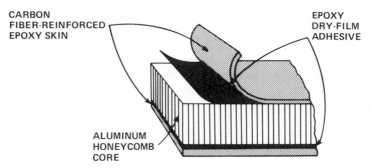

CARBON
FIBER-REINFORCED
EPOXY SKIN

EPOXY
DRY-FILM
ADHESIVE

ALUMINUM
HONEYCOMB
CORE

**FIG. 7.86**   Racing-car shell construction. *(Mod. Plast., May 1981, p. 38.)*

**FIG. 7.87**   Kv–Ep kayak.[30] *(Du Pont Co.)*

**FIG. 7.88**   Kv–Ep canoe. *(Du Pont Co.)*

**FIG. 7.89** Kv–Ep racing scull and oars.[30] *(Du Pont Co.)*

**FIG. 7.90** Kv–Ep hull, decking, and cordage on a sailing yacht.[30] *(Du Pont Co.)*

**FIG. 7.91**   Kv–Ep components in snowmobile. *(Du Pont Co.)*

**FIG. 7.92**   Kv–Ep tracks. *(Du Pont Co.)*

The composite frame weighs 3.5 lb (1.59 kg) compared with the steel frame of 5 lb (2.27 kg). Another manufacturer has produced bicycle frames of Gr–Ep tubing weighing 3.1 lb (1.41 kg). The tubes are joined up into frames with adhesives and steel lugs. In 1976 graphite bikes won 43 out of 70 major cycling events. In Sweden prototype bicycle components were injection-molded. The glass-polyester plastic components included

| | |
|---|---|
| Wheels | Pedal-crank arms |
| Frame | Package carrier |
| Forks | Fenders |
| Handlebars | |

Caliper hand brakes, gears, axles, pedals, and chains remained steel. The bike is 20% lighter than a steel bike.

## Boats

Kevlar, glass, and graphite in epoxy or polyester matrixes are now as common for sailboats, catamarans, kayaks, canoes, and power boats as wood was many years ago. Boat hulls constructed with Kv–Ep can offer significant weight savings and improved stiffness compared with Gl–Ep and offer superior damping vibration and good impact resistance. For example, a kayak made with Kevlar and polyester resin weighs about 18 lb (8.2 kg), while a comparable one made with glass fabric weighs over 30 lb (13.6 kg). The use of Kevlar in sailboats has resulted in lighter and faster boats, and stiffness and vibration damping mean smoother and more precise sailing performance. Lines and standing rigging last longer and are up to 75% lighter than their equivalents in steel wire. Less weight means faster, more responsive boats. In larger boats, the high-modulus fibers give stiffer hulls; this means reduced "power hooks" in power boats and in sailboats less tendency of the hull to bend under forces from the fore- and backstays.

Offshore racing boats, canoes, and kayaks have all benefited from composites, especially Kevlar. Canoes are 20 to 35% lighter than fiber glass or aluminum as Kevlar makes them easier to carry, launch, and maneuver. Olympic-style kayaks use Gr–Ep for hull and deck construction designed for turbulent waters, and the weight of the kayak is reduced 12%. A feature of a catamaran is its rigid wing in place of a conventional sail. The ribs and other internal parts of the wing are wood and Gr–

FIG. 7.93   Gr–Ep archery bows. *(Hercules, Inc.)*

Ep. The slats and control flaps have a framework of aluminum and Gr–Ep tubing, and in place of two aluminum tubes connecting the two hulls now there are beams of Gr–Ep.

Other applications include Gr–Ep drive shafts in large yachts to reduce vibration, Gr–Ep for long wing spars in sailplanes, and hybrid Gl–Gr–Ep in racing sculls and components for trimarans. Crossbeams in the trimaran have been fabricated of Gl–Gr–Ep-hybrid woven fabric, rovings, and weft; pultruded graphite fibers and a hybrid Gr–Gl–Ep have been used for the low section, the outriggers, deck panel, and dagger board.

Three breakthroughs for composites should be mentioned. (1) In a recent rowing championship sprint race the winning team used Gr–Ep oars. (2) Boat manufacturers have turned to RTM to supplement spray-up and hand-lay-up production. The process takes 30 min to cure a glass-polyester part the size of a bathtub in a closed mold. By comparison hand lay-up takes 2 h. RTM has been used chiefly to fabricate smaller to midsize parts like cockpit hatches, radar arches, and swim platforms. The molding technique is expected to be applied to larger items like 35- to 40-ft (10.6- to 12-m) boat hulls within 5 years. (3) An outboard-motor reed valve is a small exhaust valve that opens and closes every time the engine cycles. The single greatest cause of failure in two-cycle engines is read-valve failure from backfires. Even the highest grade of Swedish steel used for the part usually fatigues and ruins the engine. Longer-lasting Gr–Ep reed valves are in the final test phase before being implemented. The Gr–Ep valves are quieter and result in an engine that idles smoothly and runs more efficiently.

## Racing Cars, Motorcycles, and Snowmobiles

The transition to Kevlar, graphite, and fiber glass has been easy for equipment of this kind. Kevlar brake linings and clutch facings have longer wear life and are stronger than those of asbestos. Gaskets reinforced with Kevlar have been used for engine heads and manifolds. Kevlar and Gl–Ep for body panels used in racing cars can make the cars 40% lighter with no loss of strength. Another approach is combining aluminum honeycomb and Gr–Ep skins. The racing-car body was formerly constructed from welded and/or riveted aluminum panels. The new aluminum and Gr–Ep combination reduces the weight of the shell from 1700 lb (771 kg) to about 990 lb (449 kg) and permits greater design flexibility.

On motorcycles steel drive chains have been replaced with Kevlar composite material because of high energy-transfer efficiency and low noise. Other components include radiator and other hoses, timing and V-belts, and truck-chassis beams. Sulky rigs with Gr–Ep shafts are 20% lighter than the conventional wood-frame sulkies. Composite shafts break less. The *Gossamer Condor* and then the *Gossamer Albatross* used composites to prove that continued, unassisted human-powered flight is possible. Extensive use of Kevlar was made in the *Albatross* the vehicle that crossed the English Channel without help from motors, gases, ground crew, or in-air launch.

## Sports Equipment

### *Tennis Rackets*

High-strength, high-stiffness graphite fibers in tennis rackets give the player a definite edge. A properly constructed graphite racket dissipates less energy and allows the player to impart more velocity to the ball without increasing physical effort. Rackets can be tailored to fit individual players' needs. Most rackets have been manufactured from graphite prepreg, but other materials have also been used. Many graphite rackets begin with a woven graphite prepreg (shaped like a woven-cotton finger bandage). The graphite frame is pressurized internally with a rubber bladder and cured in an oven. Hybrid rackets include Gr–Kv–Ep, Gr–Gl–Ep, Gr–B–Ep, and all sorts of material combinations for frames. The fiber levels in a 10-oz. (284 g) Gr–Kv–Ep composite racket, for example, are 70% graphite and 30% Kevlar.

One interesting racket has an aluminum-and-foam frame stiffened with Gl–Ep for improved torsional rigidity. Engineers claim that torsional strength of the hybrid-composite construction is from 2.5 to 50 times greater than possible for standard sandwich or laminate designs. Strips of woven B–Ep add extra flexural stiffness to the grip. Only a small amount of boron is needed to boost performance in a graphite composite. Selective reinforcement in the throat section of a tennis racket or the highly stressed lower end of a golf-club shaft requires only a fraction of an ounce to increase performance.

In another racket a polyurethane-foam handle is molded to the frame, holes in the frame are drilled and polished, several coats of exterior finish are applied, and calfskin grips complete the racket. Other models include frames of wood and Gr–Ep laminate and graphite-filled high-pressure structural-foam core (injection-molded) laminated between frames of aluminum sheet. The unusual manufacture of a hollow tennis-racket frame is based upon an interesting use of a fusible alloy. The alloy forms a melt-out core, around which carbon-fiber reinforced nylon 6/6 is injection-molded. A tin-bismuth alloy core, corresponding in shape to the internal section of the racket frame, is low-pressure die-cast. To provide holes for the racket strings, retractable pins in the mold enter corresponding holes in the core. The polymer is rapidly injected into the mold at about 518°F (270°C). The heat is absorbed by the core so quickly that the polymer cools and hardens before the alloy can melt. After injection molding, the racket frame and core pass through an oven at 302°F (150°C), where the tin-bismuth alloy, with a sharp melting point of 281°F (138°C), is melted out and recovered for reuse. Finally, the hollow frame is filled with low-density polyurethane foam in the head to dampen racket vibration and medium-density polyurethane foam in the handle.

### Golf Clubs

In golf-club shafts lightness, response, and torsional control are key advantages of graphite. A graphite golf shaft is 40% lighter and stronger than a metal shaft. Both longer drives and increased accuracy are claimed for this composite. Less weight in the shaft means that more weight can be put in the head, where it is needed to increase distance. Torsional control is all-important and must be retained; otherwise you lose everything you gain in distance. One manufacturer wraps the graphite fibers away from the torsion to control the torque. A typical shaft would have a basic construction of 90° (1.6-rad) fiber layer for hoop strength, filament-wound layers for a combination of torsional and flexural properties, and 0° layers for final flexural patterns, stiffness, and strength.

Since the first graphite-shafted clubs had their pronounced effect on both professional and amateur golfers, B–Gr clubs have been produced. The graphite fibers are arranged around the shaft for torque control, and flexing is controlled by axial layers composed of a combination of boron and graphite. The tip is reinforced by adding several more layers of high-strength B–Ep to improve durability. The boron is slightly lighter and stronger but retains the same stiffness of a good all-graphite shaft. The boron solves the problem of getting an optimum combination of high strength and high stiffness, which cannot be solved by graphite alone.

### Fishing Rods

Gr–Ep fishing rods are lighter than Gl–Ep and even bamboo and result in better sensitivity or "feel." Since vibrations are dampened rapidly, minimizing waves in the line, casts can be longer and more accurate. Graphite rods are also stiffer and have a faster tip. Kevlar rods, which are strong and sensitive, are made by wrapping Kevlar around a mandrel with a unique resin-bonding system then adding longitudinal glass fibers to surround the Kevlar completely. The assembly is wrapped in cellophane tape and cured in an oven. The finished rod is quite a bit stronger than a fiber-glass rod and almost as strong as graphite.

One manufacturer uses 100% hand lay-up to produce the graphite rod with a balanced, special laminate wrap. This wrap puts a number of fibers in one direction and an equal number in the opposite direction, so that the rod is balanced cylindrically and longitudinally. It is believed that the rod will respond more quickly and steady more quickly since it will go right back to where it came from with a more dynamic forward thrust than a conventional rod.

For saltwater fishing a Gr–Ep rectangular rod is made of two ribbons of carbon fiber laminated above and below a lightweight core. This sandwich design has flat skins floated in space on a compatible lightweight core, giving better bending stability than standard tubular glass-fiber fishing rods. Strong handles of Gr–Ep have also been made. The handle is bonded to the rod blank using the same epoxy material as the rod. This eliminates the joining devices used with purchased handles.

S glass for fishing rods is new. Performance of these rods lies between the conventional fiber-glass and graphite rods. S-glass rods are about 25% stiffer than regular fiber-glass rods, giving much faster recovery at lower weight and greater sensitivity.

### Archery

A spiral laminate wrap design has been used to produce 100% Gr–Ep arrows in a technique similar to that used to produce fishing rods. Other composite arrows produced as a Gr–Gl–Ep hybrid have a smaller cross section (less drag) than aluminum arrows with equivalent spine. Composites give a high damping factor, significantly increasing arrow performance. Archery bows produced with graphite fibers added to other materials such as Kevlar permit 20% more energy to be stored when the bow is drawn. The high torsional stiffness of the composite bow permits extended recurve limb design while retaining the required bending stiffness. Lighter arrows can be shot accurately due to the initial lower acceleration upon release. The longer thin limbs damp shock, and little vibration is felt by the archer.

### Track and Field

Pole vaulting has changed completely since the fiber-glass pole replaced bamboo. Now, however, Gr–Ep is becoming the universal choice of athletes. In javelins graphite fibers have replaced glass fibers, providing better stiffness and improving the aerodynamic flight characteristics. The specific modulus and vibration-damping characteristics of graphite composite reduce wobble significantly, increasing distance.

### Snow Skis

Graphite fibers in skis provide better torque characteristics than glass fibers for side-hill skiing. The vibration-damping characteristics of graphite fibers are also beneficial. Other skis still contain fiber glass; one manufacturer uses bidirectional preimpregnated fiber-glass layers surrounding an aluminum-honeycomb core. Another uses a computer-controlled winding process which eliminates wood in the core, unlike some conventional metal skis. Instead, there is a reinforced urethane core, around which 5 mi of fiber glass is wound in a helix. The amount of glass fiber and the angle at which it is applied determine the flexing and torquing characteristics of the ski.

### Hockey Sticks and Ping-Pong Paddles

Hybrids of Kv–Gl–Ep in hockey sticks give better stiffness and longer life than all-fiber-glass or wood sticks. A ping-pong paddle made from three plies of wood alternately strengthened by Gr–Ep prepregs has reduced vibration, increased stability, and given more control over the ball. No rubber coating is required. Platform-tennis paddles use a glass-fiber fabric and wood laminated to achieve an optimum combination of flexing ability (to increase power) and shock damping. The paddle contains five to seven plies of domestic hardwoods, covered on each side with two layers of glass-fiber fabric. The matrix material is acrylic, which is sanded to expose the glass and roughen the surface.

### Racketball and Squash

Gr–Ep rackets are quickly overtaking conventional aluminum rackets. One racket, made by injection-molding foamed nylon and chopped graphite, has reduced vibration and can absorb energy from accidental smashes against a wall or floor. It has molded-in string holes. The next step is a Kv–Gr–Ep hybrid. Other manufacturers use carbon-reinforced high-strength polymer in the racket frame to provide a stiffer but more flexible one-piece construction for maximum feel, speed, and ball control. Other one-piece models substitute glass for carbon. Ancillary operations such as grinding and drilling had to be developed to keep pace with advances in composite fabrication. For example, development of the racketball racket required an air-controlled and air-powered production machine to drill 68 holes in 68 directions automatically in 68 s. Squash rackets are made of carbon-reinforced graphite composite or high- strength glass polymer, which provide strength, feel, and control not available with conventional rackets.

## MISCELLANEOUS

Many other industries have found what composites can do for their products (Table 7.11).

**Table 7.11** Composite Applications

| | | |
|---|---|---|
| Motorcycle helmets | Static-control accessories in | Protective clothing: |
| Electromechanical | business machines | Padding |
| cable (Fig. 7.94) | Pistons for small internal- | Shoes |
| Antenna guy wires and | combustion engines | Aprons |
| rods | High-speed centrifuge | Fragmentation vests (Fig. |
| Electrical and hoist | components | 7.100) |
| cables | Structural bearings for | Gloves (Fig. 7.99) |
| Aircraft guy wires | industrial machinery | Gears |
| Helicopter hoist cables | X-ray plates | Highway signs |
| Drive shafts (Fig. | Gaging equipment | Wheel chairs |
| 7.95) | Rotary compressor vanes | Implants[31] |
| Fume-extraction hoods | Hammer handles | Molds for casting titanium |
| Fiber-collection drums | Ladder rails | Ball valves |
| Cabs for earth-moving | Seats | Offshore drilling rigs (Fig. |
| equipment | Military helmets (Fig. | 7.101) |
| Roofs for rolling stock | 7.98) | Waste-water treatment plants |
| Housings | High-speed | Chemical plants (Fig. 7.102) |
| Diagnostic simulators | electromechanical devices | Food-processing plants |
| (Figs. 7.96 and 7.97) | for rapid and accurate | Support straps for cryogenic |
| | location and transfer of | containers |
| | recorded data in | Remote-piloted vehicles (Fig. |
| | computer systems | 7.103) |

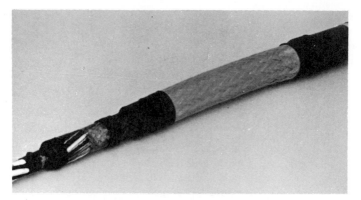

**FIG. 7.94**   Kv–Ep electromechanical cabling. *(Du Pont Co.)*

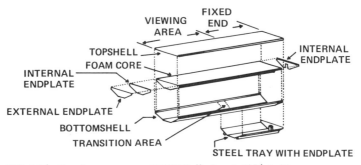

**FIG. 7.95**   Drive shafts for automotive use: steel *(below)* and Gr–Ep composite *(above)*. *(Hercules, Inc.)*

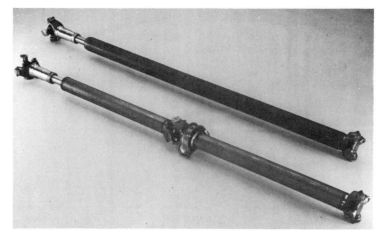

**FIG. 7.96**   Stretcher components. *(DMIC Bull., June 5, 1974, p. 3.)*

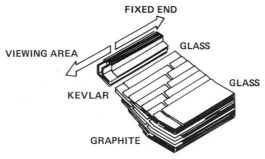

**FIG. 7.97** Section of lay-up in transition area of bottom shell (see Fig. 7.96 for location). *(DMIC Bull., June 5, 1974, p. 3.)*

**FIG. 7.98** Kv–Ep lightweight helmets. *(Du Pont Co.)*

**FIG. 7.99** Kv–Ep protective gloves. *(Du Pont Co.)*

**FIG. 7.100** Kv–Ep protective vest. *(Du Pont Co.)*

**FIG. 7.101** Cabling used for offshore rigs. *(Du Pont Co.)*

**7.96**

FIG. 7.102    Gl–Ep grating and stairways. *(Owens-Corning Fiberglas Corp.)*

FIG. 7.103    Filament-wound remote-piloted vehicle. *(Fiber Science, Inc.)*

## Textiles

Carbon and graphite fibers can be handled on textile machinery. To improve winding, braiding, weaving, and compounding the fiber is often sized with a material compatible with the desired matrix. For example, to produce easily mixed chopped fiber in compounding nylon-reinforced injection-molding compound a 3 to 5 wt % nylon-compatible sizing is placed on the fiber. Metallic yarns can also be processed into staple length, suitable for blending with natural and synthetic organic fibers on conventional textile machinery. This blend gives fabrics that resist buildup of static electricity. One of the most widely used metallic yarns for this purpose is type 304 stainless steel.

Type 347, 20Cb-3, Hastelloy X, nickel 270, tantalum, and columbium can also be fabricated into fibers. Metallic fibers are woven into surgical gowns to prevent static electricity buildup in the explosive environment of an operating room. Other applications for metallic fibers include tire cord, heating elements, flexible tubing, aerospace fabrics, corrosion- and heat-resistant fabrics, and specialized electronic applications.

The textile industry offers many potential uses for Gr–Ep composites in oscillating, reciprocating, and rotating components. One part in production is a heddle frame of a tape loom. Though material cost is 14 times that of the metal frame it replaces, overall loom cost is increased only 2.5% but loom speed is increased 50%. Gr–Ep pultrusion has been successfully used in knitting machines. The needle and sinker bars of the machine are stiffened with the composite pultrusion with 65 vol % of carbon. The pultruded strips are bound into grooves machined in the magnesium-alloy bars. The strip reinforcement reduces the coefficient of thermal expansion of the bars to that of the steel structure of the machine, reduces their weight, and increases their stiffness. These improvements enable the machine to run faster and virtually eliminate problems caused by temperature fluctuations (accurate spacing of the thousands of needles is critical).

## Medical Applications

Hybrids of Gr–Gl–Ep have been used successfully for portable iron lungs. The improved impact strength of the hybrid over all-carbon is advantageous. An experimental ankle-foot orthosis of Gr–Gl hybrid cloth weighs only 4.48 oz (127 g) and is used at $\pm 45°$ (0.79 rad) to the major axis of the leg. The lower end fits inside a shoe; the ankle section is thin and designed to allow a considerable deflection at every stride. The hybrid has proved satisfactory under repeated loading, but the glass prototype failed in flexural fatigue.

A cantilever stretcher table for a diagnostic x-ray simulator used to locate cancers for treatment was constructed of a Gr–Gl–Kv–Ep hybrid, chosen to provide the stiffness required for the cantilever structure and meet radiation-attenuation requirements [less than 0.040 in (1 mm) of aluminum]. Figures 7.96 and 7.97 show the stretcher and the selection of materials. The design of the transition between the viewing area and the fixed end of the bottom shell (Fig. 7.97) was critical because of differences in thermal expansion of the three fiber materials.

A high-impact face mask of Gl–Kv–Ep hybrid has been designed to protect the faces of epileptic and cerebral palsy patients during seizures. Similar configurations could be designed for muscular dystrophy patients, football linesmen, and riot-control police. The mask is extremely light; the lightest of the configurations (the cerebral palsy model) weighs only 4.80 oz (136 g).

## Music

Composite cellos and double basses have been made which contain chopped-glass strand mat with carbon fibers added to provide additional rigidity. The instruments are said to compare favorably in price with those made by traditional techniques while offering superior tone and appearance and sufficient robustness for school applications. Some guitars now have a compression-molded glass-fiber-reinforced one-piece body. Replacing spruce in soundboards for guitars and violins is Gr–Ep, which offers acoustical properties extremely close to those of wood but has uniform, predictable properties necessary for the mass production of instruments of consistently high quality. Sensitivity to weather variations, a severe problem with some wooden instruments, is apparently absent with composites. Gl–Ep has been substituted for wood in the shell of snare drums. Kettle drums are traditionally made from copper, but the future material is Gl–Ep for the large bowls.

## Other Applications

**Signs:** Gl–Ep composites hold advantages over aluminum, the traditional material for highway signs. They are better at withstanding collision damage, do not require costly treating to provide corrosion resistance, and are easier to repair after damage by vandalism. Gl–Ep composite panel road signs can be cut and silk-screened with the same equipment used for aluminum.

**Armor:** Glass-fiber-reinforced material has long been used as an armor protection material for governmental agencies, banks, and other security services. With the development of Kevlar, the lightweight armored product is now available for personnel, vehicle, and equipment protection.

**Precision Measurement:** Precision measuring instruments constructed from Gr–Ep composites minimize errors induced by temperature, flexure, and vibration. Compared with a metal micrometer, a Gr–Ep tubular micrometer significantly reduced the magnitude of error from all sources, including body heat. Due to flexing there is 10% less error for the micrometer made from graphite than for a metal micrometer of identical design. The graphite instrument can be handled and stored indefinitely without concern for corrosion. Large precision measuring machines that must be mounted on heavy granite bases could be moved much more easily and with less cost if they were mounted on lighter, more thermally stable graphite bases. Environmentally controlled test facilities costing thousands of dollars more than the instruments they house could be eliminated entirely by using graphite.

**Packaging Machine:** In a cigarette-packaging machine, stainless-steel arbors hold the cigarettes in place while the packages are formed. Steel satisfies requirements for strength and wear resistance but sounds like a cowbell. Arbors made with Gr–Ep composite identical in configuration to the steel arbors had sufficient strength and dimensional stability and reduced the noise level considerably. The graphite arbors are corrosion-resistant and 75% lighter than the steel arbors.

**Light Switch:** The circuit-breaker contact used as the on-off switch in fluorescent lights was made of Ag–W composite; high failure rates created the need for a new material. A Cu–Gr composite has been successfully developed as a substitute for the silver-base composite traditionally used for contacts in transformer breakers.

## REFERENCES

1. Schjelderup, H. C., and D. M. Purdy: Advanced Composites: The Aircraft Material of the Future, *AIAA 3d Aircr. Des. Oper. Meet., Seattle, July 1971.*
2. Stone, R. H.: Seven Years Experience with "Kevlar" 49 in the Lockheed L-1011 Tristar, *Symp. Des. Use "Kevlar," Aircr., Geneva, October 1980.*

3. Pride, R. A.: NASA Flight Service Programs to Evaluate Composite Structural Components, *18th Natl. SAMPE Symp. Exhib., Los Angeles, 1973.*

4. Hammer, R. H.: Composites in the Boeing 767, *Symp. Des. Use "Kevlar" Aircr., Geneva, October 1980.*

5. Vosteen, L. F.: Composite Structures for Commercial Transport Aircraft, *NASA Tech. Mem.* 78730, June 1978.

6. Ohgi, G. Y.: Advanced Composite 727 Elevator and 737 Stabilizer Programs, *23d Natl. SAMPE Symp. Exhib., Anaheim, 1978.*

7. Leonard, R. W., and D. R. Mulville: Current and Projected Use of Carbon Composites in United States Aircraft, *AGARD Avion. Panel Spec. Meet Electromagnet. Effects [Carbon] Compos. Mater. Avion. Syst., Lisbon, June 1980.*

8. August, A., R. Hadcock, and S. Dastin: Composite Materials Design from a Materials and Design Perspective, *AGARD Rep.* 639, January 1976.

9. Meade, L. E., "DOD/NASA Structural Composites Fabrication Guide," 2d ed., vol. 1, secs. 9-1 to 9-60, AFWAL, Wright-Patterson AFB, Ohio, Lockheed-Georgia Co., Marietta, Ga., Contr. F33615-77-C-5256, May 1979.

10. Petit, P. H.: An Applications Study of Advanced Composite Materials to the C-130 Center Wing Box, *NASA Contr. Rep.* CR-66979, July 1970.

11. Grosko, J. J.: Manufacturing Technology for Large Aircraft Composite Structure, First Quarterly Progress Report, *Lockheed-Georgia Co., Interim Tech. Rep., Oct. 1, 1980–Jan, 1, 1981,* IR-471-0 (1), Marietta, Ga., Contr. F33615-80-C-5076, AFWAL, Wright-Patterson AFB, Ohio.

12. Boron/Aluminum Landing Gear for Navy Aircraft, *General Dynamics Convair, Final Rep.* CASD-NADC-76-003, Contr. N62269-74-C-0619, San Diego, Calif., September 1978.

13. Hightower, D., et al.: Manufacturing Technology for Composite Launcher Shafts, *Boeing Military Airplane Co., Final Rep. September 1980–September 1981,* AFWAL-TR-81-4105, Contr. F33657-80-C-0344, Wichita, Kans., October 1981.

14. Rich, M. J.: Application of Advanced Composite Materials to Helicopter Airframe Structures, in E. M. Lenoe et al. (eds.), "Fibrous Composites in Structural Design," Plenum, New York, 1980.

15. Immen, F. H.: Army Helicopter Composites, *Natl. Def.,* November-December 1977.

16. Schwartz, M., and G. Jacaruso: A Giant Step toward Composite Helicopters, *Amer. Mach.,* March 1982, pp. 133–140.

17. Kay, B., D. Maass, and R. Kollmansberger: Airframe Preliminary Design for an Advanced Composite Design Airframe Program (ACAP), *Sikorsky Aircraft Div. United Technologies, Final Rep.* Contr. DAAK-51-79-C-0029, Army Tech. Lab., Ft. Eustis, Va., June 1980.

18. Rich, M.: Graphite Composite Transmission Support Frame, NASA Langley Research Center and Army Technical Laboratory, Ft. Eustis, Va., Sikorsky Aircraft, Division of United Technologies, Contr. NAS1-13479.

19. Thermoplastic Aerospace Structures, *Amer. Mach.,* January 1982, pp. 142–143.

20. Stonier, R. A.: Development of a Low Expansion, Composite Antenna Subreflector with a Frequency Selective Surface, *9th Natl. SAMPE Tech. Conf., Atlanta, October, 1977.*

21. Prunty, J.: Dimensionally Stable Graphite Composites for Spacecraft Structures, *9th Natl. SAMPE Tech. Conf., Atlanta, October 1977.*

22. Boeing Aerospace Co.: Design, Fabrication and Test of Graphite-Polyimide Composite Joints and Attachments for Advanced Aerospace Vehicles, *NASA Langley Res. Cen., Q. Tech. Prog. Rep.* 5, Contr. NAS1-15644, Hampton, Va., May 19, 1980.

23. Smith, G. T.: Application of Composite Materials to Turbofan Engine Fan Exit Guide Vanes, *NASA Tech Mem.* 81432, 1980.

24. Brantley, J. W., and R. G. Stabrylla: Fabrication of J79 Boron/Aluminum Compressor Blades, *General Electric–NASA–Lewis Res. Cent. Final Rep. April 1975–June 1978,* CR-159566, Contr. NAS3-18943.

25. Fannin, R. L., and P. S. O'Connell: Advanced Reinforced Titanium Blade Development, *Detroit Diesel Allison, Div. General Motors Interim 2d Q. Rep. 1981,* Indianapolis, Ind. EDR10550C, AFWAL, July 1981.

26. Darolia, R.: Feasibility of SiC Composite Structures For 1644 K (2500°F) Gas Turbine Seal Applications, *General Electric Co–NASA–Lewis Res. Cent. Final Rep.* NASA CR159597, *April 28–May 1, 1979,* November 1979.

27. Margolis, J. M.: Automotive Developments in the 1980's: Composites vs. Metals, *SAMPE J.,* July–August 1980, pp. 7–15.

28. Spaulding, K. B., Jr.: *Nav. Eng. J.,* **78:** 333 (April 1966).

29. Silvergleit, M., R. W. Deppa, and H. P. Edelstein: Potential Application of Advanced Composites for High Performance Craft, *5th Natl. SAMPE Tech. Conf. Kiamesha Lake, N.Y., Oct. 9–11, 1973.*

30. Langston, P.: Promising Recreational and Sports Equipment Applications for High Performance Composites/Kevlar, *27th Natl. SAMPE Symp., San Diego, Calif., May 4–6, 1982.*

31. McKenna, G. B., W. O. Statton, H. K. Dunn, K. D. Johnson and G. W. Bradley: The Development of Composite Materials for Orthopedic Implant Devices, *21st Natl. SAMPE Symp. Conf., Los Angeles, Apr. 6–8, 1976.*

32. Smith, C. W.: Development of Manufacturing Methods for Joining Thermoplastic Composites, General Dynamics, Convair Division, Contr. N00019-C-0227 Mod P00024, Department of the Navy, Navair, Washington, 1981.

33. Davis, G.: Composites Production Integration, *General Dynamics, Ft. Worth Division, Ft. Worth, Tex.,* AFWAL-TR-80-4018, *Final Rept.* Contr. F33615-77-C-5018, 1980.

34. McCaskill, O. K., Jr.: Composite Applications at Bell-Helicopter, *SAE Tech. Meet. Wichita, Kans. Apr. 3–6, 1979,* pap. 790578.

35. Rich, M., G. F. Ridgley, and D. W. Lowry: Application of Composites to Helicopter Airframe and Landing Gear Structures, Sikorsky Aircraft, Division of United Technologies, NASA Cr-112333, Cont. NAS1-11688, June 1973.

36. Composite Materials Find Use in Tracked Vehicle, *AMMRC Rep.* TR-79-40, May 1980, and *Mfg. Tech. Note* PB80-979380, AD72354, September 1980.

37. Zweben, C.: Advanced Composites: A Revolution for the Designer, *AIAA Meet., Long Beach, Calif., May 12-14, 1981,* AIAA-81-0894.

38. Mayer, N. J.: Composite Materials in Space Structures, *5th Natl. SAMPE Tech. Conf., Kiamesha Lake, N.Y., Oct. 9–11, 1973,* pp. 559–580.

## BIBLIOGRAPHY

Advanced Composite Aileron for L-1011 Transport Aircraft, *Lockheed Calif. Co., Q. Tech. Rep.* NASA CR162863, Dec. 22, 1977–Mar. 24, 1978, Burbank, Calif.

Advanced Manufacturing Development of a Composite Empennage Component for L-1011 Aircraft. *Lockheed Calif. Co. Q. Tech. Rep.* NASA CR162862, *Jan. 1, 1978–Mar. 31, 1978,* Burbank, Calif.

Boll, K. G.: Advanced Composite Engine Development Program, Pratt and Whitney Aircraft, Division of United Technologies, Contr. F33615-69-C-1651, AFML-TR-72-108, E. Hartford, Conn., July 1972.

CH-46 Forward Pylon Work Platform, *MTAG Meet., Orlando, Feb. 21 and 22, 1981* (Boeing-Vertol, Philadelphia, Pa., NASC Contr. N00019-79-G-0154 and N00019-80-C-0154).

Chovil, D. V., S. T. Harvey, et al.: Advanced Composite Elevator for Boeing 727 Aircraft, vol. I "Technical Summary," *Boeing Commercial Airplane Co. NASA Contr. Rep.* 3290, *May 1977–December 1979,* Seattle, Wash., Langley Research Center, Contr. NAS1-14952.

Clarke, C. A., et al.: Assessment of Risk to Boeing Commercial Transport Aircraft from Carbon Fibers, *Boeing Commercial Airplane Co., Final Rep.* NASA-CR-159211, *September 1979–January 1980,* Seattle, Wash., June 1980.

Composite F-100 Engine Augmentor Duct Program, Contr. F33615-76-C-5333, Composite Horizons, Pomona, Calif., and Contr. F33615-76-C-5429, Rohr Industries, Chula Vista, Calif.

Dastin, S. J.: Design and Concepts of Composite Structures, *3d Int. Conf. Compos. Mater., Paris, August 1980.*

Davis, J. G. Jr.: Composites for Advanced Space Transportation Systems (CASTS), NASA TM80038 *Tech. Rep. July 1, 1975–April 1, 1978,* March 1979.

Dexter, H. B.: Composite Components on Commercial Aircraft, *AGARD Meet. Effect Serv. Environ. Compos. Mater., Athens, Apr. 13–18, 1980,* NASA TM 80231.

————, et al.: Graphite Polyimide Composites, NASA CP 2079, NASA Langley Research Center, Hampton, Va., August 1979.

Friedman, G. I.: Tungsten Wire-Reinforced Superalloys for 1093°C (2000°F) Turbine Blade Applications, *TRW, Inc. NASA Contr. Rep.* 159720, Cleveland, Ohio, October 1979.

Hoffman, D. J.: 737 Graphite Composite Flight Spoiler Flight Service Evaluation, *NASA Langley Res. Cent.–Boeing Commercial Airplane Co.,* NASA CR 159362, *6th Annu. Rep., May 1979–April 1980,* Seattle, Wash., NAS1-11668.

Meade, L. E.: DoD/NASA Structural Composites Fabrication Guide, *Lockheed-Georgia Co. 7th Interim Q. Tech. Rep., Apr. 16, 1981–July 15, 1981,* IR-426-9A (VII), Contr. F33615-79-C-5125, AFWAL, Wright-Patterson AFB, Ohio.

Rich, M. J., and R. L. Foye: Low Cost Composite Airframe Structures, *NASA Tech. Mem.* X-3377, April 1976.

Schmid, T. E.: Graphite Polyimide Augmentor/Nozzle Components, *AIAA/SAE/ASME 17th Joint Propul. Conf. Colorado Springs, July 21–29, 1981,* AIAA-81-1358.

Watts, D. J.: A Study on the Utilization of Advanced Composites in Commercial Aircraft Wing Structure, *Douglas Aircraft Co.,* NASA CR 1589022, *Final Rep., July 1978,* Long Beach, Calif.

Projected structure for a helicopter fuselage.    *(Sikorsky Aircraft, Division of United Technologies.)*

# Future Potential of Composite Materials

## INTRODUCTION

The use of composite materials has progressed through several stages in the past 25 years. First, demonstration pieces were built with the idea of "let's see if we can build one." For the second stage, replacement pieces, part of the objective was to test a part designed to replace a metal part in an existing application. Examples of this are shown in Fig. 8.1, where the 727 elevator, DC-10 rudder and L-1011 aileron have been replaced by composites. The selection and ultimate fabrication (Fig. 8.2) and testing of these secondary structures led to the confidence to select and build several primary structures (Fig. 8.3), and Fig. 8.4 shows one of the initial stabilizers. The third stage is actual production pieces designed from the beginning to be fabricated wholly or in part from fiber-reinforced composite materials. The final stage is the all-composite airplane, helicopter, etc., that many people have dreamed of building. This last goal is being approached in a deliberate, conservative, multistage fashion. A substantial composite-materials technology and manufacturing base has been built and awaits further challenge.

The impact of composite materials use on jet-engine performance is substantial. Currently, with various metal alloys, thrust-to-weight ratios of 5:1 are achieved. Composites may lead to ratios as high as 16:1.

In response to today's challenges five important interacting trends are apparent:

Development of alternative materials

Very-fine-scale structural engineering of materials

**8.1**

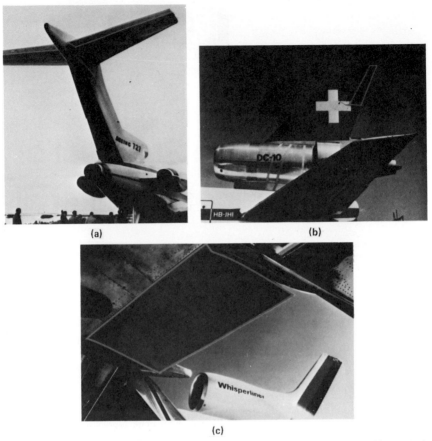

**FIG. 8.1** Composite secondary structures: (*a*) Boeing 727 elevator, (*b*) Douglas DC-10 rudder, (*c*) Lockheed L-1011 aileron. (*Langley Research Center.*)

Detailed three-dimensional process modeling

Real-time process monitoring and quality control

Control of degradative processes in materials

In the materials area, emphasis will be on developing and characterizing lower-cost material systems, e.g., pitch-based fibers, improved epoxy-resin systems with reduced sensitivity to environmental factors, cost-effective and reliable high-temperature resin systems and special product forms, e.g., thin-prepreg and hybrid-material tape and woven broad goods. Polymers represent the most rapidly growing area of materials utilization. Rapid development is anticipated in composites for structural use; in blends, where properties not found in a single polymer can be tailored more rapidly than by engineering an entirely new polymer; and in medical implants. In metals, the growth in use of metallic composites is especially high in applications where cost is secondary, e.g., those where high strength with minimum weight is critical.

The technology will continue to expand rapidly in the areas of computer-aided design and manufacturing and in the development necessary to fabricate large structures cheaply and reliably. Durability and damage tolerance will be emphasized to identify potential material limitations. Automation, including robots, material-handling systems, and micro-computer controls, will make many operations economical and feasible. Microcomputer controls in filament winding will make it acceptable to users in the 1980s. Computer control allows different winding programs to be combined, to give fiber orientation by winding different angles. The computer can teach the machine to duplicate the placement of fibers like a robot. Expansion of the process will enable continuous fibers to be directly oriented in a mold and molded into shapes.

## NEW MATERIALS COMING

Many composite materials today do not fit the production process required to produce an economical and competitive product. We must tailor the materials to go into these processes. One possibility is Gr–Ep in a thermoplastic form that can be melted and shaped very quickly and still retain most of its properties. In the automotive industries the lowest-cost material will probably be SMC, which uses discontinuous or randomly oriented short fibers. Quick-curing and quick-molding thermoplastic sheet materials can simply be laid in a stamping machine and pressed into an auto part.

**FIG. 8.2**   Upper aft rudders for DC-10. *(Langley Research Center.)*

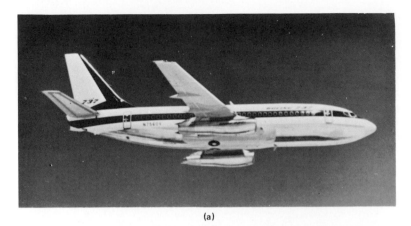

(a)

(b)

(c)

**FIG. 8.3** Composite medium-primary structures: (*a*) Boeing 747 horizontal stabilizer, (*b*) Douglas DC-10 vertical stabilizer, (*c*) Lockheed L-1011 vertical fin. (*Langley Research Center.*)

FIG. 8.4  Composite horizontal stabilizer for 737. *(Langley Research Center.)*

## Thermoplastic Polymer

A new solvent-resistant thermoplastic polymer suitable for the resinous matrix in graphite-reinforced composites has been developed. High-performance Gr–Ep composites are routinely used in aircraft, but widespread use elsewhere would be accelerated by lowering material and processing costs, improving moisture resistance, and improving damage tolerance. One class of composites that may accomplish these objectives is based on a newly developed thermoplastic with improved solvent resistance over common thermoplastic resins such as polysulfone. Simple modifications to polysulfone have given a new resin that resists fuel, lubricants, and solvents normally encountered in aircraft flight and maintenance.

Research laboratories have produced experimental sheets of thermoformable thermoplastic composites at rates that suggest cycle times of less than 1 min. Shorter cycle times are possible, depending on part thickness, geometry, and matrix material. A unique method of impregnation allows virtually any combination of fibrous materials to be combined into a thermoformable, plastic-sheet product. Data suggest that the sheets could be stored indefinitely and subsequently formed into a particular shape. Forming is achieved by heating the thermoplastic matrix material to its optimum temperature by compression-molding in matched dies. To date, composites reinforced with carbon fiber, carbon fabric, glass mat, woven roving, heavy tows, and hybrids have been prepared. Potential use for the readily formable sheets is seen in structural areas where weight savings is important and where the environment permits the use of thermoplastics, $-100$ to $300°F$ ($-73$ to $149°C$). Anticipated applications include seats for mass transportation, automotive parts, aircraft interiors, business-machine housings and frames, sporting equipment, containers, and antennas.

## Structural Thermoplastics

Highly reinforced thermoplastics are being used in load-bearing applications usually associated with metals and offering the economies of injection molding. These structural composites, with their heavy loadings of short glass or carbon fibers, provide high tensile

**Table 8.1** Contending Structural Thermoplastics[1]

| Composite identification | Reinforcement | | Thermoplastic |
| | wt % | Type | |
| --- | --- | --- | --- |
| Gl–nylon 6/6 | 50 | Glass fiber | Heat-stabilized nylon 6/6 |
| Gl–nylon 6 | 60 | Glass fiber | Heat-stabilized nylon 6 |
| Gl–nylon 6/10 | 60 | Glass fiber | Heat-stabilized nylon 6/10 |
| Gl–ST nylon | 50 | Glass fiber | Heat-stabilized supertough nylon |
| C–nylon 6/6 | 40 | Carbon fiber | Heat-stabilized nylon 6/6 |
| Gl–PC | 40 | Glass fiber | Polycarbonate |
| Gl–PPS | 50 | Glass fiber | Polyphenylene sulfide |
| C–PPS | 40 | Carbon fiber | Polyphenylene sulfide |
| Gl–PP | 50 | Glass fiber | Chemically coupled heat-stabilized polypropylene |
| Gl–PBT | 50 | Glass fiber | Thermoplastic polyester (polybutylene terephthalate) |

strength, high flexural modulus, and low creep but can readily be molded on injection equipment. They are characterized by fiber loadings in the 40 to 60% range (Table 8.1).

The European automotive industry already uses 50% glass-reinforced nylon 6. Nylon composites with 50 to 60% glass reinforcement are now finding expanded use in the United States in automotive, photographic, appliance, recreational, military, business-machine, electronic, and chemical-processing applications.[1]

## Conductive Materials

Organic polymers and intercalated graphite can be made to conduct electricity for a range of applications that include lightweight conductors, batteries, and electronic devices. Conductive polymers have a backbone like a chain of atoms or a stack of molecular poker chips. They approximate one-dimensional conductors, materials that conduct in one dimension only. Possible applications for conductive plastics include automobile ignition cables (require low currents), helical cables on telephone receivers (require flexibility), electromagnetic-interference shielding, solar photovoltaic cells (for inexpensive coverage of large surfaces), and semiconductors.

Intercalated graphite is a nonmetal with an electrical conductivity close to that of copper. Theoretically, its conductivity may even be greater than that of copper, and it is lighter and stronger than the metal. Graphite's high conductivity, however, cannot be reliably reproduced yet, although conductivity about half that of copper has been demonstrated consistently. But even this conductivity, combined with the material's light weight, makes it attractive for shielding electronic components against electromagnetic interference and for applications where weight is important, e.g., electrical wiring in aircraft.

Intercalated graphite is made by impregnating or doping high-quality graphite fiber or powder with metal-rich chemical compounds so that the compounds lodge between (intercalate) the stacked layers of carbon. It is believed that the guest compounds give electrons to the host crystal lattice, which leads to the higher conductivities. Other intercalated materials are being investigated, notably intercalated boron nitride. With electronic circuitry getting more and more sensitive and signal pollution increasing, electromagnetic/interference shielding is going to be a big market. The military is also interested in providing a way for composite aircraft to withstand lightning strikes and for grounding electronic equipment in them. The biggest possibility for commercial success is intercalated graphite wiring for use in power transmission, aircraft electronic systems, or internal carriers in transformers or motors. A composite can be formed by enclosing powdered intercalated graphite in a copper matrix (one-third copper) and swaging it into wires. This process has not yet made wire in production quantities.

## Fiber-Composite Ceramics

Brittleness can be lowered in ceramics but never eradicated altogether. A promising new material is fiber-composite ceramics. SiC fiber is embedded in a ceramic or glass-ceramic matrix. The fiber, loosely bonded to the matrix, diverts the path of any fracture, producing a fracture toughness 5 to 6 times greater than that of conventional monolithic ceramics. The drawback is that temperature capabilities are limited to 1823°F (995°C). Other fibers successfully developed include graphite with a strength level of 140 kips/in$^2$ (965 MPa) in glass, 6061 aluminum and AZ91 magnesium matrixes, and aluminum oxide.[2] Applications and production methods, including injection molding, are being investigated, for gas-turbine engines, radomes, bearings, lasers and optical mirrors, and space-system applications. Recent work in synthesizing refractory-metal glasses may give these materials the strength and thermal stability for successful construction of MMCs. For example, they might be used to reinforce an aluminum matrix.

An improved glass fiber introduced as a reinforcement for cement boosts the strength of the composite by up to 5 times. Attempts to make glass-reinforced concrete failed until it was discovered that an alkaline chemical reaction that occurs while the cement is setting destroys the glass. Further research showed that if there is some zirconia in the glass it will resist the alkaline attack. However, the properties of the composite change with time, and very rainy conditions cause some loss of strength. Now 5 years of continuing research has produced a surface treatment that reduces the reaction to alkaline conditions and increases the strength of the composite, making it resistant to weathering.

## Fiber-Reinforced Advanced Titanium

While the high specific strength, stiffness, and tailorability of advanced composites may lead to their being one of the most efficient of known structural material systems, composites lack the ductility of metal. One approach to an optimum mix of composite and metallic technologies is to combine the relatively low-cost superplastic-forming–diffusion-bonding (SPF/DB) fabrication techniques for titanium with the tailorable high strength and stiffnesses of advanced composites. The joint technology is known as fiber-reinforced advanced titanium (FRAT). The hybrid system is currently typified by the selective incorporation of a carbon-core substrate (SiC fibers or other similar high-strength, high-modulus fibers) into the face sheets of a three-sheet SPF/DB sandwich structure of Ti–6Al–4V. With fiber volume fractions of 7 to 40% local increases of 30 to 40% in strength and 100% in selective stiffness have been demonstrated.

Other metal-matrix materials investigated include CP titanium, Ti–3Al–2.5Sn, and Ti–6Al–2Sn–4Zr–2Mo. The fibers being evaluated besides SiC include B$_4$C-coated boron and Borsic. The major finding thus far is that a considerable degradation in strength can be expected after consolidation of MMC and thermal exposure [1650°F (899°C) for several hours] during SPF/DB. However, no significant degradation in modulus has been observed. The SiC laminates show a dramatic increase in compressive modulus over the matrix alloy, whereas $B_4C$-coated boron shows the highest potential for achieving both strength and stiffness after processing. Consolidation of prefabricated tapes into structures has led engineers to consider diffusion bonding (as an integral part of the SPF/DB process) and brazing of the composite laminates after SPF/DB panel fabrication. The SPF/DB process gives the structural designer additional design freedom at no additional cost. It permits joining two or more sheets in numerous ways.

Brazed MMC joints provide an attractive alternative to other joining methods for attaching the consolidated MMC preforms to the SPF/DB titanium structure. Brazing requires minimal bonding force and temperature exposure, eliminating or minimizing mechanical damage to fibers or fiber-matrix interactions.

Reinforcement of the SPF/DB structure with MMC has been accomplished with unidirectional tapes, and both continuous and selective reinforcement have been evaluated (Fig. 8.5). Cost and inspection considerations make selective reinforcement considerably more attractive. Since the basic structure is made from a high-strength alloy, the composite fibers are used unidirectionally to enhance the stringer compression resistance (local and column buckling). They may also be used to provide a higher degree of damage tolerance by reducing the stress-concentration effects of mechanical attachments and penetrations. Future development work will deal with

Scaling-up SPF/DB and selectively reinforced SPF/DB structures

Extension of selective reinforcement to titanium powdered-metal structures

Selective reinforcement of titanium with aluminum-matrix composite and refractory-metal fibers

Design and testing of damage-tolerant structures using selective reinforcements

Generation of material and structural design allowables in preferred systems

## Graphite-Polyimide

As high-temperature applications of composites increase, the use of Gr–PI and other high-temperature materials will also grow, especially in engines. Two composite applications for the augmentor and nozzle section of advanced military gas-turbine engines were designed, fabricated, and tested to determine the cost and weight-savings potential and to demonstrate the possibility of using organic matrix composites as substitutes for sheetmetal titanium structures. Figure 8.6 shows the Gr–PI flaps in the engine. Gas-turbine engines of fighter aircraft often use augmentors or afterburners to provide additional thrust for takeoff, combat maneuvers, and high-velocity flight. The augmentor is located behind the turbomachinery and is followed by a variable-area exhaust nozzle, which is used to control the turbine exhaust pressure and exhaust-gas flow rate, which affect engine performance. Since the nozzle and augmentor are aft of the engine, which in turn is typically located in the aft of the aircraft, weight savings in the augmentor and nozzle sections are highly beneficial to aircraft weight and balance. Indeed, for some fighter aircraft, every pound of weight saved in the nozzle reduces the aircraft takeoff gross weight by up to 4 lb (1.8 kg), a saving that can be used to provide increased payload or range.

The composite is a viable substitution for existing sheetmetal titanium structures. Weight savings, projected cost savings, the durability currently being demonstrated, and

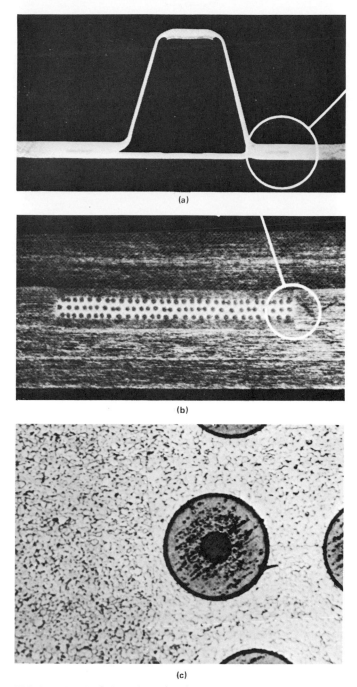

**FIG. 8.5** Details of selectively reinforced titanium structures: (*a*) end view of sectioned hat. (*b*) Placement between sheets; material is B$_4$C-coated boron in Ti–6Al-4V SPF/DB hat-stiffened section (×10). (*c*) Quality of diffusion bond (×200).[7,8]

**FIG. 8.6**  Closeup view of composite flap installation. *(Pratt & Whitney, Division of United Technologies.)*

the reduced use of a strategic element make future applications of composites to gas-turbine engines a virtual certainty. As temperatures climb and as other components are selected as potential composite components, Gr–PI and other polyimides with SiC and $B_4C$ fibers will be developed.

## NEW MATERIAL PROCESSING

### Fiber and Matrixes

A process improvement in the adhesion between graphite fibers and metal matrixes depends on depositing a thin, amorphous carbon coating on graphite fibers. This coating through the use of molten-metal baths makes it possible to apply oxide and magnesium coatings to the smooth surface of the graphite fibers.

### *Felt Fibers*

A new form of reinforcement is a rigid, bonded felt made from short fibers. Rigid-felt reinforcements can be made from almost any short metal, plastic, or ceramic fiber, e.g., metal-bonded stainless-steel fibers, epoxy-bonded aramid fibers, and ceramic-bonded SiC whiskers. The felts are formed by air-laying or vacuum-forming from a liquid suspension into a preform of flat sheets or complex shapes. After the shape has been formed, a resinous binder in a solvent is drawn through the felt and then drained of all excessive fluid. Normal surface-tension forces cause the binder to accumulate at the joints or contact points of the fibers. When the binder is cured, the felt becomes a rigid, low-density, three-dimensional trusswork of fibers. After the three-dimensional trusswork of fibers has been made, it can either be used as is (a low-density structural material) or be impregnated with a resin such as polyester or epoxy. This type of reinforcement has applications in low-density metals for modulus and strength reinforcement, in high-temperature plastic to prevent creep and raise the heat-distortion temperature, and in ceramics to improve thermal and mechanical shock resistance.

### Aramid Fibers

A new form of aramid fiber, an ultrafine, short-fiber pulp, is expected to compete successfully with glass, carbon, and steel fibers to replace asbestos in automotive brakes, clutches, gaskets, and filters and to be used in reinforced-phenolic parts for electrical applications. The pulp of very short highly fringed fibers can be handled on standard wet or dry mixing equipment. Since the aramid pulp has a high specific tensile strength, only 5 to 40% as much is needed to replace asbestos as a reinforcing material. The result is a cost-effective material, even though Kevlar fibers cost at least 7 times as much as asbestos.

The multitude of subfibers on the surface of the core fiber are curled, flattened, and branched, unlike most mineral and synthetic fibers, which are smooth, straight, and needle-like. The effective $L/D$ ratio is increased because the fibrils are very fine, less than 1 $\mu$m in diameter, so that $L/D$ is often over 500 (Fig. 8.7). The high fibrillation also improves mechanical bonding in moldings and laminates. The aramid pulp is said to be capable of handling the high temperatures generated in brakes and clutches.

### SMC and BMC Fibers

Electromagnetic interference is all around us. We cannot escape it, indoors or out, and neither can sensitive electronic equipment. Computers, business machines, communications equipment, and even the onboard microprocessors that control automobile engines can all malfunction because of electromagnetic radio-frequency interference. Many of these devices are housed in plastics, which normally provide no shielding from electromagnetic pollution. Fortunately, conductive fillers can be added to these plastics to transform them from mere electrical insulators into functional conductors, capable of shielding electronics from

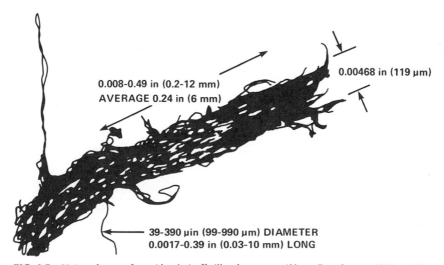

**FIG. 8.7** Unique feature of aramid pulp in fibrillated structure. (*Mater. Eng., January 1981, p. 22.*)

electromagnetic pollution. The electrically conductive additives allow plastic housings to reflect and absorb electromagnetic energy, thus combining the advantages of plastics with the shielding ability of metals.

Current research in conductive-plastics technology is aimed at determining the right type and amount of conductive filler required to shield against electromagnetic-radio-frequency interference without sacrificing part performance, particularly mechanical properties. Perhaps more important is the need to provide shielding at a cost comparable to or less than conventional methods (conductive coatings, metallization, and plating). Conductive plastics is still a new technology. Research organizations, material suppliers, and molders continue their efforts to develop better conductive plastics. The fabrication of complex conductive parts lends itself readily to a variety of plastics processes including injection molding, compression molding of BMC and SMC, structural-foam molding, and thermoforming of sheet. To provide the necessary shielding characteristics other fillers like carbon black, carbon pigments, and glass and carbon-fiber mat are part of the conductive system.[3]

New applications include an automotive SMC hood scoop to which a carbon-fiber mat imparts electromagnetic shielding, without which electromagnetic radiation from the engine, distributor, and spark plugs could reach the radio antenna and affect reception.

Conductive mats made of aluminized glass fibers are being considered for use in SMC, primarily because of lower cost. While the best material for electromagnetic shielding is the mat form, other available forms include roving, milled and chopped fibers, and woven fabrics. E-glass fibers, chemically bonded with a coating of pure aluminum, are woven into fabric. For electromagnetic shielding in aircraft, the conductive fabric called Thorstrand is used in wing-to-body fairings, wing coverings, wing-tip coverings, and other areas where both electromagnetic shielding and static bleed are necessary to protect sensitive instrumentation. The woven conductor usually forms the outside layer in a typical aircraft composite structure, which is laid up then vacuum-bagged.

Development is continuing in BMCs. Several fillers have been tried, but in BMCs, where strength is not really a consideration, flake reinforcements may be the best bet. Adding pure-aluminum flake results in more conductivity than aluminized glass fibers on a pound-for-pound basis. While aluminum-coated glass can produce some strengthening (as in SMC), it is not really necessary in BMC. Here pure aluminum ribbon or flake has an advantage in conductivity.

Stainless-steel fibers drawn from 316L stainless rod promise electromagnetic shielding at lower loadings than other conductive fillers. Both continuous strands and chopped fibers are available. Electroplated carbon fibers are yet another alternative to standard conductive fillers. A coating of silver, nickel, copper, or brass is electroplated onto graphite fibers. With an aspect ratio of 50:1, nickel-coated fibers yield electrical properties equal to those of nickel but have much lower specific gravity. Silver-plated fibers produce similar results. In recent studies nickel-coated fibers had conductivities 100 times greater than graphite fibers.

### Silicon Carbide Fibers

New MMC materials (chap. 2) are being developed for weight-critical structures and jet engines. These materials use high-performance, low-weight SiC fibers embedded in an aluminum or titanium matrix to form high-strength, high-stiffness, low-weight materials. The SiC fiber is produced by a special chemical vapor-deposition process; when it is embedded in, and metallurgically bonded to, the metallic-matrix base material, the resulting structure is stronger than steel with half its weight. For example, SiC–Al composite [with the SiC continuous fibers laid in a 0°/90° (0/1.6-rad) crosswise pattern] has a strength of over 120 kips/in$^2$ (838 MPa) and a modulus of over 20 × 10$^6$ lb/in$^2$ (138 GPa). SiC–Ti composites have somewhat higher mechanical properties, thanks to the stronger matrix, and maintain usable properties up to an exposed temperature of 1200°F (650°C).

SiC–MMC parts and components have been produced by casting and molding techniques so that the fibers are installed in the preferred orientation to carry the primary loads. Low-cost hot molding of aluminum composites is being developed so that aircraft skin panels, stringers, and beams can be molded directly to shape without having to be machined to profile. Methods for hot-pressing SiC–Ti engine components, such as blades, vanes, and shafts, are also being developed.

## Resins

New resin systems are continually being developed for applications (temperature, environment, etc.) as well as for specific industries. Two new epoxy-resin systems for aerospace applications have been recently introduced to industry. One features no tack, no flow, low bleed, and is a 350°F (177°C) cure resin for use in fabricating compression-molded parts. The second shows good tack, controlled flow, no bleed, and is self-adhesive for use in co-curing honeycomb sandwich panels and co-cured skins. Materials development continues to seek material useful to 350°F (177°C) and to develop a direct replacement for existing resin-matrix materials with improved properties that will be easier to process. Most promising at present is polybutadiene bisimide (PBBI), which resulted from development of the high-vinyl-modified epoxy (HME) resins. A composite with a PBBI matrix can be cured without an autoclave, under only vacuum-bag pressure at 15 lb/in$^2$ (103 kPa). With PBBI, there is no bleeding during curing. Thus, thick parts containing many plies can be made more easily, and co-curing becomes easier. PBBI is only slightly affected by moisture; its properties at higher temperatures exceed those of wet epoxy (of the type now used in composites) and both wet and dry HME resins.

## Self-Reinforcing Plastics

Plastics that reinforce themselves are unique polymers with extended-chain rigid-rod molecules. They can be lined up during processing into a fiber or film to give high strength and stiffness in one direction. Unlike amorphous plastics, in which the coillike molecules are looping chains like spaghetti, they need not be stretched or oriented to achieve strength or rigidity at the molecular level. The molecular composite would consist only of polymers, preferably in injection-moldable form. Any type of product or part demonstration is 5 to 10 years away.

There are several reasons for bothering with ordered polymers or molecular composites. In fiber-reinforced materials the interface between the fiber and the matrix presents problems, as do differing properties of fiber and matrix, e.g., coefficient of thermal expansion. Self-reinforced, homogeneous materials avoid this problem. The thermal stability and environmental resistance of the new polymers should be higher than those of current fiber-reinforced materials. Processability should be inherently better.

In fiber form, the rigid-rod polymers like polybenzothiazole (PBT) have essentially the properties of graphite fiber [on the order of 30 × 10$^6$ lb/in$^2$ modulus (207 GPa) and 250 to 300 kips/in$^2$ (1724 to 2068GPa) tensile strength] with an unoptimized fiber that has flaws. The properties of aluminum that could be attained in a molecular composite [10 to 11 × 10$^6$ lb/in$^2$ (69 to 76 GPa) modulus and 100 to 200 kips/in$^2$ (690 to 1380 MPa) tensile strength] make it conceivable that aluminum could eventually be replaced.

## NEW TOOLING AND CURING TECHNIQUES

### Tooling

Figure 8.8 shows an integrally heated bonding press, which can bond parts up to 10 in (254 mm) thick in only half the time of an autoclave. A unique temperature-hold feature guaranties uniformity of heating for the entire unit with a tolerance of ±5 Fahrenheit degrees (±2.8 Celsius degrees). The press can cure components up to 400°F (204°C) and can preprogram rates for temperature rise and cooldown. Heat transfer is by direct contact with heated platens (rather than convection by air, as with an autoclave), making the press more energy-efficient. Other advantages include less structure in basic bond tools, elimination of problems associated with envelope bagging, and visual inspection during the cure cycle.

The press has successfully produced fiber-glass rotor blades lasting 2½ times longer than metal blades with increased operational safety. With introduction of this blade, preimpregnated fiber-glass roving was used for the spars. It was less expensive than tape and more adaptable to machine fabrication, permitting better quality control during production of the large blades than with hand lay-up. The spar caps were wound on an orbital pin-winding machine (Fig. 8.9), a one-of-a-kind replacement for a conventional filament-winding machine. An orbital winding accessory for the machine enabled it to form spar straps.

**FIG. 8.8**  Integrally heated bonding press.[9]  *(Bell Helicopter Textron.)*

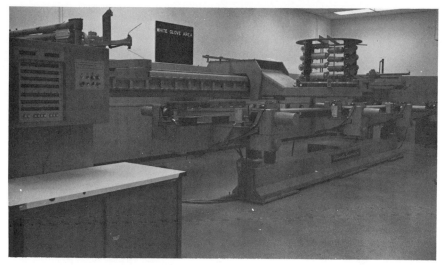

**FIG. 8.9** Orbital pin-winding machine.[9] *(Bell Helicopter Textron.)*

## Nonautoclave Curing

Nonautoclave composite curing is adaptable to large complex-shaped structures and high-rate automated production, uses less energy, eliminates component-size restrictions, enlarges the manufacturing-base flexibility of a company, and reduces current and future capital-equipment outlays. The nonautoclave processes use an oven, integrally reinforced tools, and presses. Background development work has shown most volatile material in resin-matrix laminates to be entrapped air, i.e., either the interlaminar air trapped in lay-up or the intraply air from fiber-matrix consolidation. Conventional vacuum-pressure cure with a bag in an oven provides insufficient pressure for compaction so that consolidation pressure seals entrapped air whereas autoclave pressures of 85 lb/in² (586 kPa) provide the air compaction and sufficient resin flow for adequate cure. Various physical-property-comparison curing tests and mechanical-property tests from room temperature through 250°F (20 to 121°C) have been conducted[4] (Fig. 8.10).

## Radio-Frequency Curing

The feasibility of curing Gl–Ep composites by radio-frequency heating rather than conventional conduction heating was demonstrated[5] in the early 1970s. Radio-frequency curing eliminates large energy-consuming facilities and costly flow times, and its economic feasibility for the future has been demonstrated. A conveyor can be used to pass a self-contained pressurized tool holding the uncured Gl–Ep part between flat-plate electrodes and through the emitted radio-frequency field to cure complex shapes of varying thickness and permit use of inexpensive nonmetallic tooling.

Using polystyrene in lieu of polypropylene as die material offers improved interlaminar-shear properties because polystyrene is a stronger tooling material, allowing greater pressure

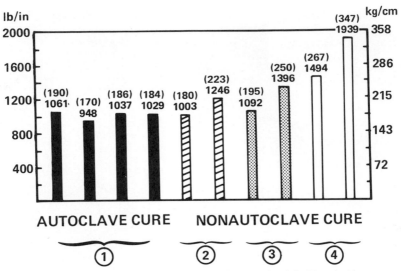

FIG. 8.10 Comparison of test results for T-shaped specimens bonded with and without auto-clave.[4] 1 = Formed in rubber–fiber-glass tool; staged and cured in autoclave. 2 = Formed and staged on thin-line rubber mold with wood tool with electrical heaters; cured in oven. 3 = Material staged in flat; formed over rubber mold with wood tool and cured in oven. 4 = Like 2 but specimen had adhesive on all sides.

during cure and capable of withstanding temperatures of 300 to 350°F (149 to 177°C). Although constant-thickness laminates cure uniformly, other shapes, e.g., wedges, require work to develop special cure cycles and parameters. Figure 8.11 shows the economics in three charts reflecting the cost savings achievable with radio-frequency curing.

## Xenon

Another potential rapid, low-energy curing system is an augmented pressure system for a xenon-flashlamp curing device.

## NEW APPLICATIONS

### Commercial Aircraft

When you look 10 years into the future of commercial aircraft (Fig. 8.12), composites will make up over 30% of two models, including the wing and fuselage sections which are primary structure. Future designers must be ready to exploit the advantages of composite materials because their use will affect the three-dimensional view of the aircraft from the very beginning. As composites are used to reduce weight in major elements, designers can use new aerodynamic shapes and smaller engines (or stay with current engine class and get better performance). The nacelles, which cover the engines, will be wrapped as shown in Fig. 8.13.

The small-commercial-aircraft business will make its major breakthrough of the decade in 1983, when the first all-composite Lear Fan aircraft is delivered. Gr–Ep composite material is used for the fuselage, airframe, wing, rudder, aileron sections, flap sections, and elevators. The propeller on this twin-engine plane is a Kv–Ep composite. It is claimed that this new aircraft can fly higher and faster than other aircraft in the general aviation field.

## Military Aircraft

Fighters and bombers in the next decade will use more composites: the B-1B bomber will include 3200 lb (1452 kg) of Gr–Ep, 400 lb (181 kg) of B–Ep, and approximately 20,000 lb (9100 kg) of Gl–Ep and Nomex honeycomb core. Initial reports indicate that over 26%

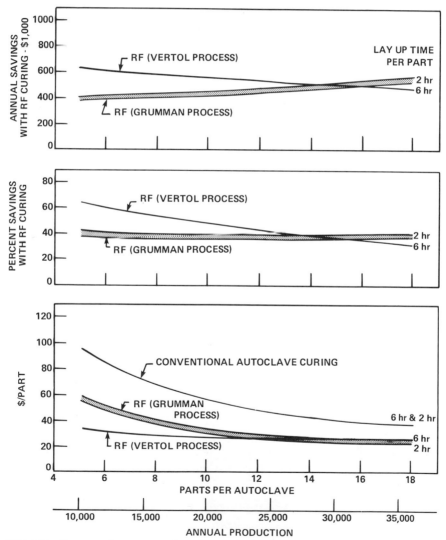

**FIG. 8.11** Representative cost comparison of conventional and radio-frequency curing (2000 autoclave cycles per year); data from Ref. 5.

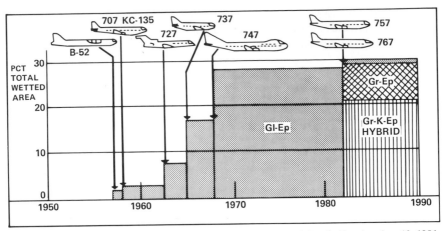

**FIG. 8.12** Evolution of composite structure in one group of commercial aircraft. *(Iron Age, Aug. 12, 1981, p. 60.)*

of the weight of the next generation of fighter aircraft design for the 1990s will be Gr–Ep. In addition to weight savings, the design exploits the unidirectional stiffness of the material. Both the wings and canards (a smaller forward set of wings) are aeroelastically tailored for increased maneuverability and performance. Both the wings and canards twist and bend in flight to achieve maximum performance for the particular flight conditions. The high-technology design is aimed at giving future fighter planes twice the maneuverability of present fighters at transonic and supersonic speeds. This is achieved in part by orienting the composite plies to control the change of the airfoil shape under load as a function of wing and canard bending, attaining the most favorable airfoil shape in terms of lift and

**FIG. 8.13** Gr–Ep nacelle of 747 airliner being filament-wound. *(Fiber Science, Inc.)*

drag. The wing and canard both use unique Gr–Ep ply orientation. The wing lamina has a ply orientation of 35° (0.61-rad) and ± 50° (0.87-rad) angles from the geometric center. For the canard, ply orientation is 15° (0.26-rad) and ± 45° (0.79 rad). The currently tested fighter of the 1990s contains 26% Gr–Ep and 3% Gl–Ep. The skins on the fuselage, wings, canards, engine inlets, vertical stabilizer, and wing and canard spars are Gr–Ep.

Large transport aircraft being modified, redesigned, and developed to meet the demands of the next decade are candidates for MMC. Two materials currently being developed under government sponsorship are SiC–Al, which is as strong and stiff as steel but as light as aluminum, and SiC-whisker-reinforced aluminum. Both materials appear to exhibit properties equal to or better than those of boron- or carbon-reinforced systems. The SiC fiber has values on the order of 18 kips/in$^2$ (124 MPa) unidirectional tensile strength and a modulus of elasticity of 28 × 10$^6$ lb/in$^2$ (193 GPa). The whisker system has an ultimate strength of more than 100 kips/in$^2$ (690 MPa) and a modulus of 14 to 16 × 10$^6$ lb/in$^2$ (97 to 110 GPa). It is isotropic, whereas the fiber material has a strength of only about 15 or 20 kips/in$^2$ (103 or 138 MPa) in the transverse direction.

The first component will be a wing box, measuring 28 by 14 by up to 3 ft (8.5 by 4 by 0.9 m), which will meet all form, fit, and function requirements for the new C-5A cargo aircraft wing.

## Engines

Composites may improve many critical areas of engine technology (Chap. 7). A wide variety of turbine-engine components have been or are being designed, fabricated, and tested to confirm the advantages of using reinforced-composite materials. Unfortunately the piecemeal development of various components on a variety of existing engines has severely limited the ability to design an optimized engine. The 1980s will be the time to exploit the advantages of applying reinforced composites to the maximum possible extent in a turbine engine, in a design that would lift the many constraints of configuration, operating-temperature variance, interface requirements, etc. Enough components have already been demonstrated to lend confidence in approaching the design, fabrication, and testing of components that have not yet been developed. A parallel development of the supporting materials, design, analysis, and testing technologies is needed. Clearly, as the energy-conscious 80s pass, it is time for the composite airplanes on the drawing board to be designed with composite engines.

## Helicopters

Composite structures as a prime element of helicopters for the decade 1985–1995 are being pursued by several companies under government funding. The program calls for development of an all-composite, nonmetallic fuselage using an existing drive train with a resultant 22% weight reduction and 17% cost reduction over conventional metal airframes. At the same time, military requirements for crashworthiness, ballistics tolerance, reliability, maintainability, and reduced radar signature must be met. If the composite technology is proved, it would pave the way for the use of composites in primary airframe structures in both helicopters and fixed-wing aircraft. Materials included in the primary structure include

Gr–Ep, Gl–Ep, Kv–Ep, B–Ep, Nomex, and polyimides for parts exposed to heat. European manufacturers have a reputation for mastery of composite materials, but United States technology exceeds the foreign competition even though implementation of composites has advanced in foreign aircraft in certain limited and restricted areas, namely, the application of composites to rotor-head design and rotor-blade design. The United States has now forged ahead in rotor-blade technology through mechanized manufacturing methods involving fiber spinning with variable geometry, but technology breakthroughs must continue.

## Ground Transportation

The government is convinced (although the automobile industry is not completely convinced) that Gr–Ep and other composite materials have a major role in ground transportation. Composites are expected to be considered as materials for the engine, drive train, chassis, suspension, body, and structure of future cars and trucks.

It should be noted that the successful replacement of one metal part often leads to secondary benefits. For example, a Gr–Ep filament-wound drive shaft has been extensively tested. Besides having excellent torsional strength, it requires less balancing, reduces the noise level, and affords a smoother ride. The use of Gr–Ep in the automobile industry will be about 5 lb (2.3 kg) per average car in 1985 but should climb to 90 lb (41 kg) per car in 1990.

With hundreds of plastic-processing machines, the auto industry is capable of handling fiber-reinforced materials. Mixing heads of injection machines can handle fibers over 2 in (50.8 mm) long. With a few adjustments to pressures, temperatures, and tolerances even the stamping machines can also be used on Gr–Ep materials. Today's 177-lb (80-kg) auto bumper will weigh 54 lb (24.5 kg) if the component is made out of 50 lb (22.7 kg) of steel and 4 lb (1.8 kg) of Gr–Ep.

## Other Innovations

To lower the cost of solar-energy systems the structural use of concrete reinforced with zirconium-sand glass fibers is being investigated. Using chopped fibers 1.5 in (38 mm) long for 5% of the total weight, researchers have been able to cast sections as thin as 0.18 in (4.6 mm). The technology promises low production costs for parabolic troughs, heliostats, and flat-plate collectors for solar and wind-power generation.

Military applications include armor (Chap. 7) and improvement in gun barrels, vehicle towbars, and missile silos.

Kv–Ep cabling is being evaluated to replace steel in suspension bridges. The Kv–Ep cables are lighter, easier to work with, and nonconductive so they will not interfere with transmission. Composite materials have been proposed to augment an all-aluminum bridge which is attached to a vehicle and can be unfolded over a river or valley, providing mechanized units with a quick, safe crossing.

Airline seats will use a Gr–Ep composite in future designs; a 35-lb (15.9-kg) seat is predicted by 1985. Fixed-gantry high-speed machines used for cutting honeycomb-core composites in aircraft fabrication (Fig. 8.14) use a tubular ram constructed of Gr–Ep com-

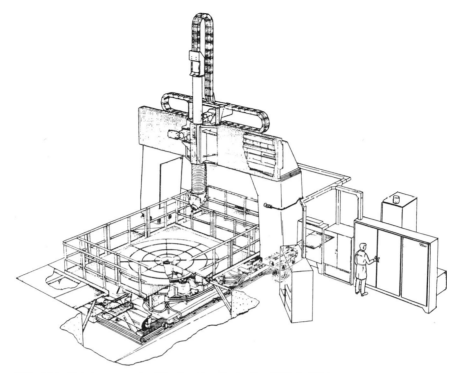

**FIG. 8.14**   Tubular ram made of Gr–Ep. *(Am. Mach., May 1981, p. 122.)*

posite material. The tube, 12 in (305 mm) OD and 0.5 in (12.7 mm) wall thickness, was made by filament winding. Future machine tools and equipment will use composites.

Composite-materials research will help develop graphite-reinforced body implants (graphite has a surprising compatibility with body tissue) and design of a sophisticated three-finger hand for robots to better mimic the tendon-in-sheath structure of the human hand.

## NONDESTRUCTIVE TESTING

By virtue of their special fabrication techniques fiber-reinforced composite materials require inspection and quality control different from those for conventional structural metals. Because the composite may vary in constitution from point to point, the removal of samples for destructive analysis would be neither permissible nor particularly useful. Of course, the quality of the fibers and matrix materials may be regulated by sampling and process control, but this leaves many significant problems with the composites themselves:

1. Determination of
   *a.* State of cure of resin matrix
   *b.* Porosity of resin matrix

    *c.* Fiber-volume fraction

    *d.* Orientation and lay-up of fibers and plies

    *e.* Fiber-matrix interface condition

2. Detection of

    *a.* Delaminations and translaminer cracks

    *b.* Foreign inclusions

    *c.* Lack of bonding of ply to adjacent ply or other structure

Existing technology is adequate for many of these problems,[6] but many problems await complete solution. Thus ultrasonic scan by immersion, jet probes, or roller probes can usually give sufficient warning of porosity or delamination. Low-voltage, high-resolution radiography and the Fokker bond tester (or similar instrument) can be applied to smaller selected areas in cases of doubt, but these techniques have their own specific difficulties. Experienced operators employing visual inspection and coin tapping are still required. There are several outstanding problems on which resources could profitably be concentrated.

## Shapes and Structures Difficult to Scan

Complex shapes and composite structures are easy to scan ultrasonically. Some structures will require special scanning rigs. Existing techniques should be supplemented by ultrasonic time-delay methods, using separate transmitter and receiver. This can be backed up by direct calculations of the time of flight for defect-free structures with given fiber distributions.

A recent development for evaluating fiber-composite panels for mechanical-strength properties and in-service strength loss combines instrumentation from two separate technologies, acoustic emission and pulse ultrasonics. The usual acoustic-emission procedure analyzes spontaneous stress waves emitted as a result of material deformation and flaw growth. The new acoustoultrasonic procedure employs ultrasonically excited elastic waves that simulate acoustic emissions. A repeating, controlled set of elastic waves is generated. They interact with material morphology and boundary surfaces like spontaneous stress waves that arise at the onset of fracture. The output waveform resembles a burst type of acoustic emission and carries substantially more information about the material than about the signal source. In fiber-composite laminates the acoustoultrasonic waveform yields correlations with ultimate-tensile and interlaminar-shear strengths. More use can also be made of microfocus [down to 0.006 in (0.15 mm)] low-energy (down to 7 keV) x-ray equipment. These techniques should also be adapted to the examination of new resin systems.

## Factors Affecting Shear Strength

Although the relationship between matrix porosity and loss of shear strength has been widely studied, more work is required on establishing the relationship between delamination size and shear strength. Shear-strength measurements should not be confined to short-beam tests but should include other measurements more directly related to true shear strength.

## Environmental and Other Effects

There is still great uncertainty about effects of environmental degradation, moisture ingress, impact and fatigue damage on mechanical performance. A unified line of approach here would also be a valuable adjunct to production testing in that large structures can be surveyed rapidly, often without physical contact. In this context optical holography and acoustic emission have already been studied with limited success, but a number of related techniques might now be studied with advantage:

Electronic speckle-pattern interferometry (ESPI)
Laser-Doppler measurement
Thermoelastic stress analysis (TSA)
Tapping analysis

## Conclusions

Without simple, low cost unambiguous nondestructive inspection methods, composite designs for aeropropulsion hardware will have to continue to rely on large design margins of safety and expensive individual component proof testing to verify the component integrity. New methods will have to be developed before low-cost structural composite hardware on gas-turbine engines becomes a reality.

Nondestructive testing involves detection of flaws and evaluation of their effect on structural performance. This requires inspection equipment and analysis techniques. Advances in mechanics will help to provide the latter, but in the meantime we must continue to rely on empirical approaches, such as the fabrication and testing of components with deliberate defects. Equipment is required for both factory and field inspection. The latter poses a particularly difficult challenge.

## FUTURE

It is increasingly apparent that composites will become economically competitive with metals for several reasons:

1. Acquisition costs will drop because of
   a. Lower material costs as volumes build
   b. Lower fabrication costs as this technology advances, aided by development of new, process-oriented resins
2. Rising fuel prices will provide an increasing premium for weight reduction
3. Material improvements will further reduce maintenance costs and extend service life

While the lead time for structural metals has increased dramatically, that for composites has been relatively constant. Although this may change as demand for composites builds, a unique advantage of composites is that complex structures are fabricated from a relatively few types of prepregs while many different forms of metals (plate, extrusions, and forgings) are required. This should help to maintain the lead-time advantage of composites.

We shall be forced to conserve our finite resources as population grows and demand increases. Composites will play a large part in helping us pay more attention to the efficient use of structural materials and fuels.

## Equipment Improvements

Processing companies will soon be able to have the technical expertise of a field-service representative right in their plant whenever they need it, through the implementation of *modem* that will allow their microcomputers to communicate directly with the supplier's staff via telephone. Diagnostic work and troubleshooting will be easier and cheaper, and programming from a remote location will also be possible.

A desk-top system being developed reportedly will provide control information to be related directly to production-scale processing. This will reduce the time needed for theoretical calculation of control data for fiber placement before keying the data into the microcomputer for precise calculation of winding patterns; it will also cut the cost of downtime incurred by using production-scale computerized machines to determine such data for winding complex shapes.

A filament-winding manufacturer has designed and built a system that incorporates four microcomputers, one for controlling each axis of a four-axis machine. The four axes are monitored and controlled by a central "supervisor" microcomputer. This machine is twice as accurate as previous machines, provides for greater memory storage of information on each axis, and supplies the computational power of a minicomputer at a fraction of the cost.

A four-axis spherical winding machine for manufacturing pressure vessels interfaces with a master shop computer, which controls at least two other filament winders and assists in developing highly sophisticated winding programs. This system exemplifies the next significant step in computer-controlled winding systems, namely, computer-aided design–computer-aided manufacturing master-control capability. In such an operation, the microcomputer of the filament winder will interface with the master control system, which, through simulated three-dimensional cathode-ray-tube display, will assist in generating both product-design plots and production-control commands.

The next major change in manufacturing will be the result of the mass application of computers to both individual processes and entire factories. Minicomputers will become sufficiently inexpensive and compact to be embedded in manufacturing equipment throughout the factory. Initially they will provide process control and automation, but soon after computers may be used to evolve smart, rather than merely automatic, processes. Flexible processes will be developed that can accommodate rapid changes in raw materials or geometric shapes. Optimum quality for each composite part produced will be ensured by computers that call upon sophisticated analytical models of material behavior and the unit processes; by extensive computerized data bases containing part, material, and process information; by sensors that can detect not only process parameters but also component property and quality parameters; and by advanced adaptive logic systems for handling vast amounts of interactive data. These closed-loop smart processes will continuously evaluate the properties of incoming raw materials, compensate by adjusting their process parameters, use their advanced logic circuits to determine process improvements in real time, continuously update the manufacturing data base so that the knowledge gained can be applied automatically to related families of parts, and effectively perform in-process inspections and quality-control measurements so that no postproduction inspection and scrappage will be necessary.

## Beyond Tomorrow

Now that the roof for the Olympic Stadium in Montreal has been built with Kevlar fabric, far greater arch spans should be possible. The city under a dome so often envisioned by futurologists is a little closer to reality. How long could a suspension bridge be if it were made with cables of Kevlar fibers instead of steel? Farther in the future is the Skyhook concept. Imagine a satellite in synchronous orbit above the earth attached to a cable fastened to the earth. (There would also have to be a counterweight cable leading away from the satellite into space.) The cable to earth could support elevator cabs running up to the satellite and beyond to launch payloads into space. The cable to the satellite would have to be over 22,000 mi (35,400 km) long, and no material can support its own weight over such a distance, even if the cable is tapered. Composites like Kevlar might conceivably meet the calculations of scientists as cable material.

Some scientists believe that Kevlar fibers can enable them to mine the moon. Robotic mechanical shovels would load bags of Kv–Ep with moon dust and place them in a giant catapult. Since the moon's gravity is only about one-sixth that of earth, the catapult would easily hurl the bags out into space, where they would be caught by a mile-wide net of Kevlar at a space station where the moon dust is processed. Skyhook and moon mining may be science fiction today, but fiction is turning into fact at an accelerating rate.

Work for space platforms and processing has been started with governmental funding to several aerospace companies. The initial platforms will be built on the ground and then be carried aloft by the space shuttle. The eventual plan is to do all the construction in space. Several companies have developed *beam builders,* machines that can be operated in space. Rolled-aluminum or composite material fed into these machines is turned into extruded 0.01-in-thick (0.25-mm) shapes strong enough in space to support any equipment needed. Ground tests have been outstanding.

Traditionally, plastics in automobiles have been used in decorative and noncritical structural applications. Expansion of plastics use beyond this will require the applications of composites in major load-bearing components. Without further technological (and economical) breakthroughs, composites growth in automotives is expected to decelerate, primarily because the number of potential new applications is decreasing. For composites to achieve further growth in the late 80s, these materials must compete for structural and semistructural load-bearing applications. Other problems which must be overcome include difficulty in repair, joining, and recycling of state-of-the-art composites.

Recycling composite components is a real need. The latest estimates over a 100,000-mi (160,900-km) vehicle life indicate that the fabrication and use of a composite part will consume about one-half the energy required for the same part made of steel and slightly less than an aluminum part. After 100,000 mi (160,900 km) industry is able to recycle most of the steel and aluminum components for reuse in some fashion. That is not true of the composite components, obviously increasing total cost and energy estimates. The practical problem of how to recycle a scrapped automobile, sort out the different kinds of polymers, and recycle the composite remains to be solved. Developments and breakthroughs in the use of composites in automobiles will increase by 12.6% a year through 1995, primarily because of anticipated improved processing techniques and rising costs of metals. At that rate, the 1995 car will carry some 160 lb (73 kg) composites, compared with about 36 lb (16.3 kg) in 1980.

The best advice I can give to readers is to broaden your knowledge of all materials, for a better understanding of the many problems involved in materials engineering and science. Composites will continue to present the designer with great opportunities and challenges. There will be at the disposal of the designer, engineer, technologist, chemist, etc., an increasing array of fibers, reinforcement forms, and matrixes with which to create a material and a structure. This requires a better understanding of material properties and behavior than is needed for metal structures. In view of the newness of the technology, material variability, and the many unknowns, composite structure designs have tended to be conservative. Improved quality assurance in manufacturing, nondestructive testing, service experience, and understanding of composite behavior will allow the designer to reduce safety margins. The result will be significant improvements in structural efficiency.

## REFERENCES

1. Banks, M. T.: Structural Thermoplastic Composites, *Mater. Eng.,* October 1980, pp. 51–53.

2. Prewo, K. M., and E. R. Thompson: Research on Graphite Reinforced Glass Matrix Composites, United Technologies Research Center, E. Hartford, Conn., NASA-CR-165711, Langley Research Center, Contr. NASl-14346, May 1981.

3. Wehrenberg, R. H., II, Today's Conductive Plastics Combine Shielding plus Strength, *Mater. Eng.,* March 1982, pp. 37–43.

4. Burroughs, W.: Nonautoclave Cure Manufacturing Technology, *MTAG Nonmet. Compos. Prog. Rev., Orlando, Feb. 23, 1981* (RI, NAAD, Los Angeles, Calif., Contr. F33615-80-C-5080).

5. Mahon, J., et al.: Manufacturing Methods for Rapidly Curing High Temperature Components, *Grumman Aerospace Corp., Air Force Mater. Lab. Tech. Rep.* 73-159, Bethpage, N.Y., August 1973.

6. Mayfield, J.: Hornets Fly on Composite Wings, *Amer. Mach.,* November 1978, pp. 107–110.

7. Lorenz, R. H., T. E. Steelman, and W. D. Padian: Selective Reinforcement of Low-Cost Titanium Components, *Rockwell International Corp.,* NAAD, AFWAL-TR-80-4014, *Final Rep. September 1978–October 1979,* Contr. F33615-78-C-5180, Air Force Materials Laboratory, Wright-Patterson AFB, Ohio.

8. Steelman, T. E., R. H. Lorenz, G. R. Martin, and R. P. Robelotto: Silicon Carbide/Titanium Material and Process Fundamentals, Rockwell International Corp. NAAD, IR-2420 (III), *Interim Rep. June 16, 1980–Jan. 15, 1981,* Contr. F33615-79-C-5012, Air Force Materials Laboratory, Wright-Patterson AFB, Ohio.

9. Anderson, R. G., and E. E. Blake: Fabrication and Demonstration of an Integrally Heated and Pressurized Mold System, *AVRADCOM Final Rep.* TR 81-F-11 (Bell Helicopter Textron, Ft. Worth, Tex., Contr. DAAG46-79-C-0032) 1979.

## BIBLIOGRAPHY

Bolinger, P. N.: Manufacturing Densification Methods for Carbon-Carbon Composite Nozzles, *General Electric Co.,* AFWAL-TR-81-4085, Contr. F33615-77-C-5207, *Final Rep. February 1978– May 1981,* Wright-Patterson AFB, Ohio.

Hoffman, R. E., W. L. Greever, and C. H. Sheppard: HME-350 Advanced Composite Resin Development, Hercules Incorporated, Magma, Utah and Boeing Co., Seattle, Wash., AFWAL-TR-81-4165, Contr. F33615-78-C-5170, *Final Rep., September 1978–August 1981,* Wright-Patterson AFB, Ohio.

Hulse, C. O.: Development of Directionally Solidified Eutectic Ceramic Matrix-Metal Composites, *United Technologies Corp. Rep.* R77-91-2587-3 and -4, Contr. N62269-76-C-0250, NADC-75282-30, E. Hartford, Conn., May 1, 1977.

Hunt, M. S.: An Introduction to the Use of Carbon Fibre Reinforced Composite Materials for Surgical Implants, National Mechanical Engineering Research Institute, Pretoria, South Africa, CSIR-ME-1689.

Monteleone, R., B. Kay, and J. Ray: Airframe Preliminary Design for an Advanced Composite Design Airframe Program (ACAP), *Sikorsky Aircraft, Div. United Technologies, Rep.* 1, July 1981; *Rep.* 2, October 1981; *Rep.* 3, January 1982, DAAK51-81-C-0017, Army Technology Laboratory, Ft. Eustis, Va.

Reynolds, W. N.: Nondestructive Examination of Composite Materials: A Survey of European Literature, NDT Cent. AFRE Harwell, Didcot, Oxon, *AVRADCOM Final Rep.* TR81-F-6, Contr. DAJA37-79-C-0553.

Ritter, L. C.: Conveyorized Radio Frequency Cure of Epoxy Glass Composites, Boeing Vertol Co., *AVRADCOM Final Rep.* TR80-F-16, Contr. DAAG46-79-C-0009.

# Glossary

*Many abbreviations are spelled out in the Glossary without a full explanation of their meaning, which should be sought through the Index.*

**ABL bottle**   An internal-pressure-test vessel about 18 in (457 mm) in diameter and 24 in (610 mm) long, used to determine the quality and properties of the filament-wound material in the vessel.

**Abhesive**   A film or coating applied to one solid to prevent (or greatly decrease) the adhesion to another solid with which it is to be placed in intimate contact, e.g., a parting or mold-release agent.

**Ablative plastic**   A material which absorbs heat (while part of it is being consumed by heat) through a decomposition process (pyrolysis) taking place near the surface exposed to the heat.

**ABS**   Acrylonitrile-butadiene-styrene.

**Accelerator**   A material mixed with a catalyzed resin to speed up the chemical reaction between the catalyst and resin; used in polymerizing resins and vulcanizing rubbers; also known as *promoter* or *curing agent*.

**Activator**   An additive used to promote the curing of matrix resins and reduce curing time. (*See also* Accelerator.)

**Additive**   Any substance added to another, usually to improve properties.

**Adherend**   A body held to another body by an adhesive.

**Adhesion**   The state in which two surfaces are held together at an interface by forces or interlocking action or both.

**Adhesion, mechanical**   Adhesion between surfaces in which the adhesive holds the parts together by interlocking action.

**Adhesive, contact**   *See* Contact adhesive.

**Adhesive film**   A synthetic resin adhesive, usually of the thermosetting type, in the form of a thin dry film of resin, used under heat and pressure as an interleaf in the production of laminated materials.

**Adhesiveness** The property defined by the adhesion stress $A = F/S$, where $F$ = perpendicular force to glue line and $S$ = surface.

**Aggregate** A hard fragmented material used with an epoxy binder, as in epoxy tools.

**Aging** The process or the effect on materials of exposure to an environment for an interval of time.

**AIMS** Advanced integrated-manufacturing system.

**Air-bubble void** Noninterconnected spherical air entrapment within and between the plies of reinforcement.

**Air locks** Surface depressions on a molded part, caused by trapped air between the mold surface and the plastic.

**Air vent** Small outlet to prevent entrapment of gases.

**Ambient** The surrounding environmental conditions, e.g., pressure or temperature.

**Anistropic** Exhibiting different properties when tested along axes in different directions.

**Anisotropic laminate** One in which the strength properties are different in different directions.

**Anisotropy of laminates** The difference of the properties along the directions parallel to the length or width into the lamination planes or parallel to the thickness into the planes perpendicular to the lamination.

**APIs** Addition-reaction polyimides.

**Arc resistance** The total time in seconds that an intermittent arc can play across a plastic surface without rendering the surface conductive.

**Ash content** The solid residue remaining after a reinforcing substance has been incinerated or strongly heated.

**Aspect ratio** The ratio of length to diameter of a fiber.

**A stage** An early stage in the polymerization reaction of certain thermosetting resins (especially phenolic) in which the material, after application to the reinforcement, is still soluble in certain liquids and is fusible; sometimes referred to as *resole*. (*See also* B stage, C stage.)

**ATLAS** Automated tape lay-up system.

**Attenuation** The process of making thin and slender, as applied to the formation of fiber from molten glass.

**Autoclave** A closed vessel for conducting a chemical reaction or other operation under pressure and heat.

**Autoclave molding** After lay-up, the entire assembly is placed in steam autoclave at 50 to 100 lb/in$^2$ (23.4 to 47.6 Pa); additional pressure achieves higher reinforcement loadings and improved removal of air.

**Automatic mold** A mold for injection or compression molding that repeatedly goes through the entire cycle, including ejection, without human assistance.

**Automatic press** A hydraulic press for compression molding or an injection machine which operates continuously, being controlled mechanically, electrically, hydraulically, or by a combination of these methods.

**Axial winding** In filament-wound reinforced plastics, a winding with the filaments parallel to the axis.

**Back draft** An area of interference in an otherwise smooth-drafted encasement; an obstruction in the taper which would interfere with the withdrawal of the model from the mold.

**Backpressure** Resistance of a material, because of its viscosity, to continued flow when a mold is closing.

**Bag molding** A technique in which the consolidation of the material in the mold is effected by the application of fluid pressure through a flexible membrane.

**Balanced design** In filament-wound reinforced plastics, a winding pattern so designed that the stresses in all filaments are equal.

**Balanced-in-plane contour** In a filament-wound part, a head contour in which the filaments are oriented within a plane and the radii of curvature are adjusted to balance the stresses along the filaments with the pressure loading.

**Balanced twist** An arrangement of twist in a plied yarn or cord which will not cause twisting on itself when the yarn or cord is held in the form of an open loop.

**Barcol hardness** A hardness value obtained by measuring the resistance to penetration of a sharp steel point under a spring load. The instrument, the Barcol Impressor, gives a direct reading on a scale of 0 to 100. The hardness value is often used as a measure of the degree of cure of a plastic.

**Bare glass** Glass (yarns, rovings, or fabrics) from which the sizing or finish has been removed or before it has been applied.

**Base** The reinforcing material (glass fiber, paper, cotton, asbestos, etc.) which is impregnated with resin in the forming of laminates.

**Batch** A measured mix of various materials. (*See also* Lot.)

**Batt** Felted fabrics; structures built by the interlocking action of fibers themselves without spinning, weaving, or knitting. (*See also* Felt.)

**Bearing area** The diameter of the hole times the thickness of the material.

**Bearing strength** The bearing stress at that point on the stress-strain curve where the tangent is equal to the bearing stress divided by $n\%$ of the bearing-hole diameter.

**Bearing stress** The applied load in pounds divided by the bearing area. (Maximum bearing stress is the maximum load in pounds sustained by the specimen during the test divided by the original bearing area.)

**Biaxial load** (1) A loading condition in which a laminate is stressed in at least two different directions in the plane of the laminate. (2) A loading condition of a pressure vessel under internal pressure and with unrestrained ends.

**Biaxial winding** In filament winding, a type of winding in which the helical band is laid in sequence, side by side, with no crossover of fibers.

**Bidirectional laminate** A reinforced plastic laminate with the fibers oriented in various directions in the plane of the laminate; a cross laminate. (*See also* Unidirectional laminate.)

**Binder** The resin or cementing constituent of a plastic compound which holds the other components together; the agent applied to glass mat or preforms to bond the fibers before laminating or molding.

**Blanket** Plies which have been laid up in a complete assembly and placed on or in the mold all at one time (flexible-bag process); also the form of bag in which the edges are sealed against the mold.

**Bleedout** In filament winding, the excess liquid resin that migrates to the surface of a winding.

**Blister** Undesirable rounded elevation of the surface of a plastic with boundaries that are more or less sharply defined, resembling in shape a blister on the human skin; the blister may burst and become flattened.

**BMC**   Bulk-molding compound.

**Bond strength**   The amount of adhesion between bonded surfaces; a measure of the stress required to separate a layer of material from the base to which it is bonded. (*See also* Peel strength.)

**Boss**   Protuberance on a plastic part designed to add strength, to facilitate alignment during assembly, to provide for fastenings, etc.

**Bottom plate**   A steel plate fixed to the lower section of a mold, often used to join the lower section of the mold to the platen of the press.

**Breathing**   (1) Opening and closing a mold to allow gases to escape early in the molding cycle (also called *degassing*). (2) Permeability to air of plastic sheeting.

**Bridging**   A region of a contoured part which has cured without being properly compacted against the mold.

**Broad goods**   Woven glass, synthetic fiber, or combinations thereof over 18 in (457 mm) wide.

**B stage**   An intermediate stage in the reaction of certain thermosetting resins in which the material swells when in contact with certain liquids and softens when heated but may not dissolve or fuse entirely; sometimes referred to as *resistol*. The resin in an uncured prepreg or premix is usually in this stage. (*See also* A stage, C stage.)

**Bubble**   A spherical internal void; globule of air or other gas trapped in a plastic.

**Buckling**   Crimping of fibers in a composite material, often occurring in glass-reinforced thermoset due to resin shrinkage during cure.

**Bulk density**   The density of a molding material in loose form (granular, nodular, etc.), expressed as a ratio of weight to volume.

**Burst strength**   Hydraulic pressure required to burst a vessel of given thickness; commonly used in testing filament-wound composite structures.

**Butt joint**   *See* Joint.

**Butt wrap**   Tape wrapped around an object in an edge-to-edge fashion.

**CAD**   Computer-aided design.

**CAM**   Computer-aided manufacturing.

**Catalyst**   A substance which changes the rate of a chemical reaction without itself undergoing permanent change in its composition; a substance which markedly speeds up the cure of a compound when added in small quantity compared with the amounts of primary reactants. (*See also* Curing agent, Hardener, Inhibitor, Promoter.)

**Catenary**   A measure of the difference in length of the strands in a specified length of roving as a result of unequal tension; the tendency of some strands in a taut horizontal roving to sag lower than the others.

**Caul**   A sheet the size of the platens used in hot pressing.

**Cavity**   (1) Depression in mold. (2) The space inside a mold into which a resin is poured. (3) The female portion of a mold. (4) That portion of the mold which encloses the molded article. (5) That portion which forms the outer surface of molded article (often referred to as the *die*). (6) The space between matched molds; depending on the number of such depressions, molds are designated as single or multiple-cavity.

**CC**   Carbonaceous heat-shield composites (C–C is carbon-carbon).

**Centerless grinding**   A technique for machining parts having a circular cross section, consisting of grinding the rod which is fed without mounting it on centers. Grinding is accomplished by working the material between wheels rotating at different speeds; the faster, abrasive wheel cuts the stock. Variations of the basic principle can be used to grind internal surfaces.

**Centrifugal casting**  A high-production technique for cylindrical composites, such as pipe, in which chopped strand mat is positioned inside a hollow mandrel designed to be heated and rotated as resin is added and cured.

**CFG iron**  Compact-flake-graphite iron.

**Charge**  The measurement or weight of material (liquid, preformed, or powder) used to load a mold at one time or during one cycle.

**Chase**  (1) The main body of the mold, which contains the molding cavity or cavities, or cores, the mold pins, the guide pins or the bushings, etc. (2) An enclosure of any shape used to shrink-fit parts of a mold cavity in place to prevent spreading or distortion in hobbing or to enclose an assembly of two or more parts of a split cavity block.

**Chill**  (1) To cool a mold by circulating water through it. (2) To cool a molding with an air blast or by immersing it in water.

**Circuit**  In filament winding (1) one complete traverse of the fiber-feed mechanism of a winding machine; (2) one complete traverse of a winding band from one arbitrary point along the winding path to another point on a plane through the starting point and perpendicular to the axis.

**Circumferential ("circ") winding**  In filament-wound reinforced plastics a winding with the filaments essentially perpendicular to the axis.

**Clamping plate**  A mold plate fitted to the mold and used to fasten the mold to the machine.

**Clamping pressure**  In injection molding and transfer molding the pressure applied to the mold to keep it closed, in opposition to the fluid pressure of the compressed molding material.

**Co-curing**  Simultaneous bonding and curing of components.

**Coefficient of elasticity**  The reciprocal of Young's modulus in a tension test.

**Coefficient of expansion**  The fractional change in dimension of a material for a unit change in temperature. Also called *coefficient of thermal expansion.*

**Coefficient of friction**  A measure of the resistance to sliding of one surface in contact with another surface.

**Coefficient of thermal expansion $\alpha$**  The change in length per unit length produced by a unit rise in temperature.

**Cohesion**  (1) The propensity of a single substance to adhere to itself. (2) The internal attraction of molecular particles toward each other. (3) The ability to resist partition from the mass. (4) Internal adhesion. (5) The force holding a single substance together.

**Cold-setting adhesive**  A synthetic resin adhesive capable of hardening at normal room temperature in the presence of a hardener.

**Compatibility**  The ability of two or more substances combined with each other to form a homogeneous composition with useful plastic properties.

**Composite**  A homogeneous material created by the synthetic assembly of two or more materials (a selected filler or reinforcing elements and compatible matrix binder) to obtain specific characteristics and properties. Composites are subdivided into the following classes on the basis of the form of the structural constituents: *fibrous:* the dispersed phase consists of fibers; *flake:* the dispersed phase consists of flat flakes; *laminar:* composed of layer or laminar constituents; *particulate:* dispersed phase consists of small particles; *skeletal:* composed of a continuous skeletal matrix filled by a second material.

**Compression mold**   A mold which is open when the material is introduced and which shapes the material by heat and by the pressure of closing.

**Compression molding pressure**   The unit pressure applied to the molding material in the mold.

**Compressive modulus** $E_c$   Ratio of compressive stress to compressive strain below the proportional limit. Theoretically equal to Young's modulus determined from tensile experiments.

**Compressive strength**   (1) The ability of a material to resist a force that tends to crush. (2) The crushing load at the failure of a specimen divided by the original sectional area of the specimen.

**Compressive stress**   The compressive load per unit area of original cross section carried by the specimen during the compression test.

**Conductivity**   (1) Reciprocal of volume resistivity. (2) The conductance of a unit cube of any material.

**Contact adhesive**   An adhesive which for satisfactory bonding requires the surfaces to be joined to be no farther apart than about 0.004 in (0.1 mm).

**Contact molding**   A process for molding reinforced plastics in which reinforcement and resin are placed on a mold, cure is at room temperature using a catalyst-promoter system or by heat in an oven, and no additional pressure is used.

**Contact-pressure resins**   Liquid resins which thicken or polymerize on heating and require little or no pressure when used for bonding laminates.

**Continuous filament**   An individual flexible rod of glass of small diameter of great or indefinite length.

**Continuous-filament yarn**   Yarn formed by twisting two or more continuous filaments into a single continuous strand.

**Cooling fixture**   A fixture used to maintain the shape or dimensional accuracy of a molding or casting after it is removed from the mold and until the material is cool enough to hold its shape.

**Core**   (1) The central member of a sandwich construction to which the faces of the sandwich are attached. (2) A channel in a mold for circulation of heat-transfer media.

**Count**   (1) For fabric the number of warp and filling yarns per inch in woven cloth. (2) For yarn the size based on relation of length and weight. Basic unit is a *tex*.

**Coupling agent**   Any chemical substance designed to react with both the reinforcement and matrix phases of a composite material to form or promote a stronger bond at the interface; a bonding link.

**CP**   (1) Cross-ply. (2) Resinous heat-shield composites.

**CPIs**   Condensation-reaction polyimides.

**Crack**   An actual separation of molding material visible on opposite surfaces of the part and extending through the thickness; a fracture.

**Crazing**   Fine cracks which may extend in a network on or under the surface of a plastic material.

**Creel**   A device for holding the required number of roving balls or supply packages in the desired position for unwinding onto the next processing step.

**Creep**   The change in dimension of a plastic under load over a period of time, not including the initial instantaneous elastic deformation; at room temperature it is called *cold flow*.

**Crimp**   The waviness of a fiber; it determines the capacity of fibers to cohere under light pressure; measured either by the number of crimps or waves per unit length or by the percent increase in extent of the fiber on removal of the crimp.

**Critical strain** The strain at the yield point.

**Critical longitudinal stress (fibers)** The longitudinal stress necessary to cause internal slippage and separation of a spun yarn; the stress necessary to overcome the interfiber friction developed as a result of twist.

**Cross-laminated** Laminated so that some of the layers of material are oriented at right angles to the remaining layers with respect to the grain or strongest direction in tension. Balanced construction above the centerline of the thickness of the laminate is normally assumed. (*See also* Parallel-laminated.)

**Crosswise direction** Refers to cutting specimens and to application of load. For rods and tubes, crosswise is the direction perpendicular to the long axis. For other shapes or materials that are stronger in one direction than in another, crosswise is the direction that is weaker. For materials that are equally strong in both directions, crosswise is an arbitrarily designed direction at right angles to the length.

**C stage** The final stage in the reaction of certain thermosetting resins in which the material is relatively insoluble and infusible; sometimes referred to as *resite*. The resin in a fully cured thermoset molding is in this stage. (*See also* A stage, B stage.)

**Cull** Material remaining in a transfer chamber after the mold has been filled. (Unless there is a slight excess in the charge, the operator cannot be sure the cavity will be filled.)

**Cure** To change the properties of a resin by chemical reaction, which may be condensation or addition; usually accomplished by the action of heat or catalyst, or both, and with or without pressure.

**Curing agent** Hardener, a catalytic or reactive agent added to a resin to cause polymerization.

**Curing temperature** Temperature at which a cast, molded, or extruded product, a resin-impregnated reinforcement, an adhesive, etc., is subjected to curing.

**Curing time** The length of time a part is subjected to heat or pressure, or both, to cure the resin; interval of time between the instant relative movement between the moving parts of a mold ceases and the instant pressure is released. (Further cure may take place after removal of the assembly from the conditions of heat or pressure.)

**Cycle** The complete, repeating sequence of operations in a process or part of a process. In molding, the cycle time is the elapsed time between a certain point in one cycle and the same point in the next.

**D glass** A high-boron-content glass made especially for laminates requiring a precisely controlled dielectric constant.

**Damping (mechanical)** Mechanical damping gives the amount of energy dissipated as heat during the deformation of a material. Perfectly elastic materials have no mechanical damping.

**Deep-draw mold** A mold having a core which is long in relation to the wall thickness.

**Daylight** The distance in the open position between the moving and fixed tables (platens) of a hydraulic press. For a multidaylight press, daylight is the distance between adjacent platens.

**Deflection temperature under load** The temperature at which a simple beam has deflected a given amount under load (formerly called *heat-distortion temperature*).

**Deformation under load** The dimensional change of a material under load for a specific time following the instantaneous elastic deformation caused by the initial application of the load; also called *cold flow* or *creep*.

**Degassing** *See* Breathing.

**Delaminate** To split a laminated plastic material along the plane of its layers. (*See also* Laminate.)

**Delamination** Physical separation or loss of bond between laminate plies.

**Denier**   A yarn and filament numbering system in which the yarn number is equal numerically to the weight in grams of 30,000 ft (9144 m) (used for continuous filaments). The lower the denier the finer the yarn.

**Dielectric**   A nonconductor of electricity.

**Dielectric constant ε**   (1) The ratio of the capacity of a capacitor having a dielectric material between the plates to that of the same capacitor when the dielectric is replaced by a vacuum. (2) A measure of the electrical charge stored per unit volume at unit potential.

**Dielectric curing**   Curing a synthetic thermosetting resin passing an electric charge from a high-frequency generator through the resin.

**Dielectric loss**   The energy eventually converted into heat in a dielectric placed in a varying electric field.

**Dielectric strength**   The ability of a material to resist the flow of an electrical current.

**Dimensional stability**   Ability of a plastic part to retain the precise shape to which it was molded, cast, or otherwise fabricated.

**Displacement angle**   In filament winding the distance of advance of the winding ribbon on the equator after one complete circuit.

**Doctor roll**   A device for regulating the amount of liquid material on the rollers of a spreader; also called *doctor bar*.

**Doily**   In filament winding the planar reinforcement applied to a local area between windings to provide extra strength in an area where a cutout is to be made e.g., port openings.

**Dome**   In filament winding the portion of a cylindrical container that forms the integral ends of the container.

**Doubler**   In filament winding a local area with extra reinforcement, wound integrally with the part or wound separately and fastened to it.

**Draft**   The taper or slope of the vertical surfaces of a mold designed to facilitate removal of molded parts.

**Draft angle**   The angle between the tangent to the surface at that point and the direction of ejection.

**Drape**   The ability of preimpregnated broad goods to conform to an irregular shape; textile conformity.

**Dry spot**   (1) Of a laminate the area of incomplete surface film on laminated plastics. (2) In laminated glass an area over which the interlayer and the glass have not become bonded. (*See also* Resin-starved area.)

**Dry winding**   Filament winding using preimpregnated roving, as differentiated from wet winding. (*See also* Wet winding.)

**Dry lay-up**   Construction of a laminate by layering preimpregnated reinforcement (partly cured resin) in a female or male mold, usually followed by bag molding or autoclave molding.

**DS**   Directionally solidified.

**Dwell**   (1) A pause in the application of pressure to a mold, made just before the mold is completely closed, to allow gas to escape from the molding material. (2) In filament winding the time the traverse mechanism is stationary while the mandrel continues to rotate to the appropriate point for the traverse to begin a new pass.

**Edgewise**   Refers to cutting specimens and to the application of load. The load is applied edgewise when it is applied to the edge of the original sheet or specimen. For compression-molded specimens of square cross section the edge is the surface parallel to the direction of motion of the molding plunger. For injection-molded specimens of square cross section this surface is selected arbitrarily; for laminates the edge is the surface perpendicular to the laminae. (*See also* Flatwise.)

**E glass**    A borosilicate glass; the type most used for glass fibers for reinforced plastics; suitable for electrical laminates because of its high resistivity. (Also called *electric glass.*)

**Ejection**    Removal of a molding from the mold impression by mechanical means, by hand, or by using compressed air.

**Ejection ram**    A small hydraulic ram fitted to a press to operate the ejector pins.

**Elastic deformation**    The part of the total strain in a stressed body which disappears upon removal of the stress.

**Elasticity**    The property of plastics materials by virtue of which they tend to recover their original size and shape after deformation.

**Elastic limit**    The greatest stress which a material is capable of sustaining without permanent strain remaining upon the complete release of the stress. A material is said to have passed its elastic limit when the load is sufficient to initiate plastic (nonrecoverable) deformation.

**Elastic recovery**    The fraction of a given deformation that behaves elastically.

$$\text{Elastic recovery} = \frac{\text{elastic extension}}{\text{total extension}}$$

$$\text{Elastic recovery} = \begin{cases} 1 & \text{for perfectly elastic material} \\ 0 & \text{for perfectly plastic material} \end{cases}$$

**Electroformed molds**    A mold made by electroplating metal on the reverse pattern on the cavity.

**Elongation**    Deformation caused by stretching; the fractional increase in length of a material stressed in tension. (When expressed as percentage of the original gage length, it is called *percentage elongation.*)

**EMC**    Elastomeric-molding tooling compound.

**End**    A strand of roving consisting of a given number of filaments gathered together (the group of filaments is considered an *end* or *strand* before twisting and; a *yarn* after twist has been applied); an individual warp yarn, thread, fiber, or roving.

**End count**    An exact number of ends supplied on a ball or roving.

**Endurance limit**    *See* Fatigue limit.

**Epoxy plastics**    Plastics based on resins made by the reaction of epoxides or oxiranes with other materials such as amines, alcohols, phenols, carboxylic acids, acid anhydrides, and unsaturated compounds.

**Equator**    In filament winding the line in a pressure vessel described by the junction of the cylindrical portion and the end dome.

**ERM**    Elastic-reservoir molding.

**Even tension**    The process whereby each end of roving is kept in the same degree of tension as the other ends making up that ball of roving. (*See also* Catenary.)

**Exotherm**    The liberation or evolution of heat during curing of a plastic product.

**Fabric**    A material constructed of interlaced yarns, fibers, or filaments, usually planar. Nonwovens are sometimes included in this classification.

**Fabricating, fabrication**    The manufacture of plastic products from molded parts, rods, tubes, sheeting, extrusions, or other form by appropriate operations such as punching, cutting, drilling, and tapping. Fabrication includes fastening plastic parts together or to other parts by mechanical devices, adhesives, heat sealing, or other means.

**Fan**  In glass-fiber forming the fan shape that is made by the filaments between the bushing and the shoe.

**Fatigue**  The failure or decay of mechanical properties after repeated applications of stress. (Fatigue tests give information on the ability of a material to resist the development of cracks, which eventually bring about failure as a result of a large number of cycles.)

**Fatigue life**  The number of cycles of deformation required to bring about failure of the test specimen under given set of oscillating conditions.

**Fatigue limit**  The stress below which a material can be stressed cyclically for an infinite number of times without failure.

**Fatigue strength**  (1) The maximum cyclic stress a material can withstand for a given number of cycles before failure occurs. (2) The residual strength after being subjected to fatigue.

**Felt**  A fibrous material made from interlocked fibers by mechanical or chemical action, moisture, or heat; made from asbestos, cotton, glass, etc. (*See also* Batt.)

**Fiber**  Relatively short lengths of very small cross section of various materials made by chopping filaments (converting); also called *filament, thread,* or *bristle.* (*See also* Staple fibers.)

**Fiber-composite material**  A material consisting of two or more discrete physical phases, in which a fibrous phase is dispersed in a continuous matrix phase. The fibrous phase may be macro-, micro-, or submicroscopic, but it must retain its physical identity so that it could conceivably be removed from the matrix intact.

**Fiber glass**  An individual filament made by attenuating molten glass. (*See also* Continuous filament, Staple fibers.)

**Fiber diameter**  The measurement of the diameter of individual filaments.

**Fiber-matrix interface**  The region separating the fiber and matrix phases, which differs from them chemically, physically, and mechanically. In most composite materials, the interface has a finite thickness (nanometers to thousands of nanometers) because of diffusion or chemical reactions between the fiber and matrix. Thus, the interface can be more properly described by the terms *interphase* or *interfacial zone.* When coatings are applied to the fibers or several chemical phases have well-defined microscopic thicknesses, the interfacial zone may consist of several interfaces. In this book "interface" is used to mean both "interphase" and "interfacial zone."

**Fiber orientation**  Fiber alignment in a nonwoven or a mat laminate where the majority of fibers are in the same direction, resulting in a higher strength in that direction.

**Fiber pattern**  (1) Visible fibers on the surface of laminates or moldings. (2) The thread size and weave of glass cloth.

**Filament**  Any fiber whose aspect ratio (length to effective diameter) is for all practical purposes infinity. i.e., a continuous fiber. For a noncircular cross section, the effective diameter is that of a circle which has the same (numerical) area as the filament cross section.

**Filaments**  Individual glass fibers of indefinite length, usually as pulled from a stream of molten glass flowing through an orifice of the bushing. In the operation, a number of fibers are gathered together to make a strand or end of roving or yarn.

**Filament weight ratio**  In a composite material, the ratio of filament weight to the total weight of the composite.

**Filament winding**  A process for fabricating a composite structure in which continuous reinforcements (filament, wire, yarn, tape, or other) impregnated with a matrix material either previously or during the winding are placed over a rotating removable form or mandrel in a prescribed way to meet certain stress conditions. Generally the shape is a surface of revolution, which may or may not include end closures. When the right number of layers has been applied, the wound form is cured and the mandrel removed.

**Fill**  Yarn oriented at right angles to the warp in a woven fabric.

**Filler**  A relatively inert material added to a plastic mixture to reduce cost, modify mechanical properties, serve as a base for color effects, or improve the surface texture. (*See also* Binder, Reinforced plastic.)

**Fillet**  A rounded filling for the internal angle between two surfaces of a plastic molding.

**Filling yarn**  The transverse threads or fibers in a woven fabric, i.e., fibers running perpendicular to the warp; also called *weft*.

**Film adhesive**  A synthetic resin adhesive usually of the thermosetting type in the form of a thin dry film of resin with or without a paper carrier.

**Fillout**  *See* Lack of fillout.

**Finish**  A material applied to the surface of fibers in a fabric used to reinforce plastics, and intended to improve the physical properties of the reinforced plastics over those obtained using reinforcement without finish.

**Flame resistance**  Ability of a material to extinguish flame once the source of heat is removed. (*See also* Self-extinguishing resin.)

**Flame retardants**  Chemicals used to reduce or eliminate the tendency of a resin to burn. (For polyethylene and similar resins, chemicals such as antimony trioxide and chlorinated paraffins are useful.)

**Flame-retarded resin**  A resin compounded with certain chemicals to reduce or eliminate its tendency to burn.

**Flame spraying**  Method of applying a plastic coating in which finely powdered fragments of the plastic, together with suitable fluxes, are projected through a cone of flame onto a surface.

**Flammability**  Measure of the extent to which a material will support combustion.

**Flash**  The portion of the charge that flows or is extruded from the mold cavity during molding; extra plastic attached to a molding along the parting line, which must be removed before the part is considered finished.

**Flash mold**  A mold designed to permit the escape of excess molding material; such a mold relies upon backpressure to seal the mold and put the piece under pressure.

**Flat lay**  (1) The property of nonwarping in laminating adhesives. (2) An adhesive material with good noncurling and nondistention characteristics.

**Flatwise**  Refers to cutting specimens and the application of load. The load is applied flatwise when it is applied to the face of the original sheet or specimen.

**Flexural modulus**  The ratio, within the elastic limit, of the applied stress on a test specimen in flexure to the corresponding strain in the outermost fibers of the specimen.

**Flexural rigidity**  (1) For fibers this is a measure of the rigidity of individual strands or fibers; the force couple required to bend a specimen to unit radius of curvature. (2) For plates the measure of rigidity is $D = EI$, where $E$ is the modulus of elasticity and $I$ is the moment of inertia, or

$$D = \frac{Eh^2}{12\,(1 - \nu)} \quad \text{in/lb}$$

where $E$ = modulus of elasticity
$h$ = thickness of plate
$\nu$ = Poisson's ratio

**Flexural strength** (1) The resistance of a material to breakage by bending stresses. (2) The strength of a material in bending expressed as the tensile stress of the outermost fibers of a bent test sample at the instant of failure. For plastics this value is usually higher than the straight tensile strength. (3) The unit resistance to the maximum load before failure by bending, usually in kips per square inch (megapascals).

**Flow** The movement of resin under pressure, allowing it to fill all parts of a mold; flow or creep is the gradual but continuous distortion of a material under continued load, usually at high temperatures.

**Foamed plastics** Resins in sponge form; may be flexible or rigid; cells may be closed or interconnected and density anywhere from that of the solid parent resin to 2 lb/ft³ (32 kg/m³).

**Foam-in-place** Foam deposition requiring the foaming machine to be brought to the work (as opposed to bringing the work to the foaming machine).

**FOD** Foreign-object damage.

**Force** (1) The male half of the mold, which enters the cavity, exerting pressure on the resin and causing it to flow (also called *punch*) (2) Either part of a compression mold (top force and bottom force).

**FP** Polycrystalline alumina fiber.

**Fracture** Rupture of the surface without complete separation of laminate.

**FRAT** Fiber-reinforced advanced titanium.

**FRP** Fibrous-glass-reinforced plastic, any type of plastic-reinforced cloth, mat, strands, or any other form of fibrous glass.

**Gage length** Length over which deformation is measured.

**Gap** In filament winding the space between successive windings, which are usually intended to lie next to each other.

**Gel** A semisolid system consisting of a network of solid aggregates in which liquid is held; the initial jellylike solid phase that develops during the formation of a resin from a liquid.

**Gelation time** For synthetic thermosetting resins the interval of time between introduction of a catalyst into a liquid adhesive system and gel formation.

**Gel coat** A resin applied to the surface of a mold and gelled before lay-up. (The gel coat becomes an integral part of the finished laminate and is usually used to improve surface appearance, etc.)

**Gel point** The stage at which a liquid begins to exhibit pseudo-elastic properties, also conveniently observed from the inflection point on a viscosity-time plot. Also called *gel time*.

**Geodesic** The shortest distance between two points on a surface.

**Geodesic isotensoid** Constant-stress level in any given filament at all points in its path.

**Geodesic-isotensoid contour**   In filament-wound reinforced plastic pressure vessels a dome contour in which the filaments are placed on geodesic paths so that the filaments will exhibit uniform tension throughout their length under pressure loading.

**Geodesic ovaloid**   A contour for end domes, the fibers forming a goedesic line. The forces exerted by the filaments are proportioned to meet hoop and meridional stresses at any point.

**Glass**   An inorganic product of fusion which has cooled to a rigid condition without crystallizing. Glass is typically hard and relatively brittle and has a conchoidal fracture.

**Glass fiber**   A glass filament that has been cut to a measurable length. Staple fibers of relatively short length are suitable for spinning into yarn.

**Glass filament**   A form of glass that has been drawn to a small diameter and extreme length. Most filaments are less than 0.005 in (0.13 mm) in diameter.

**Glass-filament bushing**   The unit through which molten glass is drawn in making glass filaments.

**Glass finish**   A material applied to the surface of a glass reinforcement to improve its effect upon the physical properties of the reinforced plastic; also called *bonding agent*.

**Glass flake**   Thin, irregularly shaped flakes of glass typically made by shattering a continuous thin-walled tube of glass.

**Glass former**   An oxide which forms a glass easily, also one that contributes to the network of silica glass when added to it.

**Glass stress**   In a filament-wound part, usually a pressure vessel, the stress calculated using only the load and the cross-sectional area of the reinforcement.

**Glass volume percent**   The product of the specific gravity of a laminate and the percent glass by weight divided by the specific gravity of the glass.

**Greige**   Fabric before finishing; yarn or fiber before bleaching or dyeing. Also called *gray goods, greige goods, greige gray*.

**Guide pin**   A pin which guides mold halves into alignment on closing.

**Guide-pin bushing**   The bushing through which the guide pin moves when the mold is closed.

**Gusset**   A piece used to give added size or strength in a particular location of an object; the folded-in portion of a flattened tubular film.

**Hand**   The softness of a piece of fabric, as determined by the touch (individual judgment).

**Hand lay-up**   The process of placing (and working) successive plies of reinforcing material or resin-impregnated reinforcement in position on a mold by hand.

**Hardener**   (1) A substance or mixture added to a plastic composition to promote or control the curing action by taking part in it. (2) A substance added to control the degree of hardness of the cured film. (*see also* Catalyst.)

**Hardness**   The resistance to surface indentation, usually measured by the depth of penetration (or arbitrary units related to depth of penetration) of a blunt point under a given load using a particular instrument according to a prescribed procedure. (*See also* Barcol hardness, Rockwell hardness number.)

**Hat**   A member in the shape of a hat.

**Heat build-up**   The temperature rise in a part resulting from the dissipation of applied strain energy as heat.

**Heat-convertible resin** A thermosetting resin convertible by heat to an infusible and insoluble mass.

**Heat resistance** The property or ability of plastics and elastomers to resist the deteriorating effects of elevated temperatures.

**High-pressure laminates** Laminates molded and cured at pressures not lower than 1 kip/in² (7 MPa) and more commonly at 1.2 to 2 kips/in² (8.3 to 13.8 MPa).

**High-pressure molding** A molding process in which the pressure used is greater than 1 kip/in² (7 MPa).

**HM** High-modulus.

**HMC** High-strength molding compound.

**HME** High-vinyl-modified epoxy.

**Honeycomb** Manufactured product of resin-impregnated sheet material (paper, glass fabric, etc.) or sheetmetal formed into hexagonal-shaped cells; used as a core material in sandwich construction.

**Hoop stress** The circumferential stress in a material of cylindrical form subjected to internal or external pressure.

**Hybrid** The result of attaching a composite body to another material such as aluminum, steel, etc., on two reinforcing agents in the matrix such as graphite and glass.

**Hydraulic press** A press in which the molding force is created by the pressure exerted on a fluid.

**Hydrophilic** Capable of adsorbing or absorbing water.

**Hydrophobic** Capable of repelling water.

**Hygroscopic** Capable of adsorbing and retaining atmospheric moisture.

**IFAC** Integrated-flexible automation center.

**ILC** Integrated laminating center.

**IM** Intermediate-modulus.

**Impact strength** The ability of a material to withstand shock loading; the work done in fracturing a test specimen in a specified manner under shock loading.

**Impregnate** In reinforced plastics to saturate the reinforcement with a resin.

**Impregnated fabric** A fabric impregnated with a synthetic resin. (*See also* Prepreg.)

**Inert filler** A material added to a plastic to alter the end-item properties through physical rather than chemical means.

**Infrared** The part of the electromagnetic spectrum between the visible-light range and the radar range; radiant heat is in this range, and infrared heaters are much used in sheet thermoforming.

**Inhibitor** A substance which retards a chemical reaction; used in certain types of monomers and resins to prolong storage life.

**Initial modulus** *See* Modulus of elasticity.

**Inorganic** Designating or pertaining to the chemistry of all elements and compounds not classified as organic; matter other than animal or vegetable, such as earthy or mineral matter. (*See also* Organic.)

**Insert** An integral part of a plastic molding consisting of metal or other material which may be molded into position or pressed into the molding after the molding is complete.

**Insert pin** A pin which keeps an inserted part (insert) inside the mold by screwing or friction; it is removed when the object is being withdrawn from the mold.

**Instron** An instrument used to determine the tensile and compressive properties of materials.

**Insulating resistance** The electric resistance between two conductors or systems of conductors separated only by insulating material.

**Insulator** (1) A material of such low electric conductivity that the flow of current through it can usually be neglected: (2) A material of low thermal conductivity.

**Interface** The junction point or surface between two different media; on glass fibers, the contact area between glass and sizing or finish; in a laminate, the contact area between the reinforcement and the laminating resin.

**Interlaminar shear strength** The maximum shear stress existing between layers of a laminated material.

**Internal stress** Stress created within an adhesive layer by the movement of the adherends at differential rates or by the contraction or expansion of the adhesive layer. (*See also* Stress.)

**IPS** Integrated process system.

**Irreversible** (1) Not capable of redissolving or remelting (2) Descriptive of chemical reactions which proceed in a single direction and are not capable of reversal (as applied to thermosetting resins).

**Isocyanate plastics** Plastics based on resins made by the condensation of organic isocyanates with other compounds. (*See also* Urethane plastics.)

**Isotropic laminate** One in which the strength properties are equal in all directions.

**Izod impact test** A destructive test designed to determine the resistance of a plastic to the impact of a suddenly applied force.

**Joint** The location at which two adherends are held together with a layer of adhesive; the general area of contact for a bonded structure. *Butt joint:* the edge faces of the two adherends are at right angles to the other faces of the adherents. *Scarf joint:* a joint made by cutting away similar angular segments of two adherends and bonding them with the cut areas fitted together. *Lap joint:* a joint made by placing one adherend partly over another and bonding together the overlapped portions.

**Lack of fillout** Characteristic of an area, occurring usually at the edge of a laminated plastic, where the reinforcement has not been wetted with resin.

**Lacquer** Solution of natural or synthetic resins in readily evaporating solvents, used as a protective coating.

**Laminate** (1) To unite sheets of material by a bonding material usually with pressure and heat (normally used with reference to flat sheets). (2) A product made by so bonding. (*See also* Bidirectional laminate, Unidirectional laminate.)

**Laminated molding** A molded plastic article produced by bonding together, under heat and pressure in a mold, layers of resin-impregnated laminating reinforcement; also called *laminated plastics.*

**Laminate ply** One layer of a product which is evolved by bonding together two or more layers of materials.

**Land** The portion of a mold which provides the separation or cutoff of the flash from the molded article.

**Lap** In filament winding the amount of overlay between successive windings, usually intended to minimize gapping.

**Lap joint** *See* Joint.

**Lay** (1) In glass fiber the spacing of the roving bands on the roving package expressed in the number of bands per inch. (2) In filament winding the orientation of the ribbon with some reference, usually the axis of rotation.

**Lay-flat** *See* Flat lay.

**Lay-up** (1) As used in reinforced plastics, the reinforcing material placed in position in the mold. (2) The process of placing the reinforcing material in position in the mold. (3) The resin-impregnated reinforcement. (4) The component materials, geometry, etc., of a laminate.

**Lengthwise direction** Refers to cutting specimens and the application of loads. For rods and tubes, lengthwise is the direction of the long axis. For other shapes of materials that are stronger in one direction than in the other lengthwise is the direction that is stronger. For materials that are equally strong in both directions lengthwise is an arbitrarily designated direction that may be with the grain, direction of flow in manufacture, longer direction, etc. (*See also* Crosswise direction.)

**LHS** Low-cost high-strength.

**LIM** Liquid injection molding.

**Linear expansion** The increase of a planar dimension, measured by the linear elongation of a sample in the form of a beam which is exposed to two given temperatures.

**Liner** In a filament-wound pressure vessel the continuous, usually flexible, coating on the inside surface of the vessel used to protect the laminate from chemical attack or to prevent leakage under stress.

**Liquidus** The maximum temperature at which equilibrium exists between the molten glass and its primary crystalline phase.

**LMC** Low-pressure molding compound.

**Load-deflection curve** A curve in which the increasing flexural loads are plotted on the ordinate and the deflections caused by those loads are plotted on the abscissa.

**Longos** Low-angle helical or longitudinal windings.

**Loop tenacity** The strength value obtained by pulling two loops, like two links in a chain, against each other in order to test whether a fibrous material will cut or crush itself; also called *loop strength*.

**Loss on ignition** Weight loss, usually expressed as percent of total, after burning off an organic sizing from glass fibers or an organic resin from a glass-fiber laminate.

**Lot** A specific amount of material produced at one time and offered for sale as a unit quantity.

**Low-pressure laminates** In general, laminates molded and cured at pressures from 0.4 kip/in$^2$ (2.8 MPa) down to and including pressure obtained by the mere contact of the plies.

**Low-pressure molding** The distribution of relatively uniform low pressure [0.2 kip/in$^2$ (1.4 MPa) or less] over a resin-bearing fibrous assembly of cellulose, glass, asbestos, or other material, with or without application of heat from an external source, to form a structure possessing definite physical properties.

**LST** Large Space Telescope.

**LWV** Lightweight vehicle.

**Macerate** (1) To chop or shred fabric for use as a filler for a molding resin. (2) The molding compound obtained when so filled.

**Mandrel** (1) The core around which paper-, fabric-, or resin-impregnated glass is wound to form pipes, tubes, or vessels. (2) In extrusion the central finger of a pipe or tubing die.

**Mat**  A fibrous material for reinforced plastic consisting of randomly oriented chopped filaments or swirled filaments with a binder, available in blankets of various widths, weights, and lengths.

**Mat binder**  Resin applied to glass fiber and cured during the manufacture of mat, used to hold the fibers in place and maintain the shape of the mat.

**Matched metal molding**  A reinforced-plastic manufacturing process in which matching male and female metal molds are used (similar to compression molding) to form the part, as opposed to low-pressure laminating or spray-up.

**Matrix**  *See* Resin.

**Matte**  A nonspecular surface having diffused reflective powers.

**Mechanical adhesion**  Adhesion between surfaces in which the adhesive holds the parts together by interlocking action.

**Melamine plastics**  Plastics based on melamine resins.

**Metallic fiber**  Manufactured fiber composed of metal, plastic-coated metal, metal-coated plastic, or a core completely covered by metal.

**M glass**  A high-beryllia-content glass designed especially for high modulus of elasticity.

**Micron**  A unit of length replaced by the micrometer ($\mu$m); 1 $\mu$m = $10^{-6}$ m = $10^{-3}$ mm = 0.00003937 in = 39.4 $\mu$in.

**Mil**  The unit used in measuring the diameter of glass-fiber strands, wire, etc. (1 mil = 0.001 in).

**Milled fibers**  Continuous glass strands hammer-milled into small modules of filamentized glass. Useful as anticrazing reinforcing fillers for adhesives.

**MMC**  Metal-matrix composite.

**Modulus**  A number which expresses a measure of some property of a material, e.g., modulus of elasticity, shear modulus, etc.; a coefficient of numerical measurement of a property. Using "modulus" alone without modifying terms is confusing and should be discouraged.

**Modulus in compression**  *See* Compressive modulus.

**Modulus in flexure**  *See* Flexural modulus.

**Modulus, initial, or Young's modulus**  *See* Modulus of elasticity.

**Modulus in shear**  *See* Shear modulus.

**Modulus in tension**  *See* Tensile modulus.

**Modulus of elasticity**  The ratio of the stress or applied load to the strain or deformation produced in a material that is elastically deformed. If a tensile strength of 2 kips/in$^2$ (14 MPa) results in an elongation of 1%, the modulus of elasticity is 2/0.01 = 200 kips/in$^2$ (1379 MPa); also called *Young's modulus*.

**Modulus of elasticity in torsion**  The ratio of the torsion stress to the strain in the material over the range for which this value is constant.

**Modulus of rigidity**  *See* Flexural rigidity.

**Modulus of rupture**  *See* Flexural strength.

**Mohs hardness**  A measure of the scratch resistance of a material; the higher the number the greater the scratch resistance (diamond is 10).

**Moisture absorption**  The pickup of water vapor from air by a material; it relates only to vapor withdrawn from the air by a material and must be distinguished from water absorption. *See also* Water absorption.

**Mold** (1) The cavity or matrix in or on which the plastic composition is placed and from which it takes form. (2) To shape plastic parts or finished articles by heat and pressure. (3) The assembly of all the parts that function collectively in the molding process.

**Mold-release agent** A liquid or powder used to prevent sticking of molded articles in the cavity.

**Mold seam** Line on a molded or laminated piece, differing in color or appearance from the general surface, caused by the parting line of the mold.

**Mold shrinkage** (1) The immediate shrinkage which a molded part undergoes when it is removed from a mold and cooled to room temperature. (2) The difference in dimensions, expressed in inches per inch (millimeters per millimeter) between a molding and the mold cavity in which it was molded (at normal temperature measurement). (3) The incremental difference between the dimensions of the molding and the mold from which it was made, expressed as a percentage of the dimensions of the mold.

**Molding** The shaping of a plastic composition in or on a mold, normally accomplished under heat and pressure; sometimes used to denote the finished part.

**Molding compounds** Plastics in a wide range of forms (especially granules or pellets) to meet specific processing requirements.

**Molding cycle** (1) The time required for the complete sequence of operations on a molding press to produce one set of moldings. (2) The operations necessary to produce a set of moldings without reference to time.

**Molding pressure** The pressure applied to the ram of an injection machine or press to force the softened plastic to fill the mold cavities completely. *See also* Compression molding pressure.

**Molding, pressure-bag** See Pressure-bag molding.

**Molding pressure, compression** *See* Compression molding pressure.

**Monofilament** (1) A single fiber or filament of indefinite length generally produced by extrusion. (2) A continuous fiber sufficiently large to serve as yarn in normal textile operations; also called *monofil*.

**Monomer** (1) A simple molecule capable of reacting with like or unlike molecules to form a polymer. (2) The smallest repeating structure of a polymer, also called a *mer*.

**Multiple-cavity mold** A mold with two or more mold impressions; i.e., a mold that produces more than one molding per molding cycle.

**Multicircuit winding** In filament winding a winding that requires more than one circuit before the band repeats by lying adjacent to the first band.

**Multifilament yarn** A multitude of fine, continuous filaments (often 5 to 100), usually with some twist in the yarn to facilitate handling. Sizes range from 5 to 10 denier up to a few hundred denier. Individual filaments in a multifilament yarn are usually about 1 to 5 denier.

**Nesting** In reinforced plastics placing plies of fabric so that the yarns of one ply lie in the valleys between the yarns of the adjacent ply (nested cloth).

**Netting analysis** The analysis of filament-wound structures which assumes that the stresses induced in the structure are carried entirely by the filaments, the strength of the resin being neglected, and that the filaments possess no bending or shearing stiffness, carrying only the axial tensile loads.

**Nol ring** A parallel filament-wound test specimen used for measuring various mechanical-strength properties of the material by testing the entire ring or segments of it.

**Nonhygroscopic** Not absorbing or retaining an appreciable quantity of moisture from the air (water vapor).

**Nonrigid plastic** A plastic which has a stiffness or apparent modulus of elasticity not over 10 kips/in$^2$ (69 MPa) at 73.4°F (23°C).

**Nonwoven fabric** A planar structure produced by loosely bonding together yarns, rovings, etc. (*See also* Fabric.)

**Notch sensitivity** The extent to which the sensitivity of a material to fracture is increased by the presence of a surface inhomogeneity such as a notch, a sudden change in section, a crack, or a scratch. Low notch sensitivity is usually associated with ductile materials and high notch sensitivity with brittle materials.

**Novolak** Trade name of a phenolic-aldehyde resin, which remains permanently thermoplastic unless a source of methylene groups is added; a linear thermoplastic B-staged phenolic resin. (*See also* Thermoplastic.)

**Offset yield strength** The stress at which the strain exceeds by a specific amount (the offset) an extension of the initial proportional portion of the stress-strain curve.

**Open-cell-foamed plastic** A cellular plastic in which there is a predominance of interconnected cells.

**Orange peel** An uneven surface resembling that of an orange peel; said of injection moldings with unintentionally rugged surfaces.

**Organic** Designating or composed of matter originating in plant or animal life or composed of chemicals of hydrocarbon origin, natural or synthetic.

**Oriented materials** Materials, particularly amorphous polymers and composites, whose molecules and/or macroconstituents are aligned in a specific way. Oriented materials are anisotropic. Orientation is generally uniaxial or biaxial.

**Orthotropic** Having three mutually perpendicular planes of elastic symmetry.

**Overcuring** The beginning of thermal decomposition resulting from too high a temperature or too long a molding time.

**Overflow groove** Small groove used in molds to allow material to flow freely to prevent weld lines and low density and to dispose of excess material.

**Overlap** A simple adhesive joint, in which the surface of one adherend extends past the leading edge of another.

**Overlay sheet** A nonwoven fibrous mat (in glass, synthetic fiber, etc.) used as the top layer in a cloth or mat lay-up, to provide a smoother finish or minimize the appearance of the fibrous pattern. *See also* Surfacing mat.

**Package** The method of supplying the roving or yarn.

**PAN** Polyacrylonitrile.

**Parallel-laminated** Laminated so that all the layers of material are oriented approximately parallel with respect to the grain or strongest direction in tension. (*See also* Cross-laminated.)

**Parameter** An arbitrary constant, as distinguished from a fixed or absolute constant. Any desired numerical value may be given as a parameter.

**Parting agent** *See* Release agent.

**Parting line** A mark on a molded piece where the sections of a mold have met in closing.

**PBBI**   Polybutadiene bisimide.

**PBT**   Polybutylene terephthalate; polybenzothiazole.

**Peel ply**   The outside layer of a laminate which is removed or sacrificed to achieve improved bonding of additional plies.

**Peel strength**   Bond strength, in pounds per inch of width, obtained by peeling the layer. (*See also* Bond strength.)

**Permanence**   The property of a plastic which describes its resistance to appreciable change in characteristics with time and environment.

**Permanent set**   The deformation remaining after a specimen has been stressed in tension a prescribed amount for a definite period and released for a definite period.

**PES**   Polyethersulfone.

**PET**   Polyethylene terephthalate.

**Phenolic, Phenolic Resin**   A synthetic resin produced by the condensation of an aromatic alcohol with an aldehyde, particularly of phenol with formaldehyde. (*See also* A stage, B stage, C stage, Novolak.)

**PI**   Polyimide.

**Pick**   (1) An individual filling yarn, running the width of a woven fabric at right angles to the warp, also called *fill, woof, weft.* (2) To experience tack. (3) To transfer unevenly from an adhesive applicator mechanism due to high surface tack.

**Pinch-off**   In blow molding a raised edge around the cavity in the mold which seals off the part and separates the excess material as the mold closes around the parison.

**Pinhole**   A tiny hole in the surface of, or through, a plastic material, usually not occurring alone.

**Pin, insert**   *See* Insert pin.

**Pit**   Small regular or irregular crater in the surface of a plastic, usually with width about the same order of magnitude as the depth.

**Planar helix winding**   A winding in which the filament path on each dome lies on a plane which intersects the dome while a helical path over the cylindrical section is connected to the dome paths.

**Planar winding**   A winding in which the filament path lies on a plane intersecting the winding surface.

**Plastic**   A material that contains as an essential ingredient an organic substance of high molecular weight, is solid in its finished state, and at some stage in its manufacture or processing into finished articles can be shaped by flow; made of plastic. A *rigid plastic* is one with a stiffness or apparent modulus of elasticity greater than 100 kips/in$^2$ (690 MPa) at 73.4°F (23°C). A *semirigid plastic* has a stiffness or apparent modulus of elasticity between 10 and 100 kips/in$^2$ (69 and 690 MPa) at 73.4°F (23°C).

**Plasticate**   To soften by heating or kneading.

**Plastic deformation**   Change in dimensions of an object under load that is not recovered when the load is removed; opposite of elastic deformation. *See also* Elastic recovery.

**Plastic flow**   Deformation under the action of a sustained force; flow of semisolids in molding plastics.

**Plasticize**   To make a material moldable by softening it with heat or a plasticizer.

**Plastic tooling**   Tools constructed of plastics, generally laminates or casting materials.

**Platens**   The mounting plates of a press, to which the entire mold assembly is bolted.

**Plied yarn**  A yarn formed by twisting together two or more single yarns in one operation.

**PMR**  Polyimides. Polymerization-of-monomer-reactant polyimides.

**Poisson's ratio** $\nu$  A constant relating change in cross-sectional area to change in length when a material is stretched;

$$\nu \approx \begin{cases} \frac{1}{2} & \text{for rubbery materials} \\ \frac{1}{4} \text{ to } \frac{1}{2} & \text{for crystals and glasses} \end{cases}$$

**Polar winding**  A winding in which the filament path passes tangent to the polar opening at one end of the chamber and tangent to the opposite side of the polar opening at the other end. A one-circuit pattern is inherent in the system.

**Polyamide**  A polymer in which the structural units are linked by amide or thioamide groupings; many polyamides are fiber-forming.

**Polyesters**  Thermosetting resins produced by dissolving unsaturated, generally linear alkyd resins in a vinyl active momoner, e.g., styrene, methyl styrene, or diallyl phthalate.

**Polyimide**  A polymer produced by heating polyamic acid; a highly heat-resistant resin [> 600°F (> 316°C)] suitable for use as a binder or an adhesive.

**Polymer**  A high-molecular-weight organic compound, natural or synthetic, whose structure can be represented by a repeated small unit (mer), e.g., polyethylene, rubber, cellulose. Synthetic polymers are formed by addition or condensation polymerization of monomers. Some polymers are elastomers, some plastics. When two or more monomers are involved, the product is called a *copolymer.*

**Polymerization**  A chemical reaction in which the molecules of a monomer are linked together to form large molecules whose molecular weight is a multiple of that of the original substance. When two or more monomers are involved, the process is called *copolymerization* or *heteropolymerization.*

**Polymerize**  To unite molecules of the same kind into a compound having the elements in the same proportion but possessing much higher molecular weight and different physical properties.

**Porosity**  The ratio of the volume of air or void contained within the boundaries of a material to the total volume (solid material plus air or void), expressed as a percentage.

**Positive mold**  A mold designed to apply pressure to a piece being molded with no escape of material.

**Postcure**  Additional elevated-temperature cure, usually without pressure, to improve final properties and/or complete the cure. Complete cure and ultimate mechanical properties of certain resins are attained only by exposure of the cured resin to higher temperatures than those of curing.

**Postforming**  The forming, bending, or shaping of fully cured, C-staged thermoset laminates that have been heated to make them flexible. On cooling, the formed laminate retains the contours and shape of the mold over which it has been formed.

**Pot life**  The length of time a catalyzed resin system retains a viscosity low enough to be used in processing; also called *working life.*

**PPS**  Polyphenylene sulfide.

**Precure**  The full or partial setting of a synthetic resin or adhesive in a joint before the clamping operation is complete or before pressure is applied.

**Preform**  (1) A preshaped fibrous reinforcement formed by distribution of chopped fibers by air, water flotation, or vacuum over the surface of a perforated screen to the approximate contour and thickness desired in the finished part. (2) A preshaped fibrous reinforcement of mat or cloth formed to the desired shape on a mandrel or mock-up before being placed in a mold press. (3) A compact "pill" formed by compressing premixed material to facilitate handling and control of uniformity of charges for mold loading.

**Preform binder**   A resin applied to the chopped strands of a preform, usually during its formation, and cured so that the preform will retain its shape and be handleable.

**Preimpregnation**   The practice of mixing resin and reinforcement and effecting partial cure before use or shipment to the user. (*See also* Prepreg.)

**Premix**   A molding compound prepared prior to, and apart from, the molding operations and containing all components required for molding, i.e., resin, reinforcement, fillers, catalysts, release agents, and other compounds.

**Prepreg**   Ready-to-mold material in sheet form, which may be cloth, mat, or paper impregnated with resin and stored for use. The resin is partially cured to a B stage and supplied to the fabricator, who lays up the finished shape and completes the cure with heat and pressure.

**Pressure**   Force measured per unit area. Absolute pressure is measured with respect to zero. Gage pressure is measured with respect to atmospheric pressure.

**Pressure-bag molding**   A process for molding reinforced plastics, in which a tailored flexible bag is placed over the contact lay-up on the mold, sealed, and clamped in place. Fluid pressure, usually compressed air, is exerted on the bag, and the part is cured.

**Primary structure**   One critical to flight safety.

**Primer**   A coating applied to a surface before the application of an adhesive or lacquer, enamel, or the like to improve the performance of the bond.

**Promoter**   *See* Accelerator.

**Proportional limit**   The greatest stress which a material is capable of sustaining without deviation from proportionality of stress and strain (Hooke's law); it is expressed in force per unit area, usually in kips per square inch (megapascals).

**Prototype**   A model suitable for use in complete evaluation of form, design, and performance.

**PS**   Polysulfone.

**Pultrusion**   Reversed extrusion of resin-impregnated roving in the manufacture of rods, tubes, and structural shapes of a permanent cross section. After passing through the resin dip tank the roving is drawn through a die to form the desired cross section.

**QCSEE**   Quick, clean, short-haul experimental engine.

**Random pattern**   A winding with no fixed pattern. If a large number of circuits are required for the pattern to repeat, a random pattern is approached; a winding in which the filaments do not lie in an even pattern.

**Reinforced molding compound**   Compound supplied by raw material producer in the form of ready-to-use materials, as distinguished from premix. *See also* Premix.

**Reinforced plastic**   A plastic with strength properties greatly superior to those of the base resin, resulting from the presence of reinforcements embedded in the composition.

**Reinforcement**   A strong inert material bonded into a plastic to improve its strength, stiffness, and impact resistance. Reinforcements are usually long fibers of glass, asbestos, sisal, cotton, etc., in woven or nonwoven form. To be effective, the reinforcing material must form a strong adhesive bond with the resin. ("Reinforcement" is not synonymous with "filler.")

**Release agent**   A material which is applied in a thin film to the surface of a mold to keep the resin from bonding to it.

**Resilience**   (1) The ratio of energy returned on recovery from deformation to the work input required to produce the deformation (usually expressed as a percentage). (2) The ability to regain an original shape quickly after being strained or distorted.

**Resin**   A solid, semisolid, or pseudo-solid organic material which has an indefinite (often high) molecular weight, exhibits a tendency to flow when subjected to stress, usually has a softening or melting range, and usually fractures conchoidally. Most resins are polymers. In reinforced plastics the material used to bind together the reinforcement material, the matrix. *See also* Polymer.

**Resin applicator**   In filament winding the device which deposits the liquid resin onto the reinforcement band.

**Resin content**   The amount of resin in a laminate expressed as a percent of total weight or total volume.

**Resin, liquid**   An organic polymeric liquid which becomes a solid when converted into its final state for use.

**Resin-rich area**   Space which is filled with resin and lacking reinforcing material.

**Resin-starved area**   Area of insufficient resin, usually identified by low gloss, dry spots, or fiber show.

**Resistivity**   The ability of a material to resist passage of electric current through its bulk or on a surface.

**Retarder**   *See* Inhibitor.

**Reverse helical winding**   As the fiber-delivery arm traverses one circuit, a continuous helix is laid down, reversing direction at the polar ends; contrasted to biaxial, compact, or sequential winding in that the fibers cross each other at definite equators, the number depending on the helix. The minimum crossover would be 3.

**Rib**   A reinforcing member of a fabricated or molded part.

**Ribbon**   A fiber having essentially a rectangular cross section, where the width-to-thickness ratio is at least 4:1.

**Rigid plastic**   *See* Plastic.

**RIM**   Reaction-injection molding.

**Rockwell hardness number**   A value derived from the increase in depth of an impression as the load on an indenter is increased from a fixed minimum value to a higher value and then returned to the minimum value.

**Room-temperature-curing adhesives**   Adhesives that set (to handling strength) within 1 h at 68 to 86°F (20 to 30°C) and later reach full strength without heating.

**Roving**   In filament winding a collection of bundles of continuous filaments either as untwisted strands or as twisted yarns. Rovings may be lightly twisted, but for filament winding they are generally wound as bands or tapes with as little twist as possible. Glass rovings are predominantly used in filament winding.

**Roving cloth**   A textile fabric, coarse in nature, woven from rovings.

**RPP**   Reinforced pyrolyzed plastic.

**RRIM**   Reinforced reaction-injection molding.

**RTM**   Resin-transfer molding.

**S glass**   A magnesia-alumina-silicate glass, especially designed to provide filaments with very high tensile strength.

**Sandwich constructions**   Panels composed of a lightweight core material (honeycomb, foamed plastic, etc.) to which two relatively thin, dense, high-strength faces or skins are adhered.

**Sandwich heating**   A method of heating a thermoplastic sheet before forming by heating both sides of the sheet simultaneously.

**SAP**   Sintered-aluminum powder.

**Satin**   A plastic finish having a satin or velvety appearance.

**Scarf joint**   *See* Joint.

**Scratch**   Shallow mark, groove, furrow or channel normally caused by improper handling or storage.

**Scrim**   A low-cost, nonwoven open-weave reinforcing fabric made from continuous-filament yarn in an open-mesh construction.

**Secondary structure**   One not critical to flight safety.

**Self-extinguishing resin**   A resin formulation which will burn in the presence of a flame but which will extinguish itself within a specified time after the flame is removed.

**Selvage**   The edge of a woven fabric finished off so as to prevent the yarns from raveling.

**Semirigid plastic**   *See* Plastic.

**Sequential winding**   *See* Biaxial winding.

**Set**   (1) To convert into a fixed or hardened state by chemical or physical action, such as condensation, polymerization, oxidation, vulcanization, gelation, hydration, or evaporation of volatiles. (2) The irrecoverable deformation or creep usually measured by a prescribed test procedure and expressed as a percentage of original dimension.

**Set up**   To harden, as in curing.

**Shear**   An action or stress resulting from applied forces and tending to cause two contiguous parts of a body to slide relative to each other in a direction parallel to their plane of contact.

**Shear edge**   The cutoff edge of the mold.

**Shear modulus $G$**   The ratio of shearing stress $\tau$ to shearing strain $\gamma$ within the proportional limit of a material.

**Shelf life**   The length of time a material can be stored under specified conditions without harmful changes in its properties, also called *storage life*.

**Shoe**   A device for gathering filaments into a strand in glass-fiber forming. (*See* Chase.)

**Short-beam shear strength**   The interlaminar shear strength of a parallel-fiber-reinforced plastic material as determined by three-point flexural loading of a short segment cut from a ring specimen.

**Shrinkage**   The relative change in dimension between the length measured on the mold when it is cold and the length on the molded object 24 h after it has been taken out of the mold.

**Silicone plastics**   Plastics based on resins in which the main polymer chain consists of alternating silicon and oxygen atoms with carbon-containing side groups; derived from silica (sand) and methyl chloride.

**Silicones**   Resinous materials derived from organosiloxane polymers, furnished in different molecular weights including liquids and solid resins and elastomers.

**Single-circuit winding**   A winding in which the filament path makes a complete traverse of the chamber, after which the following traverse lies immediately adjacent to the previous one.

**Size**   Any treatment consisting of starch, gelatin, oil, wax, or other suitable ingredient applied to yarn or fibers at the time of formation to protect the surface and facilitate handling and fabrication or to control the fiber characteristics. The treatment contains ingredients which provide surface lubricity and binding action but, unlike a finish, no coupling agent. Before final fabrication into a composite, the size is usually removed by heat-cleaning and a finish is applied.

**Sizing**   (1) Applying a material on a surface in order to fill pores and thus reduce the absorption of the subsequently applied adhesive or coating. (2) To modify the surface properties of the substrate to improve adhesion. (3) The material used for this purpose, also called *size*.

**Sizing content**   The percent of the total strand weight made up by the sizing, usually determined by burning off the organic sizing ("loss on ignition").

**Skein**   A continuous filament, strand, yarn, roving, etc., wound up to some measurable length and generally used to measure various physical properties.

**Skin**   The relatively dense material that may form the surface of a cellular plastic or sandwich.

**SMC**   Sheet-molding compound.

**S-N curve**   Stress per number of cycles to failure. *See also* Stress-strain.

**Soft flow**   The behavior of a material which flows freely under conventional conditions of molding and which, under such conditions, will fill all the interstices of a deep mold where a considerable distance of flow can be demanded.

**Specification**   A detailed description of the characteristics of a product and of the criteria which must be used to determine whether the product is in conformity with the description.

**Specific gravity**   The ratio of the weight of any volume of a substance to the weight of an equal volume of another substance taken as standard at a constant or stated temperature.

**Specific heat**   The quantity of heat required to raise the temperature of a unit mass of a substance 1 degree under specified conditions.

**Specimen**   An individual piece or portion of a sample used to make a specific test; of specific shape and dimensions.

**SPF/DB**   Superplastic-forming diffusion bonding.

**Spiral**   In glass-fiber forming the device that is used to make the strand traverse back and forth across the forming tube.

**Splice**   To join two ends of glass-fiber yarn or strand, usually by means of an air-drying glue.

**Spline**   (1) To prepare a surface to its desired contour by working a paste material with a flat-edged tool; the procedure is similar to screeding concrete. (2) The tool itself.

**Split-cavity blocks**   Blocks which, when assembled, contain a cavity for molding articles having undercuts.

**Split mold**   A mold in which the cavity is formed of two or more components, known as *splits,* held together by an outer chase.

**Split-ring mold**   A mold in which a split-cavity block is assembled in a chase to permit forming undercuts in a molded piece. The parts are ejected from the mold and then separated from the piece.

**SPMC**   Solid polyester molding compound.

**Spray**   A complete set of moldings from a multi-impression injection mold, together with the associated molded material.

**Spray-up** Techniques in which a spray gun is used as the processing tool. In reinforced plastics, for example, fibrous glass and resin can be simultaneously deposited in a mold. In essence, roving is fed through a chopper and ejected into a resin stream, which is directed at the mold by either of two spray systems. In foamed plastics, very fast-reacting urethane foams or epoxy foams are fed in liquid streams to the gun and sprayed onto the surface. On contact, the liquid starts to foam.

**Sprayed-metal molds** Molds made by spraying molten metal onto a master until a shell of predetermined thickness is achieved. The shell is then removed and backed up with plaster, cement, casting resin, or other suitable material. Used primarily as a mold in sheet-forming process.

**Spun roving** A heavy, low-cost glass-fiber strand consisting of filaments that are continuous but doubled back on each other.

**Staple fibers** Fibers of spinnable length manufactured directly or by cutting continuous filaments to relatively short lengths [generally less than 17 in (432 mm)].

**Starved area** An area in a plastic part which has an insufficient amount of resin to wet out the reinforcement completely. This condition may be due to improper wetting or impregnation or excessive molding pressure.

**Starved joint** An adhesive joint which has been deprived of the proper film thickness of adhesive due to insufficient adhesive spreading or application of excessive pressure during lamination.

**Static fatigue** Failure of a part under continued static load; analogous to creep-rupture failure in metals testing but often the result of aging accelerated by stress.

**Static modulus** The ratio of stress to strain under static conditions, calculated from static stress-strain tests, in shear, compression, or tension.

**Stiffness** The relationship of load and deformation; a term often used when the relationship of stress to strain does not conform to the definition of Young's modulus. (*See* Stress-strain.)

**Storage life** *See* Shelf life.

**Strain $\epsilon$** Although strain has several definitions, which depend upon the system being considered, for small deformations, engineering strain is applicable and is the most common definition of strain.

**Strain relaxation** *See* Creep.

**Strands** A primary bundle of continuous filaments (or slivers) combined in a single compact unit without twist. These filaments (usually 51,102 or 51,204) are gathered together in the forming operations.

**Strand count** The number of strands in a plied yarn or in a roving.

**Strand integrity** The degree to which the individual filaments making up the strand or end are held together by the sizing applied.

**Strength, flexural** *See* Flexural strength.

**Stress $\sigma$** Most commonly defined as engineering stress, the ratio of the applied load $P$ to the original cross-sectional area $A_0$.

**Stress concentration** Magnification of the level of an applied stress in the region of a notch, void, or inclusion.

**Stress corrosion** Preferential attack of areas under stress in a corrosive environment, where this factor alone would not have caused corrosion.

**Stress crack** External or internal cracks in a plastic caused by tensile stresses less than that of its short-time mechanical strength. The stresses which cause cracking may be present internally or externally or may be combinations of these stresses. (*See also* Crazing.)

**Stress relaxation**   The decrease in stress under sustained constant strain, also called *stress decay*.

**Stress-strain**   Stiffness, expressed in kips per square inch (megapascals), at a given strain.

**Stress-strain curve**   Simultaneous readings of load and deformation, converted into stress and strain, plotted as ordinates and abscissas, respectively, to obtain a stress-strain diagram.

**Structural bond**   A bond that joins basic load-bearing parts of an assembly; the load may be either static or dynamic.

**Surfacing mat**   A very thin mat, usually 0.007 to 0.020 in (0.18 to 0.51 mm) thick, of highly filamentized fiber glass used primarily to produce a smooth surface on a reinforced plastic laminate.

**Surface resistance (electric)**   The surface resistance between two electrodes in contact with a material is the ratio of the voltage applied to the electrodes to that portion of the current between them which flows through the surface layers.

**Surface resistivity (electric)**   The ratio of potential gradient parallel to the current along the surface of a material to the current per unit width of surface.

**Surface treatment**   A material applied to fibrous glass during the forming operation or in subsequent processes, i.e., size or finish.

**Surfacing mat**   *See* Overlay sheet.

**Syntactic foam**   A cellular plastic which is put together by incorporating preformed cells (hollow spheres or microballoons) in a resin matrix; opposite of foamed plastic, in which the cells are formed by gas bubbles released in the liquid plastic by chemical or mechanical action.

**Synthetic resin**   A complex, substantially amorphous, organic semisolid or solid material (usually a mixture) built up by chemical reaction of comparatively simple compounds, approximating the natural resins in luster, fracture, comparative brittleness, insolubility in water, fusibility or plasticity, and some degree of rubberlike extensibility but commonly deviating widely from natural resins in chemical constitution and behavior with reagents.

**Tack**   Stickiness of an adhesive or filament-reinforced resin prepreg material.

**Tack range**   The length of time an adhesive will remain in the tacky-dry condition after application to the adherend and under specified conditions of temperature and humidity.

**Tack stage**   The length of time deposited adhesive film exhibits stickiness or tack or resists removal or deformation of the cast adhesive.

**Tangent line**   In a filament-wound bottle, any diameter at the equator.

**Tape**   A composite ribbon consisting of continuous or discontinuous fibers that are aligned along the tape axis parallel to each other and bonded together by a continuous matrix phase.

**TD**   Thoria-dispersed.

**Tenacity**   The strength of a yarn or of a filament of a given size; equals breaking strength divided by denier.

**Tensile bar**   A compression- or injection-molded specimen of specified dimensions used to determine the tensile properties of a material.

**Tensile modulus**   The ratio of the tension stress to the strain in the material over the range for which this valve is constant.

**Tensile strength or stress**   The maximum tensile load per unit area of original cross section, within the gage boundaries, sustained by the specimen during a tension test. It is expressed as kips per square inch (megapascals). Tensile load is interpreted to mean the maximum tensile load sustained by the specimen during the test, whether this coincides with the tensile load at the moment of rupture or not.

**TFRS**   Tungsten-fiber-reinforced superalloy.

**Thermal conductivity**   Ability of a material to conduct heat; the physical constant for quantity of heat that passes through a unit cube of a substance in unit time when the difference in temperature of two faces is 1 degree.

**Thermal expansion, Coefficient of**   *See* Coefficient of thermal expansion.

**Thermoplastic**   Capable of being repeatedly softened by increase of temperature and hardened by decrease in temperature; applicable to those materials whose change upon heating is substantially physical rather than chemical.

**Thermoset**   A plastic which changes into a substantially infusible and insoluble material when it is cured by application of heat or by chemical means.

**Thixotropic**   Gel-like at rest but fluid when agitated; having high static shear strength and low dynamic shear strength at the same time.

**Thread count**   The number of yarns (threads) per inch (millimeter) in either lengthwise (warp) or crosswise (fill) direction of woven fabrics.

**TLM**   Tape-laying machine.

**TMC**   Thick molding compound.

**Toggle action**   A mechanism which exerts pressure developed by the application of force on a knee joint, used as a method of closing presses and applying pressure at the same time.

**Tolerance**   The guaranteed maximum deviation from the specified nominal value of a component characteristic at standard or stated environmental conditions.

**Torsional rigidity (fibers)**   The resistance of a fiber to twisting.

**Toughness**   The energy required to break a material, equal to the area under the stress-strain curve.

**Tow**   A large bundle of continuous filaments, generally 10,000 or more, not twisted.

**Transfer molding**   Method of molding thermosetting materials in which the plastic is first softened by heating and pressure in a transfer chamber and then forced by high pressure through suitable sprues, runners, and gates into the closed mold for final curing.

**Transfer pot**   (1) A heating cylinder. (2) Transfer chamber in a transfer mold.

**Transition temperature**   The temperature at which the properties of a material change.

**Twist**   The turns about its axis per unit of length in a yarn or other textile strand. *See also* Balanced twist.

**TZM**   Trade name of molybdenum alloy wire.

**UDC**   Unidirectional composites.

**UHM**   Ultrahigh-modulus.

**Ultimate elongation**   The elongation at rupture.

**Ultimate tensile strength**   The ultimate or final stress sustained by a specimen in a tension test; the stress at moment of rupture.

**Ultraviolet**   Zone of invisible radiations beyond the violet end of the spectrum of visible radiations. Since ultraviolet wavelengths are shorter than the visible, their photons have more energy, enough to initiate some chemical reactions and to degrade most plastics.

**UMC**   Unidirectional molding compound.

**Undercut**   Having a protuberance or indentation that impedes withdrawal from a two-piece, rigid mold; any such protuberance or indentation, depending on the design of the mold (tilting a model in designing its mold may eliminate an apparent undercut).

**Unidirectional laminate**   A reinforced plastic laminate in which substantially all the fibers are oriented in the same direction.

**Urethane plastics**   Plastics generally reacted with polyols, e.g., poly esters or poly ethers, when the reactants are joined by formation of a urethane linkage.

**Vacuum-bag molding**   A process for molding reinforced plastics in which a sheet of flexible transparent material is placed over the lay-up on the mold and sealed. A vacuum is applied between the sheet and the lay-up. The entrapped air is mechanically worked out of the lay-up and removed by the vacuum, and the part is cured; also called *bag molding.*

**Veil**   An ultrathin mat similar to a surface mat, often composed of organic fibers as well as glass fibers.

**VIM**   Vibrational microlamination.

**Virgin filament**   An individual filament which has not been in contact with any other fiber or any other hard material.

**Viscosity**   The property of resistance to flow exhibited within the body of a material expressed in terms of relationship between applied shearing stress and resulting rate of strain in shear.

**Void content**   The percentage of voids in a laminate can be calculated by the use of the formula void $\% = 100 - x$, where $x$ is usually a weight percent.

**Voids**   Gaseous pockets trapped and cured into a laminate; unfilled spaces in a cellular plastic substantially larger than the characteristic individual cells.

**Volatile content**   The percent of volatiles driven off as a vapor from a plastic or an impregnated reinforcement.

**Volatile loss**   Weight loss by vaporization.

**Volatiles**   Materials in a sizing or a resin formulation capable of being driven off as a vapor at room temperature or slightly above.

**Warp**   (1) The yarn running lengthwise in a woven fabric; a group of yarns in long lengths and approximately parallel, put on beams or warp reels for further textile processing, including weaving. (2) A change in dimension of a cured laminate from its original molded shape.

**Water absorption**   Ratio of the weight of water absorbed by a material upon immersion to the weight of the dry material. *See also* Moisture absorption.

**Weathering**   The exposure of plastics outdoors. In *artificial weathering* plastics are exposed to cyclic laboratory conditions of high and low temperatures, high and low relative humidities, and ultraviolet radiant energy, with or without direct water spray, in an attempt to produce changes in their properties similar to those observed on long continuous exposure outdoors. Laboratory exposure conditions are usually intensified beyond those in actual outdoor exposure to achieve an accelerated effect.

**Weave**   The particular manner in which a fabric is formed by interlacing yarns and usually assigned a style number.

**Web**   A textile fabric, paper, or a thin metal sheet of continuous length handled in roll form, as contrasted with the same material cut into sheets.

**Weft**   The transverse threads or fibers in a woven fabric; fibers running perpendicular to the warp; also called *filler, filler yarn, woof.*

**Wet flexural strength (WFS)**   The flexural strength after water immersion, usually after boiling the test specimen for 2 h in water.

**Wet lay-up**   The reinforced plastic which has liquid resin applied as the reinforcement is laid up; the opposite of dry lay-up or prepreg. *See also* Dry lay-up, Prepreg.

**Wet-out**   The condition of an impregnated roving or yarn wherein substantially all voids between the sized strands and filaments are filled with resin.

**Wet-out rate**   The time required for a plastic to fill the interstices of a reinforcement material and wet the surface of the reinforcement fibers; usually determined by optical or light-transmission means.

**Wet winding**   In filament winding the process of winding glass on a mandrel where the strand is impregnated with resin just before contact with the mandrel. (*See also* Dry winding.)

**Whisker**   A very short fiber form of reinforcement, usually crystalline.

**Winding, biaxial**   *See* Biaxial winding.

**Winding pattern**   (1) The total number of individual circuits required for a winding path to begin repeating by laying down immediately adjacent to the initial circuit. (2) A regularly recurring pattern of the filament path after a certain number of mandrel revolutions, leading to the eventual complete coverage of the mandrel.

**Winding tension**   In filament winding the amount of tension on the reinforcement as it makes contact with the mandrel.

**Wire**   A metallic filament.

**Working life**   The period of time during which a liquid resin or adhesive, after mixing with catalyst, solvent, or other compounding ingredients, remains usable. (*See also* Gelation time, Pot life.)

**Woven fabrics**   Fabrics produced by interlacing strands at more or less right angles.

**Woven roving**   A heavy glass-fiber fabric made by weaving roving.

**Wrinkle**   A surface imperfection in laminated plastics that has the appearance of a crease in one or more outer sheets of the paper, fabric, or other base which has been pressed in.

**XMC**   Directionally reinforced molding compound.

**Yarn**   An assemblage of twisted fibers or strands, natural or manufactured, to form a continuous yarn suitable for use in weaving or otherwise interweaving into textile materials. *See also* Continuous filament.

**Yield point**   The first stress in a material, less than the maximum attainable stress, at which an increase in strain occurs without an increase in stress. Only materials that exhibit this unique phenomenon of yielding have a yield point.

**Yield strength**   The stress at which a material exhibits a specified limiting deviation from the proportionality of stress to strain; the lowest stress at which a material undergoes plastic deformation. Below this stress, the material is elastic; above it, viscous.

**Young's modulus**   *See* Modulus of elasticity.

# Index

1

# About the Editor

**Mel M. Schwartz** is uniquely qualified to write on the subject of composite materials. With over thirty years experience in the aerospace industry, he has spent the last twelve years directing and supervising research and development in metals and nonmetals, including work in the areas of adhesive bonding and composites. He has contributed a wide variety of papers in the field of metals joining techniques. He was also editor of and contributed to the McGraw-Hill handbook *Metals Joining Manual*. Mr. Schwartz has been a speaker at many professional symposiums, including those held by the Society of Manufacturing Engineers, the American Society of Welding, and the Society of Aerospace Materials and Process Engineers. He has been a member of many industry committees as well as the National Academy of Sciences Materials Advisory Board. At present, he is Chief of Metals and Metals Processing for the Sikorsky Aircraft Division of United Technologies.